Student Study Guide/Soluti

for

General, Organic, and Biochemistry

Tenth Edition

Katherine J. Denniston
Towson University

Joseph J. Topping
Towson University

Danaè R. Quirk Dorr
Minnesota State University Mankato

Robert L. Caret
University of Massachusetts

Prepared by
Danaè R. Quirk Dorr
Minnesota State University Mankato

STUDENT STUDY GUIDE/SOLUTIONS MANUAL FOR

GENERAL, ORGANIC, AND BIOCHEMISTRY, TENTH EDITION

Published by McGraw-Hill Education, 2 Penn Plaza, New York, NY 10121. Copyright © 2020 by McGraw-Hill Education. All rights reserved. Printed in the United States of America. Previous editions © 2017, 2014, and 2011. No part of this publication may be reproduced or distributed in any form or by any means, or stored in a database or retrieval system, without the prior written consent of McGraw-Hill Education, including, but not limited to, in any network or other electronic storage or transmission, or broadcast for distance learning.

Some ancillaries, including electronic and print components, may not be available to customers outside the United States.

This book is printed on acid-free paper.

1 2 3 4 5 6 QVS 23 22 21 20 19

ISBN 978-1-260-50616-7
MHID 1-260-50616-9

Cover image: ©Tammy616/Getty Images.

All credits appearing on page or at the end of the book are considered to be an extension of the copyright page.

The Internet addresses listed in the text were accurate at the time of publication. The inclusion of a website does not indicate an endorsement by the authors or McGraw-Hill Education, and McGraw-Hill Education does not guarantee the accuracy of the information presented at these sites.

mheducation.com/highered

Contents

Contents

1 Chemistry: Methods and Measurement

Learning Goals

1. Outline a strategy for learning general chemistry.
2. Explain the relationship between chemistry, matter, and energy.
3. Discuss the approach to science, the scientific method, and distinguish among the terms *hypothesis*, *theory*, and *scientific law*.
4. Distinguish between *data* and *results*.
5. Describe the properties of the solid, liquid, and gaseous states.
6. Classify matter according to its composition.
7. Provide specific examples of physical and chemical properties and physical and chemical changes.
8. Distinguish between intensive and extensive properties.
9. Identify the major units of measure in the English and metric systems.
10. Report data and calculate results using scientific notation and the proper number of significant figures.
11. Distinguish between *accuracy* and *precision* and their representations: *error* and *deviation*.
12. Convert between units of the English and metric systems.
13. Know the three common temperature scales, and convert values from one scale to another.
14. Use density, mass, and volume in problem solving, and calculate the specific gravity of a substance from its density.

Introduction

The subject matter of chemistry deals with all material substances as well as all the changes that these materials undergo. The measurement of properties of materials and careful observation and recording of data are essential. The goal of this chapter is to enable you to further your understanding of how chemistry works and develop the skills needed to represent and communicate data and results from scientific inquiry.

1.1 Strategies for Success in Chemistry

The Science of Learning Chemistry

LG 1

The **Study Cycle** is a plan for learning. It involves previewing materials prior to each class, attending class as an active participant, reviewing your notes, studying in short intense sessions each day, and assessing your learning.

Learning General Chemistry

The first nine chapters of this book focus on principles of **general chemistry.** There are several strategies that you can use to help facilitate your understanding of these concepts that connect most aspects of chemistry.

1. Preview materials prior to each class.
2. Actively participate in class.
3. Review your notes after class.
4. Design your own flash cards.
5. Identify the big ideas.
6. Organize the big ideas and look for connections between concepts.
7. Make concept maps.
8. Solve the in-chapter and end-of-chapter questions and problems.

1.2 The Discovery Process

Chemistry

Chemistry is the study of matter, its chemical and physical properties, the chemical and physical changes it undergoes, and the energy changes that accompany those processes. **Matter** is anything that has mass and occupies space. **Energy** is nonmaterial and is defined as the ability to do work (to accomplish some change). The changes that matter undergoes always involve either the gain or loss of energy. Thus, a study of chemistry involves matter, energy, and their interrelationship.

The Scientific Method

The **scientific method** consists of six interrelated processes:

1. **Observation**. The description of the physical properties of a substance is a result of observation. The measurement of the temperature of a liquid, or the size and mass of a solid result from observation.
2. **Formulation of a question**. The observation is the basis of the question on why and how things work.
3. **Pattern recognition**. Discerning a cause-and-effect relationship may give rise to a generalized explanation describing the behavior observed.
4. **Theory development**. Observation of a phenomenon calls for some explanation. The process of explaining observed behavior begins with a **hypothesis**—an educated guess. If this hypothesis survives extensive testing, it may attain the status of a **theory**. A theory is a hypothesis supported by testing (experimentation) that explains scientific facts and is capable of predicting new facts.
5. **Experimentation**. The heart of the scientific method is the verification of theories. This verification process results from conducting carefully designed experiments intended to reinforce or refute the model system, the theory, or the hypothesis. A scientific experiment produces **data**. Each piece of data is the result of a single measurement. Examples include the mass of a sample, or the time required for a chemical reaction to occur. Mass, length, volume, time, temperature, and energy are

LG 4

the most common types of data obtained from chemical experiments. **Results** are the outcome of an experiment. Data and results may be identical, but more often several pieces of data are combined to produce a result.

6. **Information summarization.** Many phenomena have common causes and explanations. A **scientific law** summarizes and clarifies large amounts of information.

Models in Chemistry

A model of a chemical unit or system is often used to make ideas more clear. A good model is based on everyday experience and provides a great deal of information in a simple fashion. Throughout the chapters in the textbook, you will be learning models for concepts, just keep in mind the scientific definition of model.

1.3 The Classification of Matter

Chemists studying matter look for similarities in **properties** among different types of materials. By categorizing matter based on their similarities the subject is simplified and predictions can be made about new substances. Two common ways to categorize matter are by *state* and by *composition*.

States of Matter

Three states of matter exist: the gaseous state, the liquid state, and the solid state.

1. **The gaseous state.** Gases have a very low mass in a certain amount of volume (as compared to the other states) because the individual particles that comprise the gas are separated by large distances. Consequently, gases can be compressed (pushed into a smaller volume) or expanded to a larger volume. Gases take on the size and shape of their container; that is, they have no definite shape or volume.
2. **The liquid state.** Particles that make up a liquid are much closer together than the particles that make up a gaseous state. Thus, liquids expand and contract only slightly, and the mass of a certain amount of volume of a liquid is much greater than that of gases. Liquids have a definite volume but no definite shape.
3. **The solid state.** Solids are characterized by particles that are very close together. Attractive forces between the particles are strong enough to provide a rigid shape or structure to the bulk material. In other words, they have a definite shape and volume. The proximity of particles prevents significant expansion or compression.

In most cases, as a substance moves from a solid to a liquid to a gas, the molecules are separated by more space, with gases having the greatest distance between molecules.

Composition of Matter

LG 6

All matter can be classified as either a pure substance or a mixture. A **pure substance** is a form of matter that has identical 1) composition, and 2) physical and chemical properties throughout. Pure substances may also be subcategorized as elements or compounds. An **element** is a pure substance that cannot be converted into a simpler form of matter by any chemical reaction. A **compound** is a substance resulting from the combination of two or more elements in a definite, reproducible fashion.

A **mixture** is a combination of two or more pure substances in which the combined substances retain their identity. A mixture may be either homogeneous or heterogeneous matter. A **homogeneous mixture** has uniform composition. Its particles are well mixed, or thoroughly intermingled. On visual inspection a person cannot distinguish the different substances present in the mixture. A **heterogeneous mixture** has a nonuniform composition. Upon viewing it, a person can see distinct regions. For example, a mixture of salt and pepper can be mixed very thoroughly but you can always see the different substances present.

Example 1: Classification of Matter

Classify each of the following according to the classification scheme above.

1. vegetable soup
2. tap water
3. aluminum
4. carbon dioxide
5. blood

Solution:

1. Vegetable soup generally has chunks of vegetables in it. It is a mixture and more specifically a heterogeneous mixture.
2. Tap water is not pure water. It is a mixture containing many other substances such as dissolved oxygen, fluoride (if you live in a region that fluorinates its water) and chlorine (if you live in a region that uses chlorine to purify the water). Visually it looks uniform throughout, and thus is a homogeneous mixture.
3. Aluminum is element number 13 on the periodic table. It is a pure substance.
4. Carbon dioxide is a compound. Its name gives evidence of the fixed ratio or two oxygen atoms (the prefix di- means two) to one carbon atom. As such, it is a pure substance.
5. Blood appears uniform when viewed by the naked eye. It consists of many components such a red blood cells, platelets, plasma (which itself is a mixture of chemicals). It is a homogeneous mixture.

Physical Properties and Physical Change

LG 7

Water is the most common example of a substance that may exist in all three states over a reasonable temperature range. The conversion of ice to liquid water, or liquid water to the gaseous state, steam, is an example of a **physical change**. A physical change does not alter the composition or identity of the substance undergoing change, but does alter the appearance of the substance.

Properties (characteristics) of matter may be classified as either physical or chemical. **Physical properties** enable us to identify different kinds of matter without changing the identity (chemical composition) of the sample. Examples of physical properties include color, odor, taste, melting and boiling temperatures, and compressibility.

Chemical Properties and Chemical Change

Chemical properties result in a change in composition and can only be observed through chemical reactions. A **chemical reaction**, or **chemical change**, is the conversion of a chemical substance into one or more different substances. When doing so the starting material is used up (consumed) while completely new materials are formed. Think about the burning of a piece of paper or the baking of a cake. The starting components do not remain after the process has occurred.

The terms *rust*, *digest*, and *combustion* all indicate chemical change. The terms *melt*, *boil*, *freeze*, and *evaporate* all indicate a physical change.

Example 2: Chemical vs. Physical Change
Which of the following are physical changes?

1. An iron nail rusts.
2. A block of ice melts.
3. Water on the floor evaporates.
4. Food is digested in the small intestine.
5. Gasoline undergoes combustion.

Solution:
To be a physical change, the identity (or chemical composition) of the material cannot change. Think about each change and ask yourself, "Is the original material still present after the change occurs?"

1. As iron rusts, the iron with its metallic characteristics is turned into a red flaky material that is not shiny, or malleable like the iron is. It cannot be easily converted back to the iron nail. This is NOT a physical change, this is a chemical change.
2. Most students know that water has the formula H_2O. Whether in the solid or liquid state, it still has the same composition, that is, H_2O. Additionally, liquid water could be refrozen into the original block of ice. This is a physical change.
3. Again, we have water converting to the gas phase (H_2O vapor.) This is a physical change.
4. As food is digested, the materials that make up the food are broken down to similar substances which can be used by the body. The identity of many of the molecules that make up the food actually changes. Digestion involves a chemical change.
5. The combustion reaction is the burning of the material in the presence of oxygen. You cannot take the substances coming out of the exhaust pipe and readily convert them back to the gasoline from which it came. The products are mostly carbon dioxide and water, definitely different substances than gasoline. This is a chemical change.

Intensive and Extensive Properties

Properties of substances can be classified as either intensive or extensive. An **intensive property** is independent of the quantity of the substance. For example, the boiling point of water is intensive. Whether you boil one cup of water or a large pot of water, the water will boil at 100°C. An **extensive property** depends on the quantity of a substance. The more matter present, the greater the value. For example, volume is extensive. Certainly, a 12 oz. glass of tea has more matter than a 6 oz. glass of tea.

Example 3: Intensive vs. Extensive Properties
Determine if the following properties are extensive or intensive.

1. My dog is 90 kg.
2. The patient's temperature is 98.6°F.
3. The shirt is red.
4. The aluminum cube is 3.0 cm^3.

Solution:

1. The larger the dog, the greater the mass. This is extensive.
2. The patient can have a normal temperature of 98.6°C whether a large man or a small child. This is an intensive property.
3. A shirt can be red whether sized as small or extra-large, making it intensive.
4. Cubic centimeters is a volume measurement and the larger the cube, the greater the volume. Therefore, this is an extensive property.

1.4 The Units of Measurement

Chemistry involves measurement. Common quantities that are measured are mass, volume, time, and so on. However, reporting a measurement as simply a number has no meaning until it is accompanied by a unit. A **unit** defines the basic quantity of the substance being measured.

In the English system of measurement, used primarily in the United States, units are often unrelated. The English system is not used in scientific work primarily because of the difficulty involved in converting from one unit to another.

TABLE 1.1 Some Common Relationships Used in the English System

Quantity	Relationship
Weight	1 pound = 16 ounces
	1 ton = 2000 pounds
Length	1 foot = 12 inches
	1 yard = 3 feet
	1 mile = 5280 feet
Volume	1 gallon = 4 quarts
	1 quart = 2 pints
	1 quart = 32 fluid ounces

The metric system is a decimal-based system; it is inherently simpler and less ambiguous. In the metric system there are three basic units. Any subunit or multiple unit contains one of these units preceded by a prefix. This prefix indicates the power of ten by which the base unit is to be multiplied to form the subunit or multiple unit. The most common metric prefixes are shown in Table 1.2 and should be memorized.

TABLE 1.2 Some Common Prefixes Used in the Metric System

Prefix	Meaning	Decimal Equivalent
mega (M)	10^6	1,000,000.
kilo (k)	10^3	1,000.
deci (d)	10^{-1}	0.1
ceni (c)	10^{-2}	0.01
milli (m)	10^{-3}	0.001
micro (μ)	10^{-6}	0.000001
nano (n)	10^{-9}	0.000000001

Mass

Mass describes the quantity of matter in an object. The terms *weight* and *mass*, in common usage, are often considered synonymous. In fact, they are not. **Weight** is the manifestation of the force of gravity on an object. Mass is independent of gravity.

$$\text{weight} = \text{mass} \times \text{acceleration due to gravity}$$

Length

Length is the measurement of the distance between two points. The standard metric unit of length is the meter. A meter is similar in length to the English yard, yd (1 yd = 0.91 m). Large distances are measured conveniently in kilometers, and smaller distances are measured in millimeters or centimeters (according to the metric prefixes shown in Table 1.2 above).

Volume

Volume is the space occupied by an object. The standard metric unit of volume is the liter. A liter is the volume occupied by 1000 grams of water at 4° Celsius. The English unit quart, qt, is similar in size to the liter (1 qt = 0.946 L). Volume can also be reported as a length cubed. By multiplying together the length, width and depth of an object the volume can be calculated. For example, the value, 50 m^3, is a volume unit. To convert between the volume units in liters and the volume units in length-cubed, use the relationship: 1 mL = 1 cm^3. (A cm^3 is sometimes called cubic centimeter and abbreviated cc.)

Time

The standard metric unit of **time** is the second. The need for accurate measurement of time by chemists is necessary in many applications.

1.5 The Numbers of Measurement

Measurements are not merely numbers; they must have both a number and a unit. Providing one without the other does not represent a meaningful measurement. However, when reporting a measurement or a calculation it is also important to indicate how much uncertainty or doubt there is regarding the measurement or calculation. Modern pocket calculators often generate more digits than are merited, given the data used in the calculation. Reporting all the digits often portrays a greater amount of certainty about the measurement than is merited. Proper use of significant figures, scientific notation, and rounding-off is essential as this conveys the amount of uncertainty or doubt about a measurement.

Significant Figures

LG 10

Significant figures are all digits in a number representing data or results that are known with certainty, *plus the first uncertain digit.*

Consider the diagram below. The ruler is marked off in mm (that is, 1/10 of a cm). To measure the length of the pencil, we see that it is on or close to the mark "4.7 cm." We can then estimate one more place, by imagining 10 subdivisions between the 4.7 and 4.8. I'll estimate the pencil as just past the 4.7 mark and call it 4.71 cm. You might say, "It looks right on the line." If so, you would call it 4.70 cm. The first two digits are known with certainty and the final one is an estimate and thus uncertain. The measurement will be reported to 3 significant figures.

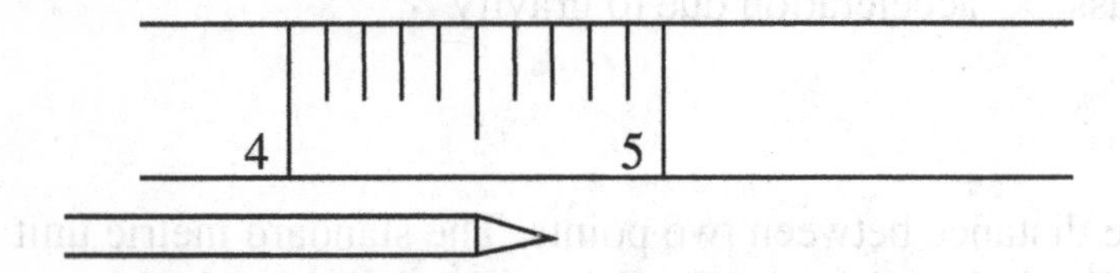

The number of significant figures associated with a measurement is determined by the measuring device. Conversely, the number of significant figures reported is an indication of the precision of the measurement itself.

Recognition of Significant Figures

Only significant digits should be reported as data or results. The six rules enumerated below describe the assignment of significant figures.

Rule 1: All nonzero digits are significant. The length 154.5 mm has 4 significant figures.

Rule 2: The number of significant digits is independent of the position of the decimal point. The length 15.45 cm also has 4 significant figures.

Rule 3: Zeros located between nonzero digits are significant. The volume 808 mL has 3 significant figures.

Rule 4: Zeros at the end of a number (often referred to as trailing zeros) are significant or not significant depending upon the existence of a decimal point in the number.

- If there *is* a decimal point, any trailing zeros are significant. The values 45.10 and 40.00 both have 4 significant figures.
- If the number *does not* contain a decimal point, trailing zeros are not significant. The number 40,000 only has one significant figure.

Rule 5: Zeros to the left of the first nonzero integer are not significant; they serve only to locate the position of the decimal point. The measurement, 0.000049 L, has 2 significant figures.

Scientific Notation

LG 10

Very large or very small numbers may be represented with the proper number of significant figures using **scientific notation**. Scientific notation, involves the representation of a number as a power of ten; the usual convention shows the decimal point in *standard position*—to the right of the first non-zero digit.

To convert a number greater than 1 to scientific notation, the original decimal point is moved x places to the left (locating it after the 1st non-zero digit), and the resulting number is multiplied by 10^x. The exponent (x) is a positive number equal to the number of places the original decimal point was moved.

To convert a number less than 1 to scientific notation, the original decimal point is moved x places to the right (locating it after the first non-zero digit), and the resulting number is multiplied by 10^{-x}. The exponent ($-x$) is a negative number equal to the number of places the original decimal point was moved.

Example 4: Conversion to Scientific Notation
Convert the following to scientific notation. Keep all significant figures in your answer.

1. 310,000
2. 0.0000450

Solution:

1. Start by placing the decimal point after the 1st non-zero digit. The decimal point is moved from the end of the number (it isn't written in but it is there) to the left by 5 places. This gives the value: 3.1×10^5.
2. Here the decimal place must be moved from its location to just after the 4. This requires it to move 5 places to the right. Keep the zero at the end because it is a significant digit. The number in scientific notation is 4.50×10^{-5}.

Accuracy and Precision

LG 11

Error is the difference between the true and measured value. **Accuracy** is the degree of agreement between the true and measured value. High levels of accuracy correlate with small values of **error. Precision** is how close together a set of measurements are to each other. Just as accuracy is measured in terms of error, precision is represented by **deviation**, the amount of variation present in a set of replicate measurements. The goal of all experiments is to have both high accuracy and high precision, but keep in mind that these do not always go hand in hand. It is possible to have high precision with low accuracy and vice versa.

Exact (Counted) and Inexact Numbers

Inexact numbers have uncertainty, illustrated by the doubt in the final significant figure. In contrast, *exact numbers*, a result of counting or definition, have no uncertainty and do not limit the number of significant figures when used in calculations.

Rounding Numbers

A generally accepted rule for rounding off states that if the first digit dropped is 5 or greater, we raise the last significant digit to the next higher number. If the first digit dropped is 4 or less, the last significant digit remains unchanged.

Example 5: Rounding
Round off the following to two significant figures.
1. 16.00468 grams   2. 0.004360 liters

Solution:
1. The third digit (the first to be dropped if rounding to two significant figures) is a zero. Therefore the digits kept remain unchanged. Thus, the answer is 16 grams.
2. After the 0.0043 is the digit, 6. This means the 3 rounds up to a 4 giving 0.0044 liters.

Significant Figures in Calculation of Results

The rules for determining the number of significant figures in a calculation differ according to the mathematical operation.

In the process of addition or subtraction, the position of the decimal point in the quantities being combined determines the number of significant figures in the answer. The answer must have as many places to the right of the decimal point as the original number with the least number of places to the right. See the following example (note that when adding numbers in scientific notation one can either convert all the numbers to standard form, as is done in example 5, or convert to numbers which have the same exponents).

Example 6: Calculations and Significant Figures – Addition and Subtraction
Report the answer to the following calculation to the correct number of significant figures.

$$3.2 \times 10^{-2} + 2.2 \times 10^{-3}$$

Solution:
When adding and subtracting, the answer must have the same number of places to the right of the decimal place as the original numbers with the LEAST number of places to the right. At first glance, it looks like they both have 1 place to the right of the decimal point. However, the numbers cannot be compared until they are written either in standard form or with the same exponent (that is, both with 10^{-2} or both with 10^{-3}.)

Example 6 Continued

If we write both numbers in standard form:

$$\begin{array}{r} 0.032 \\ +\ 0.0022 \\ \hline 0.0342 \end{array}$$

One of the values has 3 places to the right of the decimal point, the other has four. The answer must only have 3 digits to the right, giving 0.034 and can also be written in scientific notation as 3.4×10^{-2}.

If we convert to numbers that have the same exponent: $2.2 \times 10^{-3} = 0.22 \times 10^{-2}$

$$\begin{array}{r} 3.2 \times 10^{-2} \\ +\ 0.22 \times 10^{-2} \\ \hline 3.42 \times 10^{-2} \end{array}$$

However, one of the values has 1 place to the right of the decimal place while the other has 2. The answer must only have 1 digit to the right, therefore the answer is 3.4×10^{-2}.

In multiplication and division, the decimal point position is irrelevant. What is important is the number of significant figures. *The answer can be no more precise than the least precise number from which the answer is derived.* So the answer must have as many significant figures as the original number with the fewest significant figures.

Example 7: Calculations and Significant Figures – Multiplication and Division.
Perform the following multiplication and division problem using scientific notation. Round off the answer to the proper number of significant digits using scientific notation.

$$\frac{\left(6.85\times10^{-6}\right)\times\left(1.690\times10^{3}\right)}{3.6}$$

Solution:
A scientific calculator can manage this problem quite nicely. Use the exponent key to enter the numbers which are in scientific notation. When entered, the calculator gives the value: 0.003215694. Your calculator knows nothing about significant figures, so you must determine which digits are significant. There are 3 significant figures in 6.85×10^{-5}, 4 significant figures in 1.690×10^{3} and only 2 significant figures in 3.6. Your answer can only have 2 significant figures giving the answer of 3.2×10^{-3}.
Comment:
The number of significant figures is the same whether reported in scientific notation or in standard form as 0.0032.

1.6 Unit Conversion

A conversion factor is a mathematical relationship comparing values with different units. Conversion factors are used to in the factor-label method (dimensional analysis) to convert from one unit to another 1) within the same system, or 2) convert units from one system to units of another system.

Conversion of Units within the Same System

The relationships in Table 1.1 are common within the English system. Any of these relationships can be set up as conversion factors. For example, the relationship 1 pound = 16 ounces (from Table 1.1) can be used to convert between pounds and ounces. Conversion factors for this relationship can be written as:

$$\frac{1\ \text{lb}}{16\ \text{oz}} \quad \text{or} \quad \frac{16\ \text{oz}}{1\ \text{lb}}$$

If a conversion must be made within the metric system, the relationships in Table 1.2 are can be used as conversion factors. For example, the relationship 1 mg = 10^{-3} g can be written as:

$$\frac{1\ \text{mg}}{10^{-3}\ \text{g}} \quad \text{or} \quad \frac{10^{-3}\ \text{g}}{1\ \text{mg}}$$

The Factor-Label Method

The use of these conversion factors is referred to as the *factor-label method* or *dimensional analysis*. It is important to note that the method used is the same regardless of whether you are working to convert units *within* the same unit system or *between* the unit systems.

When converting from one unit to another, the unit in the data given must be the same unit that is in the denominator of the conversion factor selected. That way, when the data given is multiplied by the conversion factor, the unwanted unit will cancel and leave the unit required in the desired result. Consider the conversion of 4 g to mg:

$$4\cancel{g} \times \frac{1\ \text{mg}}{10^{-3}\ \cancel{g}} = 4 \times 10^{3}\ \text{mg}$$

Data Given x Conversion Factor = Desired Result

Conversion of Units Between Systems

It will often be important to convert from the English system to the metric system. Table 1.3 gives commonly used English system units and their relationship to the metric system unit. Your instructor *may* require that you memorize the relationships.

For example, look at the pound – ounce relationship in Table 1.1. Conversion factors for this relationship can be written as:

$$\frac{1\ \text{lb}}{454\ \text{g}} \quad \text{or} \quad \frac{454\ \text{g}}{1\ \text{lb}}$$

Since 1 lb and 454 g both represent the same mass, this ratio is equivalent to one. These conversion factors can be used to convert from one unit to another.

TABLE 1.3 Relationships Between Common English and Metric Units

Quantity	English		Metric
Mass	1 pound	=	454 grams
Length	1 inch	=	2.54 centimeters
	1 yard	=	0.91 meters
Volume	1 quart	=	0.946 liters
	1 gallon	=	3.78 liters

Example 8: Conversion between units
Convert 5.5 inches into centimeters.

Solution:
Looking at Table 1.3, we see that 1 in = 2.54 cm. These relationships can be written as conversion factors.

$$\frac{1 \text{ in}}{2.54 \text{ cm}} \text{ or } \frac{2.54 \text{ cm}}{1 \text{ in}}$$

To begin, write the quantity you know (Data Given), 5.5 inches. Then align the conversion factor so that the unit you wish to eliminate is in the denominator of conversion factor. In the work below, we start with inches and want to cancel that unit, so we place it on the denominator. This way the inches will cancel each other out (anything divided by itself equals one.)

$$5.5 \cancel{\text{in}} \times \frac{2.54 \text{ cm}}{1 \cancel{\text{in}}} = 13.97 \text{ cm} = 14 \text{ cm}$$

Data Given x Conversion Factor = Desired Result

After performing the calculation, the final steps are to check significant figures and check that the answer makes sense. Our original measurement, 5.5 in, has 2 significant figures and therefore our final answer must have 2 significant figures. Using the common convention for rounding, the answer would be 14 cm. Finally, we need to check that the answer makes sense, since centimeters are a smaller unit of measurement than inches (it takes more than two centimeters to make an inch) it makes sense that the number of centimeters in 5.5 inches is more than double 5.5.

For many of the conversions you will need to do, there will not be a single conversion factor which relates the two units and it will be necessary to use more than one conversion factor. See Examples 9 and 10.

Example 9: Conversion between units in the English system
Convert 8.62 gallons into pints.

Solution:
Notice that there is no relationship given in Table 1.1 between gallons and pints. We do see a relationship between gallons and quarts and a relationship between quarts and pints.

Example 9 Continued

We can perform this calculation step-wise:

$$8.62\ \cancel{gal} \times \frac{4\ qt}{1\ \cancel{gal}} = 34.48\ qt$$

$$34.48\ \cancel{qt} \times \frac{2\ pt}{1\ \cancel{qt}} = 68.96\ pt = 69.0\ pt$$

Or we can perform this calculation in a single step by arranging the factors in a chain:

$$8.62\ \cancel{gal} \times \frac{4\ \cancel{qt}}{1\ \cancel{gal}} \times \frac{2\ pt}{1\ \cancel{qt}} = 69.0\ pt$$

Data Given x Conversion Factor x Conversion Factor = Desired Result

Notice that in each case, the units of gallons cancel, and the units of quarts also cancel, leaving pints as the final unit. The 4 and 2 in the conversion factors are exact numbers and do not limit the number of significant figures in your final answer; your final answer is shown with 3 significant figures. Since gallons are a larger unit of volume than pints, it makes sense that our answer is fairly large.

Example 10: Conversion from English to metric numbers.
A patient's weight is 129 pounds. What is the patient's weight in kilograms?

Solution:
In Table 1.1 we have a relationship which compares pounds to grams. Table 1.2 provides the relationship between kilograms and grams. Beginning with the data given, 129 lbs, the conversion factors can be aligned as follows.

$$129\ \cancel{lb} \times \frac{454\ \cancel{g}}{1\ \cancel{lb}} \times \frac{1\ kg}{1000\ \cancel{g}} = 58.6\ kg$$

Data Given x Conversion Factor x Conversion Factor = Desired Result

If a unit is squared or cubed, the conversion factor used must also be squared or cubed. Example 11 shows this process.

Example 11: Conversion of area
A rectangle has an area of 12 in^2. What is the area in m^2?

Solution:
If we were converting from inches to meters we would use the conversion factors:

$$\frac{1 \text{ in}}{2.54 \text{ cm}} \quad \text{and} \quad \frac{0.01 \text{ m}}{1 \text{ cm}}$$

We will use these same conversion factors; however, we must modify them by squaring the unit *and* the numbers.

$$12 \cancel{\text{in}^2} \times \frac{2.54^2 \cancel{\text{cm}^2}}{1^2 \cancel{\text{in}^2}} \times \frac{0.01^2 \text{ m}^2}{1^2 \cancel{\text{cm}^2}} = 7.7\times10^{-3} \text{ m}^2$$

Data Given x Conversion Factor x Conversion Factor = Desired Result

Again it is important to double check that the units have correctly cancelled out. In^2 will cancel out with in^2, however, in^2 will not fully cancel out with in, and thus the calculation will be wrong if the conversion factor is not squared.

1.7 Additional Experimental Quantities

LG 13

Temperature

Temperature is the degree of "hotness" of an object. Many substances, such as mercury, expand as their temperature increases, and this expansion provides us with a way to measure temperature and temperature changes. The height of the mercury in a thermometer is proportional to the temperature. A mercury thermometer may be calibrated, or scaled, in different units, just like a ruler. Three common temperature scales are Fahrenheit (°F), Celsius (°C) and Kelvin (K). Two convenient reference temperatures used to calibrate a thermometer are the freezing and boiling temperatures of water. Conversion from one temperature scale to another may be accomplished as shown below.

Fahrenheit to Celsius: $T_{°C} = \dfrac{T_{°F} - 32}{1.8}$ Celsius to Fahrenheit: $T_{°F} = 1.8 \text{ x } T_{°C} + 32$

Celsius to Kelvin: $T_K = T_{°C} + 273.15$

Example 12: Temperature conversions
What value, in °F, corresponds to 310.0 Kelvin?

Solution:
First convert from Kelvin to Celsius, then convert from Celsius to Fahrenheit using the equations given above.

Example 12 Continued

$$T_K = T_{°C} + 273.15$$

or

$$T_{°C} = T_K - 273.15$$

$$T_{°C} = 310.0 - 273.15 = 36.85°C$$

$$T_{°C} = 37.0\ °C$$

The rules for significant figures (for addition and subtraction) require that the answer has the number of places to the right of the decimal point as the number with the least. The value 310.0 has one place to the right while 273.15 has two. The answer must only have one.

$$T_{°F} = 1.8 \times T_{°C} + 32$$

$$T_{°F} = 1.8 \times 37.0 + 32$$

$$T_{°F} = 98.6°F$$

The 32 and 1.8 are exact numbers and do not limit the number of significant figures in the answer.

Energy

Energy, the ability to do work, may be categorized as **kinetic energy**, the energy of motion, or **potential energy**, the energy of position, and can be thought of as the energy stored in the substance.

Energy may also be classified according to form. The principal forms of energy include light, heat, electrical, mechanical, and chemical energy. Remember the following characteristics of energy:

1. Energy cannot be created or destroyed.
2. Energy may be converted from one form to another.
3. Energy conversion always occurs with less than 100% efficiency.
4. All chemical reactions involve either a gain or loss of energy.

Energy absorbed or liberated in chemical reactions is often in the form of heat energy. Heat energy may be represented in units of calories or Joules, their relationship being

1 calorie (cal) = 4.18 Joules (J)

One calorie is defined as the amount of heat energy required to change the temperature of 1 gram of water by 1°C. The nutritional calorie, that is, the one you "count" when on a diet, is actually equivalent to 1000 calories (or 1 kcal.) The nutritional calorie is written with an upper case C. The topic of energy will be discussed more thoroughly in Chapter 7.

1000 cal = 1 kcal = 1 Cal

Concentration

Concentration is a measure of the number of particles of a substance, or the mass of those particles, that are contained in a specific volume. Concentration is used to describe mixtures of different substances, such as the number of red blood cells in a specified volume of blood. In Chapter 6, concentration will be covered in detail.

LG 14

Density and Specific Gravity

Density is the ratio of mass to volume.

$$\text{density} = \frac{\text{mass}}{\text{volumn}} = \frac{m}{V}$$

Density is typically reported in units of g/mL and is independent of the amount of material. In other words, it is an intensive property. Each substance has a unique density, for example 1 mL of water and 1 mL of gold have very different masses.

Values of density are often related to a standard, well-known reference, the density of pure water. At 4°C this value is 1.00 g/mL. This "referenced" density is the **specific gravity**. Note that specific gravity has no units.

$$\text{specific gravity} = \frac{\text{density of object (g/mL)}}{\text{density of water (g/mL)}}$$

Example 13: Calculation of Density

A solid block that has a mass of 1267.4 grams was found to have the following measurements: length = 9.86 cm, width = 46.6 mm, and height = 0.224 m. What is its density in units of g/mL? What is the specific gravity?

Solution:

To determine the density of the block, mass and volume are needed. The mass is given in the problem. The volume must be determined using the formula V = length × width × height. Before using the formula, convert each measurement into centimeters:

length = 9.86 cm
width = 46.6 mm = 4.66 cm
height = 0.224 m = 22.4 cm

Next, multiply the length, width, and height to determine the volume:

$$V_{solid} = 9.86 \text{ cm} \times 4.66 \text{ cm} \times 22.4 \text{ cm} = 1.03 \times 10^3 \text{ cm}^3$$

Since 1 mL = 1 cm^3, the volume can also be written as 1.03×10^3 mL.

Now the density formula can be used to determine density knowing mass and volume of the block:

$$d = \frac{m}{V} = \frac{1267.4 \text{ g}}{1.03 \times 10^3 \text{ mL}} = 1.23 \text{ g/mL}$$

The specific gravity $= \dfrac{1.23 \text{ g/mL}}{1.00 \text{ g/mL}} = 1.23$. Note, the specific gravity equals the density, but without units.

Example 14: Density as a conversion factor.
If the density of carbon tetrachloride is 1.59 g/mL, what is the mass (in grams) of 2.00 liters of carbon tetrachloride?

Solution:
When given density, you are given a conversion factor that will convert between the mass of a substance and its volume. A density of 1.59 g/mL can be written as:

$$\frac{1.59\ \text{g}}{1\ \text{mL}} \text{ or } \frac{1\ \text{mL}}{1.59\ \text{g}}$$

To work the problem, first convert from liters to milliliters so that the above relationship can then convert the value from milliliters to grams.

$$2.00\ \cancel{\text{L}} \times \frac{1\ \cancel{\text{mL}}}{10^{-3}\ \cancel{\text{L}}} \times \frac{1.59\ \text{g}}{1\ \cancel{\text{mL}}} = 3.18\times10^{3}\ \text{g}$$

Example 15: Density as a conversion factor.
Calculate the volume in mL of a 5.555 g ethyl alcohol. The density of ethyl alcohol is 0.789 g/mL.

Solution:
Since density is given, we have the following conversion factor:

$$\frac{0.789\ \text{g}}{1\ \text{mL}}$$

If you have an option, do not start with the conversion factor. Here we will start with the given mass.

$$5.555\ \cancel{\text{g}} \times \frac{1\ \text{mL}}{0.789\ \cancel{\text{g}}} = 7.04\ \text{mL}$$

Density in not an exact number but is derived from measurements. Consequently, you *must* consider the number of significant figures in density when determining the number of significant figures in the answer. Therefore, we round off the answer to three significant figures.

Alternate Solution:
Using the formula:

$$d = \frac{m}{\text{V}}$$

algebraically solve for the variable V, giving:

$$\text{V} = \frac{m}{d}$$

Example 15 Continued

Substitute in the value of mass and density, giving:

$$V = \frac{5.555\ g}{0.789\ g/mL} = 7.04\ mL$$

Note: the first solution is usually preferred because if you are working on a factor-label problem that has many conversion factors, you do not have to stop the conversion, solve for the variable and then resume the factor-label problem.

Self Test for Chapter One

1. According to the Study Cycle, what should you do right after you attend class?
2. What can you use as your own personal quiz in order to prepare for in class exams?
3. How is a hypothesis different from a theory?
4. What are the three states of matter?
5. Which state of matter will expand to fill any container?
6. Which state of matter has a definite shape and volume?
7. Which of the following are physical properties of an object?

   a. combustibility
   b. color
   c. melting point
   d. phase
   e. chlorine combining with hydrogen forming HCl
   f. density
   g. volume
   h. sulfur burning in air making acid rain
8. Is mass an intensive or extensive property?
9. Which of the following are mixtures?

   a. NaCl
   b. salt water
   c. a soft drink
   d. tap water
   e. blood
   f. a cake mix
10. Label each of the following mixtures as homogeneous or heterogeneous.

    a. salt and pepper
    b. ice water
    c. freshly brewed hot tea.
11. Which system of measurement is a decimal-based system?
12. A square is 3.0 inches by 3.0 inches. What is the area of the square in m^2?
    2.54 cm = 1 in.
13. How many significant figures are in each of the following?

    a. 0.0040
    b. 1.490
    c. 4.00×10^5

14. Convert the following to scientific notation. Keep the same number of significant figures.

    a. 5,640,000
    b. 0.000490
    c. 0.10090

15. Round off 0.00369865 to two significant figures.

16. Express the answer to the following calculations with the correct number of significant figures.

    a. $\dfrac{6.80\times10^{-2}}{4.61\times10^{-1}}$

    b. $\dfrac{1.06\times10^{-3}}{6.4\times10^{-4}}\times1.64\times10^{-6}\times\dfrac{1.11\times10^{3}}{1.00\times10^{3}}$

    c. 14.5 + 89.6

    d. (14.5 + 89.6) × 1.001

17. How many centimeters are in a kilometer?

18. If one atom of uranium is 238.0 atomic mass units, and one atomic mass unit is equal to 1.66×10^{-24} gram, what is the mass, in grams, of one atom of uranium?

19. Convert 50.0 miles/hour to cm/second. Use the following relationships: 5280 feet = 1 mile; 12 inches = 1 foot; 2.54 centimeters = 1 inch.

20. What is the temperature on the Fahrenheit scale corresponding to – 40.0°C?

21. What is the temperature on the Celsius scale corresponding to – 4.6°F?

22. What is the temperature on the Kelvin scale corresponding to 17.98°F?

23. Calculate the density, in units of g/mL, of a liquid with a volume of 348 mL and a mass of 0.3546 kg.

24. The density of carbon tetrachloride is 1.59 g/mL. What is the mass of 2.65 liters of carbon tetrachloride in grams?

25. What is the specific gravity of an object that weighs 13.35 g and has a volume of 25.00 mL?

26. The density of aluminum is 2.70 g/cm^3. What is the volume of 15.05 grams of aluminum?

27. What term describes the ability to do work?

28. How many calories are in 1500 Calories (nutritional calories)? Express your answer using scientific notation.

2 *The Structure of the Atom and the Periodic Table*

Learning Goals

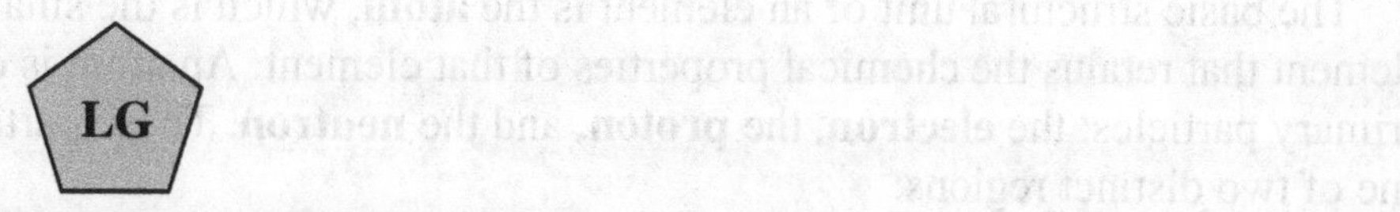

1. Describe the properties of protons, neutrons, and electrons.
2. Interpret atomic symbols, and calculate the number of protons, neutrons, and electrons for atoms.
3. Distinguish between the terms *atom* and *isotope,* and use isotope notations and natural abundance values to calculate atomic masses.
4. Summarize the history of the development of atomic theory, beginning with Dalton.
5. Describe the role of spectroscopy and the importance of electromagnetic radiation in the development of atomic theory.
6. State the basic postulates of Bohr's theory, its utility, and its limitations.
7. Recognize the important subdivisions of the periodic table: periods, groups (families), metals, and nonmetals.
8. Identify and use the specific information about an element that can be obtained from the periodic table.
9. Describe the relationship between the electronic structure of an element and its position in the periodic table.
10. Write electron configurations, shorthand electron configurations, and orbital diagrams for atoms and ions.
11. Discuss the octet rule, and use it to predict the charges and the numbers of protons and electrons in cations and anions formed from neutral atoms.
12. Utilize the periodic table trends to estimate the relative sizes of atoms and ions, as well as relative magnitudes of ionization energy and electron affinity.

Introduction

In this chapter, we will learn some of the properties of atomic particles and the experiments on which we base our current understanding of atomic structure. The atomic structure of each element is unique, but there exist structural similarities among these elements.

The periodic table is an organized "map" of the elements that relates their structure to their chemical and physical properties. The chemical and physical properties of elements follow directly from the electronic structure of the atoms that make up these elements. Familiarity with the periodic table allows prediction of the structure and properties of the various elements and serves as the basis for understanding chemical bonding.

2.1 Composition of the Atom

Electrons, Protons, and Neutrons

The basic structural unit of an element is the **atom**, which is the smallest unit of an element that retains the chemical properties of that element. An atom is composed of three primary particles: the **electron**, the **proton,** and the **neutron**. These particles are located in one of two distinct regions:

1. The **nucleus** is a small, dense, positively charged region in the center of the atom. The nucleus is composed of positively charged protons and uncharged neutrons.
2. Surrounding the nucleus is a diffuse region of negative charge populated by electrons, the source of the negative charge. Electrons are tiny in comparison to the protons and neutrons. The properties of these particles are summarized in Table 2.1.

TABLE 2.1 Selected Properties of the Three Basic Subatomic Particles

Name	Charge	Mass (amu)	Mass (grams)
Electron (e^-)	−1	5.486×10^{-4}	9.1094×10^{-28}
Proton (p^+)	+1	1.00	1.6726×10^{-24}
Neutron (n)	0	1.00	1.6750×10^{-24}

The number of protons determines the identity of the atom. In an atom, the number of protons is equal to the number of electrons. The atom is neutral because the overall charge of the atom is equal to the sum of the charges of the particles which make up the atom.

The **atomic number (Z)** is equal to the number of protons in the atom and is the order in which elements are listed on the periodic table. For example, sodium (Na) has an atomic number of 11. It has 11 protons, and the atom therefore will have 11 electrons. The **mass number (A)** is equal to the sum of the protons and neutrons (the mass of the electrons is so small as to be insignificant). The mass number cannot be identified from the periodic table. It is either given as a superscript prior to the element symbol or after an element name. An atom of sodium, for example may have 12 neutrons. It would have a mass number of 23 (11 protons + 12 neutrons) and would be written either as ^{23}Na or sodium-23.

Example 1: Atomic Number and Mass Number

Consider an atom of oxygen-16. Determine the following for the atom.

1. atomic number
2. number of protons
3. number of electrons
4. mass number
5. number of neutrons

Solution:

1. The atomic number is given for oxygen on the periodic table. The atomic number is 8.
2. The atomic number gives the number of protons. There are 8 protons in this atom.
3. In an atom, the number of protons equals the number of electrons. There are 8 electrons.

Example 1 Continued

4. The mass number is located after the atom name. The mass number is 16.
5. Since,

   mass number (A) = number of protons (atomic number Z) + number of neutrons,

   the number of neutrons is given by A – Z or 16 – 8 = 8. There are 8 neutrons.

Isotopes

Isotopes are atoms of the same element which have different masses due to different numbers of neutrons (different atomic mass). Hydrogen is the only element for which the different isotopes have distinctly different names. Hydrogen-1 is named hydrogen (and makes up the vast majority of the naturally occurring element.) Hydrogen-2 is deuterium. Hydrogen-3 is tritium.

Certain isotopes of elements emit particles and energy (radioactivity, discussed in Chapter 9) that may be useful to trace the behavior of biochemical systems. These isotopes otherwise behave identically to any other isotope of the same element.

A sample of chlorine is principally composed of two isotopes, $^{35}_{17}Cl$ and $^{37}_{17}Cl$, in approximately a 3:1 ratio. Note: in the above representation of the two isotopes, the mass number is the superscript and the subscript is the atomic number. The subscript must be 17 because 17 is the atomic number defining chlorine. It is often included in writing an isotope but is not necessary.

Besides the atomic number, the other common value listed on the periodic table is the **atomic mass**. The atomic mass (measured in atomic mass units, amu) is obtained by averaging the masses of the isotopes that make up the element, giving a greater weighting (in the average) to the element that is more abundant. This value for chlorine is 35.45 amu. Since there is three times more chlorine-35 than chlorine-37, the average is closer to 35 than 37.

Example 2: Isotopes

For the isotopes of carbon: carbon-12 and carbon-14

1. Write the symbols for the isotopes.
2. Calculate the number of protons, neutrons, and electrons for each.

Solution:

1. The numbers 12 and 14 are the mass numbers (A) for the isotopes. The mass number is written as a superscript prior to the element symbol. The atomic number (Z) is found on the periodic table and for carbon is 6. The symbols are therefore:

$$^{12}_{6}C \quad ^{14}_{6}C$$

2. The atomic number of 6 is the number of protons. The atoms are neutral and must have the same number of electrons, that is, 6. As in Example 1, the number of neutrons = Z – A. For carbon-12 the number of neutrons is 12 – 6 or 6. For carbon-14 the number of neutrons is 14 – 6 or 8.

In summary: For $^{12}_{6}C$, there are 6 protons, 6 electrons, and 6 neutrons.

Example 2 Continued

For $^{14}_{6}C$, there are 6 protons, 6 electrons, and 8 neutrons.

2.2 Development of the Atomic Theory

Dalton's Theory

LG 4

The first experimentally based theory of atomic structure was proposed by John Dalton in the early 1800s. Dalton postulated that:

1. All matter consists of tiny particles called atoms.
2. Atoms cannot be created, divided, destroyed, or converted to any other type of atom.
3. Atoms of a particular element have identical properties.
4. Atoms of different elements have different properties.
5. Atoms combine in simple, whole-number ratios.
6. Chemical change involves joining, separating, or rearranging atoms.

Postulates 1, 4, 5, and 6 are presently regarded as true. The discovery of the processes of nuclear fusion, fission, and radioactivity (Chapter 9) has disproved the postulate 2 that atoms cannot be created or destroyed. Postulate 3 that states the atoms of a particular element are identical was disproved by the later discovery of isotopes.

Evidence for Subatomic Particles: Electrons, Protons, and Neutrons

Although Dalton pictured atoms as indivisible, various experiments, particularly those of William Crookes and Eugene Goldstein, indicated that the atom is composed of charged (+ and –) particles.

J. J. Thomson demonstrated the electrical and magnetic properties of cathode rays (Figure 2.3 in the textbook). Crookes observed rays, which he called cathode rays, emanating from the cathode (– charge) of an evacuated (vacuum) tube. Further experiments showed that the ability to produce cathode rays is a characteristic of all materials. In 1897, Thomson announced that cathode rays were streams of negative particles of energy. These particles are electrons.

Similar experiments, conducted by Goldstein, led to the discovery of protons, particles equal in charge to the electron but opposite in sign. Protons were also found to be much heavier than electrons.

The third fundamental atomic particle is the neutron, a particle that has a mass virtually equal to that of the proton and zero charge. The neutron was first postulated in the early 1920s, but it was not until 1932 that James Chadwick demonstrated its existence.

Evidence for the Nucleus

In the early 1900s, it was believed that protons and electrons were uniformly distributed throughout the atom. However, an experiment in 1911 by Hans Geiger led Ernest Rutherford to propose that the majority of the mass and positive charge of the atom was actually located in a small, dense region, the nucleus, and that the small, negatively charged electrons spread across a much larger, diffuse area outside of the nucleus.

2.3 Light, Atomic Structure, and the Bohr Atom

Electromagnetic Radiation

LG 5

The study of the interaction of light and matter is termed **spectroscopy**. Light, or electromagnetic radiation, travels at a speed of 3.0×10^8 m/s (the **speed of light**). **Electromagnetic radiation** is made up of many wavelengths. This range of wavelengths is the electromagnetic spectrum. We often think of light as that which we can see, but the electromagnetic spectrum includes many wavelengths outside of the visible region. Light has a dual nature, having the properties of both waves and particles.

Photons

The particles of light, **photons**, have energies inversely proportional to their wavelength. Radio waves and microwaves have much longer wavelengths than visible light. Longer wavelength causes lower energy radiation and therefore radio waves and microwaves are lower in energy than visible light. X-rays and gamma rays have much shorter wavelengths than visible light. The short wavelengths have much higher energy associated with them. Samples of elements emit certain wavelengths of light when an electrical current is passed through the sample. Different elements emit different patterns (different wavelengths) of light.

The Bohr Atom

Bohr and his contemporaries were puzzled by the emission spectrum of hydrogen. He developed a hypothesis to explain his observations. This evolved into Bohr's atomic theory.

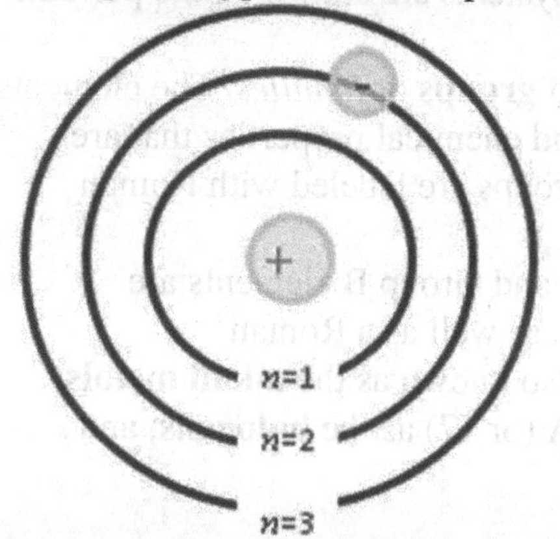

Figure 2.1 The Bohr Atom. One electron is shown, located in the $n = 2$ energy level.

- Electrons are found only in these allowed **energy levels**.
- The allowed energy levels are quantized energy levels, or orbits.
- Atoms absorb energy by excitation of electrons to higher energy levels, farther from the nucleus.
- Atoms release energy by relaxation of electrons to lower energy levels.
- Energy that is emitted upon relaxation, as the electron moves from a higher energy level to a lower energy level, is observed as a single wavelength of light.
- The observed spectral lines, wavelengths of light, are a result of electron transitions between allowed levels in the atom.
- Energy differences may be calculated from the wavelengths of light emitted.

An atom is in the **ground state** when electrons of the atom are in the lowest possible energy levels. When the electron absorbs energy and is promoted to a higher energy level, it is elevated to an **excited state**.

Modern Atomic Theory

Once we consider a system with more than one electron, that is, move beyond the hydrogen atom, the simplicity of the Bohr model does not hold. We now speak of the *probability* of finding an electron in a *region* of space within the principal energy level, referred to as an **atomic orbital**. In certain regions of space where **electron density** is greater, there is a higher probability of finding the electron at any point in time. These atomic orbitals are part of the principal energy levels and they are referred to as sublevels. In Chapter 3, we will see that the orbital model of the atom can be used to predict how atoms can bond together to form compounds. Furthermore, electron arrangement in orbitals enables us to predict various chemical and physical properties of these compounds.

2.4 The Periodic Law and the Periodic Table

By 1869, two scientists, Dmitri Mendeleev and Lothar Meyer, found that if the elements were arranged in order of increasing atomic mass there was regular variation (or periodicity) about their properties (**periodic law**). The periodic table represents this regular variation and we now know that these similar chemical and physical properties correlate with the electronic structure of the atoms.

Numbering Groups in the Periodic Table

LG 7

Two systems currently exist to number the columns (called groups) in the periodic table. The older system uses Roman numerals followed by the letter A or the letter B. The newer system numbers the groups from 1 to 18. Both numbering systems are shown on the periodic table below.

The columns of elements in the periodic table are called **groups** or *families*. The elements of a particular family share many similarities in physical and chemical properties that are related to similarities in electronic structure. The various groups are labeled with Roman numerals, and each is subtitled with the letter A or B.

Group A elements are called **representative elements**, and Group B elements are **transition elements.** Certain families have common names as well as a Roman numeral-letter designation. Group IA (or 1) elements are also known as the **alkali metals**; Group IIA (or 2) as the **alkaline earth metals**; Group VIIA (or 17) as the **halogens**; and Group VIIIA (or 18) as the **noble gases.**

Periods

A horizontal row of elements in the periodic table is referred to as a **period**. The periodic table consists of seven periods, six of which contain 2, 8, 8, 18, 18, 32 and 32 elements. Note that the lanthanide series is a part of period six and the actinide series is a part of period seven.

Alkali Metals — Alkaline Earth Metals — Halogens — Noble Gases

1 IA	2 IIA	3 IIIB	4 IVB	5 VB	6 VIB	7 VIIB	8	9 VIIIB	10	11 IB	12 IIB	13 IIIA	14 IVA	15 VA	16 VIA	17 VIIA	18 VIIIA
1 **H** 1.008																	2 **He** 4.003
3 **Li** 6.941	4 **Be** 9.012											5 **B** 10.81	6 **C** 12.01	7 **N** 14.01	8 **O** 16.00	9 **F** 19.00	10 **Ne** 20.18
11 **Na** 22.99	12 **Mg** 24.31											13 **Al** 26.98	14 **Si** 28.09	15 **P** 30.97	16 **S** 32.07	17 **Cl** 35.45	18 **Ar** 39.95
19 **K** 39.10	20 **Ca** 40.08	21 **Sc** 44.96	22 **Ti** 47.88	23 **V** 50.94	24 **Cr** 52.00	25 **Mn** 54.94	26 **Fe** 55.85	27 **Co** 58.93	28 **Ni** 58.69	29 **Cu** 63.55	30 **Zn** 65.39	31 **Ga** 69.72	32 **Ge** 75.59	33 **As** 74.92	34 **Se** 78.96	35 **Br** 79.90	36 **Kr** 83.80
37 **Rb** 85.47	38 **Sr** 87.62	39 **Y** 88.91	40 **Zr** 91.22	41 **Nb** 92.91	42 **Mo** 95.94	43 **Tc** (98)	44 **Ru** 101.1	45 **Rh** 102.9	46 **Pd** 106.4	47 **Ag** 107.9	48 **Cd** 112.4	49 **In** 114.8	50 **Sn** 118.7	51 **Sb** 121.8	52 **Te** 127.6	53 **I** 126.9	54 **Xe** 131.3
55 **Cs** 132.9	56 **Ba** 137.3	57 **La** 138.9	72 **Hf** 178.5	73 **Ta** 180.9	74 **W** 183.9	75 **Re** 186.2	76 **Os** 190.2	77 **Ir** 192.2	78 **Pt** 195.1	79 **Au** 197.0	80 **Hg** 200.6	81 **Tl** 204.4	82 **Pb** 207.2	83 **Bi** 209.0	84 **Po** (209)	85 **At** (210)	86 **Rn** (222)
87 **Fr** (223)	88 **Ra** (226)	89 **Ac** (227)	104 **Rf** (257)	105 **Db** (260)	106 **Sg** (263)	107 **Bh** (262)	108 **Hs** (265)	109 **Mt** (266)	110 **Ds** (281)	111 **Rg** (272)	112 **Cn** (277)	113 **Nh** (284)	114 **Fl** (285)	115 **Mc** (288)	116 **Lv** (289)	117 **Ts** (294)	118 **Og** (294)

Lanthanide series	58 **Ce** 140.1	59 **Pr** 140.9	60 **Nb** 144.2	61 **Pm** (147)	62 **Sm** 150.4	63 **Eu** 152.0	64 **Gd** 157.3	65 **Tb** 158.9	66 **Dy** 162.5	67 **Ho** 164.9	68 **Er** 167.3	69 **Tm** 168.9	70 **Yb** 173.0	71 **Lu** 175.0
Actinide series	90 **Th** 232.0	91 **Pa** (231)	92 **U** 238.0	93 **Np** (237)	94 **Pu** (242)	95 **Am** (243)	96 **Cm** (247)	97 **Bk** (247)	98 **Cf** (249)	99 **Es** (254)	100 **Fr** (253)	101 **Md** (256)	102 **No** (254)	103 **Lr** (257)

Metals and Nonmetals

A bold zig-zag line runs from top to bottom of the table beginning below boron (B) and ending between tennessine (Ts) and oganesson (Og). This line acts as the boundary between **metals**, to the left, and **nonmetals**, to the right. Metals tend to lose electrons during chemical change, typically conduct heat and electricity, and have a characteristic luster. Nonmetals tend to gain electrons and are poor conductors. Elements straddling the boundary between metals and nonmetals, such as Ge and As (excluding Al), have properties intermediate between metals, and nonmetals and are often termed **metalloids**.

Information Obtained from the Periodic Table

The atomic number and the atomic mass of each element are available from the periodic table. More detailed periodic tables may provide information such as electron arrangement, relative sizes of atoms, and most probable ion charges.

Example 4: Periodic Table and Classification of the Elements

For each of the following symbols, provide the name of the element and classify the element in as many of the classification schemes as you can from the discussion above.

1. Li	3. Mg	5. Cu	7. I
2. Rn	4. As	6. Al	8. H

Example 4 Solution:

1. Li: lithium. This element is in the IA (or 1) group, an alkali metal. The "A" associated with the group number lets us know it is a representative element. Since it lies to the left of the bold diagonal boundary line, lithium is a metal. Additionally, lithium has an atomic number of 3 and an atomic mass of 6.94.
2. Rn: radon. This element is in the VIIIA (or 18) group, a noble gas. Again, the "A" lets us know it is a representative element. It lies to the right of the boundary line and is thus a nonmetal. Radon has an atomic number of 86 and an atomic mass of 222.02.
3. Mg: magnesium. This element is in the IIA (or 2) group, an alkaline earth metal. It too is a representative element and is a metal. Magnesium has an atomic number of 12 and an atomic mass of 24.31.
4. As: arsenic. This element is in the VA (or 15) group and is a representative element. It directly lies on the boundary line and is therefore a metalloid. Arsenic has an atomic number of 33 and an atomic mass of 74.92.
5. Cu: copper. This element is a IB (or 11) element. It is a metal and (because of the B designation) a transition element. Because all transition elements are metals, these are also correctly called *transition metals.* Copper has an atomic number of 29 and an atomic mass of 63.55.
6. Al: aluminum. This element is in the group IIIA (or 13) and is a representative element. Even though it borders the boundary line, it is *not* a metalloid. It is a metal. Aluminum has an atomic number of 13 and an atomic mass of 26.98.
7. I: iodine. This element is a VIIA (or17) element also named the halogen group. It is a representative element and a nonmetal. Iodine has an atomic number of 53 and an atomic mass of 126.90.
8. H: hydrogen. This is in the IA (or 1) group but is not considered an alkali metal, nor is it a metal. It is a nonmetal as it is the exception to the boundary rule. It is, however, a representative element. Hydrogen has an atomic number of 1 and an atomic mass of 1.01.

2.5 Electron Arrangement and the Periodic Table

LG 9

The most important factor in chemical bonding is the arrangement of the electrons in the atoms that are combining. The periodic table provides us with a great deal of information about the electron arrangement, or **electron configuration**, of atoms.

The Quantum Mechanical Atom

Due to the failings of the Bohr model for atoms with more than one electron, another model was developed. Erwin Schröedinger described electrons in atoms in probability terms and developed equations that emphasize the wavelike character of electrons. His theory, often described as quantum mechanics, incorporates Bohr's principal energy levels ($n = 1, 2, 3...$), which are made up of one or more sublevels – not part of Bohr's model. Each sublevel contains one or more atomic orbitals.

Principal Energy Levels, Sublevels, and Orbitals

The **principal energy levels** are designated $n = 1, 2, 3$, and so forth with higher values of n representing a greater average distance from the nucleus. These principle energy levels are regions where electrons may be found. These energy levels may contain **sublevels**. The number of possible sublevels in a principal energy level is also equal to n. In the first energy level, when $n = 1$, there can be only one sublevel. The second energy level, $n = 2$, allows two sublevels; $n = 3$ allows 3 sublevels and so forth.

The sublevels are designated with the names, *s, p,* and *d.* (Additional higher energy sublevels such as *f* sublevels, exist, however, are only important for the heaviest elements and will not factor into our discussion in this book). These sublevels increase in energy

$$s < p < d$$

Both the principal energy level and type of sublevel are specified when describing the location of an electron. For example: *1s*, 2*s*, 2*p*. The first principal energy level ($n = 1$) has one possible sublevel, 1*s*. The second principal energy level ($n = 2$) has two possible sublevels: 2*s* and 2*p*. The third principal energy level ($n = 3$) has three possible sublevels: 3*s*, 3*p*, and 3*d*. Figure 2.2 below shows a schematic of the electronic structure of an atom. The top half shows the level and sublevel designation.

An **atomic orbital** is a specific region of a sublevel containing a maximum of two electrons. The *s* sublevel contains only one orbital, the *p* sublevel contains three orbitals, and the *d* sublevel contains five orbitals. Each orbital may be empty, may contain one electron, or maybe filled, containing two electrons. The *s*, *p*, and *d*, sublevels have maximum capacities of 2, 6, and 10 electrons respectively. The bottom half of Figure 2.2 gives the summary of the orbital and electron information.

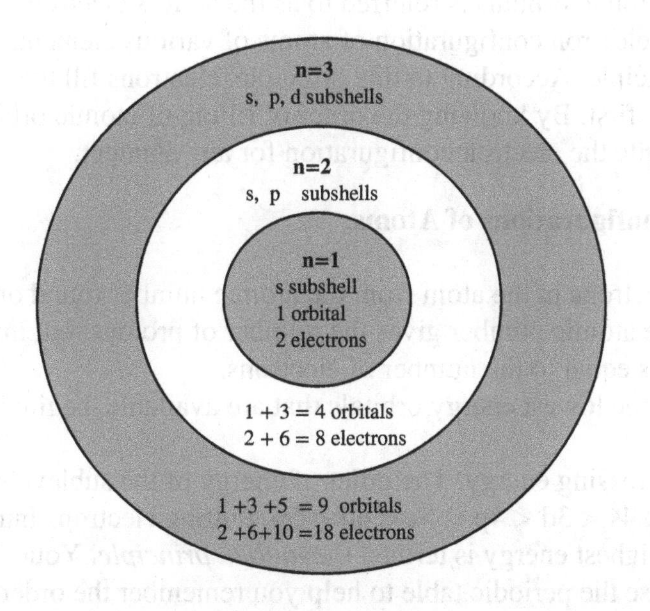

Figure 2.2 Schematic of the principal levels and sublevels. Top half of the schematic gives sublevel information, and the bottom half gives orbital and electron information.

These concentric circles *do not* represent orbits (as in Bohr's model of the atom) but regions in space where the s, p, and d sublevels are located.

> Note: You need to know all the information contained in the diagram, that is, the number of sublevels in each level and the number of orbitals and electrons in each sublevel.

Each type of orbital has its own characteristic shape:

- The *s* orbital is spherically symmetrical; a model appears as a Ping-Pong ball with its center corresponding to the intersection of imaginary *x, y,* and *z* coordinates.
- There exist three kinds of *p* orbitals, each identical in shape (often modeled as a dumbbell). They differ only in their orientation in space, along the hypothetical *x*-axis, (p_x), *y*-axis, (p_y), and *z*-axis, (p_z).
- The *d* orbitals are more complex; the textbook focuses almost exclusively on *s* and *p* orbitals in subsequent discussion.

Each atomic orbital has a maximum capacity of two electrons. The electrons are perceived to *spin* on an imaginary axis, and the two electrons in the same orbital must have opposite spins, clockwise and counterclockwise. Two electrons in one orbital that possess opposite spins are referred to as *paired* electrons.

Example 5: Principal Levels and Sublevels
Write the symbols of all the possible sublevels found in the third principal energy level ($n = 3$). Write the symbols in order of increasing energy.

Solution:
Since n = 3, there are three possible sublevels. Put the number 3 with each of the names of the sublevels, giving, in order of increasing energy: 3*s*, 3*p*, and 3*d*

Electron Configurations

LG 10

The arrangement of electrons in atomic orbitals is referred to as the atom's electron configuration. We may represent the electron configuration of atoms of various elements using the aufbau, or building-up, principle. According to this principle, electrons fill the lowest-energy orbital that is available first. By knowing the order of filling of atomic orbitals, lowest to highest energy, you may write the electron configuration for any element.

Guidelines for Writing Electron Configurations of Atoms

1. Obtain the total number of electrons in the atom from the atomic number found on the periodic table. Remember, the atomic number gives the number of protons, but in an atom, the number of protons is equal to the number of electrons.
2. Electrons in the atom occupy the lowest energy orbitals that are available, beginning with 1s.
3. Fill subshells according to increasing energy. The order of energy of the sublevels are: 1s < 2s < 2p < 3s < 3p < 4s < 3d < 4p < 5s < 4d < 5p. Putting electrons into their orbitals from lowest to highest energy is termed the *aufbau principle*. Your textbook describes a way to use the periodic table to help you remember the order.
4. Remember:
   - The s sublevel has one orbital and can hold two electrons.
   - The *p* sublevel has three orbitals. The electrons will half-fill before completely filling the orbitals for a maximum of six electrons [Recall that each orbital can

contain no more than 2 electrons (*Pauli exclusion principle*) and since orbitals in a given sublevel are equal in energy, electrons half-fill each of the orbitals in the sublevel before filling with the second electron (*Hund's rule*).]

- The *d* sublevel has five orbitals. Again, the electrons will half-fill before completely filling the orbitals for a maximum of ten electrons.

The electron configuration has the following form:

Principal energy level ⟶ $1s^2$ ⟵ Number of electrons in sublevel

Sublevel and orbital

Orbital diagrams are an alternate and more detailed way of describing the arrangement of electrons in an atom. Each orbital is represented as a box with electrons represented by arrows (up and down arrow representing the spins of the electrons.)

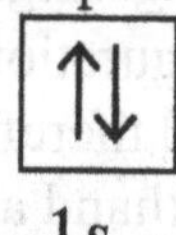

1*s*

Shorthand Electron Configurations

Electron configurations may be abbreviated by substituting the symbol for a noble gas (in brackets, [ ]) for the part of the electron configuration represented by that noble gas. To do this, locate the element on the periodic table and find the nearest noble gas preceding the element. Put the noble gas in square brackets. Move to the left of the periodic table and place electrons in the next available sublevels, finish at the element in question. For example, the electron configuration of strontium, Sr, would be quite long: $1s^2 2s^2 2p^6 3s^2 3p^6 4s^2 3d^{10} 4p^6 5s^2$. However, the noble gas that is located before Sr on the periodic table is Kr, which has the following electron configuration: $1s^2 2s^2 2p^6 3s^2 3p^6 4s^2 3d^{10} 4p^6$. The electron configuration of Sr, then, can be abbreviated as:
Sr: $[Kr]5s^2$.

Example 6: Aufbau Principle
Name the two elements that have electrons only in the first principal energy level.

Solution:
The first principal energy level is $n = 1$. There is only one sublevel (1*s*) in this level. There is only one orbital in the 1*s* sublevel, and it can only have 2 electrons in the orbital. The two elements which have no more than 2 electrons are hydrogen and helium.

Example 7: Electron Configuration
Give the electron configuration for each of the following elements:

1. He
2. O
3. Na
4. Ar
5. Zn

Example 7 Solution:

1. He has an atomic number of 2. Therefore, an atom of helium has 2 electrons. Starting with the lowest energy sublevel ($1s$), this sublevel can hold two electrons giving the electron configuration $1s^2$.
2. O has 8 electrons. $1s$ holds two electrons, ($1s^2$) leaving 6 more electrons to assign. The next available sublevel is $2s$. It too can hold 2 electrons, giving: $1s^2 2s^2$. For the remaining 4 electrons, we will write the next higher sublevel of $2p$. This sublevel can hold a maximum of 6 electrons. We only have 4 left to place. The electron configuration is therefore, $1s^2 2s^2 2p^4$.
3. Na has 11 electrons to place in orbitals. Filling according to the order (keeping in mind the maximum number of electrons each sublevel can hold), the configuration is $1s^2 2s^2 2p^6 3s^1$. Since the noble gas neon has an electron configuration of $1s^2 2s^2 2p^6$. The electron configuration of sodium could also be written in shorthand as $[Ne]3s^1$.
4. Ar has 18 electrons to place in orbitals, giving: $1s^2 2s^2 2p^6 3s^2 3p^6$.
5. Zn has 30 electrons. The electron configuration is $1s^2 2s^2 2p^6 3s^2 3p^6 4s^2 3d^{10}$. Remember, there are 5 orbitals in the d sublevel and therefore it can hold a maximum of 10 electrons. This could be written in shorthand as $[Ar]4s^2 3d^{10}$.

Example 8: Orbital Diagram

What is the orbital diagram for oxygen?

Solution:

As we see in number 2 from the above example, the electron configuration of O is $1s^2 2s^2 2p^4$. Represent the orbitals in the sublevels as boxes. Remember, there is one orbital in the s sublevel and three orbitals in the p sublevel. The electrons are placed in the orbitals as follows:

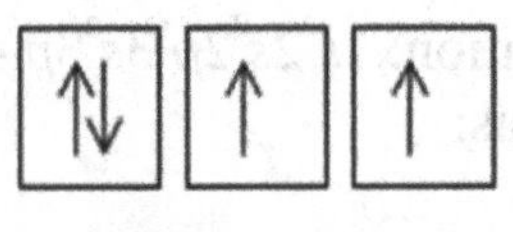

$1s$ $2s$ $2p_x$ $2p_y$ $2p_z$

The three p orbials in the $2p$ sublevel differ only in their orientation in 3-deminsional space. The orbitals are perpendicular to each other with x, y and z subscripts denoting their different orientations.

2.6 Valence Electrons and the Octet Rule

Valence Electrons

Outermost electrons in an atom, which are electrons in the highest principal energy level, are referred to as **valence electrons**. For representative elements, the number of valence electrons in an atom corresponds to the number of the group in which the atom is found, (the Roman numeral value). The exception is helium, whose group number is VIIIA (or 18). Since helium only has two electrons total, it cannot have eight valence electrons. The number of valence electrons in an atom tells us a lot about the properties of the element and how it reacts to form compounds. You will find that you refer to this number often.

The principal energy level of where these valence electrons are located is given by the period number. For example, nitrogen is in the second period (2nd horizontal row of the periodic table). The valence electrons are therefore in the $n = 2$ level. Additionally, nitrogen is in the Group VA (or 15) family and therefore has 5 valence electrons.

LG 11

The Octet Rule

Elements in the last family, the noble gases, have either two valence electrons (helium) or eight valence electrons (neon, argon, krypton, xenon, and radon). Their most important property is their extreme stability. A full $n = 1$ energy level (as in helium) or an outer octet of electrons (two in the *s* and 6 in the *p* sublevel) is responsible for this unique stability.

Atoms of elements in other groups are more reactive than the noble gases because they are, in the process of chemical reaction, trying to achieve a more stable "noble gas" configuration by gaining or losing electrons. This is the basis of the **octet rule**. In chemical reactions atoms will typically gain, lose, or share the minimum number of electrons necessary to achieve a more stable energy state (the octet of valence electrons). An advantage to the shorthand electron configuration, in addition to the brevity, is that the valence electrons are clearly evident.

Ions

Ions are charged particles that result from a gain of one or more electrons by the parent atom (forming negative ions, or **anions**) or a loss of one or more electrons from the parent atom (forming positive ions, or **cations**). For example, a hydrogen atom may lose an electron and thus become a hydrogen cation. This would be written as H^+. If a hydrogen atom were to gain an electron and form the anion, it would be written as H^-.

To represent the process of losing or gaining electrons, equations can be used. The following are examples of atoms losing electrons to form cations.

$Mg \rightarrow 2e^- + Mg^{2+}$ • Two electrons were lost by the magnesium atom.

$K \rightarrow e^- + K^+$ • One electron was lost by the potassium atom.

A negative ion on the other hand, is formed by a gain of one or more electrons by the given atom. This is represented as follows:

$S + 2e^- \rightarrow S^{2-}$ • Two electrons were gained by the sulfur atom.

$Cl + e^- \rightarrow Cl^-$ • One electron was gained by the chlorine atom.

Example 9: Ions and the Number of Protons and Electrons.

Determine the number of protons and electrons in each of the following ions.

1. S^{2-}
2. Ba^{2+}
3. H^+
4. F^-

Example 9 Solution:

1. First determine the number of protons. This is determined from the atomic number on the periodic table. Sulfur is element 16. It has 16 protons. To be negatively charged, electrons are gained. There must be 2 more electrons than protons to give an 2^- charge to the ion. Thus, the ion has 18 electrons.
2. Barium is element 56, and therefore it has 56 protons. For the 2^+ charge of the ion it has to have two fewer electrons than protons. Ba^{2+} has 54 electrons.
3. An atom of hydrogen has one proton and one electron. For the H^+ ion, one electron is lost giving one proton and no electrons. The H^+ ion is often called a "proton" because that is all that is there.
4. The F^- ion must have one more electron than the fluorine atom. It has 9 protons and 10 electrons.

Ion Formation and the Octet Rule

LG 11

Metallic elements tend to form cations. These ions are more stable than their corresponding neutral atoms. For the representative elements, the ion is **isoelectronic** (that is, it has the same electron configuration) with its nearest noble gas neighbor and has an octet of electrons in its outermost energy level.

Nonmetallic elements tend to gain electrons to become isoelectronic with the nearest noble gas element, forming anions.

The transition metals, like the representative metals, tend to lose electrons forming cations. However, the transition elements are characterized as "variable valence" elements; depending upon the type of substance they react with, they may form more than one stable ion. For example, iron has two stable ionic forms: Fe^{2+} and Fe^{3+}. The charges of the transition metals cannot easily be determined by looking at the periodic table.

Example 10: Predicting the Charge of a Metal Ion

Use the octet rule to predict the charge on an ion of sodium. Write the equation representing the process of forming a sodium ion from the atom.

Solution:

Sodium is a Group IA (or 1) metal and has the shorthand electron configuration $[Ne]3s^1$. Both the group number and the electron configuration show that the atom has one electron in its valence shell. To obtain a noble gas configuration, sodium will lose one electron giving the Na^+ ion. The equation is written as follows:

$$Na \rightarrow Na^+ + e^-$$

Example 11: Predicting the Charge of a Nonmetal

Use the octet rule to predict the charge on an ion of oxygen. Write the equation representing the process of forming a oxygen ion (oxide) from the atom.

Example 11 Solution:

Oxygen is in Group VIA (or 16). It has 6 valence electrons and an electron configuration of [He]$2s^2 2p^4$. Oxygen will gain two electrons to obtain an octet giving O^{2-}. The gaining of electrons can be represented by the equation:

$$O + 2e^- \rightarrow O^{2-}$$

2.7 Trends in the Periodic Table

LG 12

Atomic Size

The size of the atom will be determined principally by two factors:

1. The principal energy level (*n*-level) in which the outermost electrons are located increases as we go down a group (recall that the outermost *n*-level correlates with period number). Consequently, atomic size increases down a group.
2. As the magnitude of the positive charge of the nucleus increases, the nucleus' ability to "pull" on all of the electrons increases, and the electrons are drawn closer to the nucleus. As a result, atomic size decreases from left to right across a period.

Ion Size

Three generalizations can be made about the size of ions:

1. Positive ions (cations) are smaller than the parent atom.
2. Negative ions (anions) are larger than the parent atom.
3. The greater the positive charge on a cation (with the same parent atom), the smaller the radius. $Cu^{2+} < Cu^{+}$ and $Fe^{3+} < Fe^{2+}$.

Ionization Energy

The energy required to remove an electron from an isolated atom in the gas phase is the **ionization energy**. The magnitude of the ionization energy correlates with the strength of the attractive force between the nucleus and the outermost electron.

1. As we go down a group, the ionization energy decreases since the atom's size is increasing. The outermost electron is progressively farther from the nuclear charge and hence easier to remove.
2. As we go across a period, atomic size decreases, since the outermost electrons are closer to the nucleus, more tightly held, and more difficult to remove. Therefore, the ionization energy must increase.

A correlation does indeed exist between trends in atomic size and ionization energy. Atomic size decreases and ionization energy increases from bottom to top of a group. Likewise, atomic size decreases and ionization energy increases from left to right in a period. Note also that ionization energies are highest for the noble gases; this accounts for the extreme stability and nonreactivity of the noble gases.

Electron Affinity

The energy change when a single electron is added to a neutral atom in the gaseous state is known as the **electron affinity**. Electron affinity is a measure of the ease of forming negative ions. A large value of electron affinity (a large release of energy) indicates that the atom becomes more stable as it becomes a negative ion (through the process of gaining an electron).

Periodic trends for electron affinity are as follows:

1. Electron affinities generally decrease as we go down a group.
2. Electron affinities generally increase as we go across a period.

Be aware that exceptions to these general trends do exist.

Self Test for Chapter Two

1. Which subatomic particle of an atom has the least mass?
2. How many protons and electrons are in a magnesium atom?
3. An isotope contains 16 protons, 16 neutrons and 16 electrons. Write the isotope symbol (including mass number and atomic number).
4. Symbolically represent the three isotopes of the element hydrogen.
5. Which postulates of Dalton's theory of atoms are still believed to be true?
6. William Crookes and J.J. Thomson ran experiments showing that streams of negative particles of energy could emanate from materials. How did this help scientists understand the model of the atom?
7. Radio waves have relatively long wavelengths and X-rays have relatively short wavelengths. Which is higher in energy?
8. As an electron in an excited state decreases in energy to the ground state, how is the energy released?
9. Determine which of the following are metalloids: aluminum, copper, boron, silicon, carbon, phosphorus, nickel.
10. What are the possible sublevels in the $n = 2$ principal energy level?
11. List the sublevels in order of increasing energy.
12. How many electrons can be contained in the second principal energy level, $n = 2$?
13. Which orbital has a spherical shape?
14. How many orbitals are in each of the s, p, and d sublevels.
15. What is the relationship of the spins of two electrons in the same orbital?
16. Write the electron configuration of carbon.
17. Draw the orbital diagram for boron.
18. Write the shorthand electron configuration of barium.

19. How many valence electrons are in each of the representative groups (IA – VIIIA or 1, 2 and 13-17)?
20. How many valence electrons are in sulfur?
21. Name the only noble gas with two valence electrons.
22. What is the group number of the halogen family?
23. What do we call an ion that has the same electronic arrangement as its nearest noble gas neighbor?
24. Use the octet rule to determine the charge on an aluminum ion.
25. How many protons and electrons are in a chloride ion, Cl^-?
26. Give the number of protons and electrons found in a lithium atom and a lithium ion, Li^+. Write the equation showing how a lithium atom can form a lithium ion.
27. Write the equation representing the conversion of the phosphorus atom to the phosphorus ion.
28. What is the charge of the ion of any alkaline earth element?
29. Rank the following from smallest to largest atomic radius: Mg, P, S.
30. Which is larger, S or S^{2-}?
31. Which element has the highest ionization energy, K or Br?
32. Define electron affinity.

3 Structure and Properties of Ionic and Covalent Compounds

Learning Goals

1. Draw Lewis symbols for representative elements and their respective ions.
2. Classify compounds as having ionic, polar covalent, or nonpolar covalent bonds.
3. Write the formula of a compound when provided with the name or elemental composition of the compound.
4. Name inorganic compounds using standard naming conventions, and recall the common names of frequently used substances.
5. Predict differences in physical state, melting and boiling points, solid-state structure, and solution chemistry that result from differences in bonding.
6. Draw Lewis structures for covalent compounds and polyatomic ions.
7. Explain how the presence or absence of multiple bonding relates to bond length, bond energy, and stability.
8. Use Lewis structures to predict the geometry of molecules.
9. Describe the role that molecular geometry plays in determining the polarity of compounds.
10. Use polarity to determine solubility, and predict the melting and boiling points of compounds.

Introduction

This chapter describes the role of electrons in bond formation between atoms. Systems of naming the resultant compounds are discussed as well as the procedure for writing formulas based on the names of the compounds. The chemical and physical properties of these compounds are related to structure and bonding.

3.1 Chemical Bonding

When two atoms are joined together to make a chemical compound, the force of attraction between the two species is referred to as a **chemical bond**. Interactions involving valence electrons are responsible for the chemical bond.

Lewis Symbols

LG 1

The **Lewis symbol** is a convenient way of representing atoms singly or in combination. Its principal advantage is that only valence electrons are shown. This results in simpler structures and greater clarity. The chemical symbol of the atom is written. This symbol represents the nucleus and all of the lower-energy nonvalence electrons, which do not directly participate in bonding. The valence electrons are indicated by dots (•) arranged around the atomic symbol. To draw Lewis symbols, determine the number of valence electrons an atom has and place that number of dots around the element symbol. Begin by placing one dot on each of the four sides and then pair up the dots until all the valence electrons are represented by a dot. Figure 3.1 gives two examples.

•Mg•   •N̈•

Figure 3.1 Magnesium is in the group IIA and has two valence electrons. This is represented by two dots around the Mg symbol. Nitrogen is in group VA and therefore has five valence electrons.

Principal Types of Chemical Bonds: Ionic and Covalent

LG 2

Ionic bonding is characterized by an electron transfer process occurring prior to bond formation. In *covalent bonding*, electrons are shared between atoms in the bonding process.

The essential features of an **ionic bond** (the attraction of opposite charges) are as follows:

- Metals tend to form cations because they have low ionization energy and low electron affinity.
- Nonmetals tend to form anions because they have high ionization energy and high electron affinity.
- Ions are formed by the transfer of electrons
- The oppositely charged ions formed are held together by an electrostatic force.
- Reactions between metals and nonmetals tend to result in ionic bonds.

To write the formula of an ionic compound based on the identities of the two component ions, determine the charge on each ion (this can usually be determined by the group in the periodic table) and then combine the cation and anion in such a way that the resulting formula has no net charge (the total negative charge equals the total positive charge).

Although we often refer to ionic compounds as ion pairs, in the solid state these ion pairs do not actually exist as individual units. Positive and negative ions arrange themselves in a regular, three-dimensional, repeating array known as a **crystal lattice**.

When electrons are shared rather than transferred, the shared electron pair is referred to as a **covalent bond**. Compounds characterized by covalent bonding are called covalent compounds. Covalent bonds tend to form among atoms with similar tendencies to gain or lose electrons. **Molecules** are covalently bonded compounds and are made up of discreet groups of atoms bonded together, unlike the massive three-dimensional crystals of ionic compounds. The most obvious examples are the diatomic molecules H_2 as well as N_2, O_2, F_2, Cl_2, I_2, and Br_2. Bonding in these molecules is totally covalent because there is no net tendency for electron transfer between identical atoms.

When covalent bonds are formed, the sharing occurs in most cases to satisfy the octet rule for the elements involved. This can be shown using **Lewis structures**, see Figure 3.2. Lewis structures show the location of both bonding electrons and **lone pair** (unshared) electrons.

$$:\ddot{\underset{\cdot\cdot}{F}}\cdot + \cdot\ddot{\underset{\cdot\cdot}{F}}: \rightarrow :\ddot{\underset{\cdot\cdot}{F}}:\ddot{\underset{\cdot\cdot}{F}}:$$

Figure 3.2 The fluorine atoms on the left side of the equation each have only seven electrons. By sharing the electrons, each fluorine atom is surrounded by eight electrons.

Two atoms do not have to be identical in order to form a covalent bond. Compounds such as hydrogen fluoride (HF), water (H_2O), methane (CH_4), and ammonia (NH_3) are common examples. When nonmetals are bonded to nonmetals, they will form covalent bonds. Look again at the examples of compounds which contain covalent bonds above, and notice that each is a compound containing only nonmetals.

Polar Covalent Bonding and Electronegativity

LG 2

Polar covalent bonds, like covalent bonds, are based on the concept of electron sharing; however, the sharing is unequal and based on the electronegativity difference between joined atoms. **Electronegativity** is a measure of the tendency of an atom in a molecule to attract shared electrons.

In a molecule composed of two atoms of the same element, as in F_2 (see Figure 3.2) the sharing is completely equal and the covalent bond is nonpolar. However, if the molecule is composed of two different elements, the electronegativities of the two elements will be different. The sharing will be unequal with the more electronegative atom being more electron rich (partially negative, δ^-) and the less electronegative atom being electron deficient (partially positive, δ^+).

Electronegativity is represented by a scale derived from the measurement of energies of chemical bonds. Refer to Figure 3.4 in the textbook for the electronegativity values of the elements. The most electronegative element is fluorine, F, with a value of 4.0, and the least electronegative elements are Cs and Fr, each of which has a value of 0.7. The periodic trends for electronegativity, which increases from left to right and decreases from top to bottom, are similar to both ionization energy and electron affinity. If the electronegativity difference between two elements is less than 0.5 the bond is considered nonpolar covalent. If the electronegativity difference lies between 0.5 and 2.0 the bond is considered polar covalent. Ionic bonds form when the electronegativity difference exceeds 2.0.

Example 1: Classify Bond Polarity
Use electronegativity values (see Figure 3.4 in the textbook) to classify the bonds in these compounds as ionic, nonpolar covalent or polar covalent.

1. LiF
2. O_2
3. HF
4. NH_3

Example 1 Solution:

1. Lithium has an electronegativity value of 1.0. F has an electronegativity value of 4.0. The electronegativity difference (4.0 - 1.0 = 3.0) is greater than 2.0. Therefore, the bond is an ionic bond.
2. Oxygen has an electronegativity value of 3.5. Since the bond is between the same element, the electronegativity difference (3.5 - 3.5 = 0) is less than 0.5. Therefore, the bond is nonpolar. Bonds between identical atoms are always nonpolar covalent.
3. Hydrogen has an electronegativity value of 2.1. F has an electronegativity value of 4.0. The electronegativity difference (4.0- 2.1 = 1.9) is greater than 0.5 and less than 2.0. Therefore, the bond is polar covalent.
4. In NH_3, the three hydrogen atoms are each bonded to the nitrogen atom. Nitrogen has an electronegativity value of 3.0, and hydrogen has an electronegativity value of 2.1. The electronegativity difference (3.0- 2.1 = 0.9) is greater than 0.5 and less than 2.0. Hence, the bond is polar covalent.

3.2 Naming Compounds and Writing Formulas of Compounds

LG 3

LG 4

Proper use of **nomenclature**, the assignment of a correct and unambiguous name to every chemical compound, is fundamental to the study of chemistry. The student must be able to write the name of a compound given the formula and the formula of a compound given the name. The naming system is different for ionic compounds and covalent compounds.

Ionic Compounds

The "shorthand" symbol for a compound is its **formula**. Examples are: NaCl, $MgBr_2$, and Li_2O. The formula identifies the number and type of the various atoms that compose the compound. The number of like atoms is denoted by a subscript. The presence of one atom is implied when no subscript is present.

The names given to ionic compounds are based upon their formulas, with the name of the cation appearing first, followed by the anion name. The positive ion is simply the name of the element, while the negative ion is named as the stem of the element's name joined to the suffix *-ide*.

Example 2: Naming Ionic Compounds.

Give the correct name for each of the following ionic compounds.

1. NaCl	3. $AlCl_3$	5. Ba_3N_2
2. Li_2S	4. CaO	6. MgO

Solution

In each case, name the metal first, then follow it with the nonmetal using the *–ide* suffix.

1. sodium chloride	3. aluminum chloride	5. barium nitride
2. lithium sulfide	4. calcium oxide	6. magnesium oxide

If the cation and anion exist in only one common charged form, there is no ambiguity between formula and name. Sodium chloride consists of Na^+ and Cl^- and therefore must be NaCl. Lithium sulfide is made of Li^+ and S^{2-} and must be Li_2S, so that the sum of positive and negative charges is zero. With many elements, such as the transition metals, several ions of different charge may exist. Fe^{2+}, Fe^{3+} and Cu^+, Cu^{2+} are common examples. Clearly, an ambiguity exists if we use the name iron for both Fe^{2+} and Fe^{3+} or copper for Cu^+ and Cu^{2+}. Two systems have been developed that avoid this problem: the Stock system and the common nomenclature system.

In the Stock system for naming an ion (the systematic name), a Roman numeral indicates the magnitude of charge of the cation. In the older common nomenclature, the suffix *-ous* indicates the lower of the ionic charges, and the suffix *-ic* indicates the higher ionic charge. Systematic names are preferred; they are easier and less ambiguous.

Example 3: Naming Ionic Compounds Containing Transition Metals Using the Stock System

Give the correct systematic name for each of the following ionic compounds that contains transition metal ions in its structure.

1. $FeCl_3$ 2. $FeCl_2$

Solution:

When a transition metal is found in an ionic compound, we must first find the charge on the transition metal before we can name the compound.

1. To determine the charge on the metal, consider the charge of the anion it is connected to. In $FeCl_3$, the iron is connected to three Cl^- ions, for a total of 3 negative charges.

   The amount of positive charge *must* equal the amount of negative charge. Therefore the charge on Fe is 3^+.

   Use the Roman numeral (III) after the metal name, giving the name iron(III) chloride.

2. In this compound, the iron is connected to two chloride ions (Cl^-) for a total negative charge of 2^-. The iron ion must therefore have a 2^+ charge to make the compound neutral.

   Use the Roman numeral (II) after the metal name, iron(II) chloride.

Ions consisting of only a single atom are said to be **monatomic ions**. In contrast, **polyatomic ions**, such as the hydroxide ion, OH^-, are composed of two or more atoms bonded together. The polyatomic ion has an overall positive or negative charge. Some common polyatomic ions are listed in Table 3.1 below (a more inclusive list is in Table 3.3 of the textbook.) Your instructor may suggest that several of the most common polyatomic ions be committed to memory.

TABLE 3.1 Common Polyatomic Cations and Anions

Ion	Name
H_3O^+	hydronium
NH_4^+	ammonium
NO_3^-	nitrate
SO_4^{2-}	sulfate
OH^-	hydroxide
CN^-	cyanide
PO_4^{3-}	phosphate
CO_3^{2-}	carbonate
HCO_3^-	bicarbonate
CH_3COO^- or $C_2H_3O_2^-$	acetate

Example 4: Naming Ionic Compounds with Polyatomic Ions

Name the following ionic compounds. Use Roman numerals when needed.

1. $Ca_3(PO_4)_2$
2. $Mg(C_2H_3O_2)_2$
3. CsOH
4. $FeSO_4$

Solution:

1. The cation is calcium, a Group IIA (or 2) metal that always has a 2^+ charge as an ion. There is no need for a Roman numeral. The anion PO_4^{3-} is phosphate. Therefore, the name is calcium phosphate.
2. The cation is magnesium, again a Group IIA (or 2) metal that always has a 2^+ charge as an ion. The anion is the acetate ion. The ionic compound is named magnesium acetate.
3. The cation is cesium, a Group IA (or 1) metal which will not need a Roman numeral. The anion is hydroxide. Hence, the compound is cesium hydroxide.
4. The cation is iron and is a transition metal. You will need to give the charge of the metal as a roman numeral. It is connected to the SO_4^{2-} ion (note: you will need to know the charge of the anion to determine the correct name.) Since sulfate has a 2^- charge, iron has a 2^+ charge; the name is iron(II) sulfate.

It is equally important to write the correct formula when given the compound name. It is essential to be able to predict the charge of monatomic ions from the periodic table and *know* the charge and formula of polyatomic ions by memory. Remember, the relative number of positive and negative ions in the unit must result in a unit (compound) charge of zero.

Example 5: Determine the Formula for an Ionic Compound from the Name

Write the formulas of the following ionic compounds:

1. sodium chloride
2. magnesium phosphate
3. iron(III) hydroxide
4. ammonium carbonate

Example 5 Solution:
To determine the formula from the name, first write the charge of each ion, then combine the ions to make a neutral compound.

1. Na is a Group IA (or 1) metal, and thus the ion is Na^+. Cl is in Group VIIA (or 17)and therefore is Cl^-. To make the compound neutral, the formula must be 1:1, NaCl.

2. Mg is a Group IIA (or 2) metal and is therefore Mg^{2+}. Phosphate is the polyatomic ion PO_4^{3-}.To make the compound neutral, determine the least common multiple of 2 and 3, which is the number 6. Three Mg^{2+} ions are needed to make a total positive six charge. Two PO_4^{3-} are needed to make negative six (2 x 3^-). The formula is therefore $Mg_3(PO_4)_2$.

   Note: parentheses are needed around polyatomic ions when there is more than one in the formula.

3. The charge of Fe is not found from the periodic table; it is given in the compound name as a Roman numeral (III), indicating a 3^+ charge, hence, Fe^{3+}. Hydroxide is OH^-. A neutral compound is formed if there are three OH^- for every Fe^{3+}, giving $Fe(OH)_3$.

4. This compound contains two polyatomic ions, NH_4^+ and CO_3^{2-}.
   To balance the charges, two NH_4^+ are needed for each CO_3^{2-}, giving $(NH_4)_2CO_3$.

   Note: since there is only one CO_3^{2-}, no parentheses are needed around it.

Covalent Compounds

LG 4

Covalent compounds are formed by the reaction of nonmetals. The convention used for naming covalent compounds is as follows:

1. The names of the elements are written in the order in which they appear in the formula (the less electronegative atom is always listed first).
2. A prefix indicating the number of each kind of atom found in the unit is placed before the name of the element. (See Table 3.2 below.)
3. If only one atom of a particular kind is the first element present in the molecule, the prefix *mono-* is usually omitted from that first element.
4. The stem of the name of the last element is used with the suffix *-ide*.
5. The final vowel in a prefix is often dropped before *oxide*.

TABLE 3.2 Prefixes Used to Denote Number of Atoms in a Compound

Prefix	No. of Atoms	Prefix	No. of Atoms
Mono-	1	Hexa-	6
Di-	2	Hepta-	7
Tri-	3	Octa-	8
Tetra-	4	Nona-	9
Penta-	5	Deca	10

Common names are often used. For example, H_2O is water and NH_3 is ammonia. Organic compounds have their own nomenclature system. For example, C_2H_5OH (ethanol) is ethyl alcohol and $C_6H_{12}O_6$ is glucose. You will begin learning the organic system in Chapter 10. In this chapter we will focus on the conventions for naming covalent compounds.

Example 6: Naming Covalent Compounds
Name the following covalent compounds (oxides):
1. CO_2 3. SO_2 5. N_2O_4
2. CO 4. SO_3 6. N_2O_3

Solution:
A prefix must be used to indicate the number of atoms of each nonmetal in the covalent compound.

1. Carbon dioxide. Mono- is usually not used to designate one of the element, especially if there is only one of the first element in the formula.
2. Carbon monoxide. The prefix mono- is used here. It is more commonly used to designate one of the second element listed in formula.
3. Sulfur dioxide
4. Sulfur trioxide
5. Dinitrogen tetroxide. Here is an example where the –a suffix of tetra- is dropped.
6. Dinitrogen trioxide.

Example 7: Obtaining the Formula from the Name of Covalent Compounds
Give the formula for the following compounds
1. dintirogen tetroxide 2. phosphorus trichloride

Solution:
The process of giving the formula from the name of a covalent compound is simply a matter of using the prefix to give the subscripts for the elements.

1. Di- mean 2, tetra- (the a- was dropped) means 4. The formula is N_2O_4.
2. Since phosphorus has no prefix, we assume one phosphorus. Tri- means 3. The formula is PCl_3.

3.3 Properties of Ionic and Covalent Compounds

The differences in ionic and covalent bonding account for the different properties of ionic and covalent compounds. Covalently bonded molecules are discrete units, and they have fewer tendencies to form an extended, three-dimensional structure in the solid state. Ionic compounds do not have definable units but form a crystal lattice composed of hundreds of billions of positive and negative ions in an extended three-dimensional network.

Major differences in the properties of these compounds are summarized below.

Physical State

All ionic compounds are solids at room temperature, while covalent compounds may be solids (sugar, silicon dioxide), liquids (water, ethyl alcohol), or gases (carbon monoxide, carbon dioxide).

Melting and Boiling Points

The **melting point** is the temperature at which a solid is converted to a liquid, and the **boiling point** is the temperature at which a liquid is converted to a gas. Considerable energy is needed to break apart an ionic crystal lattice and convert an ionic solid to a liquid or a gas. Therefore, the melting and boiling temperatures for ionic compounds are generally higher than those of covalent compounds (which have weaker interactions between molecules).

Structure of Compounds in the Solid State

Ionic solids are crystalline, characterized by a regular structure, whereas covalent solids may be either crystalline or amorphous (no regular structure).

Solutions of Ionic and Covalent Compounds

Many ionic solids dissolve in water. If soluble, an ionic solid will **dissociate** in solution to form positive and negative ions (ionization). Because these ions are capable of conducting electric current, these compounds are **electrolytes**, and the solution is termed an electrolytic solution. Covalent solids in solution are neutral and are **nonelectrolytes**. The solution is not an electrical conductor.

3.4 Drawing Lewis Structures of Molecules and Polyatomic Ions

Lewis Structures of Molecules

Lewis structures show the location of all valence electrons in molecules and polyatomic ions. To write a Lewis structure for the compounds, follow these steps:

Step 1: Use chemical symbols for the various elements to write the skeletal structure of the compound. Typically, the first element given in the formula is the central atom. The least electronegative atom will usually be the central atom. Hydrogen and fluorine cannot ever be a central atom.

Step 2: Determine the number of valence electrons associated with each atom; combine them to determine the total number of valence electrons in the compound. Add additional electrons for polyatomic anions, and subtract electrons for polyatomic cations.

Step 3: Connect the central atom to each of the surrounding atoms with **single bonds**.

Step 4: Place electrons as lone pairs around the *terminal atoms* (non-central atoms) first to complete the octet for each. Remember, hydrogen only needs two electrons. Keep a count of how many electrons you have used, and do not exceed the number of available electrons.

Step 5: If the octet rule is not satisfied for the central atom, move one or more electron pairs from the surrounding atoms to create **double** or **triple bonds** until all atoms have an octet.

Step 6: After you are satisfied with the Lewis structure that you have constructed, perform a final electron count verifying that the total number of electrons used and the number around each atom are correct.

Example 8: Drawing Lewis Structures.
Draw the Lewis structure of carbon tetrafluoride.

Solution:

Step 1: Skeletal structure: Carbon is the central atom surrounded by four fluorine atoms.

Step 2: Count valence electrons: Carbon has 4; each fluorine has 7. Hence, $4 + 4(7) = 32$.

Step 3: Connect each fluorine to the central carbon with a single bond. Either use pairs of electrons or a dash to represent two electrons.

```
      F
      |
  F — C — F
      |
      F
```

Step 4: Distribute electrons around the terminal atoms in pairs to satisfy the octet rule. Do not use more than 32 electrons. Before we put any lone pair electrons in, we have already used 8 for bonding pairs.

Example 8 Continued

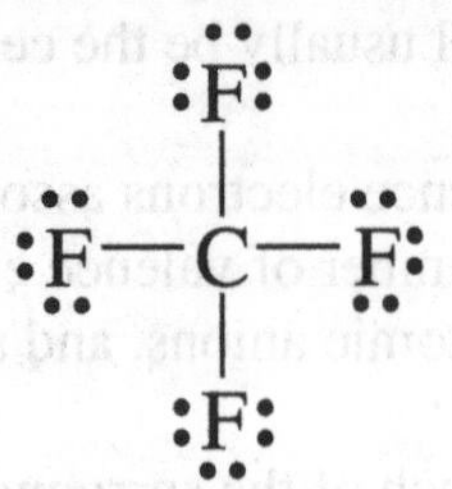

```
     ..
    :F:
     |
  .. | ..
 :F—C—F:
  .. | ..
     |
    :F:
     ..
```

Step 5: All 32 electrons have been used, and the octet rule is satisfied for each atom, so no movement of electron pairs is needed.

Step 6: One final count. There are 8 bonding electrons connecting the F's to the C's. There are 24 nonbonding electrons. Hence, 8 + 24 = 32.

Example 9: Drawing Lewis Structures

Draw the Lewis structure of SO_3.

Solution:

Step 1: Sulfur is less electronegative and is the central atom. The connectivity is

```
   O

O  S  O
```

Step 2: Count the valence electrons. For oxygen, each atom has 6 valence electrons. 3×6=18. The one sulfur atom has 6 additional electrons. Hence, 18 + 6 = 24.

Step 3: Connect the oxygen atoms to the sulfur atom with single bonds. This uses 6 of the valence electrons.

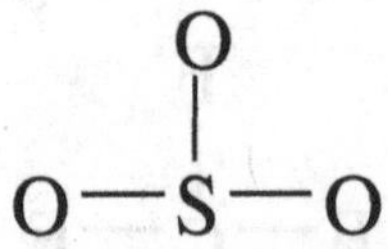

```
    O
    |
O—S—O
```

Step 4: Place electrons around terminal atoms to complete the octet for each.

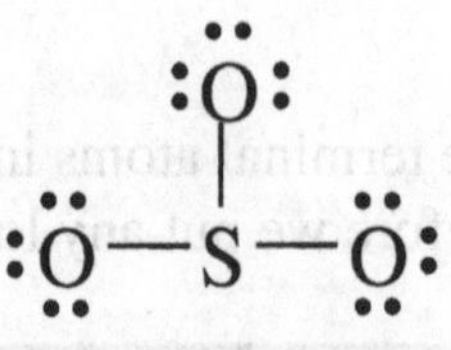

```
      ..
     :O:
  ..  |  ..
 :O—S—O:
  ..     ..
```

This uses 24 electrons. However, the octet is not yet satisfied for the sulfur atom.

Example 9 Continued
Step 5: Shift one of the lone pairs on one of the oxygens in to make a double bond. The structure below shows the bottom two electrons on the left-hand oxygen being moved to make the double bond.

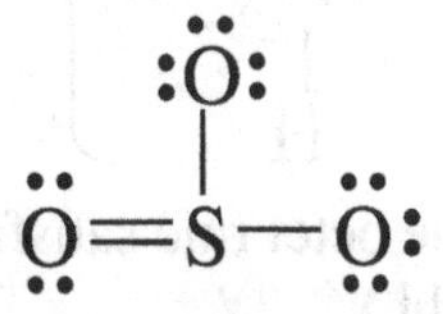

:Ö:
|
Ö=S—Ö:

Step 6: Each atom has the octet rule satisfied and uses 24 electrons (8 bonding + 16 lone pair electrons).

Lewis Structures of Polyatomic Ions

LG 6

The charge on the ion must be accounted for while computing the total number of valence electrons when writing Lewis structures of ions. For negative ions, one valence electron is added for each unit of negative charge, and we subtract one valence electron for each unit of positive charge in a positive ion. All the rules for drawing Lewis structures apply whether for a molecule or polyatomic ion.

Example 10: Drawing Lewis Structures of a Polyatomic Ion
Draw the Lewis structure of the ammonium ion.

Solution:
Step1: Ammonium is NH_4^+. Nitrogen is the central atom. The square brackets are used for polyatomic ions with the charge of the ion outside the bracket.

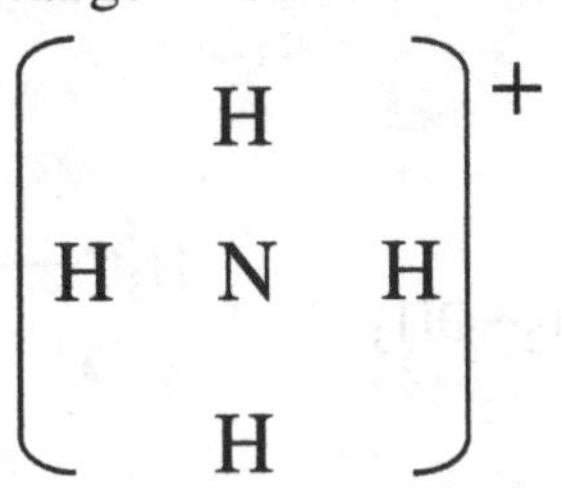

$$\left[\begin{array}{ccc} & H & \\ H & N & H \\ & H & \end{array}\right]^+$$

Step 2: Count valence electrons, which are determined by the group number. Nitrogen has 5 valence electrons and each hydrogen has one valence electron. Hence, 5 + 4(1)=9 electrons. Because the ion has a 1^+ charge, one electron must be removed giving a total of 8 valence electrons.

Step 3: Connect the hydrogen atoms to the nitrogen with single bonds.

Example 10 Continued

$$\left[\begin{array}{c} H \\ | \\ H-N-H \\ | \\ H \end{array}\right]^{+}$$

Step 4: At this point, each atom has the octet rule satisfied. Remember, hydrogen only needs two valence electrons (not eight.)

Steps 5 & 6: All eight electrons are used, and the octet rule is satisfied.

Lewis Structure, Stability, Multiple Bonds, and Bond Energies

LG 7

The order of stability of a molecule often parallels the bond order (the number of bonds between adjacent atoms). Stability is also related to the bond energy, where the **bond energy** is defined as the amount of energy in kilocalories required to break a bond holding two atoms together. The magnitude of the bond energy (and thus stability) decreases in the order triple bond > double bond > single bond. The bond length decreases in the order single bond > double bond > triple bond. That is, the more bonds there are, the shorter and stronger they are.

Isomers

Isomers are compounds that share the same molecular formula but have different structures. For example, the hydrocarbon C_4H_{10} has two different isomers that are shown in Figure 3.3. Because isomers have difference structures, isomers also have different physical properties.

$$CH_3-CH_2-CH_2-CH_3 \qquad \begin{array}{c} CH_3 \\ | \\ H_3C-C-CH_3 \\ H \end{array}$$

Figure 3.3 Two isomers of C_4H_{10}.

Lewis Structures and Resonance

More than one Lewis structure may satisfy the octet rule for a particular compound. When a compound has two or more Lewis structures that contribute to the real structure, the compound displays **resonance**. The contributing Lewis structures are *resonance forms*. The true structure, a hybrid or mixture of the resonance forms, is known as a *resonance hybrid*. The more resonance forms that exist for a structure, the greater the stability of that molecule. Look back at Example 9. If, instead of shifting an electron pair from the left-hand oxygen in to make a double bond, we moved an electron pair from the top or right-hand oxygen, we would have a different Lewis structures which equally obey the octet rule. See Figure 3.4 below.

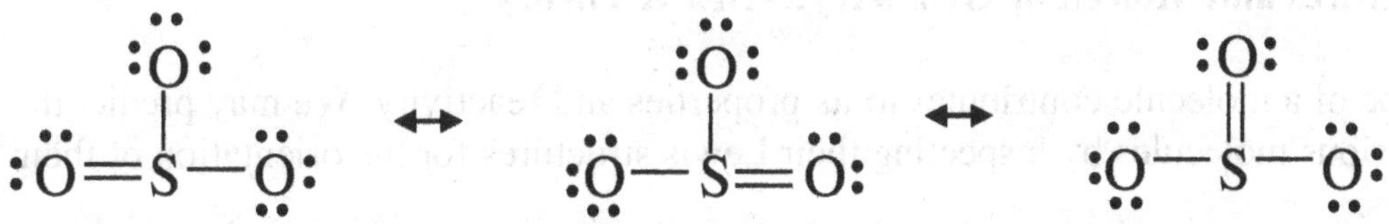

Figure 3.4 The three resonance structures of SO_3. All three structures obey the octet rule and together represent the bonding of the element.

Lewis Structures and Exceptions to the Octet Rule

Molecules with fewer than four and as many as five or six electron pairs around the central atom also exist. They are exceptions to the octet rule. In general, if one atom in a molecule (or polyatomic ion) must be an exception to the octet rule, it will be the central atom. The three common exceptions are:

1. **Incomplete Octet**. Some compounds have fewer than eight electrons surrounding the central atom. BeH_2 only has four valence electrons, and hence cannot have eight electrons surrounding Be. Compounds containing boron with three atoms connected to it, such as BF_3, only have six valence electrons. It is possible to follow the steps and create a Lewis structure for BF_3 which obeys the octet rule; however, you simply need to remember that it is an example of an incomplete octet. See Figure 3.5.

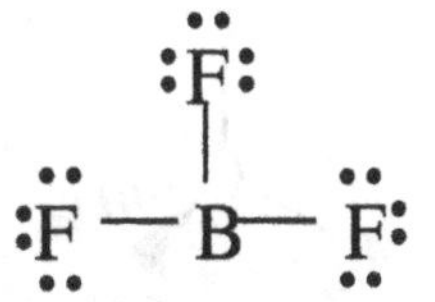

Figure 3.5 Boron trifluoride with an incomplete octet on boron.

2. **Odd Electron**. If after the valence electrons is determined, there are an odd number of electrons, then it is impossible to have each atom with eight electrons. One atom will be left with seven electrons.
3. **Expanded Octet**. If the central atom is in the 3rd period or below on the periodic table, it is possible to have more than eight electrons around the central atom. When drawing the Lewis structure, if you have all atoms with an octet and there are still more electrons in your electron count of Step 2, add these extra electrons around the central atom. See Figure 3.6.

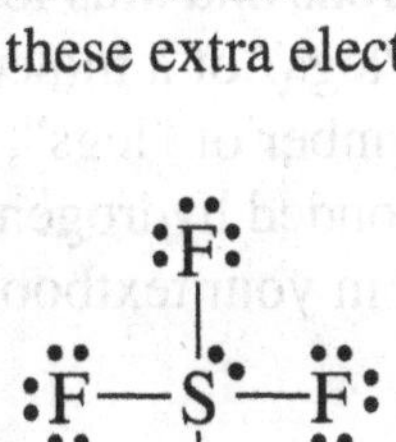

Figure 3.6 Sulfur tetrafluoride has 34 valence electrons. After drawing single bonds and

satisfying the octet rule for each atom, only 32 electrons are used. Therefore, the extra two electrons are placed on the central atom. This gives sulfur 10 electrons, an expanded octet.

Lewis Structures and Molecular Geometry; VSEPR Theory

LG 8

The shape of a molecule contributes to its properties and reactivity. We may predict the shapes of various molecules by inspecting their Lewis structures for the orientation of their electron pairs.

Electron pairs (both bonding electrons and lone pair electrons) around the central atom of the molecule arrange themselves to minimize repulsion; the electron pairs are as far as possible from each other. This is termed the **valence-shell electron pair repulsion theory (VSEPR Theory)**. VSEPR allows us to determine the three-dimensional shape of the molecule. To determine the shape first draw the Lewis structure, then examine the central atom and count the number of electron pairs around the atom (count multiple bonds as one pair.)

Two electron pairs around the central atom lead to a **linear** arrangement of the attached atoms, since this 180° arrangement is the farthest apart the two electron pairs can be. An example is BeH_2 which has a Lewis structure of H–Be–H. There are no lone pairs and two bonds to the beryllium.

Three electron pairs around the central atom lead to a **trigonal planar** arrangement. Refer to Figure 3.5 above. Boron has 3 bonding pair and no lone pair electrons on the central atom. These three bonded atoms will evenly arrange themselves with a 120° bond angle.

Four electron pairs result in a **tetrahedral** geometry. In Example 8, the Lewis structure of CF_4 is shown. Carbon has 4 bonds and no lone pairs of electrons. See Figure 3.7.

Figure 3.7 Tetrahedral geometry of CF_4. F-C-F bond angles are 109.5°. The bottom right fluorine is in front of the plane of the paper (represented by the solid wedge), and the top right fluorine is behind the plane of the paper (represented by the hashed bond line).

The examples above have no lone pair electrons on the central atom. However, lone pair electrons can occupy any of the "legs." The repulsion between lone pair-bond pair electrons is greater that bond pair-bond pair electrons and thus lone pair electrons occupy a slightly larger area of space than the bond pair "legs" of a molecule. The geometric shape is defined by the arrangement of atoms, not the number of "legs". In the Lewis structure of NH_3, there is one lone pair of electrons and three bonded hydrogen atoms. As a result, the structure or shape is **trigonal pyramidal.** Table 3.5 in your textbook gives a summary of the geometries.

Periodic Molecular Geometry Relationships

The periodic similarity of group members is also useful in predictions involving bonding. Consider oxygen, sulfur, and selenium (Group VIA or 16). Each has six valence electrons and needs two more electrons to complete its octet. When they react with hydrogen, H_2O, H_2S, and H_2Se form. These are **bent** molecules with similar Lewis structures and shapes. This logic also applies to other representative elements.

Lewis Structures and Polarity

LG 9

We earlier discussed that a covalent bond could be considered polar or nonpolar. Here we consider all the bonds in a molecule to determine if the molecule as a whole is polar or nonpolar. A molecule is polar if it has two "poles" (or ends) where one pole is more negative and one pole more positive. When placed in an electric field, these types of molecules will align themselves with the field. Nonpolar molecules do not have two poles and thus do not align with an electric field.

A molecule containing only nonpolar bonds will be a nonpolar molecule and all electrons will be equally shared between the atoms. If the molecule contains polar bonds (elements with different electronegativities), it may or may not be polar. To determine if a molecule is polar, first draw the Lewis structure then use the following guidelines:

1. Molecules are *nonpolar* which have
   - no lone pair on the central atom and
   - all the same terminal atoms
2. Molecules with one lone pair on the central atom are *polar.*
3. Molecules with two or more lone pairs on the central atom are *usually polar*, with exceptions.
4. Molecules that are made up of only carbon and hydrogen are *nonpolar.*

3.5 Properties Based on Molecular Geometry and Intermolecular Forces

The effects of molecular polarity result from the strength of attractive forces *between* individual molecules of a compound. These attractions are **intermolecular forces**. Be careful not to confuse intermolecular forces with intramolecular forces. **Intramolecular forces** are the attractive forces *within* molecules (ionic and covalent bonds). It is the intermolecular forces that determine such properties as solubility, melting point, and boiling point.

Solubility

Solubility is defined as the maximum amount of solute that dissolves in a given amount of solvent at a specific temperature. (The solute is present in lesser quantity, and the solvent is present in the greater amount.) Polar molecules are most soluble in polar solvents, and nonpolar molecules are most soluble in nonpolar solvents. This is the rule of *"like dissolves like."*

Boiling Points of Liquids and Melting Points of Solids

Boiling is the conversion of a liquid to a vapor. Such a process requires energy to overcome the intermolecular attractive forces between molecules in the liquid. The energy required is related to the magnitude of the boiling point and depends upon the strength of the intermolecular attractive forces in the liquid. These attractive forces are directly related to polarity. Molecular size is also an important consideration. The larger or heavier the molecule, and the more polar the molecule, the more difficult it becomes to convert to the gas phase (therefore the higher the boiling point).

The melting points of solids may be described on the basis of intermolecular forces as well. As a general rule, polar compounds have strong attractive (intermolecular) force, and their boiling and melting points tend to be higher than for nonpolar substances.

Self Test for Chapter Three

1. What type of chemical bond results from the transfer of electrons from one atom to another?
2. Describe how atoms are held together in a covalent bond.
3. What particles are found in ionic compounds?
4. Use electronegativity values to classify the following bonds as ionic, polar covalent, or nonpolar covalent:

   a. P and O b. K and Br c. Ni and C d. F and F
5. Provide the systematic name for each of the following:

   a. Na_2O b. Li_2S c. $FeSO_4$ d. NCl_3
6. Write the formula for each of the following:

   a. copper(I) oxide c. ammonium sulfide

   b. copper(II) nitrate d. sulfur hexafluoride
7. A substance is at gas a room temperature. Predict whether the compound is ionic or covalent.
8. A certain substance is either salt (NaCl) or sugar ($C_{12}H_{22}O_{11}$). It is dissolved in water, and the solution conducts electricity. Determine the identity of the substance.
9. Draw the Lewis structure of H_2O. How many bonding electrons and nonbonding electrons are present?
10. Draw the Lewis structure for the hydronium ion, H_3O^+. How many bonding electrons and nonbonding electrons are present?
11. Draw the Lewis structure of the nitrate ion, NO_3^-? How many single and double bonds are present?
12. List the three causes of exceptions to the octet rule and give an example of each.

13. How many electrons are shared in the hydrogen molecule? Is this considered an incomplete octet?
14. Does a triple bond have more or less bond energy than a double bond?
15. Is a single bond longer or shorter than a double bond?
16. What is the geometry of BCl_3?
17. What is the geometry of NH_4^+?
18. Which of the following is linear? H_2O, NO_2^-, BeH_2
19. What are the bond angles in a trigonal planar molecule and in a tetrahedral molecule?
20. Is HF more soluble in a polar solvent or a nonpolar solvent?
21. Is I_2 more soluble in a polar solvent or a nonpolar solvent?
22. Which type of compound has the higher melting point, covalent on ionic?
23. When H_2O is compared to C_2H_4, which has the higher boiling point?
24. When Cl_2 is compared to ICl, which has the higher boiling point?

Calculations and the Chemical Equation

Learning Goals

1. Calculate the mass of an atom using the atomic mass unit.
2. Use the relationship between Avogadro's number and the mole to perform calculations.
3. Determine molar mass, and demonstrate how it is used in mole and mass conversion calculations.
4. Use chemical formulas to calculate the formula mass and molar mass of a compound.
5. Describe the functions served by the chemical equation, the basis for chemical calculations.
6. Classify chemical reactions by type: combination, decomposition, or replacement.
7. Balance chemical equations given the identity of products and reactants.
8. Write net ionic equations, and use solubility rules to predict the formation of a precipitate.
9. Distinguish between an acid and a base.
10. Describe oxidation-reduction, and identify what is being oxidized and reduced.
11. Use a chemical equation and a given number of moles or mass of a reactant or product to calculate the number of moles or mass of a reactant or product.
12. Calculate theoretical and percent yields.

Introduction

In this chapter you will learn to determine the quantity of a substance produced from a chemical reaction if given the amount of starting material. You will also be able to calculate how much of a substance would be required to produce a desired amount of a compound in a chemical reaction.

These calculations of chemical quantities are termed stoichiometry and are based on the chemical equation. The chemical equation provides all of the needed information: the combining ratio of elements or compounds that are necessary to produce a particular product or products.

The chemical equation is also used to classify reactions according to their unique pattern and characteristics.

4.1 The Mole Concept and Atoms

Atomic masses have been experimentally determined for each of the elements. Their unit of measurement is the **atomic mass unit**, abbreviated amu.

$$1 \text{ amu} = 1.661 \times 10^{-24} \text{ grams}$$

LG 1

The Mole and Avogadro's Number

A more practical unit for defining a "collection" of atoms is the **mole**.

$$1 \text{ mole of atoms} = 6.022 \times 10^{23} \text{ atoms of an element}$$

This number is **Avogadro's number**.

Note: the mole is abbreviated mol (not m).

The mole and the amu are related. The **atomic mass** of a given element corresponds to the average mass of a single atom in amu *and* the mass of a mole of atoms in grams. For example, the average mass of one copper atom is 63.54 amu, while the mass of one *mole* of copper atoms is 63.54 grams. This mass is called the **molar mass,** the mass of one mole of the element. Again, the molar mass and atomic mass of an element are numerically equivalent, but have different units. The value is obtained from the periodic table. Look at carbon on a periodic table. The atomic mass of carbon is 12.01 amu (the mass of one atom in amu). The molar mass of carbon is 12.01 g/mol (the mass of one mole of carbon in grams). One mole of atoms of *any element* contains the same number, Avogadro's number, of atoms.

Calculating Atoms, Moles, and Mass

LG 2

At this point, we will revisit the concept of the factor-label method. You learned in Chapter 1 how to convert between units, allowing the units to guide you through the process. Here we will learn how the new conversion factors of Avogadro's number and molar mass are used to proceed from the information *provided* in the problem (Data given) to the information *requested* by the problem (Desired result).

Avogadro's number as a conversion factor: $\dfrac{6.022\times10^{23}\text{ atoms}}{1\text{ mol}}$

Molar mass as a conversion factor, with carbon used as example: $\dfrac{12.01\text{ g C}}{1\text{ mol C}}$

LG 3

Before you begin any calculations, map out a pattern for the required conversion. You may be given the number of grams and need the number of atoms that corresponds to that mass. Begin by tracing a path to the answer. Figure 4.1 can be used as a guide.

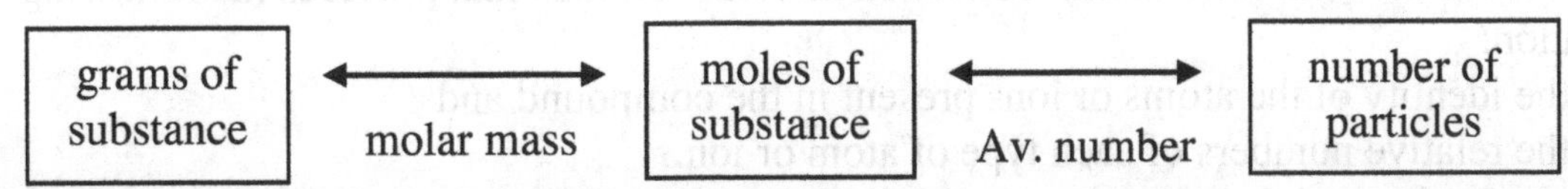

Figure 4.1 Use the factor-label method to convert between the grams, moles and particles.

Examples 1 and 2 demonstrate the process of using the factor-label method to convert between the amounts above. You will also be required to apply conversions you already learned in Chapter 1

Example 1: Conversions from Atoms to Mass.
Calculate the mass, in kilograms, of 1.5×10^{24} atoms of iron.
Solution:
Map out your work. Refer to Figure 4.1 for assistance. Start with the data given and determine each step that will take you to the quantity you want to determine.

atoms Fe $\xrightarrow{1}$ mol Fe $\xrightarrow{2}$ g Fe $\xrightarrow{3}$ kg Fe

For Step 1, use the definition of Avogadro's number; Step 2, use the molar mass of iron; and Step 3, use the definition of kilo (learned in Chapter 1). These three steps can be put together into one chain of factor-label conversions. Be sure to be very meticulous in writing the units.

$$1.5\times10^{24}\ \text{atoms Fe}\times\frac{1\ \text{mol Fe}}{6.022\times10^{23}\ \text{atoms Fe}}\times\frac{55.85\ \text{g Fe}}{1\ \text{mol Fe}}\times\frac{1\ \text{kg Fe}}{1000\ \text{g Fe}}=0.14\ \text{kg Fe}$$

Example 2: Conversions from Mass to Atoms.
Calculate the number of sulfur atoms in 14.0 grams of sulfur.

Solution:
Map of the conversions: grams S $\xrightarrow{1}$ moles S $\xrightarrow{2}$ atoms S. For Step 1, use the molar mass of S. Step 2, use the definition of Avogadro's number.

$$14.0\ \text{g S}\times\frac{1\ \text{mol S}}{32.07\ \text{g S}}\times\frac{6.022\times10^{23}\ \text{atoms S}}{1\ \text{mol S}}=2.63\times10^{23}\ \text{atoms S}$$

Note: when using molar mass, the value from the periodic table *always* has the unit of mass and the number 1 *always* is associated with the unit of mol.

4.2 The Chemical Formula, Formula Mass, and Molar Mass

The Chemical Formula

LG 4

Chemical compounds are represented by their **chemical formula**, a combination of symbols of the various elements that make up the compounds. The chemical formula is based upon the **formula unit**, the smallest collection of atoms on ions that provides the following information:

1. the identity of the atoms or ions present in the compound and
2. the relative numbers of each type of atom or ion.

Compounds containing one or more water molecules as an integral part of their structure are termed **hydrates.**

The term *formula* may be used in reference to any ionic or covalent compound. In Chapter 3, compounds were classified as nonpolar covalent, polar covalent, or ionic, depending upon the type of bonding that holds the unit together.

Formula Mass and Molar Mass

LG 4

Formula mass is the sum of the atomic mass of the atoms which make up the formula of the compound. It is the mass, in amu, of one molecule (or one unit of an ionic compound). The formula mass is calculated by adding the masses of all the atoms of which the unit is composed. To calculate the formula mass, the formula unit must be known. The *molar mass* is the mass, in grams, of one mole of the compound. Molar mass is numerically equivalent to formula mass, but with different units. Once you can calculate the formula mass and molar mass of a compound, you will be able to do calculations for compounds that are similar to the ones for elements (Examples 1 and 2).

Example 3: Calculation of Formula Mass and Molar Mass
Calculate the formula mass of carbon dioxide. What is the molar mass?

Solution:
Carbon dioxide has the formula CO_2. From the periodic table, we find that one atom of carbon has a mass of 12.01 amu, and one atom of oxygen has a mass of 16.00 amu.

1(12.01 amu) + 2(16.00 amu) = 44.01 amu. This is the formula mass, the mass of one molecule of CO_2.

The molar mass is 44.01 g/mol. (The mass, in grams, of one mole of CO_2 molecules.)

4.3 The Chemical Equation and the Information It Conveys

LG 5

A Recipe for Chemical Change

The **chemical equation** describes all of the substances that react, called **reactants,** to produce the **products.** The chemical equation also describes the physical state of the reactants and products as solid, liquid, or gas. It tells us whether the reaction occurs, and identifies the solvent and experimental conditions. Most importantly, the relative number of moles of reactants and products appears in the equation. According to the **law of conservation of mass,** matter cannot be either gained or lost in the process of a chemical reaction. In other words, we must have a balanced chemical equation.

Features of a Chemical Equation

1. The identity of products and reactants must be specified using chemical symbols.
2. Reactants are written to the left of the reaction arrow and products to the right.
3. The physical state of reactants and products is shown in parentheses.
4. The symbol Δ over the reaction arrow means that energy is necessary for the reaction to occur.
5. The equation must be balanced.

The Experimental Basis of a Chemical Equation

The chemical equation represents a real chemical transformation. One or more substances will be converted into different substances, and this transformation is evident by the following:

- Gas formation. This will be evident as bubbles are formed.
- Solid (or precipitate) formation from solution. Two clear solutions become cloudy when mixed.
- Heat production. If an acid and a base are mixed, there may be no other evidence than that the mixture will be warm to the touch.
- Change in color.
- Light absorption or emission. Light sticks are an example of the production of light in a chemical reaction.
- Change in magnetic and electrical properties. For example, the reactant may be attracted by a magnet while the products are not.

Strategies for Writing Chemical Reactions

LG 6

Spontaneous chemical reactions occur for a variety of reasons, linked by the tendency to achieve the lowest (most stable) electronic energy state. Strong electrolytes will react to form weak (less dissociated) electrolytes, if possible. Reactions forming gaseous products or solid products from solution are favored.

Chemical reactions tend to fit into one of a few simple patterns:

Combination reactions: A + B → AB. Combination reactions involve the joining of two or more atoms or compounds, producing a product of different composition.
Water is produced by this type of reaction: $2H_2(g) + O_2(g) \rightarrow 2H_2O(l)$

Decomposition reactions: AB → A + B. Decomposition reactions produce two or more products from a single reactant. An example of a decomposition reaction involves the heating of mercury(II) oxide: $2HgO(s) \rightarrow 2Hg(l) + O_2(g)$

Replacement reactions: Replacement reactions are subcategorized as:

1) **Single-replacement reactions: A + BC → AC + B.** Single-replacement reactions occur between atoms and compounds; one atom replaces another in the compound, producing a new compound and the replaced atom. Placing a piece of copper into a solution containing silver ions is an example: $Cu(s) + 2AgNO_3(aq) \rightarrow 2Ag(s) + Cu(NO_3)_2(aq)$
2) **Double-replacement reactions: AB + CD → AD + CB.** Double-replacement reactions, on the other hand, involve two compounds exchanging atoms to produce two new compounds. Acids typically react with bases in this type of reaction. Hydrochloric acid, HCl, and the base sodium hydroxide give an example: $HCl(aq) + NaOH(aq) \rightarrow NaCl(aq) + H_2O(l)$.

Example 4: Patterns of Chemical Reactions

Label each reaction as combination, decomposition, single-replacement or double-replacement.

1. $CaCO_3(s) \rightarrow CaO(s) + CO_2(g)$
2. $2Na(s) + 2H_2O(l) \rightarrow 2NaOH(aq) + H_2(g)$
3. $2Li(s) + Cl_2(g) \rightarrow 2LiCl(s)$
4. $NaCl(aq) + AgNO_3(aq) \rightarrow AgCl(s) + NaNO_3(aq)$

Solution:

1. The $CaCO_3$ breaks apart into two substances. This is a decomposition reaction.
2. The Na replaces an H in the water molecule. This is a single-replacement reaction.
3. Li combines with Cl_2 to yield a single product. This is a combination reaction.
4. The ionic compounds are exchanging atoms, yielding two new ionic compounds. This is a double-replacement reaction.

4.4 Balancing Chemical Equations

Not only does a chemical equation give the reactants and products involved in a reaction, it will also give the relative amount, in moles, of reactants and products.

Balancing Chemical Equations

LG 7

The number of moles of each product and reactant is indicated by placing a whole number *coefficient* before the formula of each substance in the chemical equation.

Many equations are balanced by trial and error. The following steps provide a method for correctly balancing a chemical equation:

Step 1: Count the number of atoms of each element on both product and reactant sides.
Step 2: Determine which elements are not balanced.
Step 3: Balance one element at a time using coefficients.
Step 4: After you believe that you have successfully balanced the equation, check, as in Step 1, to be certain that mass conservation has been achieved.

Example 5: Balancing Chemical Equations

Write the balanced equation of an aqueous solution of solid barium reacting with water to produce barium hydroxide and hydrogen gas.

Solution:

1. First write the correct symbols and formulas needed for the reactants and products:

$$H_2O \; + \; Ba \rightarrow Ba(OH)_2 \; + \; H_2$$

2. Next, count the number of atoms of each element on the left and right side of the reaction arrow:

H	O	Ba	‖	H	O	Ba
2 atom	1 atom	1 atom	‖	4 atoms	2 atoms	1 atom

Example 5 Continued

Notice that the numbers of atoms on each side are not equal. The equation is **not** balanced.

3. After the correct formulas of all reactants and products are given, you can now use coefficients alone to balance the equation. By adding a 2 in front of H_2O, the number of atoms for H and O will be doubled giving the equation:

$$2H_2O + Ba \rightarrow Ba(OH)_2 + H_2$$

4. Check how many of each atom you have in your new equation:

H	O	Ba	H	O	Ba
4 atom	2 atom	1 atom	4 atoms	2 atoms	1 atom

The equation is now balanced.

4.5 Precipitation Reactions

LG 8

Precipitation reactions involve the conversion of soluble reactants into one or more insoluble products (a **precipitate**). Table 4.1 gives information to help you predict which ionic compounds are insoluble. To predict if a precipitation reaction will occur between two ionic compounds, write the products of a double-replacement reaction and determine if one of the products is insoluble according to the rules. If one is, the precipitation reaction will occur. The insoluble salt is the precipitate.

TABLE 4.1 Solubility Rules for Common Ionic Compounds in Water

Soluble compounds contain:	Exceptions
alkali metal ions (Li^+, Na^+, K^+, Rb^+, Cs^+) or the ammonium ion (NH_4^+)	
nitrate (NO_3^-), bicarbonate (HCO_3^-), or chlorate (ClO_3^-)	
halides (Cl^-, Br^-, I^-)	Compounds which contain halides of Ag^+, Hg_2^{2+}, and Pb^{2+}
sulfate (SO_4^{2-})	Compounds which contain sulfate of Ag^+, Ca^{2+}, Sr^{2+}, Ba^{2+}, Hg_2^{2+}, or Pb^{2+}
Insoluble compounds contain:	**Exceptions**
carbonate (CO_3^{2-}), phosphate (PO_4^{3-}), chromate (CrO_4^{2-}), or sulfide (S^{2-})	Compounds which contain alkali metal ions or NH_4^+
hydroxides (OH^-)	Compounds containing alkali metal ions or the Ba^{2+} ion

Example 6: Predicting if a Precipitation Reaction Will Occur

Predict if a precipitation reaction will occur when a solution of sodium sulfide and iron(II) nitrate are mixed.

Example 6 Solution:

Precipitation reactions are double-replacement reactions in which the ions exchange atoms and one of the new compounds is insoluble according to the rules in Table 4.1.

1. Write the possible reaction by writing the correct formulas for the given compounds and swapping the ions to give the products. Use the charges to be sure the reactants and products have the correct formula.

$$Na_2S(aq) + Fe(NO_3)_2(aq) \rightarrow FeS(?) + 2NaNO_3(?)$$

Note: the coefficient of 2 is placed in front of $NaNO_3$ to balance the equation.

2. Use the solubility rules to determine if either product is insoluble. If there is an insoluble product, it will receive an (*s*) after it, and that solid is the precipitate.
   FeS: Table 4.1 states that sulfides are insoluble and Fe^{2+} is not one of the exceptions. Therefore FeS is insoluble and an (*s*) is placed after it.
   $NaNO_3$: Table 4.1 states that all alkali metal compounds are soluble; therefore, $NaNO_3$ is soluble. The balanced precipitation reaction is given by:

$$Na_2S(aq) + Fe(NO_3)_2(aq) \rightarrow FeS(s) + 2NaNO_3(aq)$$

4.6 Net Ionic Equations

The final answer to Example 6 gives the reaction equation as a molecular equation. However, the aqueous ionic compounds are not combined in solution. The ions are actually dissociated from each other. Ions which were in their dissociated form as reactants and are still in the dissociated form as products (that is, they have (*aq*) written after them) are known as spectator ions. In Example 6, the spectator ions are Na^+ and NO_3^-. Removing the spectator ions from the reaction equation will leave what is known as the **net ionic equation**. By writing the chemical equation this way we highlight the elements that are undergoing a chemical change during the reaction. For the reaction in Example 6, the net ionic equation is $Fe^{2+}(aq) + S^{2-}(aq) \rightarrow FeS(s)$.

4.7 Acid-Base Reactions

LG 9

Acid-base reactions involve the transfer of a hydrogen ion, H^+, from one reactant (the acid) to another (the base). A **neutralization reaction** involves the reaction of an acid and a base, resulting in a neutral aqueous salt solution.

4.8 Oxidation - Reduction Reactions

Oxidation and Reduction

LG 10

Oxidation-reduction reactions involve the transfer of a negative charge (one or more electrons) from one reactant to another. Oxidation and reduction are complementary processes and are often termed *redox reactions*. The atom that loses electrons is oxidized, the atom that gains electrons in reduced.

Oxidation-reduction processes will be discussed more fully in Section 8.5.

4.9 Calculations Using the Chemical Equation

General Principles

In doing calculations involving chemical reactions, we apply the following rules:

1. The chemical formula of all reactants and products must be known.
2. The equation must be balanced because the conservation of mass must be obeyed.
3. The conversion between measured quantities of one substance to measured quantities of another substance in the balanced equation must occur by way of moles.

Using Conversion Factors

Since the mol is the basis of chemical calculations, most measurements of chemicals are done in g (or kg) so it is important to be proficient at the interconversion between moles and grams. Just as the conversion between moles and grams of an element requires the atomic mass conversion factor (see Example 3), the conversion of moles to grams of a compound requires the formula mass conversion factor. It is important to remember all previous conversion factors learned in this and preceding chapters.

Example 7: Conversion from Moles to Grams for Compounds
How many moles of carbon dioxide are contained in 440.1 g of carbon dioxide?

Solution:

The molar mass of carbon dioxide (CO_2) must be calculated first:

$$1(12.01\ \text{g/mol}) + 2(16.00\ \text{g/mol}) = 44.01\ \text{g/mol}$$

Writing this as a conversion factor allows moles to be converted to grams in a single step.

$$440.1\ \cancel{\text{g}\ CO_2} \times \frac{1\ \text{mole}\ CO_2}{44.01\ \cancel{\text{g}\ CO_2}} = 10.00\ \text{mol}\ CO_2$$

Example 8: Conversion from Volume to Molecules
Calculate the number of molecules in 25 mL of H_2O. (The density of water is 1.00 g/mL.)

Solution:

Map of the conversions: mL $\xrightarrow{1}$ g $\xrightarrow{2}$ mol $\xrightarrow{3}$ molecules

Step 1 requires the use of density as the conversion factor: 1.00 g/ml or $\dfrac{1.00\ \text{g}\ H_2O}{1\ \text{mL}\ H_2O}$

Step 2 requires the molar mass of $H_2O = 2(1.008\ \text{g/mol}) + 1(16.00\ \text{g/mol}) = 18.02\ \text{g/mol}$

or $\dfrac{18.02\ \text{g}\ H_2O}{1\ \text{mol}\ H_2O}$

Example 8 Continued

Step 3 will use Avogadro's number. Note: Avogadro's number was used for atoms in Examples 1 and 2. Here we apply the same conversions to molecules.

$$25\ \cancel{\text{mL } H_2O} \times \frac{1.00\ \cancel{\text{g } H_2O}}{1\ \cancel{\text{mL } H_2O}} \times \frac{1\ \cancel{\text{mol } H_2O}}{18.02\ \cancel{\text{g } H_2O}} \times \frac{6.02 \times 10^{23}\ \text{molec. } H_2O}{1\ \cancel{\text{mol } H_2O}} = 8.4 \times 10^{23}\ \text{molec. } H_2O$$

Notice how the units are very specific. Not only are the units of g and mol included, but also the substance in included. Get into the habit of being very specific with the units and you will make fewer errors in your calculations.

The balanced equation will provide another important conversion factor for chemical calculations. Consider the balanced equation $2H_2(g) + O_2(g) \rightarrow 2H_2O(l)$. Using the coefficients, the following conversion factors can be obtained from the equation:

$$\frac{2\ \text{mol } H_2}{1\ \text{mol } O_2} \quad \text{or} \quad \frac{2\ \text{mol } H_2}{2\ \text{mol } H_2O} \quad \text{or} \quad \frac{1\ \text{mol } O_2}{2\ \text{mol } H_2O}$$

Indicating that 2 moles of H_2 reacts with 1 mole of O_2; 2 moles of H_2 produces 2 moles of H_2O; and 1 mole of O_2 produces 2 moles of H_2O. Example 9 demonstrates the use of these conversion factors.

Example 9: Mole to Mole Conversions

How many moles of O_2 are required to react with 1.5 moles of H_2 to form water?

Solution:

Write the balanced equation: $2H_2 + O_2 \rightarrow 2H_2O$. Use the coefficients of the balanced equation to convert between moles of O_2 and H_2.

$$1.5\ \cancel{\text{mol } H_2} \times \frac{1\ \text{mol } O_2}{2\ \cancel{\text{mol } H_2}} = 0.75\ \text{mol } O_2$$

You may be able to do this problem in your head. However, practice using the conversion tool so you can work it into a series of calculations as in the next example. Map out your conversions, and remember to convert the given quantity of a substance (A) to moles, then convert from moles of A to moles of the unknown (B), then to the requested unit of B.

Given quantity of A →1→ Moles of A →2 (Use coefficients)→ Moles of B →3→ Desired quantity of B

Figure 4.2 Step 1 requires conversion from the given quantity of the know substance (A) to moles. This may involve several conversion factors, represented by the two arrows. Step 2 requires the conversion from moles of A to moles of the unknown (B) using the coefficients of the balanced equation. Step 3 requires the conversion of moles of B to whatever unit is asked for in the problem. Again, this may require several steps.

Example 10: Conversion from Grams of Reactant to Grams of Product
The reaction of propane (C_3H_8 is found in bottled gas) with oxygen produces carbon dioxide and water. The balanced equation is given below:

$$C_3H_8 + 5\,O_2 \rightarrow 3\,CO_2 + 4\,H_2O$$

Calculate the number of grams of O_2 needed to react completely with 4.0 grams of C_3H_8.

Solution:
Start with the grams of C_3H_8 given and map out the conversions needed. Let Figure 4.2 be your guide.

Map: $g\ C_3H_8 \xrightarrow{1} mol\ C_3H_8 \xrightarrow{2} mol\ O_2 \xrightarrow{3} g\ O_2$.

Step 1: $g\ C_3H_8 \rightarrow mol\ C_3H_8$ requires the molar mass of C_3H_8
3(12.01 g/mol) + 8(1.01 g//mol) = 44.04 g/mol
Step 2: $mol\ C_3H_8 \rightarrow mol\ O_2$ uses the coefficients of the balanced equation.
Step 3: $mol\ O_2 \rightarrow g\ O_2$ requires the molar mass of O_2. 2(16.00 g/mol) = 32.00 g/mol

$$4.0\ g\ C_3H_8 \times \frac{1\ mol\ C_3H_8}{44.04\ g\ C_3H_8} \times \frac{5\ mol\ O_2}{1\ mol\ C_3H_8} \times \frac{32.00\ g\ O_2}{1\ mol\ O_2} = 15\ g\ O_2$$

Theoretical and Percent Yield

The **theoretical yield** is the *maximum* amount of product that can be produced if the reactants are completely converted to products. This is the calculated yield. In the real world it is difficult to produce the amount calculated as the theoretical yield.

A **percent yield**, the ratio of the actual and theoretical yields multiplied by 100%, is often used to show the relationship between predicted and experimental quantities. Thus

$$\%\ yield = \frac{actual\ yield}{theoretical\ yield} \times 100\%$$

The percent yield for most chemical reactions is less than 100%. This is due, in part, to experimental limitations in isolation, transfer, and recovery of the product. Many reactants do not completely convert to products. We will study these reactions in detail in Chapter 8.

In many cases, more of one reactant is added than is needed. This results in one reactant being used up while the other reactant is left over at the end of the reaction. The reactant that is used up limits the amount of product obtained and is termed the limiting reagent. It is the limiting reagent which is used to calculate the theoretical yield.

Self Test for Chapter Four

1. What is the mass of one atom of fluorine in amu? What is the mass of 1 mol of fluorine atoms, in grams?
2. How many iron atoms are present in 2.5 mole of iron?
3. How many grams of sulfur are found in 0.150 mole of sulfur?

4. What is the mass of one mole of H_2O?
5. What is the formula mass of one molecule of H_2O?
6. What is the molar mass of iron(II) nitrate?
7. What is the term that represents the starting materials in a chemical reaction?
8. Classify each of the following reactions as a combination reaction, decomposition reaction, single-replacement, or double-replacement reaction?
   a. $2\ Na(s) + 2\ H_2O(l) \rightarrow 2\ NaOH(aq) + H_2(g)$
   b. $C(s) + O_2(g) \rightarrow CO_2(g)$
   c. $AgNO_3(aq) + NaCl(aq) \rightarrow AgCl(s) + NaNO_3(aq)$
9. What is the coefficient of HCl when the following equation is balanced?
   $Ca + HCl \rightarrow CaCl_2 + H_2$
10. Balance the equation: $H_3PO_4(aq) + NaOH(aq) \rightarrow Na_3PO_4(aq) + H_2O(l)$
11. Write the balance chemical equation for the reaction: methanol (CH_3OH) reacts with oxygen to produce carbon dioxide and water.
12. Select the ionic compound(s) that is(are) insoluble in water: $SrSO_4$, Na_2S, $Fe(NO_3)_2$, AgBr, Li_2CrO_4.
13. What precipitate results from mixing aqueous solutions of $BaCl_2$ and $CuSO_4$? Write the molecular and net ionic equation.
14. What precipitate results from mixing aqueous solutions of $AgNO_3$ and KBr? Write the molecular and net ionic equation.
15. How many moles of hydrogen are needed to react with oxygen to form two moles of water? Note: begin by writing the balanced chemical equation.
16. How many moles of Cl_2 are in 6.02×10^{23} molecules Cl_2?
17. How many molecules are in 12.0 g nitrogen trichloride?
18. What is the volume in milliliters of 3.0 moles of ethanol (C_2H_5OH). The density of ethanol is 0.789 g/mL.
19. How many grams of sodium hydroxide will react with 73.00 grams of hydrochloric acid? $NaOH + HCl \rightarrow NaCl + H_2O$
20. Calculate the number of grams of water produced when 46.85 grams of C_2H_6 reacts with oxygen to produce carbon dioxide and water.
21. Iron reacts with oxygen to form iron(III) oxide (Fe_2O_3). How many grams of product will be formed from 5.00 grams of Fe?
22. What is the theoretical yield of NaCl when 3.25 g Na reacts with an excess of Cl_2 to produce NaCl?
23. In the laboratory, 3.25 g Na reacted with an excess of Cl_2, and 6.92 g NaCl was produced. Calculate the percent yield.
24. Do most reactions have a percent yield equal to 100%? Explain your answer.

5 States of Matter: Gases, Liquids, and Solids

Learning Goals

1. Perform conversions between units of pressure.
2. Describe the major points of the kinetic molecular theory of gases.
3. Explain the relationship between the kinetic molecular theory and the physical properties of measurable quantities of gases.
4. Describe the behavior of gases expressed by the gas laws: Boyle's law, Charles's law, combined gas law, Avogadro's law, the ideal gas law, and Dalton's law.
5. Use gas law equations to calculate conditions and changes in conditions of gases.
6. Use molar volume and standard temperature and pressure (STP) to perform calculations.
7. Discuss the limitations to the ideal gas model as it applies to real gases.
8. Describe properties of the liquid state in terms of the properties of the individual molecules that comprise the liquid.
9. Describe the processes of melting, boiling, evaporation, condensation, and sublimation.
10. Describe the dipolar attractions known collectively as van der Waals forces.
11. Describe hydrogen bonding and its relationship to boiling and melting temperatures.
12. Relate the properties of the various classes of solids (ionic, covalent, molecular, and metallic) to the structure of these solids.

Introduction

The major differences between solid, liquid, and gaseous substances lie in the following properties:

1. The average distance of separation of particles in each state
2. The strength of the attractive forces between the particles
3. The degree of organization of particles

The behavior of the states of matter, their properties, and their interconversion are considered in this chapter.

5.1 The Gaseous State

Ideal Gas Concept

An **ideal gas** is a model that describes how molecules or atoms behave at the atomic/molecular level. The ideal gas model has been used to determine the interrelationship of temperature (T), volume (V), pressure (P), and quantity in mass (m) or number of moles (n) in gaseous systems. These relationships are summarized in the ideal gas law and are described in this chapter.

Measurement of Properties of Gases

Measurement of the temperature, volume, or mass of a gas is familiar to us and we have already used those properties. **Pressure** is a measurement of force per unit area.

Pressure can be measured with a mercury **barometer**, in which a long tube filled with mercury (Hg) is inverted into a dish of mercury. The atmospheric pressure pushing on the surface of the mercury in the dish supports the column of mercury in the tube. So, the height of the mercury in the tube is proportional to the atmospheric pressure; "mm Hg" can be used as a unit of pressure.

In addition to *mm Hg (millimeters of mercury)*, other common units of pressure are: *atmosphere (atm)* and *torr*. Still other units of pressure include the *pascal (Pa)* or *kilopascal (kPa)*, *lb/in^2 (pounds per square inch,* or *psi)*, and *in Hg (inches of mercury)*. These units are related as follows:

$$1 \text{ atm} = 760 \text{ mm Hg} = 760 \text{ torr} = 14.7 \text{ lb/in}^2 = 29.9 \text{ in Hg} = 1.01 \times 10^5 \text{ Pa} = 101 \text{ kPa}$$

As you work with pressure units in this chapter, it is important to be able to convert between pressure units.

Example 1: Conversion between Pressure Units.
Convert each of the following into atmospheres.

1. 77.0 mm Hg
2. 30.2 in Hg
3. 16.8 psi (The abbreviation psi means pounds per square inch.)
4. 800.00 torr

Solution:
The relationships between the different pressure units can each be treated as a conversion factor. Here are a couple examples:

$$\frac{1 \text{ atm}}{760 \text{ mm Hg}} \qquad \frac{101 \text{ kPa}}{29.9 \text{ in Hg}}$$

1. The conversion of mm Hg to atm requires the knowledge that 1 atm = 760 mm Hg.

$$77.0 \text{ mm Hg} \times \frac{1 \text{ atm}}{760 \text{ mm Hg}} = 0.101 \text{ atm}$$

2. The conversion of mm Hg to atm requires the knowledge that 29.9 in Hg = 1 atm.

$$30.2 \text{ in Hg} \times \frac{1 \text{ atm}}{29.9 \text{ in Hg}} = 1.01 \text{ atm}$$

Example 1 Continued

Alternately, if you have not memorized this relationship, you can use the relationship between centimeters and inches which was covered in Chapter 1.

$$30.2\ \cancel{\text{in Hg}} \times \frac{2.54\ \cancel{\text{cm Hg}}}{1\ \cancel{\text{in Hg}}} \times \frac{1\ \cancel{\text{m Hg}}}{100\ \cancel{\text{cm}}\ \text{Hg}} \times \frac{1000\ \cancel{\text{mm Hg}}}{1\ \cancel{\text{m Hg}}} \times \frac{1\ \text{atm}}{750\ \cancel{\text{mm Hg}}} = 1.01\ \text{atm}$$

3. Use the relationship that 14.7 psi = 1 atm.

$$16.8\ \cancel{\text{psi}} \times \frac{1\ \text{atm}}{14.7\ \cancel{\text{psi}}} = 1.14\ \text{atm}$$

4. Use the relationship that 760 torr = 1 atm.

$$800.0\ \cancel{\text{torr}} \times \frac{1\ \text{atm}}{760\ \cancel{\text{torr}}} = 1.053\ \text{atm}$$

Kinetic Molecular Theory of Gases

The fundamental model of particle behavior in the gas phase is the **kinetic molecular theory**. This theory describes an ideal gas and is summarized as follows:

1. Gases are made up of small atoms or molecules that are in constant, random motion. The particles are moving linearly.
2. The distance of separation among these atoms or molecules is very large in comparison to the size of the individual atoms or molecules. In other words, a gas is mostly empty space.
3. All of the atoms and molecules behave independently. No attractive or repulsive forces exist between atoms or molecules in a gas.
4. Atoms and molecules collide with each other and with the walls of the container without losing energy. The energy is transferred from one atom or molecule to another.
5. The average kinetic energy of the atoms or molecules increases or decreases in proportion to absolute temperature.

Properties of Gases and the Kinetic Molecular Theory

The kinetic molecular theory explains many of the properties observed of gases. Table 5.1 gives a list of a few properties and a corresponding explanation of the property using the kinetic molecular theory.

Gases behave most ideally when they are at low pressure and high temperatures. That is, for gases that are not about to become a liquid. This allows for the greatest distance between gas particles and minimizes attractive forces.

TABLE 5.1 Properties of Gases and Kinetic Molecular Theory

Property of Gas	Kinetic Molecular Theory Explanation
Gases are easily compressible.	The distance between atoms or molecules in a gas is very large. (Statement 2)
Gases expand to fill the container.	Molecules can move freely because they have no attractive forces between molecules holding them together. (Statement 3).
Gases diffuse through each other.	Gases are in continuous random motion (Statement 1) and the large distance between gas particles (Statement 2) allows for many available paths.
Gases have low density.	Density is mass per volume. Since the gas is mostly empty space (Statement 2), the mass is very small.
Gases exert a constant pressure on their container. This is the property that keeps a balloon inflated.	Pressure is a measure of force per area. As the molecules collide with the walls of the container (Statement 4) they exert the force necessary to create the pressure. The pressure is constant because the molecules do not lose energy as they collide (Statement 4).

Boyle's Law

Robert Boyle found that the volume of a gas varies inversely with the pressure exerted by the gas, if the number of moles and the temperature of the gas are held constant. This relationship is known as **Boyle's law**.

Mathematically, the product of pressure (P) and volume (V) is a constant:

$$PV = k_b$$

Boyle's law is often used to calculate the volume resulting from a pressure change or vice versa. We consider $P_iV_i = k_b$ the initial condition, and $P_fV_f = k_b$ the final condition. Since (PV), initial or final, is constant and is equal to k_b,

$$P_iV_i = P_fV_f$$

Note that the calculation can be done with volume units of mL or L. It is only important that the units be the same on both sides of the equation. This is also true of pressure; however, the most commonly used unit of pressure measurement is the atmosphere.

Example 2: Boyle's Law
A given mass of carbon dioxide at 25°C occupies a volume of 500.0 mL at 2.00 atm. What pressure must be applied to compress the gas to a volume of 50.0 mL, assuming no temperature change?

Example 2 Solution:

1. Because pressure and volume are given at a constant number of moles and temperature, use Boyle's law:

$$P_iV_i = P_fV_f$$

2. From the given information, we have the following:
   P_i = 2.00 atmospheres V_i = 500.0 mL (the initial volume)
   P_f = ? (not known yet) V_f = 50.0 mL (the final volume)

3. Solve the equation for the unknown variable.

$$P_f = \frac{P_iV_i}{V_f}$$

4. We can thus find the pressure needed to compress the 500.0 mL of gas to 50.0 mL.

$$P_f = \frac{(2.00 \text{ atm})(500.0 \cancel{\text{mL}})}{50.0 \cancel{\text{mL}}} = 20.0 \text{ atm}$$

20.0 atmospheres of pressure will be needed to compress the gas.

Charles's Law

Jacques Charles found that the volume of a gas varies directly with the absolute temperature (K) if pressure and the number of moles of the gas are held constant. This relationship is **Charles's law.**

Mathematically, the ratio of volume (V) and temperature (T) is a constant:

$$\frac{V}{T} = k_c$$

In a way that is analogous to Boyle's law, Charles's Law is expressed as:

$$\frac{V_i}{T_i} = \frac{V_f}{T_f}$$

Example 3: Charles's Law
A balloon filled with helium has a volume of 5.00 liters at 0.00°C. What would be the balloon's volume at 25.00°C if the pressure surrounding the balloon remained constant?

Example 3 Solution:

1. Use Charles's law: $\frac{V_i}{T_i}=\frac{V_f}{T_f}$.

2. It is very important (and easy to forget!) to change the °C values into Kelvin. Use the following equation:

$$T_K = T_{°C} + 273.15$$

$$T_i = 0.00°C; \text{ in K, } T_i = 0.00°C + 273.15 = 273.15 \text{ K}$$

$$T_f = 25.00°C; \text{ in K, } T_f = 25.00°C + 273.15 = 298.15 \text{ K}$$

3. Solve for V_f.

$$\frac{V_i}{T_i}=\frac{V_f}{T_f}$$

$$(V_f)(T_i) = (V_i)(T_f)$$

$$V_f = \frac{(V_i)(T_f)}{T_i}$$

4. $V_f = \frac{(5.00 \text{ L})(298.15 \cancel{K})}{273.15 \cancel{K}} = 5.46 \text{ L}$

Combined Gas Law

Often a sample of gas (a fixed number of moles of gas) undergoes change involving volume, pressure, and temperature simultaneously. It would be useful to have one equation that describes such processes.

A **combined gas law** is such an equation. It can be derived from Boyle's law and Charles's law and takes the form:

$$\frac{P_iV_i}{T_i}=\frac{P_fV_f}{T_f}$$

Avogadro's Law

The relationship between the volume and number of moles of a gas is known as **Avogadro's law**. This law states that equal volumes of a gas contain the same number of moles, if measured under the same conditions of temperature and pressure.

Mathematically, the ratio of volume (V) and number of moles (n) is a constant:

$$\frac{V}{n} = k_a$$

A relationship comparing initial and final conditions, similar to Boyle's and Charles's laws, may be written as:

$$\frac{V_i}{n_i} = \frac{V_f}{n_f}$$

Problems involving the combined gas law or Avogadro's law are solved similarly to Examples 2 and 3. The key to working any of these gas law problems is to look for the variables which are changing and the variables which are held constant in order to write the correct gas law equation. Example 4 will give you practice in determining the equation needed to solve the problem.

Example 4: Determination of the Appropriate Gas Law Equation.
For each of the following, determine the equation needed to work the problem.

1. A balloon filled with air has a volume of 2.0 L at 25.0°C. A constant pressure of 1.0 atm is exerted on the balloon. What is the volume if the balloon is cooled to 1.0 °C?
2. A gas cylinder contains 500.0 mL of gas at 100°C at 2.0 atm. What is the pressure in the cylinder if the gas is compressed to 100.0 mL at 200°C?
3. A balloon contains 1.0 L at some temperature and pressure. The number of moles of gas is doubled while the pressure and temperature remain constant. What is the new volume of the balloon?

Solution:

1. The variables which are changing are: volume and temperature. The experiment does not allow for a change in the amount of gas or a change in pressure; that is, *n* and *P* are constant. This is Charles's Law:

$$\frac{V_i}{T_i} = \frac{V_f}{T_f}$$

2. Only *n* is held constant. This is noted by the fact that there is nothing to infer the adding or removing of a gas. *P*, *V*, and *T* are all changing. Therefore, the Combined Gas Law applies.

$$\frac{P_i V_i}{T_i} = \frac{P_f V_f}{T_f}$$

3. "The number of moles of gas is doubled" lets us know that *n* is *not* held constant. Volume is also changing. The constants are *P* and *T*. These are the conditions for Avogadro's Law.

$$\frac{V_i}{n_i} = \frac{V_f}{n_f}$$

Molar Volume of a Gas

The volume occupied by 1 mole of any gas is referred to as its **molar volume**. The basis for this relationship is Avogadro's law. At **standard temperature and pressure** (STP), the molar volume of any gas is 22.4 L. STP conditions are defined as follows:

$$T = 273\ K\ (\text{or } 0°C) \qquad P = 1\ atm$$

Thus, 1 mole of N_2 (28.0 g), O_2 (32.0 g), H_2 (2.02 g), and He (4.00 g) all occupy the same volume, 22.4 L at STP.

Gas Densities

It is also possible to compute the density of various gases at STP if one recalls that density is the mass/unit volume:

$$d = \frac{m}{V}$$

Example 5: Density of a Gas.
Determine the density of N_2 and O_2 at standard temperature and pressure.

Solution:
For N_2 at STP, the mass of one mole is 28.0 g (from the molar mass of N_2), and the volume of one mole = 22.4 L (molar volume at STP). Therefore the density of N_2 is

$$d = \frac{28.0\ g}{22.4\ L} = 1.25\ g/L$$

For O_2 at STP the volume of one mole of the gas is not different than for N_2; however, the mass of one mole is different.

$$d = \frac{32.0\ g}{22.4\ L} = 1.43\ g/L$$

Notice that the mass is greater for O_2; therefore, the density is greater.

The Ideal Gas Law

Boyle's law (relating volume and pressure), Charles' law (relating volume and temperature) and Avogadro's law (relating volume to the number of moles) may be combined into a single expression relating all four terms. This expression is the **ideal gas law**:

$$PV = nRT$$

where R is the ideal gas constant.

$$R = 0.0821\ \frac{L \cdot atm}{mol \cdot K}$$

To use this value of R, the units of atmospheres, liters, moles, and Kelvin must be used.

This equation is useful when a sample of gas in *not* undergoing a change of condition (that is, the temperature is *not* changing, or the pressure is *not* increasing…), but instead you know three of the four variables of a gas (*P, V, n* and *T*) and you need to solve for the fourth variable.

Example 6: Using the Ideal Gas Law.
Calculate the number of grams of helium in a 1.00-liter balloon at 27°C and under 1.00 atmosphere of pressure.

Solution:
Notice that you are given three of the variable (*V, T* and *P*). You are asked to calculate the mass. By solving the ideal gas law for moles (*n*), you will have a quantity that can then be easily converted to mass using the molar mass of helium (4.003 g/mol).

1. Use the ideal gas law:
   $PV = nRT$

2. Convert 27.0°C into Kelvin units:

   $$T_K = T_{°C} + 273.15$$

   $$T_K = 27.°C + 273.15 = 300.\ K$$

3. Rearrange $PV = nRT$ so we can find the number of moles (*n* in the equation) of gas present:

   $$n = \frac{PV}{RT}$$

4. $$n_{helium} = \frac{(1.00\,atm)(1.00\,L)}{\left(\frac{0.0821(L)(atm)}{(K)(mole)}\right)(300\,K)}$$

   $n_{helium} = 4.06 \times 10^{-2}$ moles of helium are present.

5. Convert moles of helium to grams of helium using its molar mass, 4.003 g/mole.

   $$4.06 \times 10^{-2}\ mol\ He \times \frac{4.003\ g\ He}{1\ mol\ He} = 0.163\ g\ He$$

Dalton's Law of Partial Pressures

A mixture of gases exerts a pressure that is the sum of the pressure that each gas would exert if it were present alone under similar conditions. This is known as **Dalton's law** of partial pressures.

$$P_t = p_1 + p_2 + p_3 + \bullet\bullet\bullet\bullet$$

in which P_t = total pressure and p_1, p_2, p_3, and so on are the **partial pressures** of the component gases.

Air, for example, is a mixture of mostly $N_2(g)$ and $O_2(g)$, and the total atmospheric pressure is the sum of the pressure of the two gases.

Ideal Gases vs. Real Gases

LG 7

We have assumed so far, both in theory and calculations, that all gases behave as ideal gases. However, in reality, there is no such thing as an ideal gas. Interactive forces, even between the widely-spaced particles of gas, are not totally absent in any sample of gas.

Attractive forces are particularly significant in gases composed of polar molecules. Calculations using ideal gas equations that involve polar gases such as HF, NO, and SO_2 are approximate at best. However, at low pressures, the approximations are useful to make.

5.2 The Liquid State

Molecules in the liquid state are close to one another. Attractive forces are large enough to keep the molecules together, in contrast to gases, in which cohesive forces are so low that a gas expands to fill any volume. These attractive forces in a liquid are not large enough to restrict movement. The molecules can move past each other.

Compressibility

Liquids are almost incompressible. In fact, the spacing between molecules is so small that even the application of many atmospheres of pressure does not significantly decrease the volume of a liquid.

Viscosity

The **viscosity** of a liquid is a measure of its resistance to flow. Viscosity is a function of the attractive forces present between molecules as well as the molecular geometry. Complex molecules, which do not "slide" smoothly past each other, as well as polar molecules, tend to have higher viscosity than less structurally complex, polar liquids. Liquids tend to become less viscous at higher temperatures.

Surface Tension

The **surface tension** of a liquid is a measure of the attractive forces at its surface. Intermolecular attraction is stronger at the surface of a liquid. This increased surface force is responsible for the spherical shape of drops of liquid. Substances known as **surfactants** may be added to a liquid to decrease surface tension.

Vapor Pressure of a Liquid

According to the kinetic theory, the molecules of a liquid are in continuous motion, with their average kinetic energy proportional to the temperature. Although the average kinetic energy is too small to allow all the molecules to "escape" from liquid to vapor, a few high-energy, surface molecules possess sufficient energy to escape from the bulk liquid.

At the same time, a fraction of these vaporized molecules lose energy (perhaps by collision with the walls of the container) and return to the liquid state. The process of conversion of liquid to vapor, at a temperature too low to boil, is **evaporation**. The reverse process, conversion of the gas to the liquid state, is **condensation**. After some period of time, the rates of evaporation and condensation become equal, and this constitutes a dynamic equilibrium between liquid and vapor states. The **vapor pressure of the liquid** is therefore defined as the pressure exerted by the vapor at equilibrium.

As temperature increases, the vapor pressure of the substance will increase. The boiling point of a liquid is defined as the temperature at which the vapor pressure of the liquid becomes equal to the prevailing atmospheric pressure. The "normal" atmospheric pressure is 760 torr, or one atmosphere, and the **normal boiling point** is the temperature at which the vapor pressure of the liquid is equal to one atmosphere.

van der Waals Forces

In Chapter 3, we discussed how polar molecules are attracted to other polar molecules resulting in intermolecular forces. If a polar molecule is attracted to another polar molecule this attractive force is called a **dipole-dipole interaction.**

In 1930, Fritz London demonstrated that he could account for a weak attractive force between any two molecules, whether polar or nonpolar. He postulated that the electron distribution in molecules is not fixed; electrons are in continuous motion, relative to the nucleus. So, for a short period of time a nonpolar molecule could experience an instantaneous dipole, a short-lived polarity due to a temporary dislocation of the electron cloud. These temporary dipoles could interact with other temporary dipoles, just as permanent dipoles interact in polar molecules. We now call these intermolecular forces **London dispersion forces**.

London dispersion forces and dipole-dipole interactions are collectively known as **van der Waals forces**. London dispersion forces exist among polar and nonpolar molecules because electrons are in constant motion in all molecules. Dipole-dipole attractions occur only among polar molecules. In the next section, we will see a third type of intermolecular force, the *hydrogen bond*.

Hydrogen Bonding

Molecules in which a hydrogen atom is bonded to the highly electronegative atom of nitrogen, oxygen, or fluorine exhibit **hydrogen bonding**. This arrangement of atoms produces a very polar bond, often resulting in a polar molecule with strong intermolecular attractive forces. Hydrogen bonding is a type of dipole-dipole attraction but is so much stronger than a typical dipole-dipole attraction that it gets a category all of its own.

Example 7:
Illustrate the hydrogen bonding found between a water molecule and its closest three neighboring water molecules.

Example 7 Solution:

1. The polar nature of water is caused by the large electronegativity difference between oxygen and hydrogen atoms.
2. One water molecule has the following polar structure:

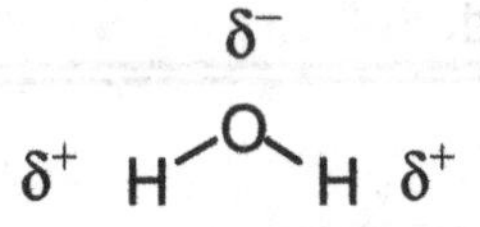

$$\overset{\delta^-}{\ } \\ \delta^+ \ H{-}O{-}H \ \delta^+$$

Notice that the oxygen end is slightly negative, and the two hydrogen ends are slightly positive.

3. The hydrogen bonding in liquid water is due to the attraction of the slightly negative oxygen atom to a slightly positive hydrogen in another water molecule. In similar fashion each slightly positive hydrogen atom attracts an oxygen atom in two other water molecules.
4. Thus, one water molecule has three hydrogen bonds with three other water molecules.

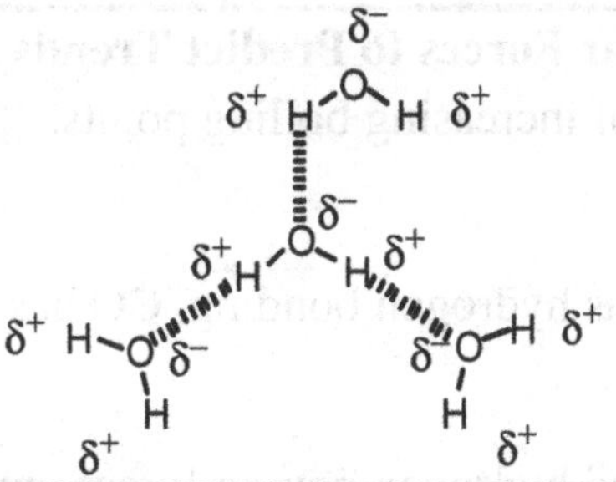

Example 8: Identifying Intermolecular Forces

For each of the following substances classify the type of intermolecular force which would occur between the molecules.

1. NH_3 3. CO_2
2. CO 4. CH_3OH

Solution:

1. NH_3 meets the criteria for hydrogen bonding. It has a hydrogen atom directly bonded to a nitrogen atom.
2. Carbon monoxide does not meet the criteria for hydrogen bonding; you must determine if the molecule is polar (and thus has dipole-dipole forces) or nonpolar (and thus has only London forces). Carbon monoxide is a molecule with only one bond. The bond is polar; therefore, the molecule is polar. Polar molecules have dipole-dipole forces.
3. Carbon dioxide also does not meet the criteria for hydrogen bonding. Refer to Chapter 3 of the study guide to review the process for determining if a molecule with more than two atoms is polar or nonpolar. We will first draw a Lewis Structure of CO_2:

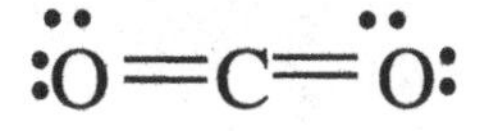

$$:\ddot{O}=C=\ddot{O}:$$

Example 8 Continued

The central atom of carbon has two of the same atoms bonded to it (oxygen atoms) and no lone pairs. The molecule is nonpolar and has London forces only.

4. Methanol (CH_3OH) has a hydrogen atom directly bonded to an oxygen atom. This meets the criteria for the hydrogen bond.

Learn the following two trends:

1. As the strength of intermolecular forces increases, the vapor pressure decreases. This trend will make sense if you realize that the stronger the attraction between molecules the less likely they will be able to overcome the attractions and evaporate, leaving fewer molecules in the vapor phase.
2. As the strength of the intermolecular forces increases, the boiling point will increase. It takes a higher temperature (more kinetic energy) to make the molecules leave the liquid phase.

Example 9: Using Intermolecular Forces to Predict Trends in Boiling Points

Rank NH_3, CO and CO_2 in order of increasing boiling points.

Solution:

In Example 8, we saw that NH_3 has hydrogen bonding, CO has dipole-dipole attractions and CO_2 has London forces.

London forces are the weakest, and hydrogen boning forces are the strongest of the intermolecular forces. Since boiling point increases with strength of intermolecular forces, the boiling point increases as follows:

$$CO_2 < CO < NH_3$$

5.3 The Solid State

The close packing of the particles of a solid results from attractive forces that are strong enough to restrict motion. The particles are "locked" together in a defined and often highly organized fashion. The result is a fixed shape and volume (recall that gases have no fixed shape or volume, and liquids have a fixed volume but no fixed shape).

Properties of Solids

Solids are virtually incompressible, due to the small distance between particles.

Melting point is the temperature at which a solid is converted to the liquid phase. Much like the trend for boiling point, melting point increases with increasing intermolecular forces.

Solids may be **crystalline solid**, having a regular repeating structure, or **amorphous solid**, having no organized structure.

Types of Crystalline Solids

LG 12

Crystalline solids may exist in one of four general groups:

1. **Ionic solids.** The units that make up an **ionic solid** are positive and negative ions. Electrostatic forces hold the crystal together. Ionic solids are brittle and have very high melting points.
2. **Covalent solids.** The units that make up a **covalent solid** are atoms held together by covalent bonds (joining nonmetals to nonmetals) throughout the entire crystal. These too have very high melting points.
3. **Molecular solids.** The units composing **molecular solids** are molecules held together by intermolecular forces (dipole-dipole, London or hydrogen bonds). These have comparatively low melting points
4. **Metallic solids** Metal atoms held together by metallic bonds make up **metallic solids.** Metallic bonds are formed by the overlap of orbitals of metal atoms, resulting in regions of high electron density surrounding the positive metal nuclei. Electrons in these regions are extremely mobile, resulting in the high conductivity (ability to carry electric current) exhibited by many metallic solids. Metallic solids have a wide range of melting point, from mercury (Hg), which is a liquid at room temperature, to tungsten (W), which is used as the filament in an incandescent light bulb.

Sublimation of Solids

The process where a solid converts directly to a gas (without going through a liquid state) is referred to as **sublimation**. Dry ice (solid carbon dioxide) and naphthalene (moth balls) are common examples of this.

Self Test for Chapter Five

1. Complete each of the following statements of the kinetic molecular theory of gases:
   a) Gases are made up of small atoms or molecules that are in constant, …
   b) The distance of separation among these atoms or molecules is very large in comparison to the size of the individual atoms or molecules. In other words, a gas is …
   c) All of the atoms and molecules behave independently. No attractive …
   d) Atoms and molecules collide with each other and with the walls of the container without …
   e) The average kinetic energy of the atoms or molecules increases or decreases in …
2. Convert 525 mm Hg to atm.
3. Convert 29.96 in Hg; that is, inches of mercury, to torr.
4. A given mass of gas at 20.0°C occupies a volume of 250.0 mL at 1.00 atmosphere of pressure. What pressure must be applied to compress the gas to a volume of 50.0 mL, assuming no temperature change?
5. A balloon filled with air has a volume of 2.00 liters at 24.0°C. What would be the balloon's volume at 37.0°C if the pressure surrounding the balloon remained constant?
6. If the temperature of 10.0 liters of gas is increased from 273 K to 546 K, what will be

the new volume of that gas? Assume that pressure is held constant.

7. Calculate the number of moles of helium in a 0.500-liter balloon at 25.0°C and 0.960 atmospheres of pressure.
8. If, at constant temperature and pressure, 2.5 moles of the gas occupy a volume of 50.0 L, how much gas would be required to occupy a volume of 110 L?
9. What is the volume of 7.0 g carbon dioxide gas at 25.0°C and 5.0 atm?
10. The total pressure of our atmosphere is primarily equal to the sum of the pressures of what two gases?
11. Define viscosity.
12. Define the terms *evaporation* and *vapor pressure.*
13. How is vapor pressure related to boiling point?
14. Why is the boiling point of a liquid not constant?
15. What term describes the attractive forces at the surface of a liquid?
16. What type of substances may be added to a liquid to decrease its surface tension?
17. What type of intermolecular force exists between molecules of CH_4?
18. Show the hydrogen bonds between two hydrogen fluoride molecules. The polar structure of HF is: $\delta^+ H-F \delta^-$.
19. Use your understanding of intermolecular forces to rank the following in order of increasing boiling point: CH_4, HF, HCl
20. What is the relationship between the strength of intermolecular forces and vapor pressure?
21. What term describes the temperature at which a solid is converted into a liquid?
22. Are ionic solids characterized by high or low melting points?
23. Match the property with the type of crystalline solid.

___ Characterized by relatively low melting points.	a. Ionic crystal
___ Made up of cations and anions and has very high melting points.	b. Covalent crystal.
___ Made up of all nonmetals and has very high melting points.	c. Molecular crystal
___ Made up only of metals and has a wide range of melting points.	d. Metallic crystal

6 *Solutions*

Learning Goals

1. Distinguish among the terms *solution, solute,* and *solvent.*
2. Describe the properties and composition of various kinds of solutions.
3. Explain which factors influence the degree of solubility, and use trends to make predictions.
4. Describe the relationship between solubility and equilibrium.
5. Use Henry's law to calculate equilibrium solubility values for gases.
6. Calculate solution concentration in mass/ volume percent, mass/mass percent, parts per thousand, and parts per million.
7. Determine the quantity of solute or solution from the concentration of solution.
8. Calculate the molarity of solution from mass or moles of solute.
9. Perform dilution calculations.
10. Describe and explain concentration-dependent solution properties.
11. Perform calculations involving colligative properties.
12. Describe why the chemical and physical properties of water make it a truly unique solvent.
13. Interconvert molar concentration of ions and milliequivalents/liter.
14. Explain the role of electrolytes in blood and their relationship to the process of dialysis.

Introduction

Almost all of the reactions that take place in your body take place in solution, and more specifically with water as the solvent. Not only do nearly all biological reactions occur in solution, many other reactions we have studied take place in solution. Precipitation reactions, acid-base reactions and many oxidation-reduction reactions take place in solution.

6.1 Properties of Solutions

A **solution** is a homogeneous mixture of two or more substances. A solution is composed of one or more **solutes** dissolved in a **solvent**. The solute is the component of the mixture present in smaller quantities; while the solvent is the component of the mixture present in larger quantities. When the solvent is water we refer to the homogeneous mixture as an **aqueous solution**.

A liquid solution has a liquid as the solvent, and the solute may be a solid, liquid, or gas. Since a solution is a homogeneous mixture, there are also solutions which are gaseous or solid. Air is a homogenous mixture of gases (mostly nitrogen and oxygen) and therefore is a solution. Gold jewelry is an example of a solid solution. It is a homogeneous mixture of gold, silver and copper.

General Properties of Liquid Solutions

Liquid solutions are clear and transparent. They may be colored or colorless, depending upon the properties of the solute and solvent. Solutions may be classified as **electrolytic** or **nonelectrolytic**. Electrolytic solutions are formed from ionic compounds or acids and bases that dissociate in solution to produce ions. Electrolytic solutions are good conductors of electricity. Nonelectrolytic solutions are formed from nondissociating molecular solutes (nonelectrolytes), and these solutions are nonconducting. Aqueous sodium chloride (NaCl, an ionic compound) is an example of an electrolytic solution; sugar water ($C_{12}H_{22}O_{11}$, a molecular compound) is a common nonelectrolytic solution.

A **true solution** is a homogeneous mixture with uniform properties throughout. In a true solution the solute particles cannot be isolated through filtration and will not "settle out" over time.

True Solutions, Colloidal Dispersions and Suspensions

At first glance, a colloidal suspension looks like nothing more than a solution. However, a **colloidal dispersion** is not a true solution; the colloid particles are not identical in size nor homogeneously distributed throughout the solution. Particles with diameters between one nanometer and two hundred nanometers are colloids. Smaller particles form a true solution. Larger particles are **precipitates**.

A colloidal dispersion is characterized by its light-scattering property, the Tyndall effect. If a light from a laser or flashlight is directed through a colloidal suspension, the beam is visible. Conversely, if the light is passed through a solution, the beam is not visible.

A **suspension** is a heterogeneous mixture that contains particles larger than a colloidal dispersion. As a result, the larger particles may settle out over time; thus it is not a true solution, it is a precipitate.

Degree of Solubility

The degree of **solubility**, that is, how much solute can dissolve in a given volume of solvent, is difficult to predict. Below are general trends for solid and liquid solutes. Gases dissolved in liquids are discussed later.

1. *The magnitude of difference between polarity of solute and solvent.* The more similar the polarity of the two substances, the more soluble they are. The adage "like dissolves like" can be used to remember this trend. For example, carbon tetrachloride, CCl_4 (a nonpolar liquid) is not soluble in water (a polar liquid). However, it is soluble in benzene, C_6H_6 (a nonpolar liquid.)

2. *Temperature.* An increase in temperature usually increases solubility. The trend must be determined experimentally for the mixture to know for certain if solubility increases with temperature or decreases with temperature.

3. *Pressure.* Pressure has little effect on the solubility of solids and liquids in liquids. However, the solubility of gases in solution is extremely pressure dependent, as we will see later.

When a solution contains all the solute that can be dissolved at a particular temperature, it is a **saturated solution**. Cooling a saturated solution often results in the formation of a precipitate. Occasionally however, on cooling, the excess solute may remain in solution for a period of time. Such a solution is described as a **supersaturated solution**. Syrup and honey are examples of a supersaturated solution of sugar in water.

Solubility and Equilibrium

When an excess of solute is brought into contact with a solvent, the resulting dissolution establishes a dynamic equilibrium. Once a saturated solution is achieved, solute molecules are still dissolving. However, every time one molecule of the solute dissolves, one molecule of the solute will precipitate. There is a continual exchange of solute molecules between the dissolved and undissolved phase. The concepts explaining solution/solute equilibria will be discussed more thoroughly in the next chapter.

Solubility of Gases: Henry's Law

Henry's law states that the number of moles of a gas dissolved in a liquid at a given temperature is proportional to the partial pressure of the gas. In other words, the gas solubility is directly proportional to the pressure of that gas in the atmosphere that is in contact with the liquid. Henry's law can be expressed by the mathematical equation: $M = kP$; where M is the molar concentration of the gas in the liquid in units of moles/liter, P is the pressure (in atm) of the gas over the solution at equilibrium, k is a constant for a given gas dependent on temperature in units of mol/(L·atm).

Henry's Law and Respiration

Henry's law explains the process of respiration. In the lungs, where the pressure of oxygen is relatively high, oxygen can easily dissolve into the blood. As the blood travels through the bloodstream the pressure of oxygen decreases and the pressure of carbon dioxide increases, causing the oxygen to leave the blood stream, and dissolve the carbon dioxide. As the blood travels back to the lungs where the pressure of CO_2 is low and O_2 is high, the CO_2 is released and the O_2 is absorbed again.

Solubility of a gas increases with decreasing temperature. Large amount of hot water released into a river or lake can kill fish due to the fact that the solubility of the oxygen will decrease as the temperature of the water increases. The fish can suffocate from lack of oxygen.

6.2 Concentration Based on Mass

The amount of solute dissolved in a given amount of solution is defined as the solution **concentration**.

$$\text{concentration} = \frac{\text{amount of solute}}{\text{amount of solution}}$$

The concentration of a solution has a profound effect on the properties of a solution, both physical (melting and boiling points) and chemical (solution reactivity). The units of the numerator and denominator in the equation above define many different concentration units. Concentration units based on mass of solute are covered in this section.

Mass/Volume Percent

LG 7

Mass/volume percent defines the amount of solute in terms of grams and the amount of solution in terms of milliliters.

$$\%\left(\frac{m}{V}\right) = \frac{\text{grams of solute}}{\text{mL of solution}} \times 100\%$$

You can use the equation to determine the % (m/V) concentration. Or, if given a % (m/V) solution, you can use the value as a conversion factor in calculations. For example, if a solution of sugar in water is 1.5% (m/V) this would mean that there are 1.5 grams of sugar dissolved in every 100 mL of solution, giving the conversion factor:

$$\frac{1.5 \text{ g sugar}}{100 \text{ mL solution}}$$

Example 1: Calculating % (m/V)
Calculate the % (m/V) of a sodium chloride solution that was made by mixing 15.0 grams of NaCl with enough distilled water to prepare 500.0 mL of solution.

Solution:

1. Use the equation for % (m/V): $\%\left(\frac{m}{V}\right) = \frac{\text{grams of solute}}{\text{mL of solution}} \times 100\%$
2. The solute is NaCl, so grams of solute = 15.0 grams of NaCl.
3. $\%\left(\frac{m}{V}\right) = \frac{15.0 \text{ g}}{500 \text{ mL}} \times 100\% = 3.00\ \%$

Example 2: Using % (m/V) as a Conversion Factor.
How many grams of glucose are found in 1.00 liter of 5.00% (m/V) glucose solution?

Solution:

1. The 5.00% (m/V) glucose solution tells us that there are 5.00 grams of glucose per 100.00 mL of total solution:

$$\frac{5.00 \text{ g glucose}}{100.00 \text{ mL solution}}$$

Example 2 Continued

2. Map out your calculation starting with the given quantity of 1.00 liter of solution:

   $1.00 \text{ L solution} \xrightarrow{1} \text{mL solution} \xrightarrow{2} \text{g glucose.}$

   Step 1: Use the definition of milli.
   Step 2: Use the %(m/V) conversion factor.

3. $1.00 \text{ L sol'n} \times \dfrac{1000 \text{ mL sol'n}}{1 \text{ L sol'n}} \times \dfrac{5.00 \text{ g glucose}}{100.00 \text{ mL sol'n}} = 50.0 \text{ g glucose}$

In 1.00 liter of 5.00% (m/V) glucose solution, there are 50.0 grams of glucose.

Example 3: Using %(m/V) as a Conversion Factor.

If a patient is to receive 2.00 grams of a specific drug, and the drug is supplied as a 2.50% (m/V) drug solution, how many milliliters of the drug solution must the patient be given?

Solution:

1. The 2.50% (m/V) drug solution means that there are 2.50 grams of the drug per 100.00 mL of the drug solution:

$$\frac{2.50 \text{ g drug}}{100.00 \text{ mL solution}}$$

2. Map out your calculation starting with 2.00 grams of the drug.

   $2.00 \text{ g drug} \rightarrow \text{mL solution.}$

   The conversion factor listed above is used for this one step conversion.

3. $2.00 \text{ g drug} \times \dfrac{100.00 \text{ mL sol'n}}{2.50 \text{ g drug}} = 80 \text{ mL drug}$

The patient must take 80.0 mL of the 2.50% (m/V) drug solution to receive 2.00 grams of the specific drug.

Note: You do not want to start with the conversion factor in a calculation, if at all possible, because you will not know which way to use the factor (as seen above or "flipped over") until you see what units you need to cancel.

Mass/Mass Percent

Similar to the mass/volume percent concentration, **mass/mass percent** defines the solution in terms of mass instead of volume.

$$\%\left(\frac{m}{m}\right)=\frac{\text{grams of solute}}{\text{grams of solution}}\times 100\%$$

This is more often used for solid solutions. Similar calculations can be performed using this unit of concentration as with the % (m/V). See Examples 4 and 5.

Example 4: Calculating % (m/m)
Calculate the % (m/m) of gold in a wedding band that has a total mass of 12.0 grams and was found to contain 8.97 g of gold.

Solution:

1. Use the equation for % (m/m): $\%\left(\frac{m}{m}\right)=\frac{\text{grams of solute}}{\text{grams of solution}}\times 100\%$

2. $\%\left(\frac{m}{m}\right)=\frac{8.97\text{ g Au}}{12.0\text{ g sol'n}}\times 100\%=74.8\%$

The wedding band is 74.8 % (m/m) gold.

Example 5: Using % (m/m) as a Conversion Factor.
How many grams of iron are in 5.0 kg of ore which is 15.0 % (m/m) iron?

Solution:

1. The ore has 15.0 grams of iron in every 100.0 grams of iron ore:
$$\frac{15.0\text{ g Fe}}{100.0\text{ g ore}}$$

2. Map out your calculation beginning with the given quantity of 5.0 kg ore

   $5.0\text{ kg ore}\xrightarrow{1}\text{g ore}\xrightarrow{2}\text{g Fe}$

   Step 1: use the definition of kilo.
   Step 2: use the conversion fact obtained by the defined % (m/m).

3. $5.0\text{ kg ore}\times\frac{1000\text{ g ore}}{1\text{ kg ore}}\times\frac{15.0\text{ g Fe}}{100.0\text{ g ore}}=750\text{ g Fe}$

There are 750 grams of iron in 5.0 kg of ore.

Parts per Thousand and Parts per Million

Parts per thousand (ppt) and **parts per million** (ppm) are used for solutions that have very low concentration. These units are similar to the percent concentration units we learned above, but instead of multiplying by 100% the ratios will be multiplied by 1,000 or 1,000,000.

$$\text{ppt} = \frac{\text{grams of solute}}{\text{grams of solution}} \times 10^3 \text{ ppt} \quad \text{and} \quad \text{ppm} = \frac{\text{grams of solute}}{\text{grams of solution}} \times 10^6 \text{ ppm}$$

Example 6: Calculating ppm
Calculate the ppm of chloride in solution if there are 0.10 mg of chloride ions in 1.00 L of aqueous solution.

Solution:
1. Convert the solute and solution to the needed units of grams. The solution is so dilute that the density of water (1.00 g/mL) can be used for the density of the solution in order to convert to grams of solution from the given volume.

$$0.10 \text{ mg Cl}^- \times \frac{1 \text{ g Cl}^-}{1000 \text{ mg Cl}^-} = 1.0 \times 10^{-4} \text{ g Cl}^-$$

$$1.00 \text{ L sol'n} \times \frac{1000 \text{ mL sol'n}}{1 \text{ L sol'n}} \times \frac{1.00 \text{ g}}{1.00 \text{ mL}} = 1.0 \times 10^3 \text{ g sol'n}$$

2. Use the definition of ppm: $\text{ppm} = \frac{\text{grams of solute}}{\text{grams of solution}} \times 10^6 \text{ ppm}$

3. $\text{ppm} = \frac{1.0 \times 10^{-4} \text{ g Cl}^-}{1.0 \times 10^3 \text{ g sol'n}} \times 10^6 \text{ ppm} = 0.10 \text{ ppm}$

If ppt or ppm is given in a problem, the value can be used as a conversion factor as we did with percent. For example, if a solution is 1.5 ppt NaCl in water, this can be written as:

$$\frac{1.5 \text{ g NaCl}}{1000 \text{ g sol'n}}$$

6.3 Concentration Based on Moles

Molarity

Another common concentration unit used in chemistry is **molarity**, symbolized *M*. Molarity is defined as the number of moles of solute per liter of solution:

$$\text{Molarity} = M = \frac{\text{moles solute}}{\text{L solution}}$$

Example 7: Calculating Molarity

Calculate the molarity of a sodium chloride solution made by mixing 3.51 grams of solid NaCl in enough water to prepare 500.0 mL of total solution.

Solution:

1. Use the molarity equation: $M = \frac{\text{moles solute}}{\text{L solution}}$

2. The amount of solute (NaCl) must be converted to moles. To convert from grams to moles, use the molar mass.

$$3.51 \text{ g NaCl} \times \frac{1 \text{ mol NaCl}}{58.5 \text{ g NaCl}} = 0.0600 \text{ mol NaCl}$$

3. The amount of solvent must be converted to liters.

$$500.0 \text{ mL sol'n} \times \frac{1.000 \text{ L sol'n}}{1000 \text{ mL sol'n}} = 0.5000 \text{ L sol'n}$$

4. We can now use the molarity equation, since we know all the needed values:

$$M = \frac{0.600 \text{ mol NaCl}}{0.5000 \text{ L sol'n}} = 0.120 \ \frac{\text{mol}}{\text{L}} \quad \text{or} \quad 0.120\ M$$

Example 8: Using Molarity as a Conversion Factor

Calculate the number of grams of solid silver nitrate needed to prepare 250.0 mL of a 0.100 *M* $AgNO_3$ solution.

Solution:

1. A 0.100 *M* $AgNO_3$ solution can be written as the following conversion factor:

$$\frac{0.100 \text{ mol AgNO}_3}{1.00 \text{ L sol'n}}$$

2. Starting with the given quantity of 250.0 mL of solution, map out your conversion.

   $250.0 \text{ mL sol'n} \xrightarrow{1} \text{L sol'n} \xrightarrow{2} \text{mol AgNO}_3 \xrightarrow{3} \text{g AgNO}_3$

   Step 1: use the definition of milli
   Step 2: use molarity as a conversion factor
   Step 3: use the molar mass of $AgNO_3$.

Example 8: Continued

3. $250.0 \text{ mL sol'n} \times \frac{1.00 \text{ L sol'n}}{1000 \text{ mL sol'n}} \times \frac{0.100 \text{ mol } AgNO_3}{1 \text{ L sol'n}} \times \frac{169.91 \text{ g } AgNO_3}{1 \text{ mol } AgNO_3}$

$= 4.25 \text{ g } AgNO_3$

We need to add 4.25 grams of solid $AgNO_3$ to prepare 250.0 mL of a 0.100 molar $AgNO_3$ solution.

LG 9

Dilution

The technique of dilution is often used to prepare less-concentrated solutions. The approach to such a calculation is outlined below:

M_1 = molarity before dilution. M_2 = molarity after dilution.

V_1 = volume before dilution. V_2 = volume after dilution.

The formula used for dilution problems is:

$$(M_1)(V_1) = (M_2)(V_2)$$

Knowing any three of these four variables enables one to calculate the fourth. Note that the number of moles of solute remains unchanged during the dilution. This is because dilution only involves the addition of extra solvent.

The dilution equation is valid with any concentration units, such as % (m/V) as well as molarity. Be certain to use the same units for both initial and final concentration values. Only in this way will proper unit cancellation occur.

Example 9: Dilution

How many mL of 12.0 *M* HCl solution do we need to add to distilled water so we can prepare 500.0 mL of 2.50 *M* HCl solution?

Solution:

1. Use $(M_1)(V_1) = (M_2)(V_2)$
   V_1 is the starting volume; we do not know this value yet.
   M_1 is the molarity of the solution we have available to us. M_1=12.0 *M*
   M_2 is the molarity of our final solution. M_2=2.50 *M*.
   V_2 is the final volume of our diluted solution. V_2 = 500.0 mL

2. Rearrange the equation to find V_1:

   $(M_1)(V_1) = (M_2)(V_2)$

Example 9 Continued

$$V_1 = \frac{(M_2)(V_2)}{M_1} = \frac{(2.50M)(500.0\text{ mL})}{12.0\,M} = 104\text{ mL}$$

To prepare 500.0 mL of a 2.50 M HCl solution by dilution, measure out 104 mL of the 12.0 *M* HCl solution and then add distilled water to obtain 500.0 mL of total solution.

6.4 Concentration-Dependent Solution Properties

Solution properties that are dependent upon the concentration of the solute particles, rather than the identity of the solute, are referred to as **colligative properties**. There are four colligative properties of solutions.

Vapor Pressure Lowering

Raoult's law states that when a solute is added to a solvent the vapor pressure of the solvent decreases in proportion to the concentration of solute particles. As a result of this, boiling point is directly affected by the concentration of solute particles in a solution.

Freezing-Point Depression and Boiling-Point Elevation

When a nonvolatile solid is added to a solvent, the freezing point of the resulting solution decreases (a lower temperature is required to form the solid state), and the boiling point increases (requiring a higher temperature to form the gaseous state).

The magnitude of the freezing-point depression (ΔT_f) is proportional to the solute concentration over a limited range of concentration.

$$\Delta T_f = k_f \times \text{(solute concentration)}$$

Note: k_f and k_b are constants associated with the solvent.

The boiling point elevation is also proportional to the solute concentration.

$$\Delta T_b = k_b \times \text{(solute concentration)}$$

If the value of the k_f or k_b is known, the magnitude of the freezing-point depression or boiling-point elevation may be calculated for a solution of known concentration of particles.

We have already worked with one mole-based unit, molarity. A second mole-based concentration unit is molality, which is more commonly used in these types of situations. **Molality** (symbolized by *m*) is defined as the number of moles of solute per kilogram of solvent in a solution:

$$m = \frac{\text{moles solute}}{\text{kg solvent}}$$

Molality does not vary with temperature, whereas molarity is temperature dependent. This property of molality makes it a better choice for use for freezing-point depression and boiling-point elevation calculations.

With colligative properties, it is important to consider the number of particles in solution. When dissolving one mole of a nonelectrolyte (such as glucose, $C_6H_{12}O_6$), one mole of dissolved particles will be present in the solution. However, when dissolving an electrolyte, the electrolyte will break apart into the ions. For example when one mole of sodium chloride is dissolved in water, one mole of sodium ions and one mole of chloride ions are produced, resulting in a total of two moles of particles.

Calculating Freezing Points and Boiling Points of Aqueous Solutions

LG 11

The values of the constants, k_f and k_b are dependent on the solvent used. For aqueous solutions, the values are $k_f = 1.86°C/m$ and $k_b = 0.52°C/m$. If the molality is known for a solution, the freezing-point and boiling-point of the solution can be determined.

For freezing-point, solve for ΔT and *subtract* this value from 0.0°C (the normal freezing-point of water.) For boiling-point, solve for ΔT and *add* this value to 100.0°C (the normal boiling-point of water.

Example 10: Boiling-point Elevation
Calculate the boiling-point and freezing-point of a 1.5 *m* solution of glucose ($C_6H_{12}O_6$).

Solution:
For the boiling-point calculation:
1. Use the boiling-point formula: $\Delta T = k_b \times m$.
2. $\Delta T = 0.52°C/m(1.5\ m) = 0.78°C$. The temperature of the solution is elevated by 0.78°C.
3. Add this temperature to the boiling-point of pure water. 100.00°C + 0.78°C = 100.78°C.

The boiling-point of 1.5 *m* glucose solution is 100.78°C.

For the freezing-point calculation:
1. Use the freezing-point formula: $\Delta T = k_f \times m$.
2. $\Delta T = 1.86°C/m(1.5\ m) = 2.8°C$. The temperature of the solution is depressed by 2.8°C.
3. Subtract this temperature from the freezing-point of pure water.
   0.00°C – 2.8°C = –2.8°C.

The freezing-point of 1.5 *m* glucose solution is –2.8°C.

Osmosis, Osmotic Pressure, and Osmolarity

Diffusion is the net movement of solute or solvent molecules from an area of high concentration to an area of low concentration. This region where the concentration decreases over a distance is termed a **concentration gradient**. The cell membrane is an example of a **selectively permeable membrane** because it only allows small molecules to diffuse across the barrier.

Osmosis is the movement of solvent from a dilute solution to a more concentrated solution through a **semipermeable membrane**. Semipermeable membranes are permeable to the solvent but not the solute. Pressure must be applied to the more concentrated solution to stop this flow, and the magnitude of the minimum pressure required to stop the flow is termed the **osmotic pressure**. The "driving force" for the osmotic process is the need to establish an equilibrium between the solutions on either side of the membrane.

The osmotic pressure may be treated quantitatively. The value for solution concentration of particles (iM) in conjunction with the ideal gas constant (R) and the temperature (T) of the solution can be used to calculate the osmotic pressure of a solution, symbolized by π, using the equation:

$$\pi = iM\text{RT}$$

The osmotic pressure may be calculated from the solution concentration at any given temperature. Osmosis is a colligative property; it is dependent on the concentration of solute particles.

By convention, the *molarity of particles* in solution is termed **osmolarity** for osmotic pressure calculations. Osmolarity = $i \times M$, where i = number of mol of particles/mol solute. The difference between molarity and osmolarity can be shown using a solution of an electrolyte such as 1.0 M NaCl. The concentration of NaCl is 1.0 M. However, the NaCl breaks apart according to the equation:

$$\text{NaCl}(s) \rightarrow \text{Na}^{+}(aq) + \text{Cl}^{-}(aq)$$

The concentration of *particles* in solution is two times as great as the concentration of the NaCl since NaCl breaks apart into two particles. Osmolarity = $i \times M = 2 \times 0.10\ M = 0.20$ *mol particles/L.*

Example 11: Calculating Osmotic Pressure.
What is the osmotic pressure of a 0.30 M Na_2SO_4 solution at 298 K (25 °C)?

Solution:

1. Calculate the osmolalty. Na_2SO_4 breaks apart into 3 ions when dissolved.

$$\text{Na}_2\text{SO}_4(s) \rightarrow 2\text{Na}^{+}(aq) + \text{SO}_4^{2-}(aq)$$

Osmolarity = $i \times M = 3 \times 0.30\ M = 0.90$ *mol particles/L*

2. Use osmolaity (iM) in the formula $\pi = iM\text{RT}$.

$$\pi = 0.90\ \frac{\text{mol}}{\text{L}} \times 0.0821 \frac{\text{L}\cdot\text{atm}}{\text{mol}\cdot\text{K}} \times 298\ \text{K} = 22\ \text{atm}$$

Osmosis is important as a transport mechanism to all living organisms. A living cell contains an aqueous solution, and material movement in and out of cells is based partly upon osmosis. If the concentration of the fluid surrounding red blood cells is higher than that inside the cell (a **hypertonic solution**), water flows from the cell, causing it to collapse. On the other hand, too low a concentration of this fluid relative to the solution within the blood cell (a **hypotonic solution**) will cause water to flow into the cell, causing cell rupture.

Two solutions are **isotonic solutions** if they have identical osmotic pressures. In that case, the osmotic pressure differential across the cell is zero, and no cell disruption occurs.

6.5 Aqueous Solutions

Water as a Solvent

The role of water in the solution process deserves special attention because of its many unique characteristics:

1. It is often referred to as the universal solvent.
2. In view of its small size, it has a relatively high boiling point, which enhances its usefulness as a solvent. This is due to the polarity of water.
3. Water is the principal biological solvent. Approximately 60% of the adult human body is water.

Concentration of Electrolytes in Solutions

The concentration of ions in solution may be represented as moles per liter (molarity) and **equivalents** per liter.

Molarity emphasizes the number of individual ions. A one molar solution of Na^+ contains Avogadro's number, $6.02s \times 10^{23}$ Na^+. In contrast, equivalents per liter emphasizes charge; one equivalent of Na^+ contains Avogadro's number of positive charge.

For an aqueous solution, the concentration of any ion in equivalents per liter (eq/L) can be determined using the equation:

$$\text{eq/L} = (\text{ion charge})(M)$$

where the absolute value of the ion charge is used. For example, for Na^+, the ion charge to put into the equation is 1. For Cl^-, the ion charge is 1. For S^{2-}, the ion charge is 2.

Example 12: Calculating Equivalents

Calculate the number of equivalents per liter of Mg^{2+} in a solution that is 1.0×10^{-4} M $MgCl_2$.

Solution:

Use the equation: eq/L = (ion charge)(*M*). The ion charge for Mg^{2+} is 2.

eq/L = $(2)(1.0 \times 10^{-4}\ M) = 2.0 \times 10^{-4}$ eq/L.

LG
14

Biological Effects of Electrolytes in Solutions

Proper cell function in biological systems is critically dependent on the concentration of electrolytes. The maintenance of proper muscle and nervous system functions depends on the sodium/potassium ratio inside and outside of the cell. A stable osmotic pressure in biological fluids depends on the electrolyte concentration inside and outside of the cell.

Dialysis is a process similar to osmosis, where waste materials are removed from the blood. Patients with partial or complete kidney failure must often undergo hemodialysis where a machine mimics the function of the kidney by pumping blood through a semipermeable membrane to exchange particles between the blood and a dialyzing fluid.

Self Test for Chapter Six

1. Consider a solution of sucrose (table sugar) dissolved in water. What is the solute? What is the solvent?
2. What makes something an aqueous solution?
3. Would a solution of potassium chloride be an electrolytic solution or a nonelectrolytic solution?
4. What properties make a colloidal suspension different from a homogeneous solution?
5. How will decreasing the temperature typically affect the solubility of a solid dissolved in a liquid?
6. Would you expect $CaCl_2$ to be more soluble in water or benzene (a nonpolar solvent)?
7. A saturated solution is described as being in dynamic equilibrium. What does this mean?
8. What conditions of temperature and pressure will increase the solubility of oxygen gas in water?
9. A certain ring is 58.3% (m/m) gold. If the ring has a mass of 75 grams, how many grams of gold are in the ring?
10. What is the % (m/V) of NaOH in 150.0 mL of a solution that contains 1.65 grams of sodium hydroxide?
11. How many milliliters of a 5.000% (m/V) glucose solution would contain 40.00 grams of glucose?
12. How many liters of D-5-W solution [5.00% (m/V) glucose solution] would need to be given to a patient by I.V. so that she would receive 250.0 Calories of energy from the given glucose? One gram of glucose produces 4.00 Calories of energy in the human body.
13. A highly pure sample of hydrochloric acid used in industry must have no more than 5 ppm of nitrate (NO_3^-) in solution. How many grams of nitrates are contained in a 1.0 kg solution of hydrochloric acid if the solution has the maximum allowable nitrate impurity of 5 ppm?
14. A water sample contains 3 ppm of lead. How many grams of lead are present in 5 L of the solution?
15. How many moles of KCl are present in 100.0 mL of a 0.552 molar potassium chloride solution?
16. How many grams of KCl are present in 50.0 mL of a 0.125 molar potassium chloride solution? The molar mass of KCl is 74.60 g/mol.
17. If 20.80 grams of HCl are added to enough distilled water to form 3.0 liters of solution,

what is the molarity of the solution? The molar mass of HCl is 36.46 g/mol.

18. Calculate the molarity of a solution that contains 5.60 grams of KNO_3 in 300.0 mL of solution. The molar mass of KNO_3 is 101.1 g/mol.
19. How many milliliters of 12.0 *M* HCl solution are required to prepare 250.0 mL of a 2.50 *M* HCl solution?
20. Of the following two solutions, which one has the higher vapor pressure: 0.10 *M* glucose or 0.50 *M* glucose?
21. What is the molality (*m*) of a solution prepared by dissolving 15.0 g NaOH in 500.0 grams of water? The molar mass of NaOH is 40.0 g/mol.
22. What is the freezing point of a 4.0 *m* sucrose solution? $k_f = 0.52\ ^{\circ}C/m$.
23. What is the osmolarity of a 0.15 *M* NaCl solution?
24. What is the osmotic pressure of a 2.65 *M* $Mg(NO_3)_2$ solution at 25 °C?
25. What term defines the movement of solvent through a semipermeable membrane from a dilute solution to a more concentrated solution?
26. Is a concentrated solution hypertonic or hypotonic when compared to a dilute solution?
27. Describe two properties of water that make it a unique solvent.
28. Calculate the number of equivalents/liter in a 0.5 *M* solution of Ba^{2+}.
29. How does dialysis assist patients suffering from kidney failure?

7 *Energy, Rate, and Equilibrium*

Learning Goals

1. Correlate the terms *endothermic* and *exothermic* with heat flow between a *system* and its *surroundings.*
2. Explain what is meant by *enthalpy, entropy,* and *free energy* and demonstrate their implications.
3. Describe experiments that yield thermochemical information, and use experimental data to calculate the quantity of energy involved in reactions.
4. Describe the concept of reaction rate and the role of kinetics in chemical and physical change.
5. Describe the importance of *activation energy* and the *activated complex* in determining reaction rate.
6. Predict the way reactant structure, concentration, temperature, and catalysis affect the rate of a chemical reaction.
7. Write rate laws, and use these equations to calculate the effect of concentration on rate.
8. Recognize and describe equilibrium situations.
9. Write equilibrium constant expressions, and use these expressions to calculate equilibrium constants or equilibrium concentrations.
10. Use LeChatelier's principle to predict changes in equilibrium position.

Introduction

This chapter focuses on three topics pertaining to chemical reactions.

1. *Thermodynamics* deals with energy changes associated with chemical reactions. The laws of thermodynamics enable us to predict if a reaction will occur under a certain set of conditions.
2. *Kinetics* concerns itself with the rate or speed of a chemical reaction. If a reaction is favorable according to the laws of thermodynamics, it may occur so slowly that it is never observed.
3. *Equilibrium* considerations determine the completeness of a reaction. Does the reaction proceed to form a 100% yield, or some amount less than 100%?

The concepts of energy, rate, and equilibrium can be applied to changes in state as well as to chemical reactions. The behavior of the states of matter (gas, liquid, and solid), their properties, and their interconversion are considered in this chapter.

7.1 Thermodynamics

Thermodynamics is the study of energy, work, and heat. Thermodynamics may be applied to chemical change or physical change. Three basic laws of thermodynamics exist; only the first two are of concern in Chapter 7.

The Chemical Reaction and Energy

Our current view of events at the molecular level is consistent with the first law: energy cannot be created or destroyed, only converted from one form to another. When substances react:

- molecules and atoms in a reaction mixture are in constant, random motion;
- these molecules and atoms frequently collide with each other;
- only some collisions, those with sufficient energy, will break bonds in molecules;
- when reactant bonds are broken, new bonds may form and products result.

As we study thermodynamics we will look at the transfer of energy between the **system** and the **surroundings**. The system contains the process under study. Typically this is the reaction of interest. The surroundings are the rest of the universe.

First Law of Thermodynamics.

Every chemical reaction involves a change in energy, most often in the form of **heat**. The first law of thermodynamics (also known as the law of conservation of energy) states that energy cannot be created or destroyed in the course of the reaction. Energy may only be converted from one form to another or transferred from one component of the system to another.

A reaction that absorbs energy (exhibits a net gain in energy) is termed an **endothermic reaction**. The products of the reaction possess more energy than the reactants. On the other hand, a reaction that releases energy (exhibits a net loss of energy) is termed an **exothermic reaction**. The products of the reaction possess less energy than the reactants. In order to keep energy constant, energy is absorbed from the surroundings in an endothermic reaction and released to the surroundings in an exothermic reaction.

Consider the following reaction:

$$\text{A-A} + \text{B-B} \rightarrow \text{A-B} + \text{A-B}$$

Energy is stored in the bonds which hold the atoms together. The stored energy is called chemical energy. It requires energy to break the bonds between A-A and B-B. To form the bonds A-B, energy is released.

- If more energy is required to break the bonds (A-A and B-B) than is released when the new bonds (A-B and A-B) are formed, there must be a net input of energy and the reaction is endothermic.
- If less energy is required to break the bonds (A-A and B-B) than is released when the

new bonds (A-B and A-B) are formed, there must be a net release of the extra energy and the reaction is exothermic. Figure 7.1 depicts the endothermic and exothermic process.

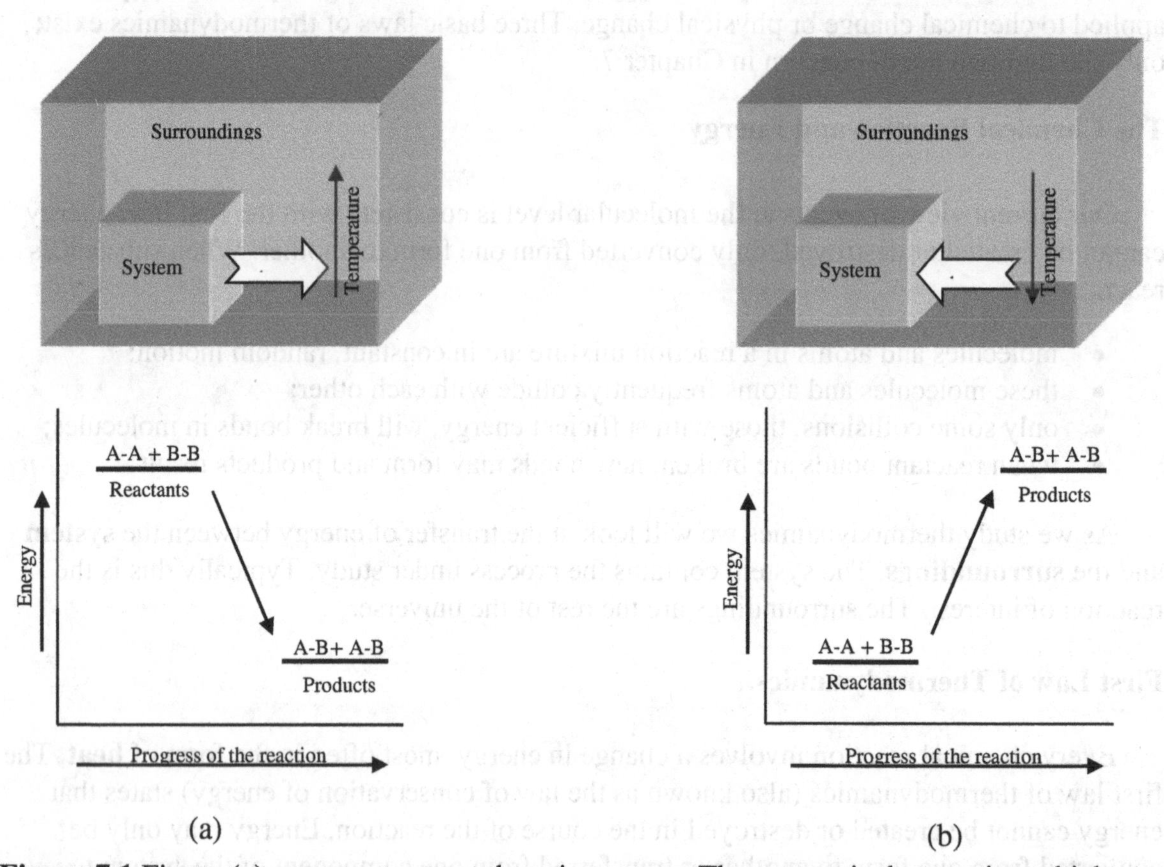

Figure 7.1 Figure (a) represents an exothermic process. The system is the reaction which releases heat to the surroundings. The products have less energy than the reactants. Figure (b) represents an endothermic process. The products have more energy than the reactants.

LG 2

Enthalpy is the term used to represent heat energy and is symbolized by H. In chemical change, we are primarily interested in the total heat gained or released by a reaction, otherwise known as enthalpy change, ΔH. For exothermic reactions where energy is released to the surroundings, the enthalpy change is negative (see Figure 7.1(a)). For endothermic reactions where energy is absorbed by the surroundings, the enthalpy change is positive (see Figure 7.1(b)). The enthalpy change is most often reported in units of joules or calories. Enthalpy change can be calculated by

$$\Delta H_{reaction} = H_{products} - H_{reactants.}$$

Spontaneous reactions are just that: they occur spontaneously under a certain set of conditions. Thermodynamics can be used to predict if a reaction will occur spontaneously.

It seems that all exothermic reactions should be spontaneous. After all, an external supply of energy does not appear to be necessary; in fact, energy is a product of the reaction. It also seems that all endothermic reactions should be nonspontaneous: energy is a reactant that we must provide. However, these hypotheses are not supported by experimentation.

Experimental measurement has shown that most, *but not all*, exothermic reactions are spontaneous; likewise, most, *but not all,* endothermic reactions are nonspontaneous. There must be some factor in addition to enthalpy that will help us to explain the less-obvious cases of nonspontaneous exothermic reactions and spontaneous endothermic reactions. This other factor is entropy.

The Second Law of Thermodynamics.

The second law of thermodynamics states that a system and its surroundings spontaneously tend toward increasing disorder (randomness). A measure of the randomness of a chemical system is referred to as **entropy** and is symbolized as S. A random, or disordered system, is characterized by high entropy. A well-ordered system is said to have low entropy. Disorder or randomness increases as we proceed from the solid to liquid to the vapor state. Solids often have an ordered crystalline structure, and liquids have a loose structure, while gas particles are virtually random in their distribution. Thus, gases have high entropy while crystalline solids have low entropy.

As was the case with enthalpy, we are usually concerned with the change in entropy, ΔS, the difference between the entropy of products and reactants. Another way to state the second law of thermodynamics is that the ΔS of the universe increases for a spontaneous process.

Reactions that are exothermic and whose products are more disordered (higher in entropy) *always* occur spontaneously, while endothermic reactions that produce products of lower entropy are *always* nonspontaneous. If they occur at all, they will require external energy input.

Example 1: Determining the Sign of ΔS.

For each process, what is the sign of ΔS?

1. Boiling water.
2. Freezing water to make ice.
3. $2\ HgO(s) \rightarrow 2\ Hg(l) + O_2(g)$
4. $2\ H_2(g) + O_2(g) \rightarrow 2\ H_2O(g)$

Solution:

1. The process of boiling water can be written as $H_2O(l) \rightarrow H_2O(g)$. Since we know that a gas has much more disorder than a liquid, the process of boiling water results in an increase in the disorder or a positive ΔS.
2. The freezing of water can be written as $H_2O(l) \rightarrow H_2O(s)$. The solid is less disordered than the liquid; therefore the process of making ice results in a decrease in the disorder, and ΔS is negative.
3. Anytime a reaction has more moles of gas on the product side than are on the reactant side, the reaction will have an increase in disorder and have a positive ΔS. In this case, there are no moles of gas on the reactant side and one mole of gas (O_2) on the product side.
4. In this reaction, there are three moles of gas on the reactant side. The reaction produces

Example 1 Continued
only two moles of gas. This is a decrease in the amount of gas and therefore a decrease in the disorder in the system. ΔS is negative.

Free Energy

In chemical change, the maximum amount of energy that can be converted to a useful form is called **free energy** (G). Free energy incorporates the two factors discussed above, the energy factor and the entropy factor, into a single expression for predicting the spontaneity of chemical change:

$$\Delta G = \Delta H - T\Delta S$$

In this expression, ΔG is the difference in free energy between products and reactants, ΔH is the difference in heat energy (enthalpy) between products and reactants, and ΔS is the difference in entropy between products and reactants. The temperature of the reaction is symbolized by T. The temperature must be in Kelvin when using this equation.

We find that a spontaneous reaction will have a negative ΔG. A positive ΔG is indicative of a nonspontaneous reaction. The following four statements can be determined using the equation above. After each statement, look at the equation and see if the statement makes sense. Keep in mind that the temperature is in Kelvin and is always positive.

- A reaction is nonspontaneous (positive ΔG) when ΔH is positive (endothermic) and ΔS is negative (decrease in disorder).
- A reaction is spontaneous (negative ΔG) when ΔH is negative (exothermic) and ΔS is positive (increase in disorder).
- Reactions which have both a positive ΔH and ΔS are spontaneous (negative ΔG) only at high temperatures.
- Reactions which have both a negative ΔH and ΔS are spontaneous (negative ΔG) only at low temperatures.

Example 2: Determining if a Reaction Is Spontaneous.
Is the following reaction spontaneous at all temperatures, nonspontaneous at all temperatures, or is spontaneity temperature dependent?

Ethanol burns in the presence of oxygen to produce carbon dioxide and water according to the following reaction:

$$C_2H_5OH(l) + 3\ O_2(g) \rightarrow 2\ CO_2(g) + 3\ H_2O(g)$$

Solution:

1. Determine the signs of ΔH and ΔS and then consider the equation $\Delta G = \Delta H - T\Delta S$ to decide the sign of ΔG. Use the four bulleted statements above as a guide.

2. The combustion process is always an exothermic process. ΔH is negative.

Example 2 Continued

3. There are three moles of gas as reactants and five moles of gas as products. An increase in the moles of gas in a reaction will result in an increase in disorder. ΔS is positive.

4. If ΔH is negative and ΔS is positive, according to the reaction ΔG = ΔH – TΔS, ΔG is negative at all temperatures and the reaction must be spontaneous.

7.2 Experimental Determination of Energy Change in Reactions

The measurement of the energy demand or energy release in a chemical reaction is **calorimetry**. This technique involves the measurement of the change in the temperature of a quantity of water (the surroundings) that is in contact with the reaction of interest (the system). The reaction and surrounding water is contained in a **calorimeter**, which is insulated in order to prevent any heat to escape beyond the calorimeter. If the reaction is exothermic, it releases heat which increases the temperature of the surrounding water. If the reaction is endothermic, it absorbs heat from the surrounding water, and the temperature decreases. A measure of the temperature change allows one to calculate the heat transferred.

The **specific heat** (*SH*) is defined as the number of calories needed to raise the temperature of 1 gram of a substance by 1 degree Celsius. For water, the specific heat is 1.00 cal/(g·°C). The specific heat, along with the mass of solution in the calorimeter (m_s) and the change in temperature (ΔT), allows one to calculate the heat, Q, absorbed by the reaction in the calorimeter.

$$Q = m_s \times \Delta T_s \times SH_s$$

Example 3: Calorimetry
When equal moles of hydrochloric acid (HCl) and sodium hydroxide (NaOH) are mixed in a calorimeter, the temperature of 50.0 grams of solution increased from 25.00°C to 38.00°C. If the specific heat of the solution is the same as the specific heat of water, which is 1.00 cal/(g•°C), calculate the quantity of energy in calories involved in this reaction. Is the reaction endothermic or exothermic?

Solution:

1. Use the relationship: $Q = m_s \times \Delta T_s \times SH_s$ to calculate the heat transferred to the water.

2. To use the equation, you must first determine the change in temperature:
   $\Delta T = 38.00°C - 25.00°C = 13.00°C$

3. Plug all values into the equation above.
   $Q = 50.0\ g \times 1.00\ cal/(g{\cdot}°C) \times 13.00°C = 650.\ cal$.

4. Since the water's temperature is raised it absorbs heat from the reaction. The reaction is exothermic.

Combustion reactions produce heat. In our bodies, many food substances, principally carbohydrates and fats, are oxidized to produce energy, and these reactions are identical to combustion reactions. In these cases, the amount of energy per gram of food is referred to as its fuel value. A special type of calorimeter, a bomb calorimeter, is useful for the measurement of the **fuel value** of foods. The **nutritional calorie** was mentioned earlier in the book as the Calorie (capital letter) and is equal to 1 kilocalorie (or 1000 calories). The fuel value is usually reported using this nutritional calorie as Cal/g.

Example 4: Calorimetry
When 1.00 gram of glucose ($C_6H_{12}O_6$) was burned in a bomb calorimeter, the temperature of 100. grams of water in the calorimeter increased from 25.00°C to 63.00°C. Calculate the fuel value of one gram of glucose. Remember that the SH for water is 1.00 cal/(g·°C).

Solution:
1. Use the equation $Q = m_s \times \Delta T_s \times SH_s$
2. $\Delta T = 63.00°C - 25.00°C = 38.00°C$
3. $Q = 100.\ g \times 1.00\ cal/(g·°C) \times 38.00°C = 3.80 \times 10^3\ cal$
4. The amount of glucose used in the problem was 1.00 grams, but this fuel value should be reported using the nutritional calorie (Calorie). 3,800 calories = 3.80 Calories.

7.3 Kinetics

Chemical Kinetics

Chemical **kinetics** is the study of the **rate** (speed) **of a chemical reaction**. This rate is measured by the disappearance of reactants or appearance of products versus time. Kinetics also proposes the step-by-step processes that take place as reactions occur. These step-by-step processes are called mechanisms.

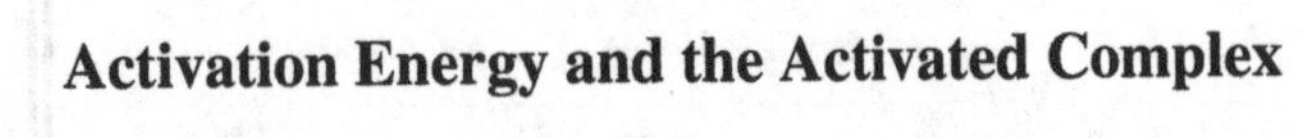

Activation Energy and the Activated Complex

For reactions to occur, the molecules must collide with each other. If the energy made available by the collision of reacting molecules exceeds the bond energies, the bonds will break, and the resulting atoms will recombine in a lower energy configuration. For complex molecules, molecules must also collide with the proper orientation in order for a reaction to occur. A collision meeting the above conditions and producing one or more product molecules is referred to as an effective collision. Only effective collisions lead to chemical reactions.

The minimum amount of energy required to produce a chemical reaction is the **activation energy** for the reaction. As implied above, a large component of the activation energy is the bond energy of the reacting molecules.

The chemical reaction may be represented in terms of the changes in potential energy that occur as a function of the time of the reaction. Several important characteristics of this relationship follow:

1. The reaction proceeds from reactants to products through an extremely unstable intermediate state that we term the **activated complex.**
2. Formation of the activated complex requires energy. The difference between the energy of reactants and activated complex is the activation energy.
3. For an exothermic reaction, the overall energy change must be a net release of energy. The net release of energy is the difference in energy between products and reactants.
4. For an endothermic reaction, the overall energy change must be a net absorption of energy.

Factors That Affect Reaction Rate

Six major experimental conditions influence the rate of a chemical reaction.

1. *The structure of the reacting species.* Reactions among ions tend to occur quickly since the ions are already dissociated in solution. Reactions between covalent compounds may occur more slowly since activation energy may be higher for bonds to be broken and new bonds formed.
2. *The molecular shape and the orientation of the molecules as they collide.* Only molecular collisions with the correct collision orientation, as well as sufficient energy, lead to product formation.
3. *The concentration of reactants.* As the concentration of the reactants increases, the rate of the reaction generally increases.
4. *The temperature of reactants.* Increasing temperature gives increased kinetic energy of the molecules and they collide with greater energy, thus making more collisions effective.
5. *The physical state of reactants.* Gases and reactions which take place in solution tend to occur at a greater rate than if the reactants are solids. Particles in the gaseous and liquid state have more free motion and thus a greater chance of colliding compared to solids.
6. *The presence of a catalyst.* A catalyst is a substance that increases the rate of a reaction. If added to a reaction mixture, the catalytic substance undergoes no net change, nor does it alter the outcome of the reaction.

Mathematical Representation of Reaction Rate

LG 7

For a generalized reaction: **A + B → C + D,** the rate of the reaction is represented by a **rate law**, which takes the form:

$$\textbf{rate} = \textbf{k}[\textbf{A}]^{n}[\textbf{B}]^{n'}$$

In this expression, the brackets, [ ], represent molar concentration, that is [A] = molar concentration of species A, [B] = molar concentration of species B.

The experimentally determined exponents, symbolized as n and n′, are the *reaction order*. They cannot be deduced from the chemical equation. The **reaction order** represents the number of molecules that are involved in the formation of the product.

The symbol k represents a constant that is unique to the reaction under study. It is termed the **rate constant**.

Knowing the form of the rate law enables chemists to design chemical reactions that will produce product in the shortest time. In the chemical industry, where the quantity of product produced and sold determines the amount of profit, a thorough understanding of the kinetics of reactions is of paramount importance.

7.4 Equilibrium

Physical Equilibrium

The most common examples of **physical equilibrium** involve interconversion between the various states of matter. For example, there may be equilibrium between the water in a lake and water in the surrounding atmosphere (as humidity).

$$H_2O(l) \rightleftharpoons H_2O(g)$$

When the temperature is 32°F (0°C), we may observe both ice and liquid water on the road (a very dangerous situation).

$$H_2O(s) \rightleftharpoons H_2O(l)$$

In both of these cases water is constantly changing states but since the forward and reverse reaction are occurring at the same rate, the overall amount of water in each state remains unchanged.

Many reactions may proceed in either direction, left to right or right to left, these are referred to as **reversible reactions**. The concentration of the various species is fixed at equilibrium because product is being consumed and formed at the same rate. In other words, the reaction continues indefinitely (dynamic) but the concentration of products and reactants is fixed (equilibrium). This is a **dynamic equilibrium**.

The equilibrium arrow, $\rightleftharpoons$, is used to indicate that [1] a reaction is reversible, [2] after a period of time will establish equilibrium, and [3] even at equilibrium the forward and reverse reaction will still be occurring, just at the same rate.

Chemical Equilibrium

Many chemical reactions do not proceed to "completion". For these reactions, after a period of time (determined by the kinetics of the reaction) the concentration of reactants no longer decreases and the concentration of products no longer increases. At this point, a mixture of products and reactants exists, and its composition would remain constant unless the experimental conditions were changed. This mixture is in a state of **chemical equilibrium**.

Chemical change, as well as physical change, can attain equilibrium. For example:

$$H_2(g) + I_2(g) \rightleftharpoons 2HI(g)$$

At equilibrium, the rate of *disappearance* of H_2 and I_2 is equal to the rate of *formation* of H_2 and I_2. The rates of the forward and reverse reactions are equal, and the concentrations of H_2, I_2, and HI are fixed. The process is dynamic, with a continuous interconversion of products and reactants. We represent this process using an **equilibrium constant.**

LG 9

The Generalized Equilibrium Constant Expression for a Chemical Reaction

The most precise description of an equilibrium process is achieved through the use of the equilibrium constant, K_{eq}. We write the general form of an equilibrium chemical reaction as

$$aA + bB \rightleftharpoons cC + dD$$

where A and B represent reactants, C and D represent products, and a, b, c, and d are the coefficients of the balanced equation. The equilibrium constant expression for this general case is

$$K_{eq} = \frac{[C]^c[D]^d}{[A]^a[B]^b}$$

Writing Equlibrium Constant Expressions

If we consider the case of H_2 and I_2 formign HI (shown above) the equilibrium constant expression would be written as:

$$K_{eq} = \frac{[HI]^2}{[H_2][I_2]}$$

Note:

1. Products of the equilibrium reaction are in the numerator, and reactants are in the denominator.
2. The exponents correspond to the coefficients of the balanced equation.
3. Brackets, [ ], represent molar concentration, *M* (moles/liter). The equilibrium constant, K_{eq} is unitless.
4. Solids and liquids *do not* appear in the equilibrium-constant expression, only gases and substances in solution are shown.

It does not matter which initial amounts (concentrations) of reactions or products we choose. When the system reaches equilibrium, the calculated value of K_{eq} will not change. The magnitude of K_{eq} can be altered only by changing the temperature; thus, K_{eq} is temperature dependent.

Example 5: Writing Equilibrium Constant Expressions
Write the equilibrium constant (K_{eq}) for each of the following reactions.

1. $N_2(g) + 3\,H_2(g) \rightleftharpoons 2\,NH_3(g)$
2. $N_2(g) + O_2(g) \rightleftharpoons 2\,NO(g)$
3. $H_2(g) + Cl_2(g) \rightleftharpoons 2\,HCl(g)$

Example 5 Solution:

The equilibrium constant (K_{eq}) for any equilibrium reaction is equal to the concentration of each product, raised to a power equal to its coefficient in the balanced equation, divided by all the reactants raised to their coefficients' powers. Do not include solids and liquids.

1. $$K_{eq} = \frac{[NH_3]^2}{[N_2][H_2]^3}$$

2. $$K_{eq} = \frac{[NO]^2}{[N_2][O_2]^3}$$

3. $$K_{eq} = \frac{[HCl]^2}{[H_2][Cl_2]}$$

Interpreting Equilibrium Constants

The reversible arrow in a chemical equation tells us that equilibrium exists, however it doesn't give any information on whether products predominate, reactants predominate, or significant concentrations of both products and reactants exist at equilibrium. The equilibrium constant provides us with this information.

- Large values of K_{eq} (larger than 10^3) indicate that mostly product is present at equilibrium (numerator larger than denominator).
- Very small values of K_{eq} (less than 10^{-3}) indicate that mostly reactant is present at equilibrium (denominator larger than numerator).
- K_{eq} values between about 10^{-3} and 10^3 indicate that significant amounts of both reactants and products are present at equilibrium.

Calculating Equilibrium Constants

Since the equilibrium constant expression is a mathematical equation, if concentrations of all reactants and products are known at equilibrium, the equilibrium constant can be calculated.

Using Equilibrium Constants

Once this equilibrium constant is known, it can be used to determine the equilibrium concentrations of one or more of the reactants or products.

LeChatelier's Principle

LeChatelier's principle states that if a stress is placed on an equilibrium system, the system will respond by altering the equilibrium in such a way as to minimize the stress.

- **Effect of Concentration**
  Addition of extra product or reactant to a fixed reaction volume is a way of saying that we have increased the concentration of product or reactant. Removal of material from a fixed volume decreases the concentration. This puts a stress on the system and the system will relieve that stress by shifting the equilibrium away from the extra added material. (See Figure 7.2) Therefore, changing the concentration of one or more components of a reaction mixture is a way to alter the equilibrium composition of an equilibrium mixture.

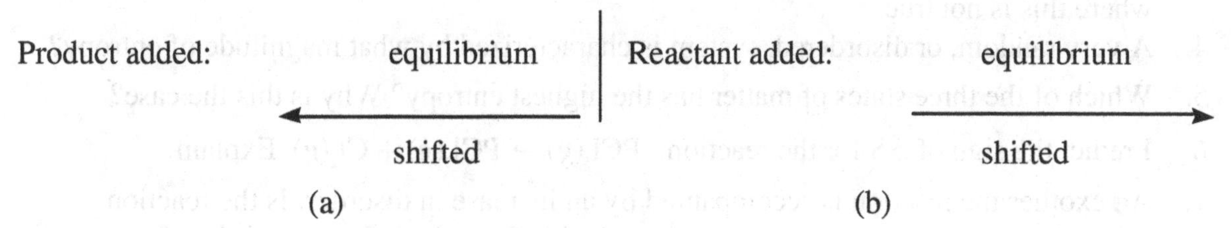

Figure 7.2 (a) Addition of a product shifts the reaction to the left (towards reactants.)
(b) Addition of a reactant shifts the reaction to the right (toward products.)

- **Effect of Heat**
  The change in equilibrium composition caused by the addition or removal of heat from an equilibrium mixture can be explained by treating heat as a product or reactant. Adding heat to an exothermic reaction is similar to increasing the amount of product, shifting the equilibrium to the left. Removing heat from an exothermic reaction shifts the equilibrium to the right. Similarly, heat is a reactant in an endothermic reaction; its removal shifts the equilibrium to the left. Adding heat favors product formation.

- **Effect of Pressure**
  Only gases are affected significantly by changes in pressure, because gases are free to expand and compress in accordance with Boyle's law. However, liquids and solids are not compressible, so their volumes are unaffected by pressure.
  Therefore, pressure changes will alter equilibrium composition only when they involve a gas or variety of gases as products and/or reactants. Note the following:
  - If the number of moles of gaseous product is <u>greater</u> than the number of moles of gaseous reactant, an increase in pressure shifts the equilibrium to the left (towards the reactants).
  - If the number of moles of gaseous product is <u>less</u> than the number of moles of gaseous reactant, an increase in pressure shifts the equilibrium to the right (towards the products).
  - If the number of moles of gaseous product and reactant is identical, pressure will have no effect on the equilibrium because there is no volume advantage.

- **Effect of a Catalyst**
  A catalyst has no effect on the equilibrium composition. A catalyst increases the rates of both forward and reverse reactions to the same extent. The equilibrium composition and equilibrium concentration do not change when a catalyst is used.

Self Test for Chapter Seven

1. For a certain reaction, heat is transferred from the reaction to the surroundings.

a. Is the reaction endothermic or exothermic?
b. Is the energy of the reactants higher or lower than the energy of the products?

2. For a certain reaction, heat is transferred from the surroundings to the reaction.
   a. Is the temperature of the surroundings increased or decreased?
   b. Is the amount of energy gained by the reaction equal to, greater than or less than the energy lost by the surroundings?
3. Are endothermic reactions *usually* spontaneous or nonspontaneous? Describe a case where this is not true.
4. A very random, or disordered, system is characterized by what magnitude of entropy?
5. Which of the three states of matter has the highest entropy? Why is this the case?
6. Predict the sign of ΔS for the reaction: $PCl_5(g) \rightarrow PCl_3(g) + Cl_2(g)$. Explain.
7. An exothermic reaction is accompanied by an increase in disorder. Is the reaction spontaneous, nonspontaneous or undeterminable from the information given?
8. For a particular reaction ΔG is calculated to be a positive value. What does that tell you about the reaction?
9. Is the reaction: $PCl_5(g) \rightarrow PCl_3(g) + Cl_2(g)$ spontaneous, nonspontaneous or undeterminable from the information given?
10. What is a calorimeter used for?
11. When two aqueous solutions containing equal moles of HCl and NaOH were mixed in a calorimeter, the temperature of the 100.0 grams of solution increased from 22.5°C to 36.5°C. If the specific heat of the aqueous solution is 1.000 cal/(g·°C), calculate the quantity of heat energy in calories that was released.
12. Chemical kinetics is the study of what property of a reaction?
13. How will the following factors influence the rate of a chemical reaction?
    a. concentration of the reactants is decreased.
    b. temperature of the reaction is increased.
    c. a catalyst is added to the reaction.
14. What is the name of an extremely unstable intermediate state of a chemical reaction?
15. Write the rate law for the reaction $2H_2(g) + S_2(g) \rightarrow 2H_2S(g)$. Use *n* and *n'* in the law. How are *n* and *n'* determined?
16. When the rate of the forward reaction equals the rate of the reverse reaction in a reversible reaction, the system is in what state?
17. Write the equilibrium constant for each of the following:
    a. $2\,NO_2(g) \rightleftharpoons N_2O_4(g)$
    b. $2\,NO(g) + 2\,H_2(g) \rightleftharpoons N_2(g) + 2\,H_2O(l)$
18. Consider the reaction at equilibrium: $N_2(g) + 3\,H_2(g) \rightleftharpoons 2\,NH_3(g)$. Which way will the reaction shift if nitrogen gas is added to the reaction vessel?
19. For the exothermic reaction at equilibrium: $2NO_2(g) \rightleftharpoons N_2O_4(g)$ + heat, which way will the equilibrium shift if the temperature is raised?
20. For the exothermic reaction at equilibrium: $2NO_2(g) \rightleftharpoons N_2O_4(g)$ + heat, which way will the equilibrium shift if the pressure is increased?

8 *Acids and Bases and Oxidation-Reduction*

Learning Goals

1. Classify compounds with acid-base properties as acids, bases, or amphiprotic.
2. Write equations illustrating the role of water in acid-base reactions.
3. Identify conjugate acid-base pairs.
4. Describe the relationship between acid and base strength and dissociation.
5. Use the ion product constant for water to solve for hydronium and hydroxide ion concentrations.
6. Calculate pH from solution concentration data.
7. Calculate hydronium and/or hydroxide ion concentration from pH data.
8. Describe the meaning and utility of neutralization reactions.
9. Use titration data to determine the molar concentration of an unknown solution.
10. Demonstrate the reactions and dissociation of polyprotic substances.
11. Describe the effects of adding acid or base to a buffer system.
12. Calculate the pH of buffer solutions.
13. Explain the role of buffers in the control of blood pH under various conditions.
14. Write oxidation and reduction half-reactions, and identify oxidizing agents and reducing agents.
15. Compare and contrast voltaic and electrolytic cells.
16. Describe examples of redox processes.

Introduction

Acids and bases include some of the most important compounds in nature. Digestion is aided by stomach acid and many biochemical processes depend on the proper level of acidity. For this reason, the level of acidity must be carefully regulated with substances called buffers.

8.1 Acids and Bases

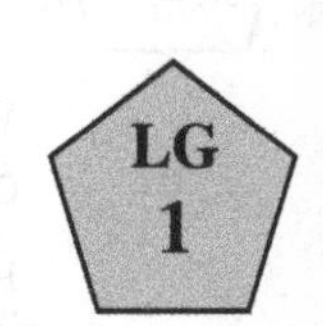

Acids have a sour taste (like vinegar), dissolve some metals, and cause vegetable dyes to change color. Bases have a bitter taste and a slippery feel (soap, for example). Bases also will cause vegetable dyes to change color. Acids and bases can be characterized by various theories, two of which are discussed in this chapter.

Acid and Base Theories

An **Arrhenius acid** dissociates in water to form hydrogen ions or **protons (H^+)**. For example, HCl dissolves in water according to the reaction:

$$HCl(g) \xrightarrow{H_2O} H^+(aq) + Cl^-(aq)$$

An **Arrhenius base** dissociates in water to form **hydroxide ions (OH^-)**. Sodium hydroxide and water dissociates according to the reaction:

$$NaOH(s) \xrightarrow{H_2O} Na^+(aq) + OH^-(aq)$$

The **Arrhenius theory** is a simple theory that is somewhat limited in its explanation of acids and bases. For a more comprehensive understanding of acids and bases, we must learn about another theory, the **Brønsted-Lowry theory**.

A **Brønsted-Lowry acid** is a proton donor and a **Brønsted-Lowry base** is a proton acceptor. Remember that H^+ is simply a proton, so a *proton donor* gives away an H^+. Examples include:

$$\text{Acid: } HCl(g) + H_2O(l) \rightarrow H_3O^+(aq) + Cl^-(aq)$$

$$\text{Base: } NH_3(aq) + H_2O(l) \rightleftharpoons NH_4^+(aq) + OH^-(aq)$$

(Note: The double arrow with ammonia indicates a reversible reaction and that an equilibrium state will be reached)

Amphiprotic Nature of Water

Water is **amphiprotic**, meaning that it possesses *both* acid and base properties. The fact that water can either accept or donate protons (H^+ ions) makes it an excellent solvent for both acids and bases.

In this reaction, water acts a base, and accepts a proton from hydrochloric acid and forms a **hydronium ion (H_3O^+)**:

$$HCl(g) + H_2O(l) \rightarrow H_3O^+(aq) + Cl^-(aq)$$

In this reaction, water acts as an acid, and donates a proton to ammonia:

$$NH_3(aq) + H_2O(l) \rightarrow NH_4^+(aq) + OH^-(aq)$$

Conjugate Acid-Base Pairs

Any acid-base reaction can be represented by the general equation

$$HA + B \rightleftharpoons BH^+ + A^-$$

where HA represents an acid and B represents a base. In the forward reaction, the acid donates a proton to the base. The products of this reaction also have acid-base properties. The protonated base (BH^+) can, in fact, donate a proton to the anion, A^-, reforming HA and B. The product acids and bases are termed *conjugate acids and bases*.

A **conjugate acid** is the species formed when a base accepts a proton.
A **conjugate base** is the species formed when an acid donates a proton.

The acid and base on the opposite sides of the equation are collectively termed a **conjugate acid-base pair**. For our general expression, B and BH^+ is a conjugate acid-base pair. Similarly, HA and A^- are a conjugate acid-base pair.

Example 1: Writing a Brønsted-Lowery Acid-Base Reaction and Identifying Conjugate Acid-Base Pairs.

1. Write an equation for the reaction of HNO_2 (a weak acid) in water.
2. Identify the conjugate acid-base pairs.

Solution

1. As an acid, HNO_2 is a proton donor and will donate the proton to water. Since the acid is weak, equilibrium is established and reversible arrows are used in the equation.

$$HNO_2(aq) + H_2O(l) \rightleftharpoons NO_2^-(aq) + H_3O^+(aq)$$

2. The conjugate acid-base pairs are HNO_2/NO_2^- and H_2O/H_3O^+.

Acid and Base Strength

The terms acid or base *strength* and acid or base *concentration* are easily confused. Strength is a measure of the degree of dissociation of an acid or base in solution, independent of its concentration. Concentration refers to the amount of acid or base per quantity of solution. An acid can be high in concentration and cause great damage to tissue if it came in contact with your skin but still be a weak acid

The strength of acids and bases in water is dependent upon the extent to which they react with the solvent, water. Acids and bases are classified as *strong* when the reaction with water is virtually 100% complete and as *weak* when the reaction is much less than 100% complete.

All strong bases are metal hydroxides. Strong bases completely dissociate in aqueous solution to produce hydroxide ions and metal cations. For example, NaOH dissociates to form Na^+ and OH^- ions in water.

Weak acids and weak bases dissolve in water principally in the molecular form. Only a small percentage of the molecules dissociate to form the hydronium or hydroxide ion. For example, the weak base ammonia, NH_3, dissolves in water primarily as NH_3, but a small percentage reacts with water to form NH_4^+ and OH^-.

The most fundamental chemical difference between strong and weak acids and bases is their equilibrium situation. A strong acid, such as HCl, *does not* exist in equilibrium with its ions, H_3O^+ and Cl^-. A weak acid, such as acetic acid (CH_3COOH), establishes a dynamic equilibrium with its ions, H_3O^+ and $C_2H_3O_2^-$. (Note that the acetate ion may be written as CH_3COO^- or as the more condensed form $C_2H_3O_2^-$.) The situation for bases is analogous. A weak base such as ammonia, will establish equilibrium with its ions NH_4^+ and OH^-. Examples of these equations are shown below. (Note that the double arrow indicates the presence of dynamic equilibrium between the reactants and products).

$$CH_3COOH(aq) + H_2O(l) \rightleftharpoons H_3O^+(aq) + CH_3COO^-(aq)$$

$$NH_3(aq) + H_2O(l) \rightleftharpoons NH_4^+(aq) + OH^-(aq)$$

Self-Ionization of Water and K_w

LG 5

Although pure water is virtually 100% molecular, a small number of water molecules do dissociate. This process occurs by the transfer of a proton from one water molecule to another, producing a hydronium and hydroxide ion. This equilibrium is shown below:

$$H_2O(l) + H_2O(l) \rightleftharpoons H_3O^+(aq) + OH^-(aq)$$

This process is the **self-ionization** of water. (The term *ionize* is synonymous with dissociate and is typically used for acids and bases.) Water is a very weak electrolyte and a poor conductor of electricity. Water has both acid and base properties; the dissociation produces both the hydronium and hydroxide ion.

Water at room temperature has a hydronium ion concentration of 1.0×10^{-7} *M*. The hydroxide ion concentration is also 1.0×10^{-7} *M*. The product of hydronium and hydroxide ion concentration is the **ion product constant for water,** symbolized K_w.

$$K_w = [H_3O^+][OH^-] = [1.0 \times 10^{-7}][1.0 \times 10^{-7}] = 1.0 \times 10^{-14}$$

The ion product is a constant because its value is not dependent upon the nature or concentration of the solute, as long as the temperature does not change. This relationship is the basis for the pH scale.

8.2 pH: A Measurement Scale for Acids and Bases

A Definition of pH

The **pH scale** correlates the hydronium ion concentration with a number, the pH, which serves as a useful indicator of the degree of acidity or basicity of a solution. The pH scale specifies "how acidic" or "how basic" a solution is. The pH scale has typical values which range from 0, for very acidic, to 14, for very basic. A pH of 7 is neutral, that is, neither acidic nor basic.

1. Addition of an acid (proton donor) to water *increases* $[H_3O^+]$ and decreases $[OH^-]$. The product of the two concentrations must give the same value of K_w (1×10^{-14} at 25°C)
2. Addition of a base (proton acceptor) to water *decreases* $[H_3O^+]$ and increases $[OH^-]$. The product of the two concentrations must give the same value of K_w (1×10^{-14} at 25°C)
3. $[H_3O^+] = [OH^-]$ when *equal* amounts of acid and base are present.

Measuring pH

The pH of a solution can be calculated if $[OH^-]$ or $[H_3O^+]$ is known and visa-versa. This is done using the ion product for water ($K_w = 1 \times 10^{-14} = [H_3O^+][OH^-]$). The pH of a solution may also be measured by a pH meter. Additionally, indicating paper (pH paper) can give an approximate value for the solution pH based on the color of the paper.

Example 2: Calculating Concentration of the Hydronium ion.
What are the hydroxide ion concentration and the hydronium ion concentration of a 0.200 *M* NaOH solution?

Example 2 Continued

Solution

1. NaOH is a strong base. It dissociates 100%, breaking apart into Na^+ and OH^- according to the equation:

$$NaOH(s) \xrightarrow{H_2O} Na^+(aq) + OH^-(aq)$$

2. Since NaOH is a base, we will first calculate the $[OH^-]$. The 0.200 *M* value of NaOH is the amount of NaOH placed in solution. After it dissociates, there is no NaOH remaining in solution. The concentrations of ions present after dissociation are: $[Na^+] = 0.200\ M$ and $[OH^-] = 0.200\ M$.

3. Next use the ion product for water.

$$K_w = [H_3O^+][OH^-]$$

4. Plug in the value of K_w and $[OH^-]$ as determined in step 2 and solve for the $[H_3O^+]$.

$1.0 \times 10^{-14} = [H_3O^+](0.200)$

$[H_3O^+] = \dfrac{1.0 \times 10^{-14}}{0.200} = 5.0 \times 10^{-14}\ M$

Calculating pH

The pH of a solution is defined as the negative logarithm of the molar concentration of the hydronium ion.

$$pH = -\log [H_3O^+]$$

The pH of a solution can be calculated if the concentration of either H_3O^+ or OH^- is known. Also, $[H_3O^+]$ or $[OH^-]$ can be calculated from the pH.

Example 3: Calculating pH of an acid solution.

Calculate the pH of a 0.010 *M* HCl solution.

Solution

1. Since HCl is a strong acid, if 1 mole of HCl dissociates, it produces 1 mole of H_3O^+ ions. Therefore, a 0.010 *M* HCl solution has $[H_3O^+] = 0.010\ M$.

2. Use the pH equation:

   $pH = -\log [H_3O^+]$ (Note: [ ] means molar.)

3. $pH = -\log (0.010)$ or $pH = -\log(1.0 \times 10^{-2})$

   $pH = 2.00$

Example 4: Calculating pH of a basic solution.

Calculate the pH of a 1.0×10^{-4} *M* NaOH solution.

Solution

1. NaOH is a strong base; thus, in our problem $[OH^-] = 1.0 \times 10^{-4}$ *M*.
2. Use $K_w = [H_3O^+][OH^-]$ to find $[H_3O^+]$, so we can use the pH equation.

$$[H_3O^+] = \frac{K_w}{[OH^-]}$$

$$[H_3O^+] = \frac{1\times10^{-14}}{1\times10^{-4}}$$

$$[H_3O^+] = 1.0\times10^{-10}\,M$$

3. Next, use the pH equation:

$pH = -\log[H_3O^+]$

$pH = -\log(1.0 \times 10^{-10})$

$pH = 10.00$

Example 5: Calculating the $[H_3O^+]$ and $[OH^-]$ from pH

What is the $[H_3O^+]$ and $[OH^-]$ of a solution when pH = 3.00. Is the solution acidic or basic?

Solution

1. Use the equation $pH = -\log[H_3O^+]$ and solve $[H_3O^+]$.

$3.00 = -\log[H_3O^+]$

$-3.00 = \log[H_3O^+]$

$\text{antilog}(-3.00) = [H_3O^+]$ or $10^{-3.00} = [H_3O^+]$

$1.0 \times 10^{-3}\,M = [H_3O^+]$

2. Use the equation $K_w = [OH^-][H_3O^+]$ and solve for $[OH^-]$

$$[OH^-] = \frac{K_w}{[H_3O^+]}$$

$$[OH^-] = \frac{1\times10^{-14}}{1\times10^{-3}}$$

$$[OH^-] = 1.0\times10^{-11}\,M$$

3. The pH < 7; therefore, the solution is acidic.

The Importance of pH and pH Control

Solution pH and pH control play a major role in many facets of our everyday lives:

1. **Agriculture: soil pH is extremely important for proper crop growth.**
2. **Physiology: the pH of our blood must be maintained within a narrow range in order for the biochemical reactions in our body to occur properly.**
3. **Industry: manufacturing processes often require careful pH control.**
4. **Municipal services: purification of drinking water and sewage treatment must be carried out at an optimum pH.**
5. **Acid rain: vehicle emissions and electric power generation can increase the amount of nitric and sulfuric acid in rain. This lowers the pH of aquatic systems and can cause problems for fish populations.**

8.3 Reactions between Acids and Bases

Neutralization

The reaction of an acid with a base to produce a salt and water is known as a **neutralization reaction.** In the strictest sense, neutralization requires equal numbers of moles of H_3O^+ and OH^- to produce a neutral solution (no excess acid or base). Since strong acids generate H^+ in solution and strong bases generate OH^- in solution, the reaction of H^+ and OH^- produce water. The neutralization reaction for any strong acid/strong base reaction can be described as:

$$H_3O^+\ (aq) + OH^-\ (aq) \rightarrow 2\ H_2O(l)$$

A neutralization reaction may be used to determine the concentration of an unknown acid or base solution. The technique of **titration** involves the addition of measured amounts of a standard solution (one whose concentration is known) to neutralize the second, unknown solution. From the volumes of the two solutions and the concentration of the **standard solution,** the concentration of the unknown solution may be determined.

To conduct an acid-base titration the following steps will occur:

1. A known volume of the unknown acid of unknown concentration is measured into a flask.
2. An **indicator** is added to the flask. An indicator is a substance that changes color as the solution reaches a certain pH.
3. A solution of base of known concentration is carefully added to the unknown solution until the indicator undergoes a color change. The volume of this base must be carefully measured, typically using a piece of glassware called a **buret**.
4. This point is referred to as the **equivalence point**, and is the point where the solution is neutral and the moles of the hydroxide ion added is equal to the moles of the hydronium ion present in the unknown acid.
5. The volume of base dispensed by the buret is measured.
6. Using the data collected of the volume of the unknown, volume of the titrant, and molarity of the titrant, the molar concentration of the unknown substance can be calculated.

This process was described for an unknown acid being titrated with a known base. However, the concentration of an unknown base could be calculated with a known acid using the same process. See Example 6 for an example of the calculations used in this process.

Example 6: Determining the Concentration of Vinegar

A 5.00 g sample of vinegar is titrated with 0.100 M NaOH. If the vinegar requires 40.0 mL of the NaOH for a complete reaction,

A. Calculate the concentration of acetic acid in the vinegar in molarity.

B. Calculate the % (m/V) of acetic acid in the vinegar.

Assume that the density of the vinegar solution is 1.00 g/mL, and that the only acidic component of the vinegar solution is acetic acid (CH_3COOH). The neutralization reaction is:

$$CH_3COOH(aq) + NaOH(aq) \rightarrow CH_3COONa(aq) + H_2O(l)$$

Solution, Part A

Three important pieces of information included in the question are:

Amount of unknown acid solution: 5.00 g vinegar

Volume of sodium hydroxide added: 40.0 mL

Concentration of sodium hydroxide: 0.100 M.

We know from the balanced equation that one mole of acetic acid reacts with one mole of sodium hydroxide.

Map of the conversion:

$$\text{mL NaOH} \xrightarrow{1} \text{L NaOH} \xrightarrow{2} \text{mol NaOH} \xrightarrow{3} \text{mol } CH_3COOH \xrightarrow{4} \text{M } CH_3COOH.$$

Step 1: use the definition of milli

Step 2: use molarity of NaOH as a conversion factor

Step 3: use the coefficients of the balanced equation

Step 4: use the density of vinegar (the solution) to convert to volume (in liters) of vinegar. Then use the formula M = mol/L to calculate the molarity of CH_3COOH in vinegar.

Steps 1 – 3 are shown here:

$$40.0 \text{ mL NaOH} \times \frac{1 \text{ L NaOH}}{1000 \text{ mL NaOH}} \times \frac{0.100 \text{ mol NaOH}}{\text{L NaOH}} \times \frac{1 \text{ mol } CH_3COOH}{1 \text{ mol NaOH}} = 0.0400 \text{ mol } CH_3COOH$$

Step 4:

$$5.00 \text{ g vinegar} \times \frac{1 \text{ mL vinegar}}{1.00 \text{ g vinegar}} \times \frac{1 \text{ L vinegar}}{1000 \text{ mL vinegar}} = 0.00500 \text{ L vinegar (the solution)}$$

$$M = \frac{0.0400 \text{ mol } CH_3COOH}{0.00500 \text{ L vinegar}} = 0.800M\ CH_3COOH$$

Example 6 Continued

Solution, Part B

Map of the conversion:

$$\text{mL NaOH} \xrightarrow{1} \text{L NaOH} \xrightarrow{2} \text{mol NaOH} \xrightarrow{3} \text{mol CH}_3\text{COOH} \xrightarrow{4} \text{g CH}_3\text{COOH} \xrightarrow{5} \%\ (\text{M/V})$$

Steps 1 – 3 are the same as in part A.
Step 4: use molar mass of CH_3COOH.
Step 5: use the definition of % (m/V).

Steps 1 – 4 are shown here:

$$40.0\ \text{mL NaOH} \times \frac{1\ \text{L NaOH}}{1000\ \text{mL NaOH}} \times \frac{0.100\ \text{mol NaOH}}{\text{L NaOH}} \times \frac{1\ \text{mol CH}_3\text{COOH}}{1\ \text{mol NaOH}}$$
$$\times \frac{60.05\ \text{g CH}_3\text{COOH}}{1\ \text{mol CH}_3\text{COOH}} = 0.240\ \text{g CH}_3\text{COOH}$$

$$\text{Step 5: }\%\ (\text{m/V}) = \frac{\text{g CH}_3\text{COOH}}{\text{mL vinegar (the solution)}} \times 100\% = \frac{0.240\ \text{g CH}_3\text{COOH}}{5.00\ \text{mL vinegar}} \times 100\% = 4.80\%$$

Polyprotic Substances

Polyprotic substances can donate or accept more than one proton per formula unit. Sulfuric acid, H_2SO_4, is a diprotic acid, and phosphoric acid, H_3PO_4, is a triprotic acid, and barium hydroxide, $Ba(OH)_2$, is a diprotic base. Acid-base reactions may occur in various combining ratios if they involve a polyprotic substance.

One mole of a diprotic acid, such as H_2SO_4, can react with two moles of NaOH.

$$H_2SO_4 + 2NaOH \rightarrow Na_2SO_4 + 2H_2O$$

Dissociation of a diprotic acid takes place in two steps.

$$H_2SO_4 + H_2O \rightarrow HSO_4^- + H_3O^+$$

$$HSO_4^- + H_2O \rightleftharpoons SO_4^{2-} + H_3O^+$$

8.4 Acid-Base Buffers

A **buffer solution** contains components that enable the solution to resist large changes in pH when acids or bases are added.

The Buffer Process

The basis of buffer action is the establishment of equilibrium between either a weak acid and its conjugate base, or a weak base and its conjugate acid.

A buffer solution functions in accordance with LeChatelier's principle (Section 7.4), which states that an equilibrium system, when stressed, will shift its equilibrium to alleviate that stress.

Addition of Base or Acid to a Buffer Solution

For the acetic acid/sodium acetate system, when a base (OH^-) is added equilibrium shifts to the right. The OH^- consumes the H_3O^+ which causes the CH_3COOH to dissociate to replace the H_3O^+, thus maintaining a pH level close to the initial level.

$$CH_3COOH + H_2O \rightleftharpoons H_3O^+ + CH_3COO^-$$

OH^- added, equilibrium shifts to the right

$$\longrightarrow$$

For the acetic acid/sodium acetate system, when an acid (H_3O^+) is added the overall amount of H_3O^+ increases which, due to LeChaletlier's principle causes equilibrium to shift back to the left, decreasing the amount of H_3O^+ and again maintaining a pH level close to the initial level.

$$CH_3COOH + H_2O \rightleftharpoons H_3O^+ + CH_3COO^-$$

H_3O^+ added, equilibrium shifts to the left

$$\longleftarrow$$

The measure of the ability of a buffer to resist large changes in pH when an acid and base are added is **buffer capacity.** The more acid in a buffer solution, the more it can resist change when a base is added to the buffer. The more base contained in the buffer solution, the more it can resist a lowering of the pH as an acid is added.

LG
12

Determining Buffer Solution pH

To calculate the pH of a buffer solution, use either the equilibrium constant expression or a rearranged version which is solved for $[H_3O^+]$. The $[H_3O^+]$ can then be used to calculate pH. The quantities needed are the equilibrium constant and the buffer solution concentrations of the weak acid and its conjugate base (or, the weak base and its conjugate acid).

A Practical Note: Consider a buffer made from acetic acid, $HC_2H_3O_2$, and its conjugate base, the acetate anion, $C_2H_3O_2^-$. It's not possible simply to add acetate anions. The acetate is added as a salt, sodium acetate, which dissolves and dissociates to form sodium cations and acetate anions. (The sodium cations won't participate in the acid-base reactions, so they can be ignored.) Since the ratio of sodium acetate to acetate anions is equal, we can use the sodium acetate concentration to represent the acetate anion concentration.

$$NaC_2H_3O_2(s) \rightarrow Na^+(aq) + C_2H_3O_2^-(aq)$$

Example 7: Calculating the pH of a Buffer

Calculate the pH of a buffer solution in which the acetic acid concentration is 0.150M and the concentration of sodium acetate is 0.090 M. The equilibrium constant, K_a, for acetic acid is 1.75×10^{-5}.

Solution:

1. Write the equilibrium constant expression for acetic acid.

$$CH_3COOH(aq) + H_2O(l) \rightleftharpoons H_3O^+(aq) + CH_3COO^-(aq)$$

$$K_a = \frac{[H_3O^+][CH_3COO^-]}{[CH_3COOH]}$$

2. Rearrange the equilibrium constant expression to solve for $[H_3O^+]$.

$$[H_3O^+] = \frac{K_a\,[CH_3COOH]}{[CH_3COO^-]}$$

Substitute the values given in the problem and calculate the concentration of H_3O^+.

$$[H_3O^+] = \frac{(1.75 \times 10^{-5})(0.150M)}{0.090M}$$

$$[H_3O^+] = 2.9 \times 10^{-5}\,M$$

3. Calculate the pH using the value of $[H_3O^+]$ determined above.

$pH = -\log[H_3O^+]$

$pH = -\log(2.9 \times 10^{-5}\ M)$

$pH = 4.54$

The Henderson-Hasselbalch Equation

The solution of the equilibrium-constant expression and the pH may be combined into one expression: the **Henderson-Hasselbalch equation**. The form of this equation is especially amenable to buffer problem calculations. pK_a is calculated by taking the negative logarithm of K_a ($pK_a = -\log K_a$).

$$\text{Henderson-Hasselbalch equation:}\quad pH = pK_a + \log\frac{[\text{conjugate base}]}{[\text{weak acid}]}$$

Example 8: Calculating the pH of a Buffer Using the Henderson-Hasselbalch Equation

Calculate the pH of a buffer solution in which the acetic acid concentration is 0.150 *M* and the concentration of sodium acetate is 0.090 *M*. (K_a for acetic acid is 1.75×10^{-5}.) Use the Henderson-Hasselbalch Equation.

Solution:

1. Write the Henderson-Hasselbalch equation for acetic acid.

$$pH = pK_a + \log\frac{[\text{acetate}]}{[\text{acetic acid}]}$$

2. Calculate the pK_a for acetic acid.

$$pK_a = -\log K_a$$

$$pK_a = -\log(1.75 \times 10^{-5}) = 4.76$$

3. Use the Henderson-Hasselbalch equation. Substitute the values given in the problem and calculate the pH.

$$pH = 4.76 + \log\frac{[0.090\text{ M acetate}]}{[0.150\text{ acetic acid}]}$$

$$pH = 4.76 + \log(0.60)$$

$$pH = 4.76 - 0.22 = 4.54$$

Control of Blood pH

LG 13

As noted earlier, it is extremely important that the pH of the blood be maintained within a very narrow range centered around 7.4. This is accomplished by a carbonic acid-bicarbonate buffer system in the blood:

$$H_2CO_3(aq) + H_2O\ (l) \rightleftharpoons H_3O^+\ (aq) + HCO_3^-(aq)$$

Carbon dioxide (CO_2) is a waste product of metabolism and returns to the lungs via the blood. CO_2 reacts with water to produce carbonic acid (H_2CO_3) which effects the buffer system in the blood:

$$CO_2(aq) + 2H_2O(l) \rightleftharpoons H_2CO_3(aq) + H_2O\ (l) \rightleftharpoons H_3O^+\ (aq) + HCO_3^-(aq)$$

Very high CO_2 concentrations in the blood shift the equilibrium to the right and lower the pH of the blood. If the blood becomes too acidic it is termed *acidosis* and can lead to numerous medical problems. Conversely, low CO_2 levels in the blood shift equilibrium to the left making the blood more basic. This is termed *alkalosis*.

8.5 Oxidation - Reduction Processes

Oxidation and Reduction

Oxidation is defined as a loss of electrons, loss of hydrogen atoms, or gain of oxygen atoms. Sodium metal is, for example, oxidized to a sodium ion, losing one electron:

$$Na \rightarrow Na^+ + e^-$$

Reduction is defined as a gain of electrons, gain of hydrogen atoms, or loss of oxygen atoms. A chlorine atom is reduced to a chloride ion by gaining one electron. The reaction for a diatomic chlorine (2 chlorine atoms bonded together) is shown below:

$$Cl_2 + 2e^- \rightarrow 2Cl^-$$

Oxidation and reduction are complementary processes. Each reaction above is called a half-reaction because neither reaction can occur without the other. The oxidation half-reaction produces the electron(s) that is(are) the reactants for the reduction half-reaction. The number of electrons lost must equal the number of electrons gained. In our example, we must double the oxidation half-reaction in order to have 2 electrons lost to equal the two electrons gained in the reduction half-reaction. The combination of two half-reactions produces the complete reaction:

$$2Na + Cl_2 \rightarrow 2Na^+ + 2Cl^-$$

Note that the electrons cancel; in the electron-transfer process, no free electrons remain.

In the reaction described above, sodium metal is the **reducing agent**; it releases electrons for the reduction of chlorine. Chlorine is the **oxidizing agent**; it accepts electrons from the sodium, which is oxidized.

The characteristics of oxidizing and reducing agents are summarized below:

Oxidizing agent	**Reducing agent**
Is reduced	Is oxidized
Gains electrons	Loses electrons
Causes oxidation	Causes reduction

Voltaic Cells

LG 15

A **voltaic cell** is an electrochemical cell that converts stored chemical energy into electrical energy. This cell consists of an **anode**, the electrode at which oxidation occurs, and a **cathode**, the site of reduction. The oxidation reaction occurring at the anode and the reduction reaction occurring at the cathode are referred to as half-reactions. The sum of the two half reactions is the cell reaction. Electrons released at the anode travel through an external circuit, where they may do useful work, to the cathode. A salt bridge containing an electrolyte such as potassium chloride allows current to flow in a continuous circuit.

The most common application of voltaic cells is in the construction of batteries, some of which are small enough and sufficiently inert to allow implantation in the human body.

Electrolysis

An **electrolysis** cell is the reverse of a voltaic cell; it uses electrical energy to cause nonspontaneous redox reactions to occur. The best-known examples of electrolysis include charging a rechargeable battery and the electroplating process (silver electroplating and chrome [chromium] electroplating).

Applications of Oxidation and Reduction

Corrosion. In one type of corrosion, elemental iron is oxidized to iron (III) oxide:

$$4\,Fe(s) + 3\,O_2(g) \rightarrow 2\,Fe_2O_3(s)$$

The oxidation state of iron changes from zero to +3, and oxygen is reduced from zero to –2.

Combustion. When fuels, such as methane, are burned in the presence of oxygen an oxidation-reduction reaction occurs:

$$CH_4(g) + 2O_2(g) \rightarrow CO_2(g) + 2H_2O(g).$$

In this reaction methane is oxidized to form CO_2 and oxygen is reduced to form water.

Bleaching. Bleaches are often oxidizing agents which oxidize the colored compounds adhering to surfaces and convert them into compounds that are not colored or compounds that can be easily removed from the surface.

Metabolism. There are many examples of biological oxidation-reduction reactions. In many metabolic redox reactions, neither the product nor the reactant carries a charge. To identify whether a particular reactant has been oxidized or reduced, alternative descriptions of these terms are useful:

Oxidation is the *gain* of oxygen or the *loss* of hydrogen.
Reduction is the *loss* of oxygen or the *gain* of hydrogen.

This description makes it much easier to recognize the conversion of ethanol to acetaldehyde as an oxidation, for example. Ethanol gains a second bond to oxygen and loses hydrogens.

$$CH_3CH_2OH \longrightarrow H_3C{-}C({=}O){-}H$$

Ethanol　　　　Acetaldehyde

(Note that this reaction is not complete and that only the oxidation half-reaction is shown. Whenever one species is oxidized, something else must be reduced. Metabolic reactions will be discussed in much more detail later in this course.)

Chemical Control of Microbes. Many antiseptics and disinfectants are oxidizing agents. At low concentrations, hydrogen peroxide destroys pathogens associated with living tissues. At higher concentrations, hydrogen peroxide solutions are used as disinfectants.

Self Test for Chapter Eight

1. According to the Arrhenius definition, what is an acid? What is a base?
2. According to the Brønsted-Lowery definition, what is an acid? What is a base?
3. For the following reaction, label the acid, base, conjugate acid, and conjugate base.

$$CH_3NH_2 + H_2O \rightleftharpoons CH_3NH_3^+ + OH^-$$

4. Write the acid-base reaction of each substance with water.
   a. HNO_2 (a weak acid)
   b. HBr (a strong acid)
   c. NH_3 (a weak base)
   d. NaOH (Remember, all metal hydroxides are strong bases.)

5. Write the reaction for the self-ionization of water.
6. What is the hydroxide concentration, $[OH^-]$, in a solution with $[H_3O^+] = 5.0 \times 10^{-5}$?
7. Is the solution in question 6 acidic or basic?
8. What is the hydronium ion concentration, $[H_3O^+]$, of a 1.0×10^{-3} *M* NaOH solution?
9. What is the measure of the strength of an acid in water?
10. Calculate the pH of a solution that contains 6.5×10^{-5} moles of hydronium ion per liter.
11. Calculate the pH of a solution that contains 2.5×10^{-9} moles of hydroxide ion per liter.
12. Calculate the hydroxide ion concentration, $[OH^-]$, of a solution with pH = 9.00.
13. Is the solution in question 12 (above) acidic or basic?
14. Calculate the hydronium ion concentration, $[H_3O^+]$, of a solution with pH = 5.00.
15. Is the solution in question 14 (above) acidic or basic?
16. Calculate the pH of a 1.0×10^{-3} molar solution of HCl.
17. What is the term for the reaction of an acid plus a base to yield a salt plus water?
18. If 14.8 mL of 0.100 M NaOH solution are needed to completely react with 25.0 mL of an unknown HCl solution, what is the molarity of the unknown HCl solution? Note: write the neutralization reaction between NaOH and HCl.
19. Calculate the pH of a buffer solution which is 0.10 *M* HNO_2 and 0.15 *M* $NaNO_2$. The K_a of HNO_2 is 4.5×10^{-4}.
20. Calculate the pH of a buffer solution which is 0.20 *M* HNO_2 and 0.15 *M* $NaNO_2$. The K_a of HNO_2 is 4.5×10^{-4}.
21. Label each of the following half-reactions as oxidation or reduction.
    a. $Mg \rightarrow Mg^{2+} + 2e^-$
    b. $Fe^{2+} \rightarrow Fe^{3+} + e^-$
    c. $O_2 + 4e^- \rightarrow 2O^{2-}$
    d. $Ag^+ + e^- \rightarrow Ag$
22. Which of the following is *not* an example of a redox reaction?
    a. $2Mg + O_2 \rightarrow 2MgO$
    b. $HCl + NaOH \rightarrow NaCl + H_2O$
    c. $2C_2H_6 + 7O_2 \rightarrow 4CO_2 + 6H_2O$
23. Label each of the following descriptions as either a voltaic cell or electrolysis (or both):
    a. Using a battery
    b. Recharging a battery
    c. Converting electrical energy to chemical energy
    d. Converting chemical energy to electrical energy
    e. Involving an oxidation-reduction reaction
24. Does a reduction reaction occur at the anode or the cathode?

9 *The Nucleus, Radioactivity, and Nuclear Medicine*

Learning Goals

1. Use nuclear symbols to represent isotopes and nuclides.
2. Enumerate characteristics of alpha, beta, positron, and gamma radiation.
3. Write balanced equations for nuclear processes.
4. Calculate the amount of radioactive substance remaining after a specified time has elapsed.
5. Explain the process of radiocarbon dating.
6. Describe how nuclear energy can generate electricity: fission, fusion, and the breeder reactor.
7. Cite examples of the use of radioactive isotopes in medicine.
8. Describe the use of ionizing radiation in cancer therapy.
9. Discuss the preparation and use of radioisotopes in diagnostic imaging studies.
10. Explain the difference between natural and artificial radioactivity.
11. Describe characteristics of radioactive materials that relate to radiation exposure and safety.
12. Be familiar with common techniques for the detection of radioactivity.
13. Interpret common units of radiation intensity and discuss their biological implications.

Introduction

This chapter considers the nucleus, nuclear properties, and their applications.

9.1 Natural Radioactivity

Radioactivity is the process by which some atoms emit high-energy particles or rays. These particles or rays are termed radiation. Nuclear radiation occurs as a result of an alteration in nuclear composition or structure, which happens because the nucleus is unstable and hence radioactive.

Nuclear symbols are analogous to atomic symbols. The symbols were introduced in Chapter 2 where isotopes were described. However, the nuclear symbols have not been seen in the chapters that follow. As a review, the nuclear symbols consist of the elemental symbol, the atomic number (equivalent to the number of protons in the nucleus), and the mass number, which is defined as the sum of the neutrons and protons in the nucleus. For example, fluorine-19 is represented as

$${}^{19}_{9}F$$

The subscript is the **atomic number**. The superscript is the mass number. Be careful not to confuse the *mass number* (the sum of neutrons and protons) with the atomic mass, which includes the contribution of electrons, and is a true mass figure. The term **nuclide** refers to any atom characterized by an atomic number and a mass number.

Only unstable nuclei undergo change and produce radioactivity. Not all atoms of a particular element undergo radioactive decay. When writing the symbols for a nuclear process, it is important to designate the particular isotope involved.

Four types of natural radiation emitted by unstable nuclei are alpha particles, beta particles, positrons, and gamma rays.

LG 2

Alpha Particles

Alpha particles (α) contain two protons and two neutrons. An alpha particle is identical to the helium ion (He^{2+}) which is a helium nucleus and is represented as:

$$^{4}_{2}He \quad \text{or} \quad \alpha \quad \text{or} \quad ^{4}_{2}He^{2+}$$

Alpha particles emitted by radioisotopes have relatively large mass (compared to other nuclear particles) and as a result, move relatively slowly (approximately 10% of the speed of light). Alpha particles are stopped by very little mass.

Beta Particles and Positrons

The **beta particle** (β) is a fast-moving electron traveling at approximately 90% of the speed of light as it leaves the nucleus. The beta particle is represented as

$$^{0}_{-1}e \quad \text{or} \quad \beta \quad \text{or} \quad ^{0}_{-1}\beta$$

The subscript –1 is written in the same position as the atomic number and, like the atomic number (number of protons), indicates the charge of the particle.

Beta particles are smaller, faster, and more energetic and penetrating than alpha particles are.

A positron has the same mass as a beta particle but carries a positive charge. **Positrons** are produced by the conversion of a proton to a neutron. The positron can be represented as

$$^{0}_{+1}e \quad \text{or} \quad ^{0}_{+1}\beta$$

Gamma Rays

Gamma rays (γ) are pure energy (compared to alpha particles, beta particles, and positrons, which are matter). Since energy has no mass or charge, the symbol for a gamma ray is simply γ.

Gamma radiation is highly energetic and is the most penetrating form of nuclear radiation.

Example 1: Types of Radiation

What is the mass number and charge of each of the following particles?

1. alpha particle
2. beta particle
3. positron
4. gamma ray

Solution:

1. An alpha particle is a helium nucleus. It has a mass number of 4 and a +2 nuclear charge.
2. A beta particle is the same as an electron. It has a mass number of 0 (with no neutrons or protons) and a –1 charge.
3. A positron is a positively charge beta particle. It has a mass number of 0 and a +1 charge.
4. A gamma ray is not a particle. It is just energy and has no mass or charge.

Properties of Alpha, Beta, Positron and Gamma Radiation

Alpha, beta, positron, and gamma radiation are collectively termed ionization radiation. Important properties of α, β, positrons, and γ radiation are summarized in Table 9.1 below.

Table 9.1 A Summary of the Major Properties of Alpha, Beta, Positron and Gamma Radiation

Name of Symbol	Identity	Charge	Mass (amu)	Velocity	Penetration
Alpha (α)	Helium nucleus	2+	4.0026	5-10% of the speed of light.	Low
Beta ($^{0}_{-1}\beta$)	Electron	1-	0.000549	Up to 90% of the speed of light	Medium
Positron ($^{0}_{+1}\beta$)	Electron	1+	0.000549	Up to 90% of the speed of light	Medium
Gamma (γ)	Radiant energy	0	0	Speed of light	High

9.2 Writing a Balanced Nuclear Equation

A **nuclear equation** represents a nuclear process such as radioactive decay. To write a balanced equation, we must remember the following:

1. The total mass on each side of the reaction arrow must be identical.
2. The sum of the atomic numbers of the reactant nuclei must be equal to the sum of the atomic numbers of the product nuclei.

In each of the following examples, note how the sum of the mass numbers and the sum of the atomic numbers are the same on either side of the nuclear equation.

Alpha Decay

The reaction shown below represents alpha decay, where an alpha particle is lost during the reaction:

$$^{226}_{88}Ra \rightarrow ^{222}_{86}Rn + ^{4}_{2}He$$

Radium-226 → Radon-222 + α particle

Beta Decay

Beta decay may be illustrated by the following:

$$^{234}_{90}Th \rightarrow ^{234}_{91}Pa + ^{0}_{-1}e$$

Thorium-234 → Protactinium-234 + β particle

Positron Emission

An example of positron emission is the conversion of a proton to a neutron:

$$^{11}_{6}C \rightarrow ^{11}_{5}B + ^{0}_{+1}e$$

Carbon-11 → Boron-11 + positron

Gamma Production

If gamma radiation were the only product of nuclear decay, there would be no change in the mass or identity of the radioactive nuclei, since a gamma ray is pure energy, possessing no mass or charge. The gamma emitter has simply gone to a lower energy state. The decay of the *metastable isotope* technetium-99m is shown:

$$^{99m}_{43}Tc \rightarrow ^{99}_{43}Tc + \gamma$$

Technecium-99m → Technecium-99 + Gama ray

Often, gamma radiation is produced along with other products during nuclear decay.

Predicting Products of Nuclear Decay

It is possible to use a nuclear equation to predict one of the products of a nuclear reaction if the others are known. We know that the mass and atomic number of the total of all products and reactants must be equal. By subtracting, we can compute the missing charge and mass that represents the unknown product and deduce its identity.

Example 2: Predicting the Missing Substance of Nuclear Decay

Complete the following nuclear equations:

1. $^{238}_{92}U \rightarrow \ ^{234}_{90}Th + ?$

2. $^{16}_{7}N \rightarrow \ ^{16}_{8}O + ?$

3. $^{40}_{19}K \rightarrow \ ^{0}_{-1}e + ?$

4. $^{197}_{79}Au + ? \rightarrow \ ^{198}_{79}Au$

Solution:

Use the following rules:

The total mass on each side of the equation arrow must be identical.
The sum of the atomic numbers on each side of the reaction arrow must be identical.

1. Mass number: $238 = 234 + A$ $A = 4$
   Atomic number: $92 = 90 + Z$ $Z = 2$
   The nuclear particle with mass number of 4 and atomic number (nuclear charge) of 2 is an alpha particle: $^{4}_{2}He$
2. Mass number: $16 = 16 + A$ $A = 0$
   Atomic number: $7 = 8 + Z$ $Z = -1$
   The nuclear particle with mass number of 0 and atomic number (nuclear charge) of –1 is a beta particle: $^{0}_{-1}e$
3. Mass number: $40 = 0 + A$ $A = 40$
   Atomic number: $19 = -1 + Z$ $Z = 20$
   The element which has a nuclear charge of 20 (or 20 protons) is calcium, giving: $^{40}_{20}Ca$
4. Mass number: $197 + A = 198$ $A = 1$
   Atomic number: $79 + Z = 79$ $Z = 0$
   The nuclear particle which has no charge but a mass of 1 must be a neutron: $^{1}_{0}n$

Example 3: Writing a Nuclear Equation

When a nucleus of an element is hit with high energy neutrons (neutron bombardment), a radioactive nucleus is produced, termed artificial radioactivity. This nucleus will then decay.

Show the two nuclear equations for the following process:

Molybdenum-98 can be converted by neutron bombardment into technetium-99 plus beta particles. The technetium produced is unstable and can be used as a gamma source for tracer applications.

Solution for Example 3:

1. From the periodic table, we find that molybdenum has an atomic number of 42, and technetium's atomic number is 43.
2. Next we set up the equation:

$$^{98}_{42}\text{Mo} + {}^{1}_{0}\text{n} \rightarrow {}^{99m}_{43}\text{Tc} + {}^{0}_{-1}\text{e}$$

Notice that the "m" following the mass number means in the above reaction equation that the isotope is metastable.

The metastable isotope of technetium then undergoes gamma emission:

$$^{99m}_{43}\text{Tc} \rightarrow {}^{99}_{43}\text{Tc} + \gamma$$

9.3 Properties of Radioisotopes

Nuclear Structure and Stability

The energy that holds the protons, neutrons, and other particles together in the nucleus is the **binding energy** of the nucleus. When an isotope decays, some of this binding energy is released.

Factors related to nuclear stability include the following:

1. Nuclear stability correlates with the ratio of neutrons to protons in the isotope. For the lighter elements, the ratio of 1:1 of neutrons to protons is stable.
2. Nuclei with large numbers of protons (84 or more) tend to be unstable.
3. Isotopes containing 2, 8, 20, 50, 82, or 126 protons or neutrons are stable. These "magic numbers" seem to indicate the presence of stable energy levels in the nucleus.
4. Isotopes with even numbers of protons or neutrons are generally more stable than those with odd numbers of protons or neutrons.
5. All isotopes (except hydrogen-1) with more protons than neutrons are unstable. However, the reverse is not true.

Example 4: Stability of Isotopes

For each of the following pairs, predict which isotope is more stable and explain why.

1. $^{12}_{6}\text{C}$ or $^{14}_{6}\text{C}$
2. $^{208}_{82}\text{Pb}$ or $^{210}_{84}\text{Po}$

Solution:

1. Carobn-12 is more stable than carbon-14 because carbon-12 has 6 protons and 6 neutrons a 1:1 ratio. Carbon-14 has 6 protons and 8 neutrons, not a 1:1 ratio.

Example 4 Continued

2. Lead-208 is more stable than polonium-210 because polonium has 84 protons, and nuclei with 84 or more protons are unstable.

Half-Life

The **half-life, $t_{1/2}$,** is the time required for one-half of a given quantity of a substance to undergo change. Each isotope has its own characteristic half-life.

The degree of stability of an isotope is indicated by the isotope's half-life. Isotopes with short half-lives decay rapidly; they are very unstable.

Decay of a radioisotope that has a reasonably short $t_{1/2}$ is experimentally determined by following its activity as a function of time. Graphing the results produces a radioactive decay curve (see Figure 9.2 in the textbook). The mass of a radioactive substance remaining after of period of time can be calculated by the following equation:

$$m_f = m_i\,(0.5)^n$$

where m_f is the final mass, m_i is the initial mass, and n is the number of half-lives.

Example 5: Understanding Half-life.

Radium-226 ($^{226}_{88}Ra$) has a half-life of 1620 years. If the numbers on a watch were coated with paint containing Ra-226 in the year 1920 to cause the numbers to glow in the dark, why is it still radioactive?

Solution:

Since only about 89 years (a small fraction of one half-life, $\frac{89 \text{ years}}{1620 \text{ years}} = 0.054$) have passed, nearly all the radium-226 used in the paint is still radioactive.

Example 6: Stability of Isotopes and Half-life.

The following half-lives are known for different isotopes of radium:

Ra-223 ....11 days
Ra-224 ....3.64 days
Ra-226 ....1620 years
Ra-228 ....6.7 years

List the isotopes from most stable to least stable.

Solution:

The longer the half-life, the more stable the isotope. Ra-226 has the longest half-life; therefore, it is the most stable. Ra-224 will be the least stable because it has the shortest half-life. The ranking from most stable to least stable is as follows:
Ra-226, Ra-228, Ra-223, Ra-224

Example 7: Predicting Amount of Isotope Remaining Given the Half-life.
How much of a 1.00-gram sample of Ra-226 would remain after three half-lives? The half-life of Ra-226 is 1620 years.

Solution:
After each half-life, the amount of Ra-226 is cut in half.

$$1.00\text{g} \xrightarrow[\text{half-life}]{\text{first}} 0.50\text{ g} \xrightarrow[\text{half-life}]{\text{second}} 0.25\text{ g} \xrightarrow[\text{half-life}]{\text{third}} 0.13\text{ g}$$

Alternatively, we could use the equation $m_f = m_i\,(0.5)^n$, using information given in the problem about the initial mass (1.00 g) and number of half-lives (3).

$$m_f = (1.00)(0.5)^3 = 0.13\text{ g}$$

Radiocarbon Dating

The approximate age of fossils and other objects of archaeological, anthropological, or historical interest may be established through **radiocarbon dating**. Radiocarbon dating is based on the measurement of the relative amounts (or ratio) of carbon-14 and carbon-12 present in an object. The carbon-14, along with the more abundant isotope of carbon-12, is converted into living plant material by the process of photosynthesis. Over time, the amount of carbon-14 slowly decreases, because carbon-14 is radioactive ($t_{1/2}$ = 5760 years); the amount of carbon-12, however, remains constant.

When the fossil is found and studied, the relative amounts of carbon-14 and carbon-12 are determined; using suitable equations involving the $t_{1/2}$ of carbon-14, it is possible to calculate (to within ± a few percent) the time that has elapsed since the formation of the object in question.

9.4 Nuclear Power

Energy Production

Einstein's equation relating mass and energy predicts that a small amount of nuclear mass is converted to a very large amount of energy when the nuclear particle breaks apart.

Einstein's equation is as follows:

$$E = mc^2$$

E = energy, m = mass, c = speed of light

When nuclear energy is released in a controlled fashion in a **nuclear reactor**, the heat from the nuclear reaction converts liquid water into steam. The steam drives an electrical generator, which produces electricity.

Nuclear Fission

Fission (meaning to split) occurs when a heavy nuclear particle is split into smaller nuclei and large amounts of energy. The splitting is initiated by a smaller nuclear particle (such as a neutron). The fission process intensifies, producing very large amounts of energy. This process of intensification is referred to as a **chain reaction.** For example, one neutron splits one atom and produces three more neutrons which can then split three atoms. The splitting of three atoms would then generate nine neutrons. Following this pattern, nine neutrons would split nine atoms and generate how many neutrons?

Nuclear Fusion

Fusion (meaning to join together) results from the combination of two small nuclei to form a larger nucleus with the concurrent release of large amounts of energy. The best example of a fusion reactor is the sun. Continuous fusion processes furnish our solar system with light and heat.

$$^{2}_{1}H + ^{3}_{1}H \rightarrow ^{4}_{2}He + ^{1}_{0}n + \text{Energy}$$

Breeder Reactors

A **breeder reactor** literally manufactures its own fuel. A perceived shortage of fissionable isotopes makes the breeder an attractive alternative to fission reactors. A breeder reactor uses uranium-238, which is abundant but nonfissionable. In a series of steps, the uranium-238 is converted to plutonium-239, which is fissionable and undergoes a fission chain reaction, producing energy.

9.5 Medical Applications of Radioactivity

The use of radiation in the treatment of various forms of cancer as well as in the newer area of **nuclear medicine** has become widespread in the past quarter-century.

Cancer Therapy Using Radiation

When high-energy radiation, such as gamma radiation, passes through a cell, it may collide with one of the molecules in the cell and cause it to lose one or more electrons, producing an ion pair. For this reason, such radiation is termed *ionizing radiation.* Ions produced in this way may cause subtle changes in cellular biochemical processes, which may result in diminished or altered cell function or, in extreme cases, the death of the cell.

An organ that is cancerous is composed of both healthy cells and malignant cells. Cells undergoing division are particularly sensitive to gamma radiation. Therefore, exposing the tumor area to controlled dosages of high-energy gamma radiation from cobalt-60 (a high-energy gamma ray source) will generally kill a higher percentage of abnormal cells than normal cells because cancer cells tend to divide at a greater rate than do healthy cells.

Nuclear Medicine

LG 9

The diagnosis of a host of biochemical irregularities or diseases of the human body has been made routine through the use of *radioactive tracers.* **Tracers** are small amounts of radioactive substances used as probes to study internal organs. Because the isotope is radioactive, its path may be followed using suitable detection devices. A "picture" of the organ is obtained, often far more detailed than is possible with conventional X-rays. **Nuclear imaging** techniques are noninvasive; that is, surgery is not required to investigate the condition of the internal organ, eliminating the risk associated with an operation. These techniques are successful because the radioactive isotope of an element has exactly the same chemical behavior as any other isotope of the same element.

Making Isotopes for Medical Applications

The radioactivity produced by unstable isotopes is described as **natural radioactivity**. If a normally stable, nonradioactive nucleus is made radioactive through bombardment with protons, neutrons, or alpha particles, the resulting radioactivity is termed **artificial radioactivity**.

The bombardment process is often accomplished in the core of a nuclear reactor, where an abundance of small nuclear particles, particularly neutrons, are available. Alternately, extremely high-velocity charged particles (such as alpha- and beta particles) may be produced in a particle accelerator, such as a cyclotron.

These product isotopes are often used in hospital laboratories as tracers in nuclear medicine.

9.6 Biological Effects of Radiation

Radiation Exposure and Safety

Safety considerations are based on the following:

1. *The magnitude of the half-life.* The shorter the half-life, the more quickly the isotope will leave the body. However, the shorter the half-life the greater the amount of nuclear radiation generated in a given amount of time. Over time (depending on the magnitude of half-life), radioactive isotopes will decay to **background radiation**.
2. ***Shielding.*** Protection from radiation can be accomplished by blocking the alpha, beta, or gamma radiation through various materials.
3. *Distance from the radioactive source.* The greater the distance one is from the radiation source, the safer one is.
4. *Time of exposure.* Less exposure means less damage.
5. *Types of radiation emitted.* Each type of radiation has a different ability to damage tissue. Beta particles are the least damaging.

Virtually all applications of nuclear chemistry create radioactive waste and, along with it, the problems of safe handling and disposal. Most disposal sites at present are considered temporary, until a long-term, safe solution can be found.

9.7 Measurement of Radiation

The changes that take place when radiation interacts with matter provide the basis of operation for various radiation detection devices.

Photographic Imaging

An isotope is administered to a patient, and the isotope begins to concentrate in the organ of interest. Photographs of that region of the body are taken at periodic intervals using a special type of film. Upon development of the series of photographs, a record of the organ's uptake of the isotope as a function of time enables the radiologist to assess the condition of the organ.

Computer Imaging

A specialized television camera, sensitive to emitted radiation from a radioactive substance administered to a patient, develops a continuous and instantaneous record of the voyage of the isotope throughout the body.

The Geiger Counter

A Geiger counter is an instrument capable of detecting ionizing radiation.

Film Badges

A film badge is merely a piece of photographic film that is sensitive to energies corresponding to radioactive emissions. The badges are used to determine exposure to radiation over time for those who work in an industry that uses radioactive materials.

Units of Radiation Measurement

Radioactivity

The **curie** (Ci) and the **becquerel** (Bq) measure of the amount of radioactivity in a radioactive source. Both the curie and the becquerel are independent of the nature of the radiation (α, β, or) as well as its effect on biological tissue. A curie is defined as the amount of radioactive material that produces 3.7×10^{10} atomic disintegrations per second. A becquerel is the amount of radioactive material that produces 1 atomic disintegration per second.

Exposure

The **roentgen** is a measure of ionizing radiation (X-ray and gamma ray) only. The roentgen is defined as the amount of radioactive isotope needed to produce 2×10^9 ion pairs when passing through 1 cubic centimeter of air at 0°C. The roentgen is a measure of radiation's interaction with air and gives no information regarding its effect on biological tissue.

Absorbed Dosage

The **rad**, or radiation absorbed dosage, provides more meaningful information than either of the previous units of measure. It takes into account the nature of the absorbing material. It is defined as the dosage of radiation able to transfer 2.4×10^{-3} calories of energy to 1 kilogram of matter. The **gray** (Gy) is also used to measure an absorbed dosage. It is defined as the absorption of 1 joule (J) of energy by 1 kg of matter.

Dose Equivalent

The **rem**, or roentgen equivalent for man, describes the biological damage caused by the absorption of different kinds of radiation by the body. The rem is obtained by multiplication of the rad by a factor called the relative biological effect, or RBE. The RBE is a function of the type of radiation (α, β, or γ). Although a beta particle is more energetic than an alpha particle, an alpha particle is approximately ten times more damaging to biological tissue. As a result, the RBE is 10 for alpha particles and 1 for beta particles. The **sievert,** Sv, describes the biological effect that results when one Gy of radiation energy is absorbed by human tissue. One Sv is equivalent to 100 rem.

An estimated lethal dose, symbolized by **lethal dose (LD_{50})**, is 500 rems. The lethal dose, LD_{50}, is defined as the dose that would be fatal for 50% of the exposed population within 30 days. Some biological effect, however, is detectable at a level as low as 25 rem.

Self Test for Chapter Nine

1. Write the nuclear symbol for an isotope of sodium with 12 neutrons.
2. Which of the types of radiation is really pure energy?
3. Which type of radiation is similar to a helium atom?
4. Which of the types of radiation is the least-penetrating form of nuclear radiation?
5. What is defined as the spontaneous decay of a nucleus to produce high-energy particles or rays?
6. Which type of radiation travels at the speed of light?
7. Between alpha and beta radiation, which causes more damage to tissue and why?
8. Which of the following isotopes would be expected to be more stable and why? $^{11}_{5}B$ or $^{13}_{5}B$
9. Complete the following nuclear reactions by supplying the missing part:

   a. $^{27}_{13}Al + ^{4}_{2}He \rightarrow ^{30}_{15}P + ?$

   b. $^{210}_{82}Pb \rightarrow ^{0}_{-1}e + ?$

10. Which of the following isotopes of lead is more stable?
    $^{214}_{82}Pb$ with a half-life of 26.8 minutes or $^{210}_{82}Pb$ with a half-life of 20.4 years.
11. How many mg of a 100.0 mg sample of Tc-99m will remain after 30 hours? The half-life of Tc-99m is 6 hours.
12. Radiocarbon dating involves estimation of the age of objects by measuring the amounts of two isotopes. Carbon-12 is one isotope. What is the identity of the other isotope?
13. Nuclear power plants use what nuclear process to produce energy? Briefly describe this process.
14. What is the term that we use for cells that undergo unnaturally rapid cell division?
15. How can radioisotopes be used in diagnostic imaging studies?
16 What is the term that describes the amount of radiation attributable to our surroundings on a day-to-day basis?
17. Why do staff members working around radiation sources often wear film badges?
18. What is the difference between natural and artificial radioactivity?
19. When working with radioactive material why would you want to have shielding and keep a maximum distance from the radioactive source and a minimum time of exposure?
20. Why is the radiation unit of rem useful when considering safety of a nuclear source?

10 An Introduction to Organic Chemistry: The Saturated Hydrocarbons

Learning Goals

1. Explain the difference between active and passive study practices.
2. Compare and contrast organic and inorganic compounds.
3. Recognize structures that represent each of the families of organic compounds.
4. Write the names and draw the structures of the common functional groups that characterize the families of organic compounds.
5. Write condensed, structural, and line formulas for saturated hydrocarbons.
6. Describe the relationship between structure and physical properties of saturated hydrocarbons.
7. Use the basic rules of the IUPAC Nomenclature System to name alkanes and substituted alkanes.
8. From the IUPAC name of an alkane or substituted alkane, be able to draw the structure.
9. Draw constitutional (structural) isomers of simple organic compounds.
10. Write the names and draw the structures of simple cycloalkanes.
11. Draw *cis*- and *trans*- isomers of cycloalkanes
12. Describe conformations of alkanes.
13. Draw the chair and boat conformations of cyclohexane.
14. Write equations for combustion reactions of alkanes.
15. Write equations for halogenation reactions of alkanes.

Introduction

Authorities estimate that one-third of all fires are arson. While investigating fires, samples are tested for the presence of accelerants, flammable substances that cause fires to burn hotter and spread more quickly. Since many common accelerants are mixtures of hydrocarbons, technology allows them to be identified using chromatography that separates these compounds by their boiling points.

In this and future chapters, we will explore the concepts of structure, properties, and chemistry of various hydrocarbons and other organic molecules.

10.1 Strategies for Success in Organic Chemistry

Although there are millions of organic compounds, organic chemistry is rational and systematic. Learning the rules of naming these compounds and the principles that determine their properties and their reactivities can lead to success.

The key to successful learning is *active* learning.

Prepare for Class

After reading the sections of the text that will be covered in lecture, outline key ideas and make note of what is unclear.

Make the Most of Class Time

In class, ask questions. Also, take notes by hand and summarize the information in your own words.

Make your study times active learning experiences by:

1. Going over your class notes and annotating them.
2. Solving problems to test your understanding. Don't look at the solutions until after you solve the problems. If you work in a study group, talk through the problems.
3. Creating and using flash cards. Visual summaries and concept maps can also be useful tools for learning.

10.2 The Chemistry of Carbon

Organic chemistry is the study of carbon-containing compounds. There are several reasons for the existence of hundreds of thousands of organic compounds:

1. Carbon atoms are able to form stable, covalent bonds with other carbon atoms. Each carbon can form up to four covalent bonds with other carbon atoms. The resulting molecules may be linear, branched, or cyclic.
2. Carbon can form stable bonds with other elements, for example nitrogen, oxygen, sulfur, and the halogens.
3. Carbon can form double and triple bonds with other carbons atoms.
4. The number of ways that these elements may combine to form unique structures is practically limitless. Compounds having the same molecular formulas but different structures (hence, different properties) are termed ***isomers***.

LG 2

Important Differences between Organic and Inorganic Compounds

The bonds in organic molecules are almost always covalent bonds, while those found in many inorganic substances are ionic bonds. When comparing organic and inorganic compounds, it helps to remember the differences between ionic and covalent bonds.

1. Ionic bonds form when one or more electrons transfer from one atom to another. Covalent bonds are formed by a sharing of electrons to form a stable orbital containing two electrons.
2. The ionic bond is electrostatic in nature. It is formed by the attraction of positive and negative ions resulting from the electron transfer process. Charge separation of covalently bonded atoms is much less extreme.
3. Ions are arranged in large, three-dimensional crystals, consisting of many positive and negative ions. Covalently bonded substances exist as discrete units called molecules.
4. Ionic compounds often dissociate in solution (electrolytes), whereas most covalently bonded molecules retain their identity in solution (nonelectrolytes). Remember, electrolytic solutions conduct electricity.

As a result of these differences the following generalizations are typically made:

Organic Compounds:	**Inorganic compounds**
Lower melting and boiling points	Higher melting and boiling points
Tend not to dissolve in water	Tend to dissolve in water
Flammable	Nonflammable

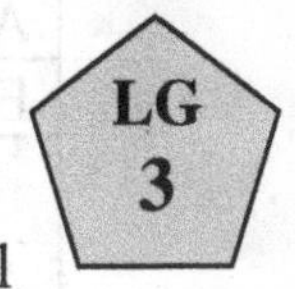

Families of Organic Compounds

Organic compounds are classified according to groups, or families. The two most general classifications are **hydrocarbons** and **substituted hydrocarbons**. Compounds that contain only carbon and hydrogen are classified as hydrocarbons. Hydrocarbons are subdivided into two principal classes: **aliphatic** and **aromatic hydrocarbons**. Aliphatic hydrocarbons are further subdivided into four families:

1. Alkanes: contain only carbon-to-carbon single bonds.
2. Cycloalkanes: contain only carbon-to-carbon single bonds and the carbon atoms link together to make a ring.
3. Alkenes: contain at least one carbon-to-carbon double bond.
4. Alkynes: contain at least one carbon-to-carbon triple bond.

Saturated hydrocarbons are composed solely of C—C and C—H single bonds. The **unsaturated hydrocarbons** are hydrocarbons that contain at least one carbon-to-carbon double bond or carbon-to-carbon triple bond. Aromatic compounds contain a benzene ring or a derivative of the benzene ring, see Figure 10.1.

Figure 10.1 Benzene, the basis for aromatic hydrocarbons.

LG 4

When one or more hydrogens on a hydrocarbon ring are substituted by functional groups a substituted hydrocarbon is formed. A **functional group** is an atom or group of atoms in a molecule principally responsible for the chemical and physical properties of that molecule. Table 10.1 lists the most common functional groups.

All compounds that contain a particular functional group, for example, the hydroxyl group (—OH), are classified as being in the same family. The hydroxyl group is the functional group of the alcohols. In the following chapters, you will study the chemical and physical properties of each of these classifications of compounds. Your instructor will probably want you to memorize the names and groups of atoms in each functional group contained in Table 10.1 Preparing and using flash cards would be a good strategy for learning this information.

TABLE 10.1 Common Functional Groups

Type of Compound	Functional Group	Structural formula*	Condensed formula*
Halide	Halogen atom	R—Cl (or any halogen)	RCl
Alcohol	Hydroxyl	R—O—H	ROH
Ether	Ether	R—O—R′	ROR′
Aldehyde	Carbonyl	R—C(=O)—H	RCHO
Ketone	Carbonyl	R—C(=O)—R'	RCOR′
Carboxylic Acid	Carboxyl	R—C(=O)—O—H	RCOOH
Ester	Ester	R—C(=O)—O—R'	RCOOR′
Amine	Amino	R—N(—H)—H	RNH_2 (Note: R_2NH and R_3N are also amines.)
Amide	Carboxamide	R—C(=O)—O—NH_2	$RCOONH_2$

*Note that in the formulas, the R and R′ represents hydrocarbons.

Example 1: Classification of Hydrocarbons

Classify each of the following hydrocarbons into the categories of aliphatic, aromatic saturated or unsaturated, alkane, cycloalkane, alkene and alkyne. More than one may apply.

1. CH_3CH_3
2. $CH_3CH{=}CHCH_3$
3. $CH{\equiv}CCH_3$

Example 1 Continued

4. [cyclohexane structural formula]

5. [benzene structural formula]

Solution:

1. Remember that each carbon will have 4 bonds. The first carbon has three hydrogen atoms bonded to it leaving a single bond to attach to the second carbon. This is a carbon-to-carbon single bond. The compound is aliphatic, saturated and an alkane.
2. The double bond is shown for this hydrocarbon (between the second and third carbon.) The compound is aliphatic, unsaturated and an alkene.
3. The triple bond is depicted between the first and second carbon. The hydrocarbon is aliphatic, unsaturated and an alkyne.
4. Only single bonds exist between the carbon atoms. It is not a benzene ring; therefore it is aliphatic. Only single bonds exist so it is saturated. The carbons are joined together in a ring so the compound is a cycloalkane.
5. This is a benzene ring; therefore, the compound is aromatic.

Example 2: Identifying Functional Groups

Identify the functional group in each of the following compounds.

1. CH_3OH
2. $H_3C-C(=O)-CH_3$
3. CH_3Br
4. $H_3C-N(CH_3)-H$

Solution:

1. The –OH group means this is an alcohol.
2. This compound takes the form of RCOR and is a ketone.
3. Bromine is a halogen. A halogen connected to a hydrocarbon is a halide.
4. Nitrogen directly bonded to a total of three carbon and/or hydrogen atoms is an amine.

10.3 Alkanes

Structure and Properties

Alkanes are hydrocarbons that contain only carbon and hydrogen bonded together through carbon-hydrogen and carbon-carbon single bonds. They have the general formula C_nH_{2n+2}.

Several different types of formulas are used to describe organic molecules. The **molecular formula** provides the atoms and number of each type of atom in a molecule, but gives no information regarding the bonding pattern. A potential downside of this type of formula is that in many cases there are multiple ways that the atoms can be arranged, and the molecular formula gives us no information about which arrangement is being referred to. For instance, there are three possible structures for a compound with the molecular formula of C_5H_{12}:

```
                                                                  H
                                                                  |
                                                              H — C — H
                                   H                          H   |   H
                                   |                          |   |   |
                               H — C — H                  H — C — C — C — H
  H   H   H   H   H            H   |   H   H                  |   |   |
  |   |   |   |   |            |   |   |   |                  H   |   H
H—C — C — C — C — C — H    H — C — C — C — C — H              H — C — H
  |   |   |   |   |            |   |   |   |                      |
  H   H   H   H   H            H   H   H   H                      H
```

Pentane — Methylbutane — Dimethylpropane

These are three completely different alkanes with the same molecular formula, as indicated by their different names. We see that these structural formulas provide more information about a molecule than does the molecular formula C_5H_{12}. The **structural formula** shows all of the atoms in a molecule and shows all bonds as lines.

A suitable compromise between the convenience of the molecular formula and the detail of the structural formula is the condensed formula. The **condensed formula** shows all of the atoms in a molecule and places them in a sequential arrangement that details which atoms are bonded to each other.

Each of the above structural formulas is shown below as condensed formulas:

$CH_3CH_2CH_2CH_2CH_3$ or $CH_3(CH_2)_3CH_3$	$CH_3CH(CH_3)CH_2CH_3$	$CH_3C(CH_3)_2CH_3$
Pentane	Methylbutane	Dimethylpropane

The most simplistic formula is the **line formula.** In this formula, we assume that there is a carbon at any location where two or more lines meet. We also assume that there is a carbon at the end of any line and that each carbon is bonded to the correct number of hydrogen atoms to give the carbon atom four bonds.

Once again, these compounds are depicted using line formulas:

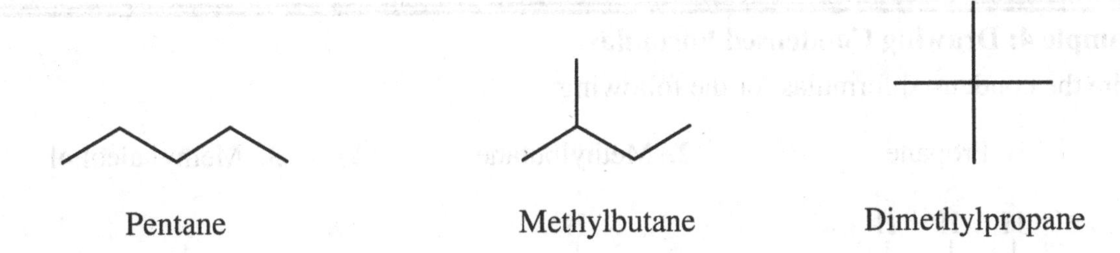

You will need to commit the names of the first ten straight-chain alkanes to memory. Refer to Table 10.2 below. For more detailed information, refer to Table 10.3 in your textbook. Use flash cards to help you learn these names.

TABLE 10.2 Names and formulas for the First Ten Straight-Chain Alkanes

Number of Carbons	Molecular Formula	Name	Number of Carbons	Molecular Formula	Name
1	CH_4	Methane	6	C_6H_{14}	Hexane
2	C_2H_6	Ethane	7	C_7H_{16}	Heptane
3	C_3H_8	Propane	8	C_8H_{18}	Octane
4	C_4H_{10}	Butane	9	C_9H_{20}	Nonane
5	C_5H_{12}	Pentane	10	$C_{10}H_{22}$	Decane

Example 3: Drawing Structural Formulas

Draw the structural formula of an alcohol and an ether which has the molecular formula C_2H_6O.

Solution:

An alcohol has an –OH group in it.

```
    H   H
    |   |
H—C—C—O—H
    |   |
    H   H
```

An ether has oxygen located between two carbon atoms.

```
    H       H
    |       |
H—C—O—C—H
    |       |
    H       H
```

Example 4: Drawing Condensed Formulas

Write the condensed formulas for the following:

1. Propane 2. Methylbutane 3. Methyl alcohol

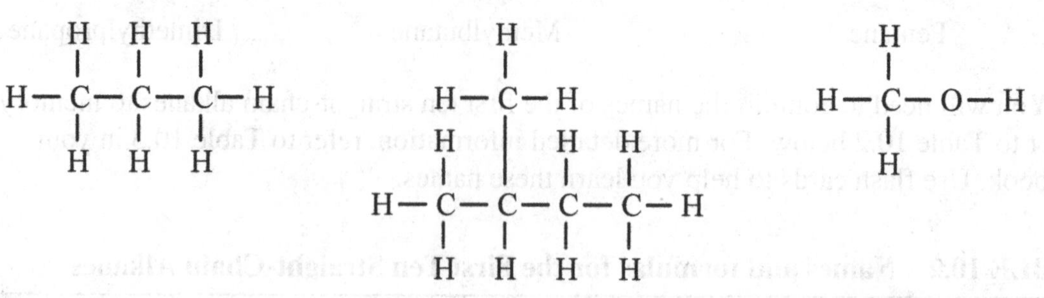

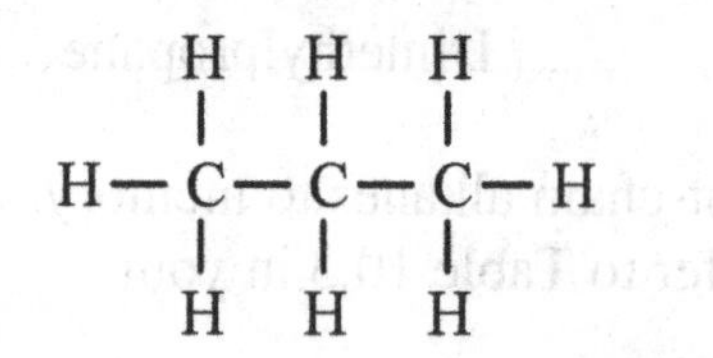

4. Acetone 5. 2-Chloropropane

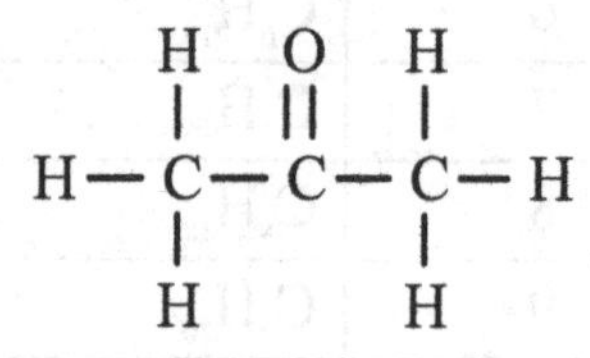

Solution:

To write the condensed formula, write each C in the chain of carbon atoms followed by each atom or groups of atoms which are directly attached to the carbon then move to the next carbon in the chain. Parentheses are sometimes placed around a group of atoms to help avoid confusion as to where that group of atoms is located.

1. $CH_3CH_2CH_3$
   Note: the hydrogen atoms are only connected to the carbon which precedes the hydrogen.

2. $CH_3CH(CH_3)CH_2CH_3$
   Notice the use of the parentheses. Without them it would be difficult to notice that the CH_3 group is attached to the second carbon and is not in the chain.

3. CH_3OH

4. CH_3COCH_3
   The ketone oxygen is attached to the second carbon so is written right after the second C. To distinguish it from an ether, you must notice that the second carbon needs a double bond to the oxygen in order to give the second carbon four bonds.

5. $CH_3CH(Cl)CH_3$
   The parentheses are placed around the Cl to help note that it, as well as the H, is connected to the second carbon.

Example 5: Molecular Formulas of Alkanes

Write the molecular formulas of the alkanes that contain 4, 8, and 40 carbon atoms.

Solution:

Use the general formula for finding the molecular formula of any alkane:
C_nH_{2n+2} where n = number of carbon atoms

4 carbons: the number of hydrogen atoms = 2(4) + 2 = 10 giving C_4H_{10};
8 carbons: the number of hydrogen atoms = 2(8) + 2 = 18 giving C_8H_{18};
40 carbons: the number of hydrogen atoms = 2(40) +2 = 82 giving $C_{40}H_{82}$.

Example 6: Drawing Line Structures.

Write a line structure for the compounds below.

1. $CH_3CH_2CH_3$
2. $CH_3CH{=}CHCH_3$
3. $CH_3CH(CH_3)CH_2CH_2CH_3$

Solution:

Remember, we assume that there is a carbon at any location where two or more lines meet. We also assume that there is a carbon at the end of any line and that each carbon is bonded to the correct number of hydrogen atoms to give the carbon atom four bonds.

1. The three carbon atoms in a chain are represented as:

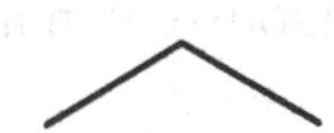

2. There are four carbon atoms in this compound with a double bond connecting the second and third carbon, giving:

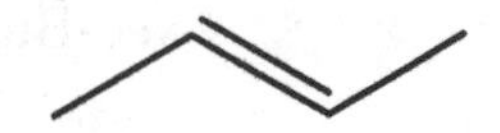

3. At first glance, it may appear to you that there are six carbon atoms in a chain. However, when you see the parentheses around the CH_3 you need to remember that the group of atoms in parentheses is not in the chain, but it is actually attached to the carbon it is written after (in this example it is after the second carbon.) Draw a five carbon chain with an extra carbon attached to the second carbon, giving:

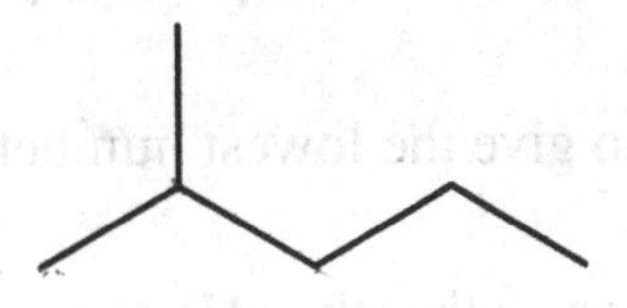

All the carbons of alkanes are attached to four other atoms, be those carbons or hydrogens.As a result, all carbons have tetrahedral geometry and 109.5° bond angles.

Physical Properties

Since all the hydrocarbons are composed of nonpolar carbon-carbon and carbon-hydrogen bonds, hydrocarbons are nonpolar molecules. Because water is a polar molecule, hydrocarbons are not water soluble, but are readily soluble in nonpolar solvents. Furthermore, virtually all of the hydrocarbons are less dense than water and have relatively low melting points and boiling points. When considering just hydrocarbons, melting points and boiling points increase with increased molar mass.

Alkyl Groups

Alkyl groups result when a hydrogen atom is removed from an alkane. The name is derived from the name of the corresponding alkane by removing the *-ane* ending and replacing it with *-yl*. For example, a two carbon atom chain is called ethane. As a group with one fewer hydrogen and attached to a longer chain, it is called an ethyl group.

Carbon atoms are classified according to the number of carbon atoms to which they are attached. A **primary (1˚) carbon** is directly bonded to one other carbon. A **secondary (2˚) carbon** is bonded to two other carbon atoms. A **tertiary (3˚) carbon** is bonded to three other carbon atoms. A **quaternary (4°) carbon** is bonded to four other carbon atoms. Similarly, Using this classification scheme, alkyl groups are also designated as primary, secondary, or tertiary based on the number of carbon atoms attached to the carbon atom that joins the alkyl group to a molecule.

Alkyl groups may linear or branched. As mentioned above, the linear alkyl groups are named based on the corresponding continuous-chain alkane. The branched-chain alkyl groups are named as shown below:

$H_3CCH—$ (with H_3C on CH)	$H_3CHCH_2C—$ (with H_3C on CH)	$H_3CH_2CCH—$ (with H_3C on CH)	$H_3C—C—$ (with CH_3 above and CH_3 below)
Isopropyl	Isobutyl	*sec*-Butyl	*t*-Butyl or *tert*-Butyl

Nomenclature

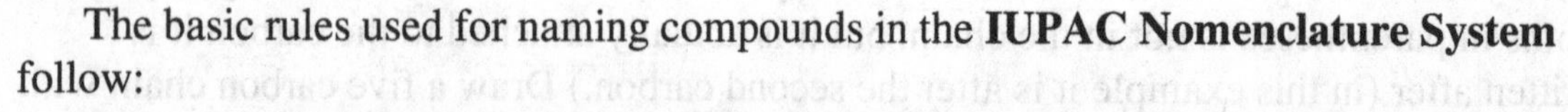

The basic rules used for naming compounds in the **IUPAC Nomenclature System** follow:

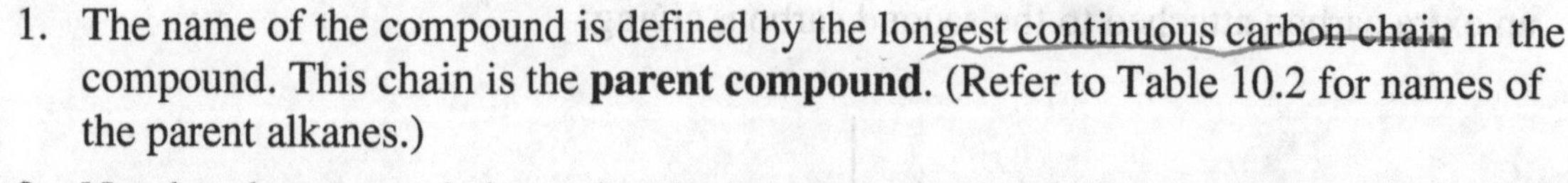

1. The name of the compound is defined by the longest continuous carbon chain in the compound. This chain is the **parent compound**. (Refer to Table 10.2 for names of the parent alkanes.)
2. Number the parent chain to give the lowest number to the carbon bonded to the first group encountered.
3. Name and number each atom (other than H) or group attached to the parent compound. The number tells you the position of the group on the main chain, and the

name tells you what type of substituent is present at that position. Halogens, as groups are named with an *–o* suffix (for example: bromine becomes bromo-), and alkyl groups are named as described above.

4. If the same substituent occurs more than once in the compound, a separate position number is supplied for each substituent, and the prefixes *di-*, *tri-*, *tetra-*, *penta-*, *hexa-*, *hepta-*, and so forth are used.
5. Place the names of the substituents in alphabetical order before the name of the parent compound. Numbers are separated from each other by commas, and numbers are separated from names by hyphens. By convention halogen substituents are placed before alkyl substituents in this priority sequence regardless of the alphabetization.

Example 7: Nomenclature

Give the IUPAC names of the following branched-chain alkanes:

1. $CH_3CHCH_2CH_3$ with a CH_3 branch on C-2

2. $CH_3CH_2\,C\,CH_2CHCH_3$ with CH_3 branches on C and CH, and Br on C

Solution:

1. The parent chain is four carbons long giving the name butane.

   Number the chain from the left-hand side of the equation because the methyl group is closest to the left-hand side. The methyl group is attached to the second carbon giving us, according to the rules, 2-methylbutane. However, this is a unique case because this is the only location that the methyl group can be attached; so no number is needed and this compound has the name methylbutane.

   Prove it to yourself: Draw a four carbon chain and attached the methyl group to each of the different carbons. Is it still a butane if it is attached to the first or last carbon? No, it is a pentane. If it is attached to the third carbon, you would just switch numbering from the opposite end and it is still on the second carbon.
2. The longest chain is six carbons long giving the parent chain of hexane.
   The methyl groups closest to one of the end carbons of the chain is on the right-hand side. Therefore, the chain gets numbered right to left. Since methyl groups are attached at carbons 2 and 4 and a bromo group is attached at carbon 4, the name is 4-bromo-2,4-dimethylhexane.

Example 8: Nomenclature

Give the IUPAC name of the following:

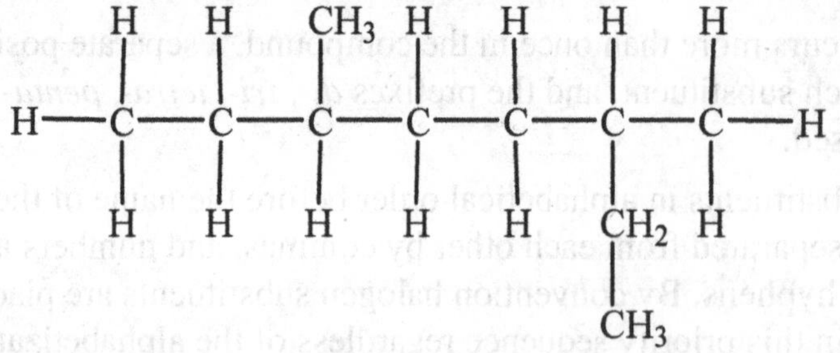

```
     H    H    CH3  H    H    H    H
     |    |    |    |    |    |    |
H—   C —  C —  C —  C —  C —  C —  C —H
     |    |    |    |    |    |    |
     H    H    H    H    H    CH2  H
                              |
                              CH3
```

Solution:

Be careful. What is the longest carbon chain? The chain can be bent and in this case it is. The longest chain is eight carbons long giving the parent name octane.

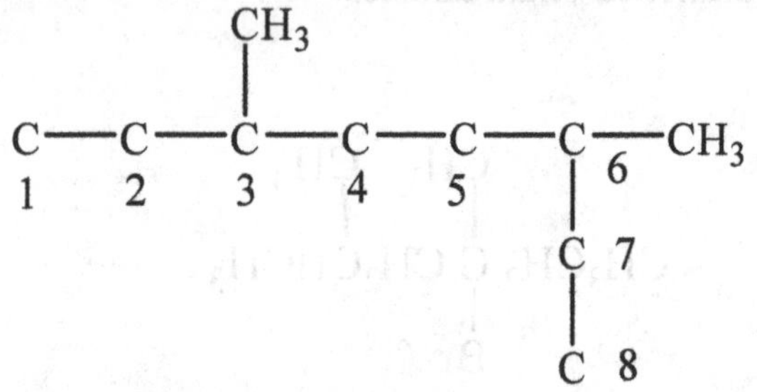

```
              CH3
              |
C —— C —— C —— C —— C —— C —— CH3
1    2    3    4    5    | 6
                         C 7
                         |
                         C 8
```

Two methyl groups are attached giving 3,6-dimethyloctane.

Example 9: Writing the Structural Formula from the Name

From the IUPAC names, draw structures of the following alkane and branched-chain alkane.

1. Heptane
2. 2,4-Dimethyloctane

Solution:

1. Draw and number the parent carbon chain.

1.

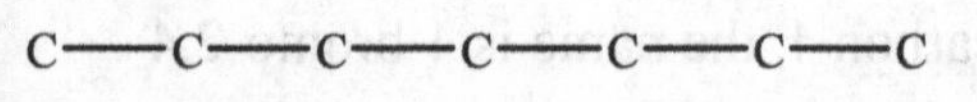

2.

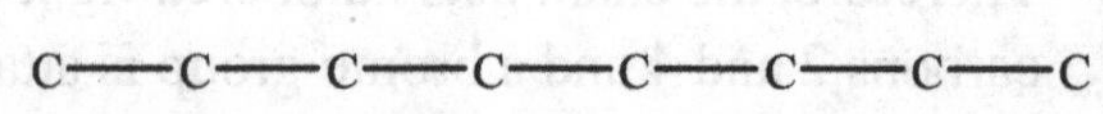

Example 9 Continued

2. Add the substituent groups to the correct carbon.

1.

```
C—C—C—C—C—C—C
1 2 3 4 5 6 7
```

2.

```
    C       C
    |       |
C — C — C — C — C — C — C — C
1   2   3   4   5   6   7   8
```

3. Add the correct number of hydrogen atoms.

1.

```
  H  H  H  H  H  H  H
  |  |  |  |  |  |  |
H—C——C——C——C——C——C——C—H
  |  |  |  |  |  |  |
  H  H  H  H  H  H  H
```

2.

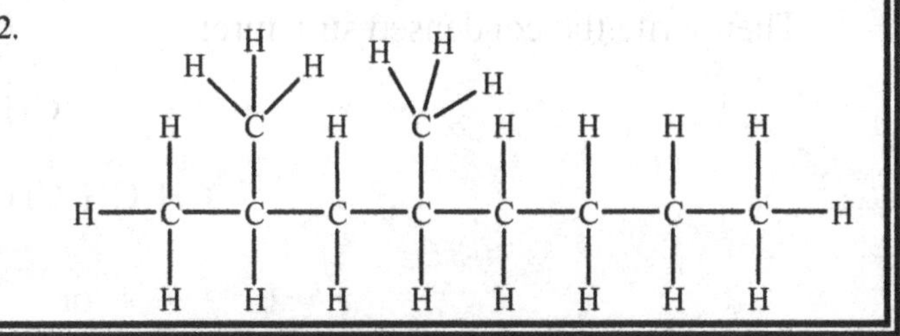

Constitutional or Structural Isomers

Two molecules having the same molecular formulas but different structures are called **constitutional** or **structural isomers**. Isomers having the same molecular formula are unique compounds due to their structural differences. They may have similar physical and chemical properties, but in many cases their properties are quite dissimilar.

Example 10: Isomers

Draw and give the IUPAC names of all the alkane isomers of C_5H_{12}.

Solution:

1. The first isomer you should always draw is the one that has all the carbons in a continuous chain structure:

$CH_3CH_2CH_2CH_2CH_3$ IUPAC name = pentane

2. Next draw four carbons in a continuous chain, and connect the fifth carbon as a methyl group to this continuous chain:

```
C— C— C— C
      |
      CH3
```

Example 10 Continued

Fill in the structure using hydrogen atoms, so that each carbon atom has four bonds.

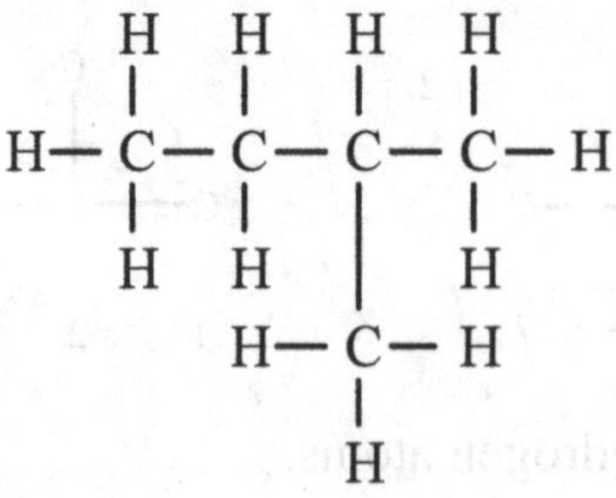

```
      H   H   H   H
      |   |   |   |
  H — C — C — C — C — H
      |   |   |   |
      H   H   |   H
              |
          H — C — H
              |
              H
```

Then write the condensed structure:

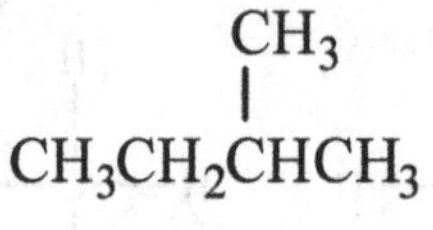

```
           CH3
           |
    CH3CH2CHCH3
```

or

$CH_3CH_2CH(CH_3)CH_3$

This compound is named methylbutane.

3. The last isomer of C_5H_{12} is found by drawing three carbons in a continuous chain and connecting the other two carbons as methyl groups to the middle carbon of the chain.

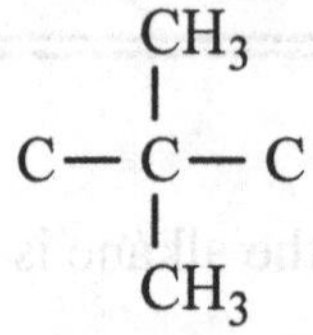

```
        CH3
        |
    C — C — C
        |
        CH3
```

Fill in the structure using hydrogen atoms as before:

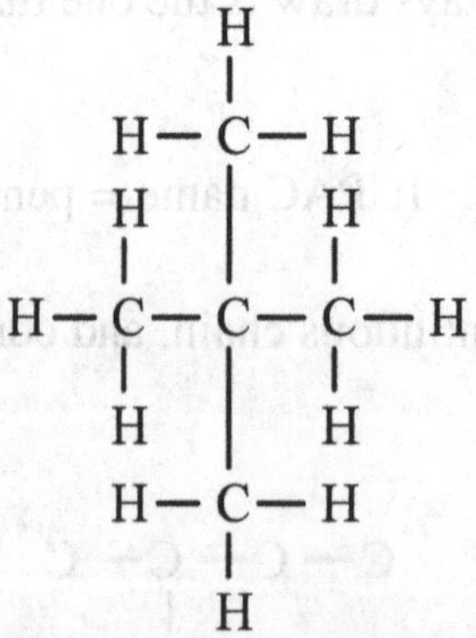

```
              H
              |
          H — C — H
          H   |   H
          |   |   |
      H — C — C — C — H
          |   |   |
          H   |   H
          H — C — H
              |
              H
```

Example 10 Continued

Then write the condensed structure:

$$\begin{array}{c} CH_3 \\ | \\ CH_3CCH_3 \\ | \\ CH_3 \end{array}$$

or

$$CH_3C(CH_3)_2CH_3$$

This compound is named dimethylpropane.

Remember that all isomers must have different IUPAC names. To check whether you have drawn duplicate isomers, name them using the IUPAC nomenclature system.

10.4 Cycloalkanes

The **cycloalkanes** are another family of hydrocarbons containing only C-C single bonds and are closely related to the alkanes. Cycloalkanes have the general molecular formula C_nH_{2n}. Note that they contain two fewer hydrogens than the corresponding alkane.

Cycloalkanes are named by adding the prefix *cyclo-* before the name of the alkane with the same number of carbon atoms. For example, cyclobutane is a cyclic alkane that has four carbon atoms. Substituted cycloalkanes are named by placing the name of the substituent before the name of the cycloalkane. No number is needed if only a single substituent is present. If more than one substituent is present, then the numbers that result in the lowest possible position numbers for the substituents are used.

cis-trans Isomerism in Cycloalkanes

Rotation around the bonds in a cyclic structure is limited by the fact that the carbons of the ring are all bonded to other ring carbons. As a result, **geometric isomers**, or ***cis-trans*** **isomers**, are produced. In the *cis-* isomer, two substituents are on the same side of the ring (either above or below). In the *trans-* isomer, the substituents are on opposite sides of the ring. See Figure 10.2.

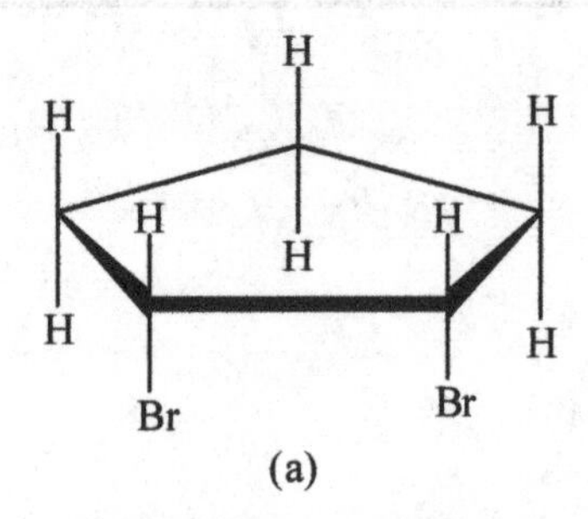

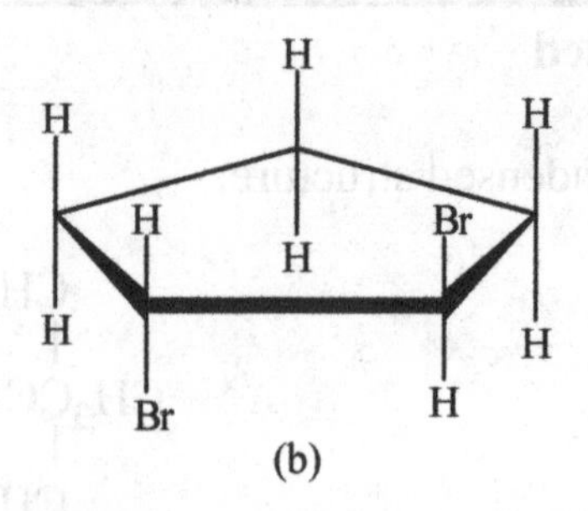

Figure 10.2 Both compounds represent 1,2-dibromocyclopentane. However, the three-dimensional location of the bromine atoms for the two compounds is different. Compound (a) is *cis*-1,2-dibromocyclopentane and compound (b) is *trans*-1,2-dibromocyclopentane.

Geometric isomers are a type of **stereoisomer**, molecules that have the same structural formulas and bonding patterns but different arrangements of atoms in space.

Recall that it is possible to look at molecules from different angles. The two 1,2-dibromocyclopentane structures shown in Figure 10.2 could also be drawn as:

Br Br	Br Br
cis-1,2-dibromocyclopentane	*trans*-1,2-dibromocyclopentane

We can still see that the left structure is *cis*, since both bromine atoms are dashed wedges and thus pointed away, and on the same side of the ring. With the *trans*- structure on the right, one bromine is pointed away (the dashed wedge) and one bromine is pointed towards us (the solid wedge) and thus are on opposite sides of the ring.

10.5 Conformations of Alkanes and Cycloalkanes

For alkanes, there is free rotation around the carbon-carbon single bond, and at room temperature, rotations occur on the order of millions of times per second. Thus, these molecules exist in an infinite variety of arrangements that are rapidly interconverting. These varying arrangements, capable of interconversion by simple rotation about a bond, are called **conformations** or **conformers.**

Alkanes

All the C-C bonds on alkanes can freely rotate 360° around the bond and thus there are an infinite number of possible conformations. These conformations are merely different forms of the same molecule and can not be separated from each other because they are rapidly interconverting. Two important conformations to consider are the staggered conformation, where the bonding electrons are farthest from each other, and the eclipsed conformation where the bonding electrons are closest to each other. As electrons move closer to each other there is increased repulsion between the electrons, and as a result, the staggered conformations are more stable than the eclipsed conformation. See Figure 10.6 in your textbook to view the staggered and eclipsed conformations of ethane and butane.

Cycloalkanes

LG 13

All cycloalkanes, except cyclopropane, also exist in different conformations. The **chair conformation**, which resembles a lawn chair, is the most stable (energetically favored) conformer for six-member ring structures since all the hydrogen atoms are staggered and the bond angle between carbons is 109.5°. The **boat conformation** resembles a rowboat and is less stable since the hydrogen atoms are more crowded. The most stable (energetically favorable) conformations of alkanes and cycloalkanes are those in which the hydrogen atoms are least crowded. See Figure 10.3.

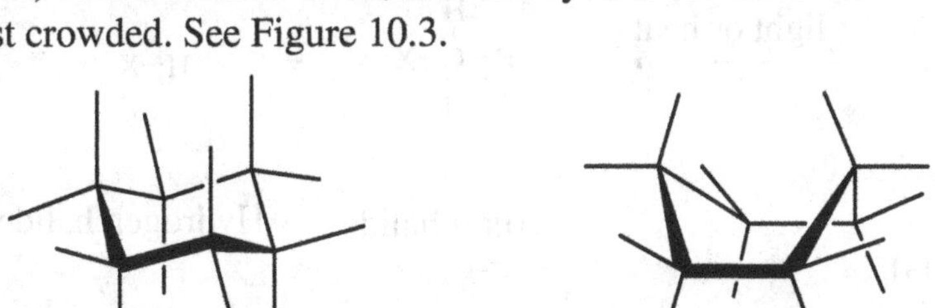

Figure 10.3 (a) The chair conformation of cyclohexane. (b) The less stable boat conformation of cyclohexane.

10.6 Reactions of Alkanes and Cycloalkanes

LG 14

Combustion

Alkanes and other hydrocarbons may be oxidized (burned in air), producing carbon dioxide and water; this reaction is called **combustion**. During combustion, a large amount of heat energy is released.

Example 11: Combustion Reaction of Alkanes
Show the complete balanced equation for the combustion of octane.

Solution:

1. Octane: The *-ane* ending tells us that the compound is an alkane. The *oct-* means that it has eight carbon atoms. By using the general molecular formula C_nH_{2n+2}, we can derive the molecular formula of octane. Thus, the molecular formula is C_8H_{18}.
2. The complete combustion of an alkane requires O_2 to react with the alkane to produce $CO_2 + H_2O$.
3. $C_8H_{18} + O_2 \rightarrow CO_2 + H_2O$ This equation is not balanced.
4. The balanced equation for the complete combustion of octane is

$$2\,C_8H_{18} + 25\,O_2 \rightarrow 16\,CO_2 + 18\,H_2O$$

LG 15

Halogenation

Halogenation of an alkane or cycloalkane is a **substitution reaction** in which a halogen atom (usually bromine or chlorine) replaces a hydrogen atom. Typically, when an alkane reacts with certain halogens in the presence of heat and/or light, a substitution reaction results. One of the C—H bonds of the alkane is broken and replaced with a C–X bond in which a halogen atom (X = Br or Cl) has substituted for a hydrogen atom. The products of this reaction are an **alkyl halide** and a hydrogen halide. A general form of this reaction is shown below where R = an alkyl chain:

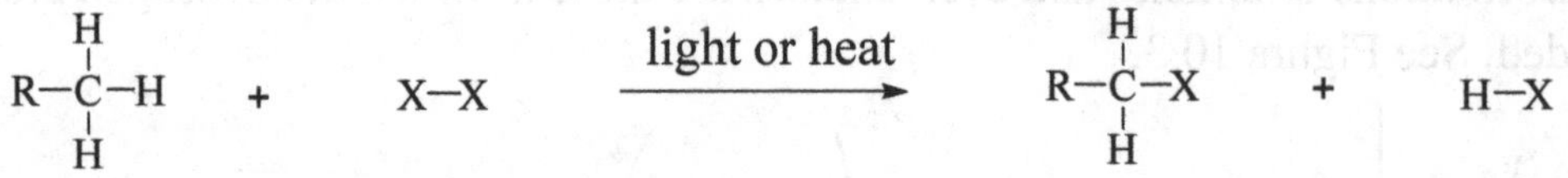

$$\underset{\text{Alkane}}{R{-}CH_2{-}H} + \underset{\text{Halogen } (Cl_2 \text{ or } Br_2)}{X{-}X} \xrightarrow{\text{light or heat}} \underset{\text{Alkyl halide}}{R{-}CH_2{-}X} + \underset{\text{Hydrogen halide}}{H{-}X}$$

In more complex alkanes, substitution can occur to some extent at all positions to give a mixture of products.

Example 12: Halogenation of Alkanes

Show the reaction for the stepwise substitution of chlorine atoms for the hydrogen atoms on methane; also, determine the IUPAC name for each organic compound produced.

Solution:

Replace one hydrogen with one chlorine.

$$CH_4 + Cl_2 \xrightarrow{UV} CH_3Cl + HCl$$

$$CH_3Cl + Cl_2 \xrightarrow{UV} CH_2Cl_2 + HCl$$

$$CH_2Cl_2 + Cl_2 \xrightarrow{UV} CHCl_3 + HCl$$

$$CHCl_3 + Cl_2 \xrightarrow{UV} CCl_4 + HCl$$

The products of these reactions are chloromethane, dichloromethane, trichloromethane, and tetrachloromethane (more commonly called carbon tetrachloride according to the inorganic nomenclature rules), respectively.

Self Test for Chapter Ten

1. Which of the following is not a normal property of an inorganic compound?

   a. flammable
   b. high boiling point
   c. soluble in water
   d. reactions are fast
   e. high melting point

2. Name the family whose simplest member is benzene.
3. Compounds that contain only carbon and hydrogen atoms are classified in what major subdivision?

   a. alkanes
   b. aliphatics
   c. aromatics
   d. heterocyclic
   e. hydrocarbons

4. What is the term used to describe the group in an organic compound that is primarily responsible for the chemical and physical properties of that compound?
5. Name the functional group found in all alkenes.
6. Name the functional group found in all alcohols.
7. Give the molecular formula of the alkane that contains nine carbon atoms.
8. Write the molecular, condensed and structural formula for the following compound:

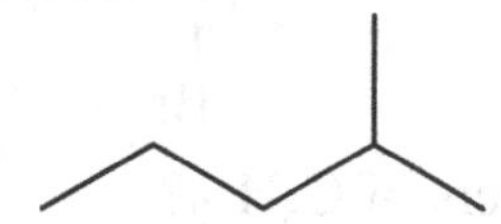

9. What is the geometry of a carbon surrounded by four single bonds?
10. Are the melting and boiling points of alkanes generally higher or lower than those of other organic compounds containing functional groups?
11. Describe the solubility of alkanes in water.
12. Give the IUPAC names of the following branched-chain alkanes:

a. $$\begin{array}{ccc} CH_3 & CH & CH_2CH_3 \\ & | & \\ & CH_3 & \end{array}$$

b. $$\begin{array}{ccc} & CH_3 & \\ & | & \\ CH_3 & C & CH_2CH_2CH_3 \\ & | & \\ & CH_3 & \end{array}$$

c. $$\begin{array}{ccccccc} & & CH_3 & & CH_3 & & \\ & & | & & | & & \\ CH_3 & - & C & - & CH & - & CH_2CH_3 \\ & & | & & & & \\ & & CH_3 & & & & \end{array}$$

13. Give the IUPAC name of the following:

$$\begin{array}{ccccc} & H & & CH_3 & \\ & | & & | & \\ CH_3 & C & CH_2 & C & CH_2Br \\ & | & & | & \\ & Cl & & H & \end{array}$$

14. Give the IUPAC name of the following:

$$
\begin{array}{c}
CH_3 \\
| \\
CH_3CCH_3 \\
| \\
Br
\end{array}
$$

15. Give the IUPAC name of the following:

$$
\begin{array}{c}
CH_2CH_3 \\
| \\
CH_3-C-CH_3 \\
| \\
CH_2CHCH_3 \\
\quad | \\
\quad Br
\end{array}
$$

16. In the IUPAC system of naming compounds, the name of the compound is defined by the longest continuous carbon chain in the compound. What do we call this continuous chain?
17. What is the IUPAC name of $CHCl_3$?
18. How many primary, secondary and tertiary carbons are in the following compound?

$$
\begin{array}{c}
CH_2CH_3 \\
| \\
CH_3-C-CH_3 \\
| \\
CH_2CHCH_3 \\
\quad | \\
\quad Br
\end{array}
$$

19. Which of the following are isomers of C_5H_{12}?
   a. $CH_3CH_2CH_2CH_3$
   b. $CH_3CH_2CH_2CH_2CH_3$
   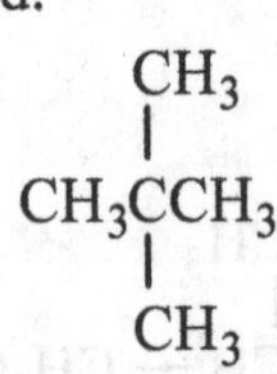
   c. $CH_3CH(CH_3)CH_2CH_3$
   d.

$$
\begin{array}{c}
CH_3 \\
| \\
CH_3CCH_3 \\
| \\
CH_3
\end{array}
$$

   e. $CH_3CH(CH_3)_2$

20. Which of the following is an isomer of 1,2-dibromoethane?
   a. Br_2CHCH_3
   b. CH_2BrCH_2Br
   c. $BrCH_2CH_2CH_3$

21. Which of the following are isomers of C_4H_{10}?
   a. benzene
   b. cyclobutane
   d. $CH_3CHBrCH_2CH_3$
   e. butane

c.

$$\begin{array}{l} CH_3CHCH_3 \\ \quad\;\; | \\ \quad\; CH_3 \end{array}$$

22. Which of the following is an isomer of CH_3OCH_3?

a. CH_3CH_2OH

b. methylbutane

c. $CH_3CH_2OCH_3$

d. oxymethylurea

e. dimethylether

23. What does the prefix *cyclo-* mean?

24. Are the following compounds constitutional isomers or stereoisomers?

25. Name each compound in question 24.

26. Why is the staggered conformation of an alkane more stable than the eclipsed conformation?

27. Is the following conformation a chair or boat conformation of methylcyclohexane?

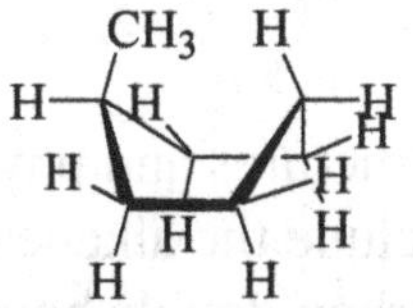

28. What is the term for the oxidation of alkanes and other hydrocarbons to carbon dioxide and water?

29. How many moles of water are produced by the complete combustion of one mole of methane?

30. Write a balanced equation representing the complete combustion of propane.

31. Write the names of the monosubstitution products that are produced when Cl_2 and propane react under UV light.

32. For methane to react with chlorine, what else is needed?

33. What are the structural formula, condensed formula and line formula of the missing alkane reactant in the following chemical equation?

$$\_\_\_\_\_\_\_\_ + Cl_2 \xrightarrow{UV} CH_3CH_2CH_2(Cl) + HCl$$

11 The Unsaturated Hydrocarbons: Alkenes, Alkynes, and Aromatics

Learning Goals

1. Describe the physical properties of alkenes and alkynes.
2. Draw the structures and write the IUPAC names of simple alkenes and alkynes.
3. Write the names and draw the structures of simple geometric isomers of alkenes.
4. Write equations predicting the products of the addition reactions of alkenes and alkynes: hydrogenation, halogenation, hydration, and hydrohalogenation.
5. Apply Markovnikov's rule to predict the major and minor products of the hydration and hydrohalogenation reactions of unsymmetrical alkenes.
6. Write equations representing the formation of addition polymers of alkenes.
7. Draw the structures and write the names of common aromatic hydrocarbons.
8. Write equations for substitution reactions involving benzene.
9. Describe heterocyclic aromatic compounds, and list several biological molecules in which they are found.

Introduction

In this chapter, we will study the structures and physical and chemical properties of the **unsaturated hydrocarbons,** which include the alkenes, alkynes, and aromatic compounds. **Alkenes** contain at least one carbon-carbon double bond, and **alkynes** contain at least one carbon-carbon triple bond. The **aromatic compounds** that we will study contain a benzene ring.

11.1 Alkenes and Alkynes: Structure and Physical Properties

As a result of multiple bonding, alkenes, alkynes and aromatic compounds contain fewer hydrogen atoms than alkanes with the same number of carbon atoms. They are referred to as unsaturated because they do not contain as many hydrogen atoms as their carbon skeleton will allow. The structures seen on the next page reveal the structural differences among alkanes, alkenes, and alkynes.

Each of the compounds in the diagram has the same number of carbon atoms but differ in the number of hydrogen atoms. Alkynes have the general formula C_nH_{2n-2} and thus contain two fewer hydrogens than the corresponding alkene (general formula C_nH_{2n}). Alkenes have two fewer hydrogens than the corresponding alkane (general formula C_nH_{2n+2}).

Alkanes contain only single bonds; alkenes have at least one carbon-to-carbon double bond; and alkynes contain at least one carbon-to-carbon triple bond. The differences in bond order result in variation in molecular geometry and chemical reactivity among these three families. Alkanes have tetrahedral carbon atoms. When carbon is bonded by one double bond and two single bonds, as in alkenes, the molecule is trigonal planar, and each bond angle is 120°. When two carbons are bonded by a triple bond, the molecule is linear, and the bond angles are 180°.

	Ethane (Ethane)	Ethene (Ethylene)	Ethyne (Acetylene)
Structural Formulas	$H_3C{-}CH_3$ (H–C–C–H with three H on each C)	$H_2C{=}CH_2$ (H₂C=CH₂ with two H on each C)	$H{-}C{\equiv}C{-}H$
Molecular Formulas	C_2H_6	C_2H_4	C_2H_2
Condensed Formulas	CH_3CH_3	$CH_2{=}CH_2$	$HC{\equiv}CH$

Like alkanes, alkenes and alkynes are nonpolar. As a result they are not water soluble but are readily soluble in nonpolar solvents and in many low-polarity organic solvents such as ether or chloroform. They are also less dense than water and have relatively low boiling points. The longer, higher molar mass alkenes and alkynes exhibit stronger London dispersion forces and therefore higher boiling and melting points.

11.2 Alkenes and Alkynes: Nomenclature

LG 2

The nomenclature of alkenes and alkynes is analogous to that of the alkanes, with the following exceptions. For alkenes, the parent name is derived from the longest continuous carbon chain containing the double bond. Then the *-ane* ending of the alkane is replaced with the *-ene* ending of an alkene. The chain is numbered to give the lowest numbers for the two carbons containing the double bond.

Alkynes are named in the same way as the alkenes, except that the *-ane* ending of the corresponding alkane is replaced with the *-yne* ending of alkynes. The rules used in numbering the alkene chain are also used in alkyne nomenclature.

Example 1: Structure and Nomenclature of Alkanes, Alkenes, and Alkynes
Name and draw the condensed formula for the alkane, alkene, and alkyne that contain three carbon atoms.

Example 1 Solution:

1. To find the molecular formula of an alkane, use the general formula C_nH_{2n+2}. The alkane with three carbon atoms has the molecular formula C_3H_8:

   $CH_3CH_2CH_3$ IUPAC name: propane

2. To find the molecular formula of an alkene, use the general formula C_nH_{2n}. The alkene with three carbon atoms has the molecular formula C_3H6:

   $CH_3CH{=}CH_2$ IUPAC name: propene

3. To find the molecular formula of an alkyne, use the general formula C_nH_{2n-2}. The alkyne with three carbon atoms has the molecular formula C_3H_4:

   $CH_3C{\equiv}CH$ IUPAC name: propyne

Example 2: Alkene Nomenclature

Give the IUPAC name of the following molecule:

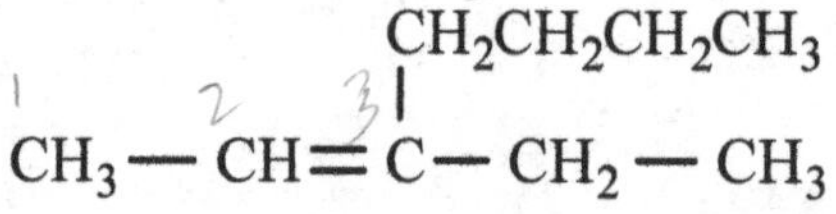

```
                 CH2CH2CH2CH3
                 |
CH3 — CH = C — CH2 — CH3
```

Solution:

1. First determine the parent compound (longest continuous chain) that contains the double bond. The parent compound is heptene. Since the *hept-* means seven carbons and the *-ene* means double bond, we have

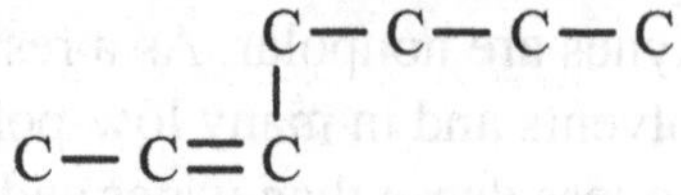

```
          C — C — C — C
          |
C — C = C
```

2. Next, label the parent to give the double bond carbons the lowest possible numbers:

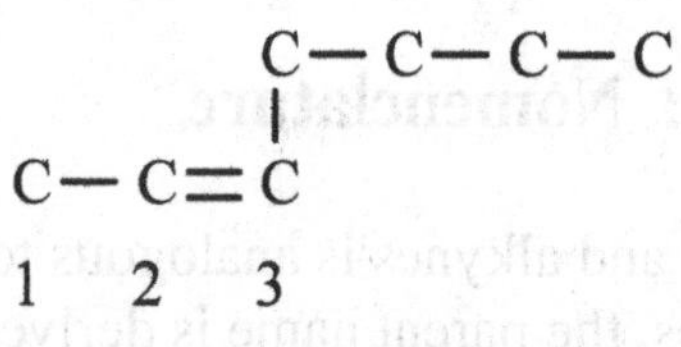

```
          4   5   6   7
          C — C — C — C
          |
C — C = C
1   2   3
```

2-heptene; the 2 tells us that the double bond is between carbons 2 and 3.

3. Finally, list the attached groups, as in naming alkanes.

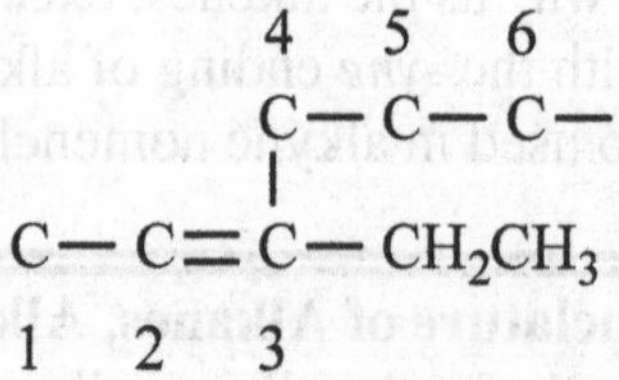

```
          4   5   6   7
          C — C — C — C
          |
C — C = C — CH2CH3
1   2   3
```

Giving, 3-ethyl-2-heptene

Example 3: Alkyne Nomenclature

Determine the IUPAC name of the following molecule:

$$CH_3-\underset{\displaystyle CH_3}{\underset{|}{CH}}-CH_2-C\equiv C-CH_3$$

Solution

1. First determine the parent compound that contains the triple bond. The parent compound is hexyne. The *hex-* means six carbons, and the *-yne* means a triple bond.

2. Label the compound so that the triple bond carbons have the lowest numbers possible.

$$\underset{6}{C}-\underset{5}{C}-\underset{4}{C}-\underset{3}{C}\equiv\underset{2}{C}-\underset{1}{C}$$

Thus, it is 2-hexyne.

3. Finally, list the attached groups as before:

$$C-\underset{\displaystyle CH_3}{\underset{|}{C}}-C-C\equiv C-C$$

Giving, 5-methyl-2-hexyne

Example 4: Determining the Structure of an Alkene and Alkyne Given the Name

Draw each of the following using structural formulas:

1. 1-Bromo-2-hexyne
2. 2-Butene

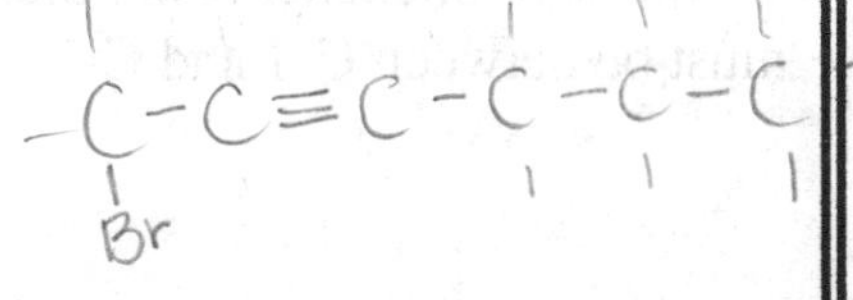

Solution:

1. 1-Bromo-2-hexyne: there are 6 carbons (*hex-*) in the parent compound and a triple bond between carbons 2 and 3 (*2-yne*). Also, there is a bromine (Br–) bonded to carbon 1.

$$\overset{\displaystyle Br}{\overset{|}{C}}-C\equiv C-C-C-C$$

Next, fill in the other bonds with hydrogen atoms.

$$H-\underset{\displaystyle H}{\underset{|}{\overset{\displaystyle Br}{\overset{|}{C}}}}-C\equiv C-\underset{\displaystyle H}{\underset{|}{\overset{\displaystyle H}{\overset{|}{C}}}}-\underset{\displaystyle H}{\underset{|}{\overset{\displaystyle H}{\overset{|}{C}}}}-\underset{\displaystyle H}{\underset{|}{\overset{\displaystyle H}{\overset{|}{C}}}}-H \quad \text{or} \quad BrCH_2-C\equiv C-CH_2CH_2CH_3$$

Example 4 Continued

2. 2-Butene: There are four carbons [*but-*] in the parent compound and a double bond between carbons 2 and 3.

$$C-C{=}C-C$$

Next, fill in the other bonds with hydrogen atoms.

$$\begin{array}{c} \quad H \quad H \quad H \quad H \\ H-\underset{H}{C}-C{=}C-\underset{H}{C}-H \end{array} \quad \text{or} \quad CH_3CH{=}CHCH_3$$

Alkenes and alkynes may have more than one double or triple bond. For example, an alkene with two double bonds is called an alkadiene. To name, identify the number for both double bonds and use the suffix, *-diene*. For example, 2,6-octadiene has the condensed structure, and line structure below.

$CH_3CH{=}CHCH_2CH_2CH{=}CHCH_3$

Cycloalkenes are numbered to give the carbon atoms on either side of the double bond C-1 and C-2. If there are groups attached to the ring, number in the direction that gives the group the lowest possible number. For example the following structure is 1-fluorocyclohexene. Note: there is no need to number where the double bond is located. It must be between C-1 and C-2.

F

11.3 Geometric Isomers: A Consequence of Unsaturation

The carbon-carbon double bond is rigid because there is no free rotation around the double bond. The rigidity of the carbon-carbon double bond in alkenes produces another class of isomers: **geometric isomers**. Similar to the discussion for cycloalkanes in section 10.4, geometric isomers are described using the prefixes *cis-* and *trans-*, which provide an easy method for naming and distinguishing between the two isomeric forms.

Remember, geometric isomers are different molecules with different physical and chemical properties. Although their properties may be similar, they are never identical.

The prefixes *cis-* and *trans-* refer to the placement of the substituents attached to the carbon-carbon double bond. When identical groups are on the same side of the double bond, the prefix *cis-* is used; when identical groups are on opposite sides of the double bond, *trans-* is the appropriate prefix.

Not every double bond will have *cis-trans* isomers. To have *cis-trans* isomerism, both carbons must have two different groups attached. For example, CH_2=CHBr, cannot have *cis-trans* isomerism. Notice that the carbon on the left-hand side has two hydrogen atoms attached. An attempt to draw the *cis-* and *trans-* isomers is shown below; notice that if you simply flip the structure on the left over, you will have the structure on the right and therefore these are not isomers.

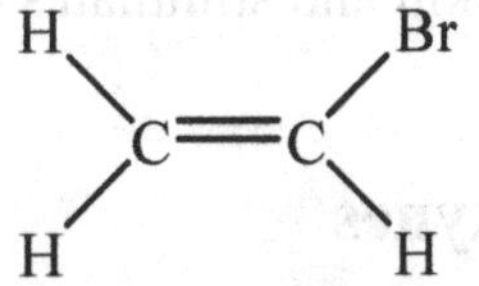

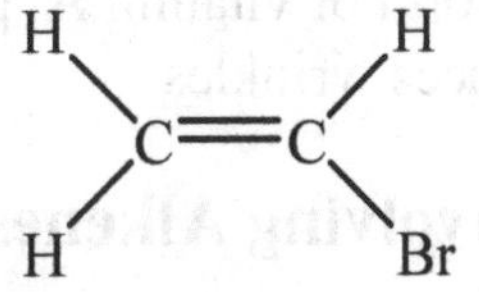

H Br C=C H H   H H C=C H Br

Example 5: Drawing Geometric Isomers of Alkenes.
Draw the structural formula for the *cis-* and *trans-* isomers of 2-butene.

Solution:

1. 2-butene has four carbon atoms with the double bond between the 2nd and 3rd carbon atom. The condensed structural formula is CH_3CH=$CHCH_3$. This condensed structure does not show the *cis-* and *trans-* isomers.

2. Examine the double bond to see what is attached to the carbon atoms on each side. In this case, there is a methyl group and a hydrogen atom on each carbon.

3. Draw the double bond with bonds to connect the attached atoms:

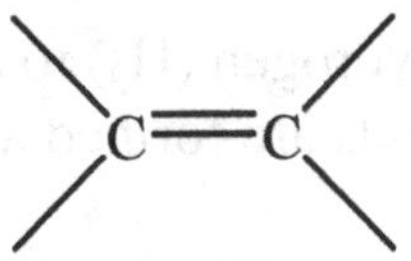

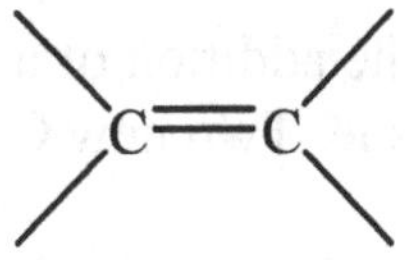

C=C   C=C

4. For the *cis-* arrangement, the methyl groups will be on the same side. For the *trans-* arrangement, the methyl groups will be on opposite sides of the formula.

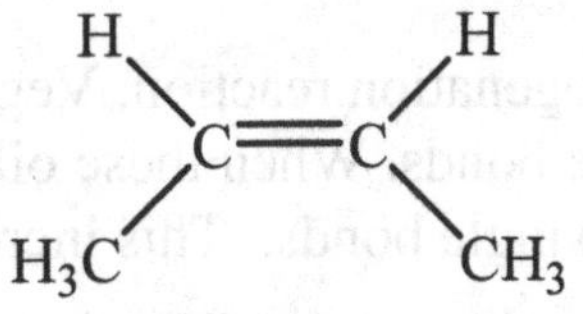

H H C=C H_3C CH_3 H CH_3 C=C H_3C H

cis-2-butene 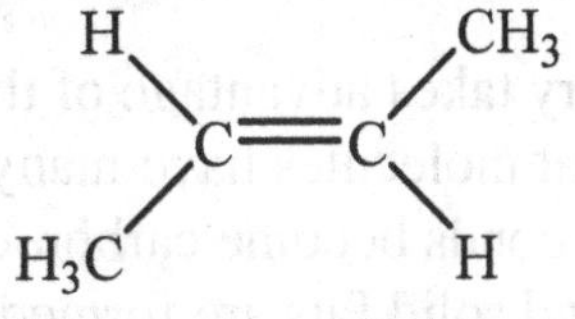   *trans*-2-butene

11.4 Alkenes in Nature

Ethene, the simplest alkene, is a plant growth substance produced by ripening fruit in areas of a plant where cell division is occurring.

There are numerous polyenes in nature. These are alkenes with multiple double bonds. The basic unit of these molecules, which include steroids, chlorophyll, and the lipid soluble vitamins A, D, E, and K, is the *isoprene* unit:

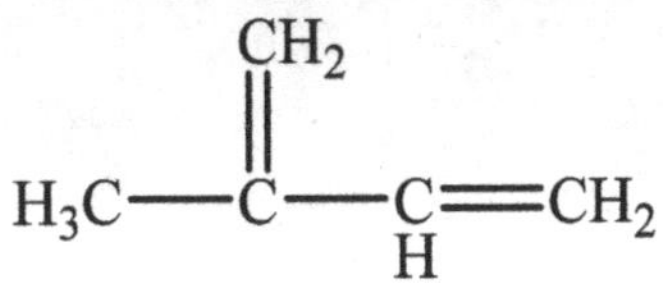

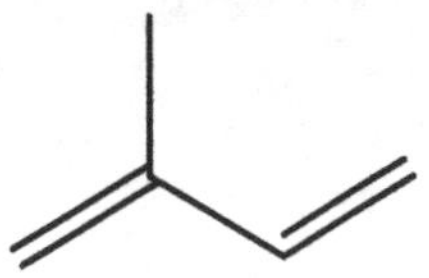

CH_2
H_3C—C—C=CH_2
H

These polyenes produced from isoprene are also called *isoprenoids* or *terpenes*. *Geraniol*, the scent of geraniums, is a natural insect repellant. *D-Limonene*, the aroma of oranges, is purified from citrus rinds. It is a useful solvent used in many cleaning products.
Myrcene is found in bayberry and has a spicy aroma used in perfumes and candles. *Farnesol*, found in many flowers, is used in skin care products marketed to increase skin elasticity. *Retinol*, a form of vitamin A, penetrates skin and stimulates collagen and elastin formation, which reduces wrinkles.

11.5 Reactions Involving Alkenes and Alkynes

LG 4

The major reactions of alkenes and alkynes involve the addition of atoms or molecules to the carbon-carbon double or triple bond. The principal kinds of **addition reactions** are hydrogenation, halogenation, hydration, and hydrohalogenation. A generalized addition reaction is shown below. The primary difference between any of the four types of addition reactions will be the identity of "A" and "B", otherwise, the reactions will follow the same pattern.

$$R_2C{=}CR_2 \; + \; A{-}B \longrightarrow R_2C(A){-}C(B)R_2$$

Hydrogenation: Addition of H_2

Hydrogenation is the addition of a molecule of hydrogen (H_2) to a carbon-carbon double bond to produce an alkane. Two new C—H single bonds are formed as the double bond is broken.

Hydrogenation generally requires heat and/or pressure. The reaction always requires a metal catalyst, such as platinum or nickel, to allow the reaction to occur at a reasonably rapid rate.

The food industry takes advantage of the hydrogenation reaction. Vegetable oils are liquid because the fat molecules have many double bonds. When these oils are hydrogenated, many of the double bonds become carbon-carbon single bonds. This increases the rigidity of the fat molecules, and solid fats are formed.

Hydrogenation of an alkyne to produce an alkane requires two moles of H_2 for every mole of alkyne.

Example 6: Writing a Hydrogenation Reaction
Write an equation representing the hydrogenation reaction for 2-butene. Indicate the conditions needed, and name all reactants and products.

Example 6 Solution:
The term *hydrogenation* means the reaction of gaseous H_2 with a specific reactant. A catalyst such as Pt, Pd, or Ni is needed to speed up the reaction. Elevated temperature and pressure are also required.

$$CH_3CH{=}CHCH_3 + H_2 \xrightarrow[\text{pressure}]{\text{Ni, heat}} CH_3CH_2CH_2CH_3$$

2-Butene → Butane

Note: the same product is formed whether the reactant is *cis*-2-butene or *trans*-2-butene.

Halogenation: Addition of X_2

Chlorine (Cl_2) or bromine (Br_2) can be added to a double bond. This reaction, called **halogenation**, proceeds readily and does not require a catalyst.

Example 7: Writing a Halogenation Reaction
Write an equation representing the halogenation of ethene to produce 1,2-dichloroethane.

Solution:
The term *halogenation* means the addition of a halogen molecule (X_2) across the double bond of an alkene. The double bond is broken and two new C–X bonds are formed.

$$H_2C{=}CH_2 + Cl{-}Cl \longrightarrow H{-}\underset{H}{\overset{H}{C}}{-}Cl \;\; / \;\; H{-}C{-}Cl$$

$$CH_2{=}CH_2 + Cl_2 \longrightarrow ClCH_2{-}CH_2Cl$$

Ethene + Chlorine ⟶ 1,2-Dichloroethane

Alkynes also react with halogens. Two moles of halogen are required for every mole of alkyne, and the product is a tetrahaloalkane.

Halogenation involving bromine (Br_2) can be used to show the presence of double or triple bonds in an organic compound. A reaction mixture of an alkane and Br_2 and a reaction mixture of an alkene and Br_2 would both be red. However, only the reaction mixture with the alkene would lose the red color as the Br_2 is used up in the halogenation reaction. Not only does this test distinguish saturated from unsaturated hydrocarbons, it can provide a measure of the degree of saturation. One mole of Br_2 is consumed per mole of alkene, and two moles of Br_2 are consumed per mole of alkyne.

Hydration: Addition of H_2O

A **hydration** reaction is the addition of a water molecule to a carbon-carbon double bond. When an alkene is reacted with water containing a trace of strong acid, an –OH group bonds to one carbon of the carbon-carbon double bond, and an –H atom bonds to the other carbon. The product of the hydration of an alkene is an alcohol.

When an alkene is unsymmetrical (carries two different groups on the double bond carbons) hydration can yield two different products. One is usually favored over the other, as explained by **Markovnikov's rule,** which can be thought of as the "rich get richer". This means that the carbon of the carbon-carbon double bond that carries the greater number of hydrogen atoms most often will receive the hydrogen atom being added to the double bond. The other carbon becomes bonded to the –OH group.

Example 8: Writing a Hydration Reaction
Write an equation representing the hydration of ethene to produce ethanol (ethyl alcohol).

Solution:
The term *hydration* means the addition of a water molecule (HOH) across the double bond of an alkene. The –H bonds to one carbon of the double bond and the –OH attaches to the other carbon of the double bond. The double bond is broken, and in its place are two new single bonds.

$$H_2C{=}CH_2 + H{-}OH \xrightarrow{H^+} H{-}CH_2{-}CH_2{-}OH$$

Ethene + Water $\xrightarrow{\text{acid}}$ Ethanol

Note: Since both carbons on either side of the double bond are identical, there is only one product.

Example 9: Writing a Hydration Reaction
Write an equation representing the hydration of propene, and draw the *major* product.

Solution:
The hydration of an unsymmetrical alkene (like propene—having different groups on each side of the double bond) favors one product over the other.

Example 9 Continued

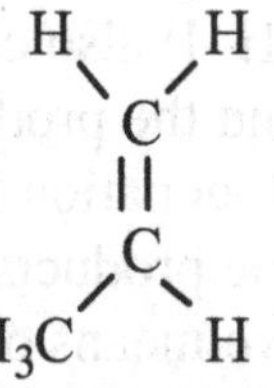

$$H_2C{=}CH{-}CH_3$$

The carbon atom of the double bond with the greater number of directly attached hydrogen atoms most often will receive the hydrogen atom being added to the double bond. The remaining carbon atom bonds to the –OH group. This is known as Markovnikov's rule.

Note that propene is unsymmetrical, since one carbon of the double bond has 2 –H atoms bonded to it while the other carbon of the double bond has only one –H atom and one $-CH_3$ group attached to it.

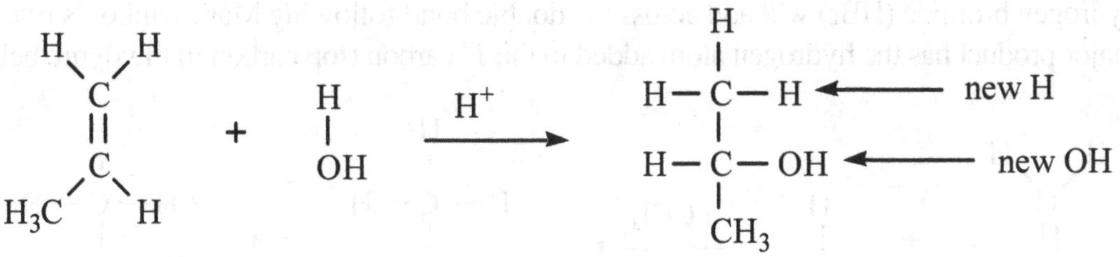

$$H_2C{=}CH{-}CH_3 + H{-}OH \xrightarrow{H^+} CH_3{-}CH(OH){-}CH_3$$

new H, new OH

Propene + Water $\xrightarrow{\text{acid}}$ 2-Propanol (major product)

Note: You will learn nomenclature of alcohols in Chapter 12.

Hydration of an alkyne produces an *enol* in which the –OH is bonded to one carbon of a carbon-carbon double bond and the –H to the other carbon. The enol is not stable and is quickly isomerized into an aldehyde or ketone as shown below.

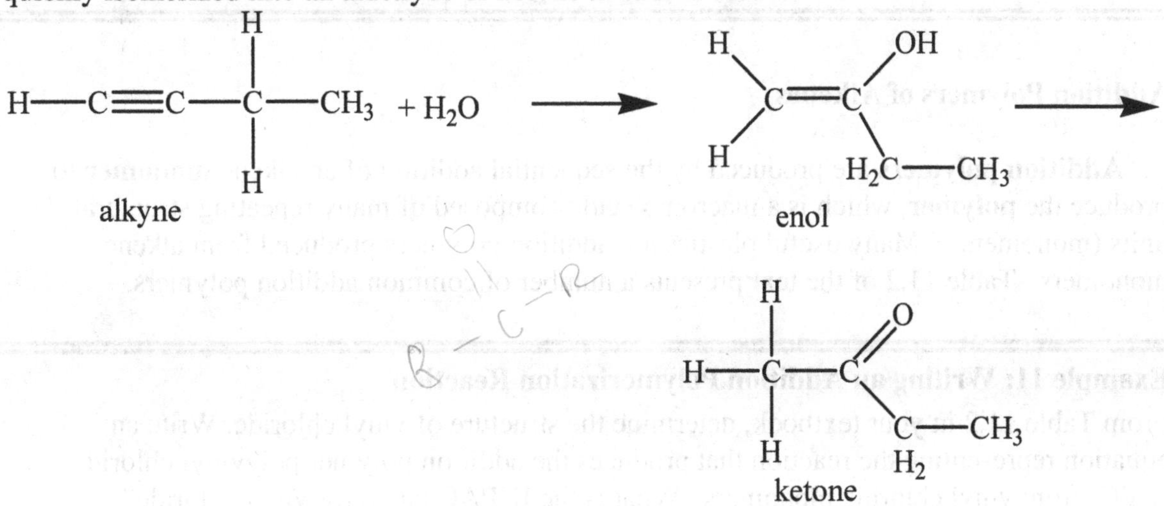

$$H{-}C{\equiv}C{-}CH_2{-}CH_3 + H_2O \longrightarrow H_2C{=}C(OH){-}CH_2{-}CH_3 \longrightarrow CH_3{-}C(=O){-}CH_2{-}CH_3$$

alkyne → enol → ketone

Hydrohalogenation: Addition of HX

A hydrogen halide (H–Br, H–Cl, or H–I) also can be added to an alkene. This addition reaction is called **hydrohalogenation,** and the product is an alkyl halide.

The reaction mechanism of hydrohalogenation is the same as that for hydration, and the nature of the major and minor alkyl halide products can be predicted by Markovnikov's rule. This reaction provides an easy, alternative means of preparing alkyl halides (recall that in Chapter 10, alkyl halides were prepared by the halogenation of alkanes in the presence of light or heat).

Example 10: Writing a Hydrohalogenation Reaction
Write an equation representing the reaction of hydrogen bromide with propene. Include the major and minor products.

Solution:
The hydrogen bromide (HBr) will add across the double bond following Markovnikov's rule. The major product has the hydrogen atom added to the 1st carbon (top carbon in the figure below.)

$$H_2C{=}CH{-}CH_3 + H{-}Br \xrightarrow{CCl_4} CH_3CH(Br)CH_3 + CH_3CH_2CH_2(Br)$$

Propene	Hydrogen bromide	2-Bromopropane (major product) or $CH_3CH(Br)CH_3$	1-Bromopropane (minor product) or $CH_3CH_2CH_2(Br)$

Addition Polymers of Alkenes

Addition polymers are produced by the sequential addition of an alkene **monomer** to produce the **polymer**, which is a macromolecule composed of many repeating structural units (monomers). Many useful plastics are addition polymers produced from alkene monomers. Table 11.2 of the text presents a number of common addition polymers.

Example 11: Writing an Addition Polymerization Reaction
From Table 11.2 in your textbook, determine the structure of vinyl chloride. Write an equation representing the reaction that produces the addition polymer polyvinyl chloride (PVC) from vinyl chloride monomers. What is the IUPAC name for vinyl chloride?

Example 11 Solution:

1. Begin by determining the structure of vinyl chloride (see Table 11.2):

$$\mathrm{H_2C{=}CHCl}$$

2. Break the double bond and link the monomers together at the carbons on either side of the previously existing double bond. Now determine the structure of the addition polymer:

$$\mathrm{H{-}CH_2{-}CHCl{-}CH_2{-}CHCl{-}CH_2{-}CHCl{-}}\ \text{etc.}$$

3. Using the simplified general form for the polymer, write the equation representing the reaction.

$$n\left[\mathrm{H_2C{=}CHCl}\right] \xrightarrow[\text{pressure}]{\text{heat}} \left[\mathrm{{-}CH_2{-}CHCl{-}}\right]_n$$

4. The IUPAC name for vinyl chloride is chloroethene.

11.6 Aromatic Hydrocarbons

Structure and Physical Properties

Aromatic compounds are all characterized by the presence of an aromatic ring within the structure. The simplest and most common aromatic compound is benzene, which consists of a six carbon ring in a planar hexagonal arrangement. Each carbon is bonded to two other carbon atoms and to a single hydrogen atom.

In 1865, Friedrich Kekulé proposed that single and double bonds alternated around the hexagonal ring. Since benzene did not decolorize bromine, he further proposed that the double and single bonds shifted position rapidly. Today, we show this structure as a resonance hybrid model.

$$C_6H_6 \longleftrightarrow C_6H_6$$

The current model of benzene structure proposes that each carbon atom is bonded to two other carbon atoms and to a hydrogen atom. The remaining six electrons are located in a cloud of electrons above and below the ring. Because the electrons are delocalized within the cloud, benzene is unusually stable and resists addition reactions typical of alkenes.

Nomenclature

Most simple aromatic compounds are named as derivatives of benzene. Others, for example phenol (a benzene ring with one hydroxyl group) and toluene (a benzene ring with one methyl group), have common historical names that must simply be memorized.

When there is only one substituent, no number is required. Simply name the group followed by benzene, as a single word. When there are two groups on the ring, there are three possible arrangements. Each arrangement is indicated by a prefix, as follows:

ortho, o: substituents are on adjacent carbon atoms (or carbons 1 and 2)

meta, m: substituents are on carbon atoms separated by one carbon atom (or carbons 1 and 3)

para, p: substituents are on carbon atoms separated by two carbon atoms (or carbons 1 and 4)

While the *ortho-, meta- and para-* prefixes are commonly used, they are not the IUPAC system. In the IUPAC system, numbers are used to indicate the location of the groups.

When more than two substituents are attached to the ring, a numerical system *must* be used in naming the compound.

Example 12: Naming Aromatic Compounds

Name the following compounds:

1. Cl

2.

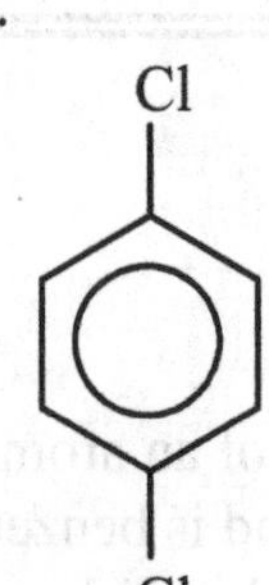

3.

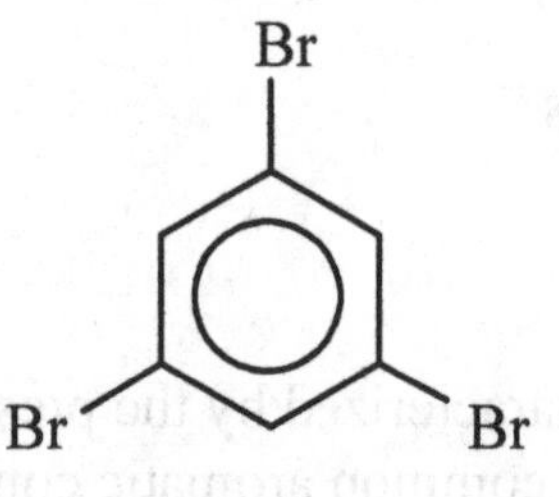

4.

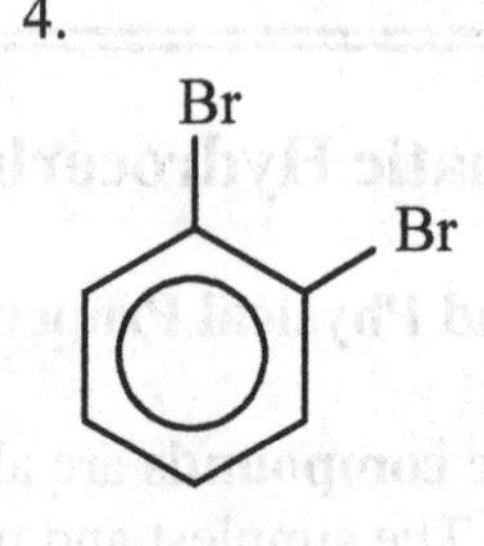

Solution:

1. There is one group attached to the benzene ring. Name the group first, and follow it with benzene to give the name chlorobenzene.
2. For two groups attached to the benzene ring, use the *ortho-*, *meta-* or *para-* prefix to designate the location of the groups. In this case, the chlorine atoms are separated by two carbon atoms giving *para*-dichlorobenzene, or 1,4-dichlorobenzene
3. With three substituents attached to the benzene ring, a numbering system is used to designate the location of the groups. This compound is 1,3,5-tribromobenzene.
4. The two bromo groups are on adjacent carbon atoms. This gives the *ortho-* arrangement: *ortho*-dibromobenzene, or 1,2-dibromobenzene.

In IUPAC nomenclature, the group that results from the removal of a single hydrogen atom from the benzene ring is called the **phenyl group**. Aromatic hydrocarbons having long aliphatic side chains are often named as phenyl-substituted hydrocarbons. For example, the following compound is 2-phenyloctane.

Polynuclear Aromatic Hydrocarbons

Polynuclear aromatic hydrocarbons are composed of two or more aromatic rings joined to one another. Naphthalene has a distinctive aroma and is used in mothballs. Benzopyrene, found in cigarette smoke, is one of the most potent of known carcinogens.

Reactions Involving Benzene

The typical reactions of benzene are **substitution reactions**. In these reactions, a hydrogen atom is replaced by another atom or group of atoms. Benzene can react with chlorine or bromine (in the presence of iron or iron halide as a catalyst).

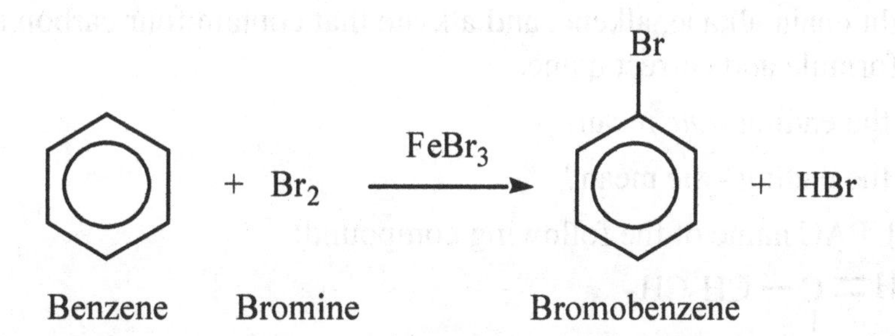

Benzene also reacts with sulfur trioxide (with concentrated sulfuric acid catalyst) to undergo substitution, as seen in the following equation:

$$\text{Benzene} + SO_3 \xrightarrow{\text{conc. } H_2SO_4} \text{Benzenesulfonic acid} + H_2O$$

Benzene | Sulfur trioxide | Benzenesulfonic acid

Benzene can also undergo nitration in a reaction with concentrated nitric acid dissolved in concentrated sulfuric acid. This reaction requires a temperature of 50-55°C.

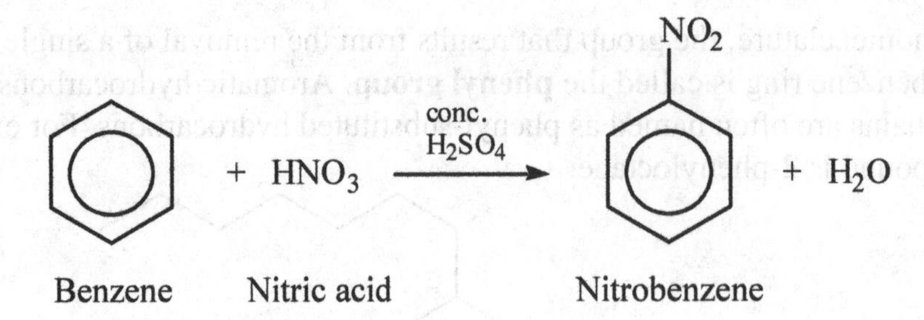

Section 11.7 Heterocyclic Aromatic Compounds

Heterocyclic aromatic compounds are those that have at least one atom other than carbon as part of the structure of the aromatic ring. Molecules such as pyridine, pyrimidines, purines, and pyrrole are heterocyclic aromatic compounds that are found in nicotine, DNA, RNA, and the porphyrin ring found in chlorophyll and hemoglobin.

Self Test for Chapter Eleven

1. Consider the molecule 1-hexene.
   a. Would you expect this compound to dissolve in water?
   b. Would you expect this compound to have a higher or lower boiling point than 1-nonene?
2. For a straight chain alkane, alkene, and alkyne that contain four carbon atoms, write the molecular formula and correct name.
3. What does the ending *-ane* mean?
4. What does the ending *-yne* mean?
5. Write the IUPAC name of the following compound:

$$CH_3-CH=\underset{\displaystyle CH_2CH_3}{\underset{|}{C}}-CH_2CH_3$$

6. Write the IUPAC name of the following compound:

$$CH_3\overset{\displaystyle Br}{\overset{|}{C}}H\underset{\displaystyle Br}{\underset{|}{C}}HC\equiv CCH_3$$

7. Draw the structure of 2-methyl-2-pentene.
8. What is the IUPAC name of the following molecule?

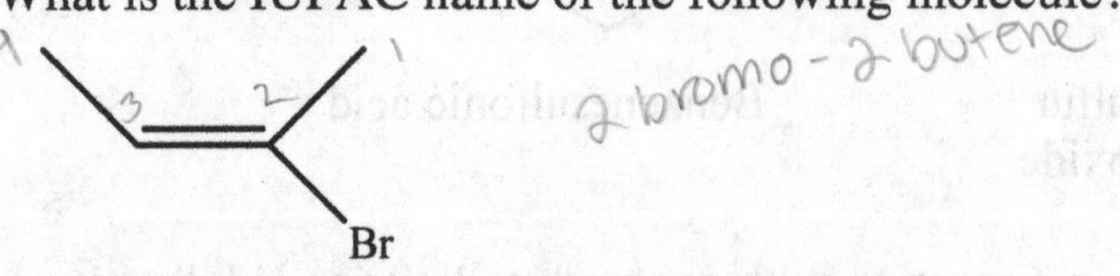

9. Draw the structure of *cis*-3-methyl-2-pentene.
10. Name the following compound using the prefixes *cis-* or *trans-*:

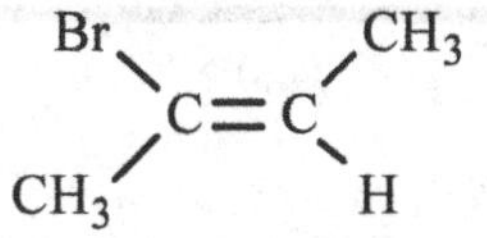

$$\begin{array}{ccc} Br & & CH_3 \\ & C{=}C & \\ CH_3 & & H \end{array}$$

11. What is the IUPAC name of the following molecule?

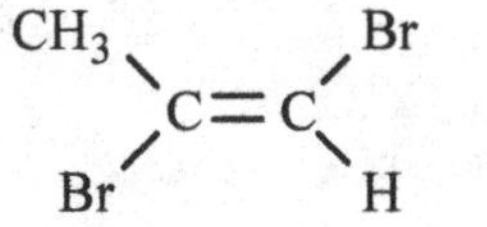

$$\begin{array}{ccc} CH_3 & & Br \\ & C{=}C & \\ Br & & H \end{array}$$

12. Alkenes are characterized by what type of reaction?
13. Which reaction is used to convert alkenes into alkanes?
14. Which three catalysts are often used in the hydrogenation of alkenes?
15. Provide the product of the reaction represented by the following equation:

$$CH_3CH_2CH{=}CH_2 + H_2 \xrightarrow{\text{Pd, heat}} ?$$

16. Write an equation showing the reaction of 1-butyne with 2 moles of hydrogen gas.
17. Write an equation for the reaction of 1-butene with Cl_2.
18. Write an equation for the hydration of 1-butene. Indicate both the major and minor products.
19. Write an equation for the reaction of hydrogen chloride with 2-methyl-2-pentene. Indicate both the major and minor products.
20. Write an equation showing the reaction of propyne with 2 moles of HBr.
21. Draw the polymer formed from the addition reaction of the styrene monomer:

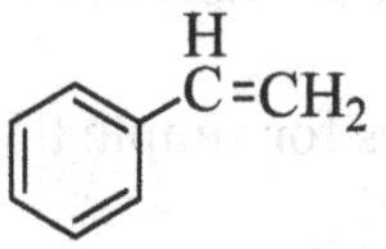

H
C=CH_2

22. Which compound makes up the basic unit of aromatic compounds?
23. Name the following aromatic compounds:

a.

b. OH

c. Cl, Cl

d. Cl, Cl, Br

e. Br, Br

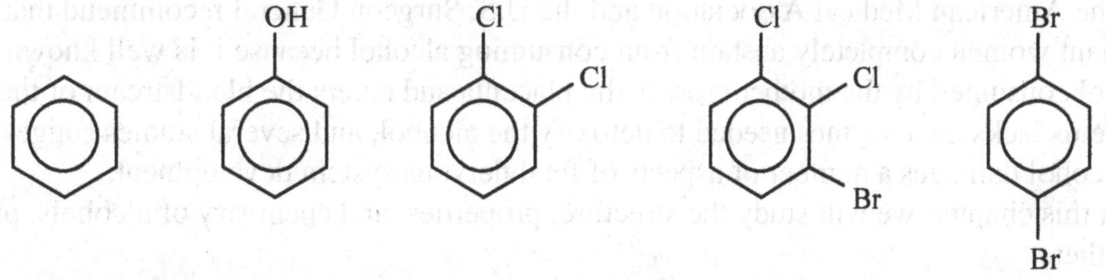

24. What is the product of the sulfuric acid catalyzed reaction between benzene and nitric acid (HNO_3)?
25. What is the product of the reaction of benzene with Cl_2? The compound $FeCl_3$ catalyzes the reaction.
26. What characteristics of a purine make it a heterocyclic aromatic compound? Where are purines used in the body?

12 Alcohols, Phenols, Thiols, and Ethers

Learning Goals

1. Classify alcohols as primary, secondary, or tertiary.
2. Rank selected alcohols by relative water solubility, boiling points, or melting points.
3. Write the names and draw the structures of common alcohols.
4. Discuss the biological, medical, or environmental significance of several alcohols.
5. Write equations representing the preparation of alcohols by the hydration of an alkene.
6. Write equations representing the preparation of alcohols by the hydrogenation (reduction) of aldehydes and ketones.
7. Write equations showing the dehydration of an alcohol.
8. Write equations representing the oxidation of alcohols.
9. Discuss the role of oxidation and reduction reactions in the chemistry of living systems.
10. Discuss the use of phenols as germicides.
11. Write names and draw structures for common ethers, and discuss their use in medicine.
12. Write equations representing the condensation reaction between two alcohol molecules to form an ether.
13. Write names and draw structures for simple thiols, and discuss their biological significance.

Introduction

The American Medical Association and the U.S. Surgeon General recommend that pregnant women completely abstain from consuming alcohol because it is well known that alcohol consumed by the mother crosses the placenta and enters the bloodstream of the fetus. The fetus lacks the enzymes needed to detoxify the alcohol, and several studies suggest that the alcohol damages a number of aspects of fetal nervous system development.

In this chapter, we will study the structure, properties, and chemistry of alcohols, phenols and ethers.

12.1 Alcohols: Structure and Physical Properties

We can think of alcohols as substituted water molecules: an H–O–H molecule in which one hydrogen atom has been replaced by an alkyl or aryl group. The functional group of the **alcohols** is the **hydroxyl group** (–OH). Alcohols have the general formula (Ar) R–O-H, in which R represents an alkyl group and Ar represents an aryl group. (An aryl group is an aromatic ring with one hydrogen removed.) The R–O–H portion of an alcohol is planar, as is H–O–H. In addition, the bond angles of both R–O–H and H–O–H are 104.5°.

Alcohols are classified as **primary (1°), secondary (2°),** or **tertiary (3°)** depending on the number of alkyl groups attached to the **carbinol carbon**; that is, the carbon that bears the hydroxyl group. If there is one alkyl group attached to the carbinol carbon, the alcohol is a primary alcohol. An alcohol with two alkyl groups bonded to the carbinol carbon is a secondary alcohol. A tertiary alcohol has three alkyl groups bonded to the carbinol carbon.

Example 1: Classification of Alcohols

Classify the following alcohols as either primary, secondary, or tertiary:

1. CH_3CH_2OH

2. CH_3CHCH_3 (with OH on the middle carbon)

3. $CH_3CH_2CHCH_3$ (with OH on the third carbon)

4. $CH_3CH_2CH_2CCH_3$ (with CH_3 and OH on the fourth carbon)

Solution:

Look at the carbon to which the -OH group is attached, the carbinol carbon, and determine the number of alkyl groups attached to that carbon.

1. Primary
2. Secondary
3. Secondary
4. Tertiary

LG 2

Alcohols are polar molecules owing to the hydroxyl group. The hydroxyl group of alcohols is very polar because the oxygen atom has a significantly higher electronegativity than the hydrogen atom. In addition, the oxygen atom has two unshared pairs of electrons. This results in a very polar bond in which the oxygen atom carries a partially negative charge (δ^-) and the hydrogen carries a partially positive charge (δ^+). This structure allows hydrogen bonds, the attractive force between a hydrogen atom covalently bonded to a small, highly electronegative atom and another atom with unshared pairs of electrons, to form. Alcohol molecules can form hydrogen bonds with other alcohol molecules and with the molecules of a polar solvent.

Both the polarity of alcohol molecules and the ability to form intermolecular hydrogen bonds exert a strong influence on the physical properties of alcohols. They have higher boiling points than hydrocarbons or ethers of similar molar mass and show greater solubility in water. Molecules with more than one hydroxyl group are even more water soluble than alcohols with a single hydroxyl group.

Example 2: Ranking Boiling Points of Compounds According to Functional Groups

For the following pairs of compounds, indicate the one that has the higher boiling point:

1. CH_4 (methane) or CH_3OH (methanol)
2. $CH_3CH_2CH_2CH_3$ (butane) or $CH_3OCH_2CH_3$ (ethyl methyl ether)
3. $CH_3CH_2OCH_3$ (ethyl methyl ether) or $CH_3CH_2CH_2OH$ (1-propanol)

Solution:

When comparing the boiling points of alkanes, ethers, and alcohols of similar molar mass, remember that the alcohols have the highest boiling points. The ethers are found to have higher boiling points than the alkanes:

alcohol > ethers > alkanes

1. Methane is an alkane, and methanol is an alcohol; therefore, methanol has the higher boiling point.
2. Ethyl methyl ether has a higher boiling point than butane, which is an alkane.
3. 1-Propanol is an alcohol and will have a higher boiling point than the ether.

Example 3: Determining Solubility of Alcohols in Water.

Predict the solubility of each of the following alcohols in water solutions:

1. $CH_3CH_2CH_2CH_2CH_2CH_2OH$ 2. CH_3CH_2OH 3. $CH_3(CH_2)_8CH_2OH$

Solution:

Alcohols of 1-4 carbons are very soluble in water. As the carbon chain length increases, the alcohol becomes less soluble. Those with more than 6 carbons are insoluble in water.

1. Not very soluble 2. Very soluble 3. Insoluble in water

12.2 Alcohols: Nomenclature

IUPAC Names

To name alcohols, determine the **parent compound**, which in this case is the longest continuous carbon chain containing the –OH group. Drop the *-e* ending of the alkane chain and replace it with *-ol*. Number the chain so that the hydroxyl group has the lowest possible number. Add substituents, named and numbered appropriately, as prefixes to the alcohol name.

If two hydroxyl groups are present, the suffix *-diol* is used; if three hydroxyl groups are present, the suffix used is *-triol*. The positions of hydroxyl groups are numbered as usual.

Example 4: IUPAC Nomenclature of Alcohols

Name the following compounds using the IUPAC method:

1. CH_3CHCH_3 with OH on C2

2. $CH_3CHCH_2CHCH_3$ with OH on C2 and CH_3 on C4

3. CH_2CH_2 with OH on each carbon

4. $CH_3CHCHCH_3$ with Br on C2 and OH on C3

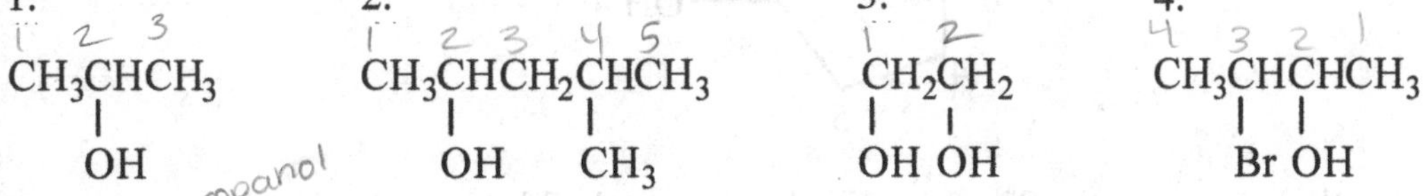

Solution:

1. The carbon chain is three carbons long giving the parent chain the name propanol. The hydroxyl group is attached to the second carbon, giving **2-propanol.**
2. The longest carbon chain has 5 carbon atoms. Number the chain from the end which has the –OH closest to it. In this case we will number the atoms from the left-hand side of the molecule.

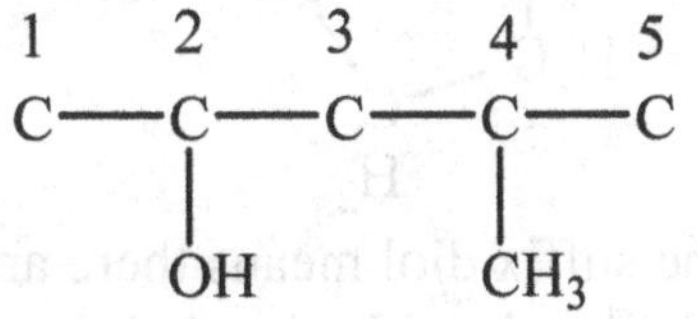

   ```
   1    2    3    4    5
   C----C----C----C----C
        |         |
        OH        CH3
   ```

   The hydroxyl group is located on the second carbon. The methyl group is located on the fourth carbon, giving **4-methyl-2-pentanol.**
3. This two carbon alcohol is a diol since it has two –OH groups. The hydroxyl groups are located on the first and second carbon, giving **1,2-ethanediol.**
4. The longest carbon chain is four carbons long giving the name butanol. Number the chain from the right hand side since this is the end the –OH group is closest to.

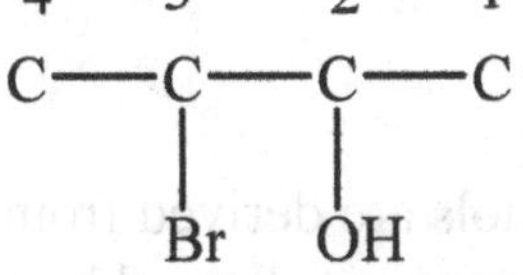

   ```
   4    3    2    1
   C----C----C----C
        |    |
        Br   OH
   ```

   The –OH is on carbon number 2, the Br is on carbon atom number 3, giving **3-bromo-2-butanol.**

Example 5: Drawing the Structural Formula of Alcohols

Draw the structural formula for the following compounds.

1. 2-methylcyclopentanol
2. 2,3-dichloro-2,3-hexanediol

Solution:

1. This compound is based on the parent compound cyclopentane. The –OH group is defined as being on carbon 1 of the ring.

Example 5 Continued

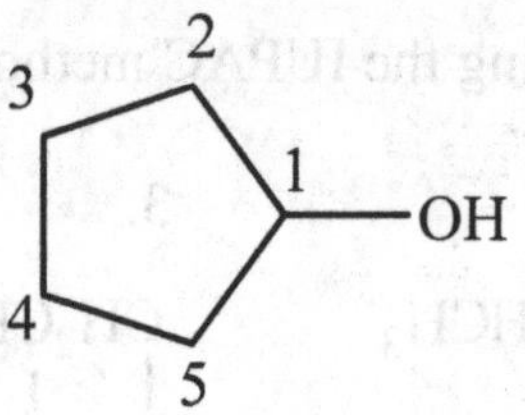

[Structure: cyclopentanol ring with carbons numbered 1–5, –OH on carbon 1]

Place the methyl group on the 2nd carbon in the ring, and add the remaining H atoms, giving:

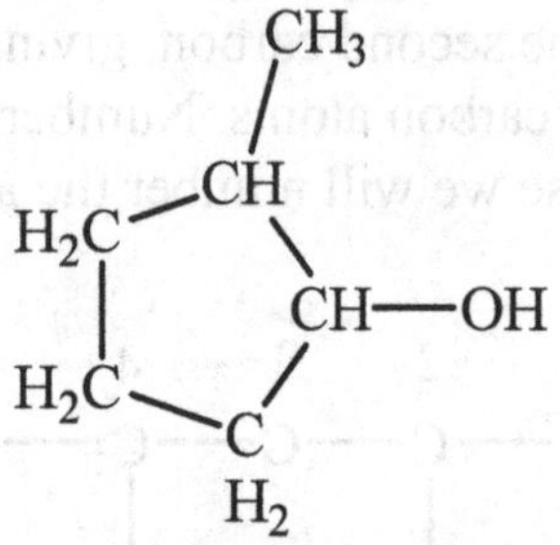

[Structure: ring of H_2C, CH (bearing CH_3), CH–OH, $C H_2$, H_2C]

2. Hexane is the parent chain. The suffix diol means there are two –OH groups. Place the hydroxyl groups on the 2nd and 3rd carbons in the chain and Cl atoms on the 2nd and 3rd carbons, giving:

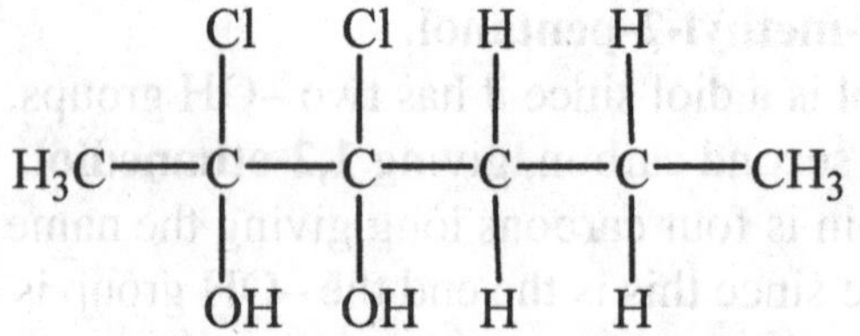

$$H_3C-C(Cl)(OH)-C(Cl)(OH)-CH_2-CH_2-CH_3$$

Common Names

The common names for the alcohols are derived from the name of the corresponding alkyl group. The name of the alkyl group is followed by the word *alcohol.*

Example 6: Naming Alcohols Using Common Names

Give the common names of each of the following alcohols:

1. CH_3OH
2. CH_3CH_2OH
3. $CH_3CH(OH)CH_3$ isopropyl
4. $CH_3CH_2CH_2CH_2OH$
5. $H_3C-C(CH_3)(OH)-CH_3$

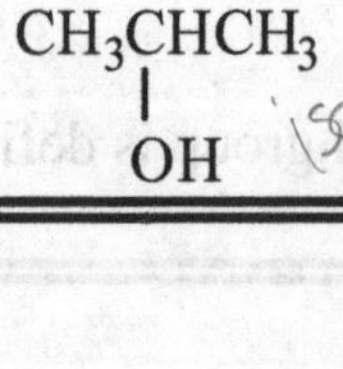

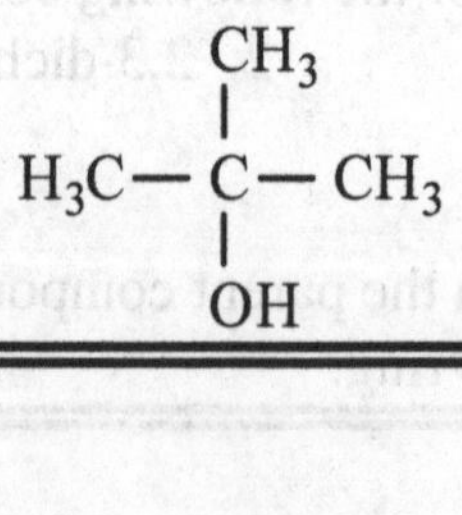

Example 6 Solution:

1. A one carbon alkane is methane. Name this as an alkyl group, and add the word alcohol, giving **methyl alcohol**.
2. A two carbon alkane is ethane. Name this as an alkyl group, and add the word alcohol, giving **ethyl alcohol**.
3. A three carbon alkane is propane, so this alcohol is a propyl alcohol. However, we must distinguish it from the linear isomer in which the -OH group is attached to the end carbon which would be called propyl alcohol. This isomer has two methyl groups attached to the end of the alkyl group and thus is called **isopropyl alcohol**. Note: isomers of propyl groups and butyl groups are summarized in Table 10.6 in your textbook.
4. This alcohol has four carbons and is thus a butyl alcohol. Because the -OH is attached to the end carbon, it is simply named **butyl alcohol**.
5. This is also a butyl alcohol but is an isomer of the alcohol in number 4. The name of the group is *tert*-butyl. Therefore, the alcohol is named ***t*-butyl alcohol**.

12.3 Medically Important Alcohols

Several of the small alcohols are important in medicine and industry. The smallest, *methanol* (methyl alcohol, wood alcohol) is a common solvent. It is extremely toxic if ingested.

Ethanol (ethyl alcohol, grain alcohol) is a solvent and disinfectant. Ethanol for human consumption is produced by **fermentation** of sugars by yeast. Denatured alcohol is ethanol to which a denaturing agent has been added to make it undrinkable.

2-Propanol (isopropyl alcohol, rubbing alcohol) is commonly used as a disinfectant and solvent. Like methanol, it is very toxic if ingested.

1,2-Ethanediol (ethylene glycol) is a common antifreeze for cars. Again, it is extremely poisonous if ingested.

1,2,3-Propanetriol (glycerol) is a component of stored fats in the body. It is used in cosmetics and pharmaceuticals.

12.4 Reactions Involving Alcohols

Preparation of Alcohols

In the laboratory, alcohols are prepared by the **hydration** of alkenes. This reaction was covered in Chapter 11. Alcohols can also be prepared by the hydrogenation (reduction) of aldehydes and ketones. The hydrogenation reaction is addition of H_2 at the C=O of the aldehyde or ketone. This reaction is very similar to the hydrogenation reaction of an alkene we learned about in Chapter 11, except it takes place at a C=O instead of a C=C site.

$$H_3C-\overset{\overset{\displaystyle O}{\|}}{C}-CH_3 + H-H \longrightarrow H_3C-\underset{\underset{\displaystyle H}{|}}{\overset{\overset{\displaystyle OH}{|}}{C}}-CH_3$$

LG 6

Oxidation-reduction reactions were covered in Chapter 8. We looked at oxidation-reduction reactions as the gain and loss of electrons. And while this definition is still true, in organic chemistry, it is more convenient to think of oxidation as the gain of oxygen or the loss of hydrogen and reduction as the gain of hydrogen or the loss of oxygen. This is why the above reaction is a reduction reaction, the carbon loses a bond to oxygen and gains a bond to hydrogen.

Example 7: Formation of Alcohols by Reduction Reactions
Show the reduction reactions for the conversions of formaldehyde (a one carbon aldehyde) and acetaldehyde (a two carbon aldehyde) into their corresponding alcohols.

Solution:

The reduction reaction is the addition of H_2. A hydrogen atom is bonded to the O and C on either side of the double bond.

1. $H-C(=O)-H + H_2 \longrightarrow H-CH_2-OH$

Formaldehyde (methanal) → Methyl alcohol (methanol)

2. $H_3C-C(=O)-H + H_2 \longrightarrow H-CH_2-CH_2-OH$

Acetaldehyde (ethanal) → Ethyl alcohol (ethanol)

Dehydration of Alcohols

Alcohols undergo dehydration reactions. A **dehydration reaction** is one in which a water molecule is lost. Dehydration is an example of an **elimination** reaction. The dehydration of an alcohol requires heat and concentrated sulfuric or phosphoric acids. The products of dehydration are an alkene and a water molecule. Quite simply, a dehydration reaction is the reverse of the hydration reactions that produce an alcohol from an alkene and water, as described in Chapter 11.

In some cases, dehydration of an alcohol may produce a mixture of products. **Zaitsev's rule** tells us that the more highly substituted alkene will be the major product. How can we predict which product will be the more highly substituted? Draw the structure of the alcohol and remove an –OH from the carbinol carbon and an –H from a neighboring carbon atom. As you look at the structure, there may be a hydrogen atom that can be removed from the carbon to the left *or* the carbon to the right of the carbinol carbon. If so, draw both possibilities. Look at the two structures. For each, examine the carbon atoms on either side of the double bond. The compound that has more alkyl groups attached to these carbons will be the major product. See Example 8, Part 2.

Example 8: Dehydration of Alcohols.

Write an equation representing the dehydration reactions for the following alcohols. If more than one possible alkene is produced, note which is the major product.

1. 1-Propanol
2. 2-Butanol

Solution:

The dehydration of an alcohol produces an alkene plus water. Remember that in some cases the dehydration of an alcohol yields a mixture of products. In those instances, it is the more highly substituted alkene that is the major product.

1. 1-Propanol is a three carbon chain with -OH on the first carbon. Remove the OH from its carbon atom (C-1) and an H from the adjacent carbon (C-2).

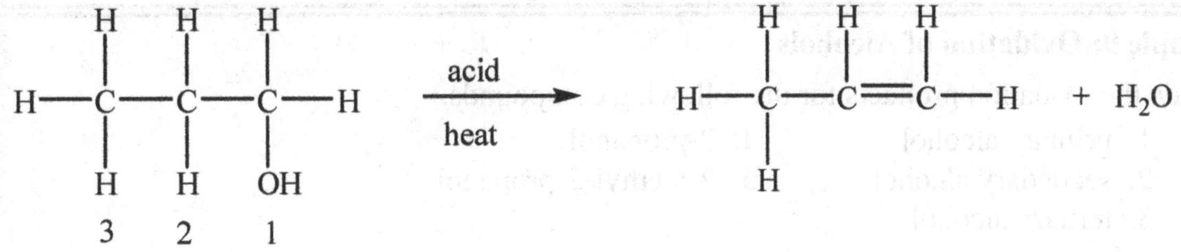

```
    H    H    H                          H    H    H
    |    |    |          acid            |    |    |
H — C —— C —— C — H   ————————►   H —— C —— C ═ C — H    +  H2O
    |    |    |          heat            |
    H    H    OH                         H
    3    2    1
      1-propanol                           propene
```

2. 2-Butanol is a four carbon chain with the hydroxyl group on the second carbon. There are two possible products. One product removes the -OH from C-2 and the H from C-1 and the other product removes the -OH from C-2 and the H from C-3.

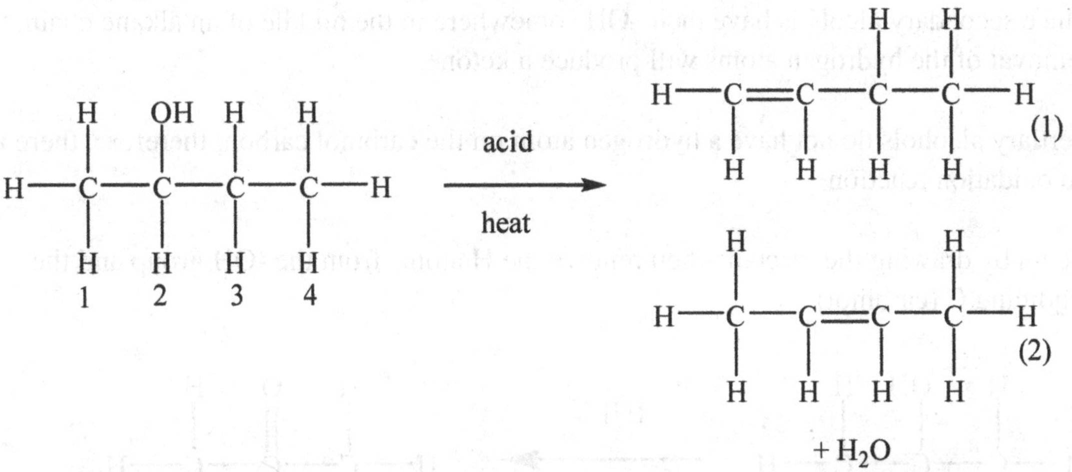

```
                                                  H    H
                                                  |    |
                                       H — C ═ C — C — C — H
    H    OH   H    H                       |   |   |   |      (1)
    |    |    |    |         acid          H   H   H   H
H — C —— C —— C —— C — H  ————————►
    |    |    |    |         heat          H             H
    H    H    H    H                       |             |
    1    2    3    4                   H — C — C ═ C — C — H
                                           |   |   |   |      (2)
                                           H   H   H   H

                                                + H2O
```

Now, we must decide, which is the major product, (1) 1-butene or (2) 2-butene. For compound (1), examine C-1 and C-2. There is only one alkyl group attached to C-1 and C-2, combined. For compound (2) examine C-2 and C-3. There are two alkyl groups attached to C-2 and C-3 combined. Since compound (2) has more alkyl groups, it is the major product.

LG 8

Oxidation Reactions

Some alcohols can be oxidized to produce aldehydes, ketones or carboxylic acids. The most commonly used oxidizing agents are basic potassium permanganate ($KMnO_4/OH^-$) and chromic acid (H_2CrO_4). Remember, oxidation is the addition of oxygen or removal of hydrogen. In this case, two hydrogen atoms are removed with one hydrogen atom removed from the hydroxyl group, and the other hydrogen atom removed from the carbinol carbon. In the process, a carbonyl is formed.

Methanol and all primary alcohols produce aldehydes, while secondary alcohols form ketones. Tertiary alcohols cannot be oxidized. This is because the carbon bonded to the hydroxyl group must contain at least one C–H bond in order for oxidation to occur. Since tertiary alcohols contain three C–C bonds on the carbinol carbon they cannot be oxidized (there is no H on the carbinol carbon that can be removed). Aldehydes can undergo further oxidation to produce carboxylic acids. This further oxidation involves the addition of oxygen (this will be discussed in more detail in section 13.4).

Example 9: Oxidation of Alcohols

Provide the oxidation products for the following compounds.

1. primary alcohol
2. secondary alcohol
3. tertiary alcohol
4. 2-propanol
5. 2-methyl-2-propanol

Solution:

1. Primary alcohols will have their -OH group on the end of a chain and produce aldehydes.

2. Since secondary alcohols have their -OH somewhere in the middle of an alkane chain, the removal of the hydrogen atoms will produce a ketone.

3. Tertiary alcohols do not have a hydrogen atom on the carbinol carbon; therefore, there is no oxidation reaction.

4. Begin by drawing the reactant, then remove the H atoms from the -OH group and the adjoining C (carbinol).

```
     H   OH   H                        H    O    H
     |   |    |          [O]           |    ‖    |
 H — C — C —  C — H   ---------->  H — C  — C  — C — H
     |   |    |                        |         |
     H   H    H                        H         H
```

The product of the reaction is a ketone—called propanone (acetone.) We will learn to name ketones in the next chapter.

Example 9 Continued

5. Begin by drawing the alcohol.

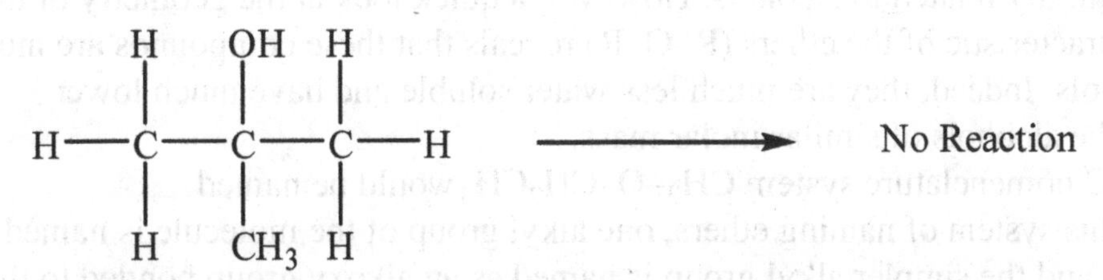

This alcohol is a tertiary alcohol. There is no hydrogen atom attached to the carbinol carbon. Therefore, there is no reaction.

12.5 Oxidation and Reduction in Living Systems

LG 9

In organic and biological chemistry, **oxidation** involves a gain of oxygen or a loss of hydrogen from a compound. A **reduction** reaction involves a loss of oxygen or gain of hydrogen. To recognize an oxidation or reduction reaction, simply count the hydrogen and oxygen atoms in the products and reactants and apply the definitions presented above.

Oxidation and reduction reactions are extremely important in the metabolic pathways that harvest energy for use by our bodies. A class of enzymes called oxidoreductases catalyzes these reactions. Coenzymes serve as acceptors, donors, and carriers of hydrogen in these cellular oxidation-reduction reactions. Nicotinamide adenine dinucleotide, NAD^+, is one such coenzyme.

12.6 Phenols

LG 10

Phenols are compounds in which the hydroxyl group is attached to a benzene ring. Owing to the polar hydroxyl group, the phenols are also polar compounds

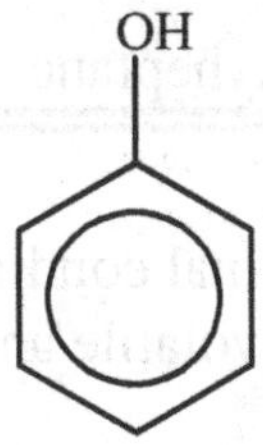

OH

The simplest member of this family is known by the common name phenol. This compound is of interest in the history of medicine. Joseph Lister, a British physician, observed that the incidence of post-surgical infections could be radically decreased if the surgical instruments and the incision were treated with an antimicrobial chemical. The agent he used was carbolic acid, a dilute solution of phenol. As a result of his observations, the use of antiseptics and disinfectants has become routine medical practice.

12.7 Ethers

Ethers are structurally related to alcohols. However, a quick look at the geometry of the functional group characteristic of the ethers (R–O–R) reveals that these compounds are much less polar than alcohols. Indeed, they are much less water soluble and have much lower boiling points than the alcohols of similar molar mass.

LG 11

Using the IUPAC nomenclature system $CH_3–O–CH_2CH_3$ would be named methoxyethane. In this system of naming ethers, one alkyl group of the molecule is named as a hydrocarbon chain and the simpler alkyl group is named as an alkoxy group bonded to that chain. The alkoxy group has the structure -OR. If needed, a number is used to indicate the position of the alkoxy group on the parent compound.

Common names for ethers are derived from the names of the alkyl groups attached to the ether oxygen. The alkyl group names are listed as prefixes before the word ether and may be ordered by size (small to large) or alphabetically. For instance, the compound $CH_3–O–CH_2CH_3$ can be called methyl ethyl ether (prefixes arranged by size of the alkyl group) or ethyl methyl ether (prefixes ordered alphabetically).

Example 10: Naming Ethers
Name the following ethers using the IUPAC nomenclature system:

a. $CH_3CH_2CH(OCH_3)CH_2CH_2CH_2CH_3$ (with OCH_3 attached to the third carbon)

b. $CH_3CH(OCH_2CH_3)CH_2CH_2CH_3$ (with OCH_2CH_3 attached to the second carbon)

Solution:

Parent compounds:	a. Heptane	b. Pentane
Position of alkoxy group:	a. Carbon-3	b. Carbon-2
Substituents:	a. Methoxy	b. Ethoxy
Name:	a. 3-Methoxyheptane	b. 2-Ethoxypentane

Ethers are chemically inert. Under normal conditions, they will not react with reducing agents, or bases. They are, however, very volatile and flammable and must be treated with care.

Ethers may be prepared by a condensation reaction between two alcohol molecules. See Example 11 below.

Example 11: Writing a Dehydration Reaction to Produce an Ether
Write an equation showing the synthesis of diethyl ether.

Example 11 Solution:

Ethers are formed by a condensation reaction between two alcohol molecules. Diethyl ether has the formula $CH_3CH_2OCH_2CH_3$. The alcohols must have the same number of carbons as the R groups of the ether.

$$\underset{\text{Ethanol}}{CH_3CH_2OH} + \underset{\text{Ethanol}}{CH_3CH_2OH} \xrightarrow{H^+} \underset{\text{Diethyl ether}}{CH_3CH_2\text{–}O\text{–}CH_2CH_3} + H_2O$$

Diethyl ether was the first general anesthetic used in medical practice. However, since ethers are highly flammable and can form explosive peroxides upon storage, diethyl ether has largely been replaced by halogenated ethers, such as desflurane, sevoflurane, and isoflurane.

12.8 Thiols

LG 13

Compounds that contain the –SH (sulfhydryl) group are known as **thiols**. Thiols are similar to alcohols in structure, but the sulfur atom replaces the oxygen atom. Thiols, and many other sulfur compounds, have nauseating aromas. They also have significantly lower melting points and boiling points than alcohols because the sulfhydryl group is much less polar than the hydroxyl group.

To name a thiol, use the suffix –thiol after the alkane name. Designate the location of the –SH group using a number.

Example 12: Naming Thiols

Name the compound:

```
          H     H     H
          |     |     |
   H3C----C-----C-----C-----CH3
          |     |     |
          SH    H    H2C----CH3
```

Solution:

Find the longest chain, and name the parent alkane. Be careful because the longest chain in this case is not in a straight line.

```
   1     2     3     4
   C-----C-----C-----C-----CH3
         |           |
         SH          C-----C
                     5     6
```

The alkane is hexane. The numbering starts on the end closest to the thiol group. The compound is 4-methyl-2-hexanethiol.

Thiols are also involved in protein structure and conformation. It is the ability of two thiol groups to easily undergo oxidation to a –S–S– (**disulfide**) bond that helps maintain the correct shape of a protein. The amino acid cysteine has a thiol group and can participate in disulfide bond formation in proteins.

Coenzyme A is a thiol that serves as a "carrier" of acetyl groups and fatty acids in cellular metabolic reactions.

Self Test for Chapter Twelve

1. In a secondary alcohol, how many hydrogen atoms are directly attached to the carbinol carbon atom?
2. From each of the following pairs of compounds, pick the one that has the higher boiling point:

   a. 1-nonanol or 1-hexanol
   b. ethane or methanol
   c. methoxyethane or 1-propanol
   d. methane or dimethyl ether

3. Which of the following alcohols has the greatest solubility in water?

   a. 1-propanol
   b. 1-butanol
   c. 1-heptanol
   d. 1-decanol

4. Why do alcohols boil at higher temperatures than pure hydrocarbons or ethers of similar molar mass?
5. Which of the following is the least polar?

   a. alcohols

   b. ethers

   c. water

6. Which of the following cannot hydrogen bond to other molecules of like kind?

   a. alcohols   b. ethers   c. water

7. Provide the IUPAC name of the following compound:

```
          Br
          |
   CH3CHCHCHCH3
        |     |
       OH    CH3
```

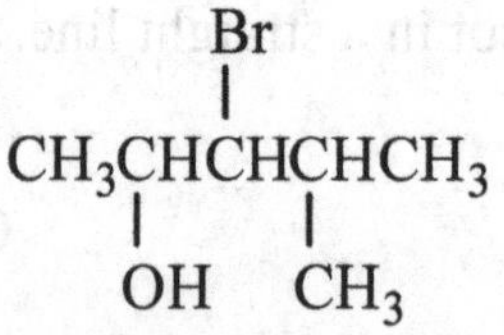

8. Provide the IUPAC name of the following compound:

```
   CH3CHCH2CH2CH2Cl
        |
       OH
```

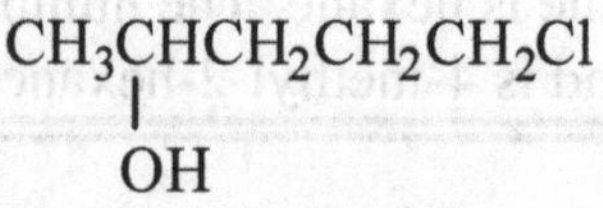

9. Determine the IUPAC name of the trialcohol called glycerol: $HOCH_2CHOHCH_2OH$

10. Provide the common names of the following:

a. $CH_3CH_2CH_2OH$

b. $CH_3CHOHCH_3$

c. $C(CH_3)_3OH$

11. If ethanol has some agent added to it to make it unfit to drink, what do we call that solution?

12. Which alcohol is often used in cosmetics and pharmaceuticals?

13. Complete the following chemical reactions:

a. 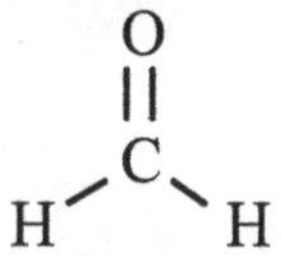 $H-C(=O)-H + H_2 \xrightarrow{\text{catalyst}} ?$

b. $CH_2=CH_2 + H_2O \xrightarrow[\text{catalyst}]{H^+} ?$

14. What are the products of the dehydration of each of the following alcohols?

a. ethanol

b. 1-propanol

c. 2-butanol

15. Complete the oxidation reactions of each of the following alcohols to form aldehydes:

a. $CH_3OH \xrightarrow{\text{oxidation}}$

b. $CH_3CH_2OH \xrightarrow{\text{oxidation}}$

16. Determine the oxidation products of the following alcohols:

a. CH_3OH

b. ethanol

c. 2-propanol

d. 2-methyl-2-propanol

17. Which kind of alcohol will not undergo oxidation under normal conditions?

18. Potassium permanganate and potassium dichromate are used with alcohols in what kind of chemical reaction?

19. Which of the following compounds is the most oxidized?

a. propane

b. 1-propanol

c. 2-propanol

d. acetic acid

e. propanone

20. Which of the following compounds is the most reduced?

a. methane | c. dichloromethane
b. chloromethane | d. trichloromethane

21. How does the structure of phenol differ from the structure of cyclohexanol?

22. Provide the common names of each of the following ethers:

a. CH_3OCH_3 | c. $CH_3OCH_2CH_3$
b. $CH_3CH_2OCH_2CH_3$

23. Which class of compounds has a nauseating aroma?

24. Give the structural formula for 2,3-dimethyl-2-heptanethiol.

25. Name the following compound:

$$\underset{\displaystyle SH}{CH_3\underset{|}{C}HCH_2CH_3}$$

13 *Aldehydes and Ketones*

Learning Goals

1. Draw the structures and discuss the physical properties of aldehydes and ketones.
2. From the structures, write the common and IUPAC names of aldehydes and ketones.
3. List several aldehydes and ketones that are of natural, commercial, health, and environmental interest, and describe their significance.
4. Write equations for the preparation of aldehydes and ketones by the oxidation of alcohols.
5. Write equations representing the oxidation of carbonyl compounds.
6. Write equations representing the reduction of carbonyl compounds.
7. Write equations for the preparation of hemiacetals and acetals.
8. Draw the keto and enol forms of aldehydes and ketones.

Introduction

Individuals have characteristic odor prints as unique as their fingerprints or DNA. Some of the molecules that have been identified in these mixtures are members of the aldehyde and ketone groups. In this chapter, you will learn about the structure, properties, and reactions of aldehydes and ketones like those that are found in human scent prints.

13.1 Structure and Physical Properties

Aldehydes and **ketones** are compounds containing a **carbonyl group** (–C = O). They differ from one another in the type of atom or atoms attached to the carbonyl carbon. In ketones, the carbonyl carbon is attached to two carbon atoms (RCOR), whereas in aldehydes the carbonyl carbon is attached to at least one hydrogen atom; the second atom attached to the carbonyl carbon in aldehydes may be either a carbon atom (RCHO) or another hydrogen atom (HCHO).

Owing to the polar carbonyl group, aldehydes and ketones are moderately polar compounds. They cannot hydrogen bond to other aldehydes or ketones. As a result, they boil at higher temperatures than do hydrocarbons of equivalent molar mass, but at temperatures lower than alcohols of similar molar mass. The figure below shows the dipole-dipole attractions between molecules as dashed lines.

H_3C CH_3 C δ^+ δ^- O O δ^- δ^+ C H_3C CH_3

Aldehydes and ketones composed of five or fewer carbon atoms are reasonably soluble in water because of the hydrogen bonding between the carbonyl group and water molecules. Larger members of these carbonyl-containing compounds are less polar, more hydrocarbon-like, and thus, more soluble in nonpolar organic solvents.

Example 1: Ranking Boiling Points of Compounds According to Functional Groups
Which member of each of the following pairs of compounds with similar molar masses has the higher boiling point?

1. butane or methoxyethane
2. methoxyethane or 1-propanol
3. propanone (a three carbon ketone) or methoxyethane

Solution:
Aldehydes and ketones are moderately polar compounds and boil at higher temperatures than hydrocarbons or ethers of similar molar masses. The aldehydes and ketones, however, boil at lower temperatures than alcohols of similar molar masses.

1. Butane is $CH_3CH_2CH_2CH_3$, a nonpolar alkane. Methoxyethane is an ether with formula $CH_3OCH_2CH_3$. The two compounds are similar in molar mass, but the ether is somewhat polar and therefore has a higher boiling point than the alkane.
2. 1-Propanol is an alcohol with the formula: $CH_3CH_2CH_2OH$. It has the same molar mass as methoxyethane (see formula in question 1 above) but is much more polar than the ether. It can also hydrogen bond with other 1-propanol molecules. 1-Propanol has the higher boiling point.
3. Though you have not yet learned ketone and aldehyde nomenclature, propanone must have the formula:

$$CH_3-\overset{\overset{\displaystyle O}{\|}}{C}-CH_3$$

These compounds are more polar than ethers of approximately the same molar mass, and therefore, the ketone has a higher boiling point.

13.2 IUPAC Nomenclature and Common Names

Naming Aldehydes

LG 2

To name aldehydes, determine the parent compound, that is, the longest continuous carbon chain containing the carbonyl group, drop the final *-e* of the parent alkane, and replace it with *-al*. The parent chain is always numbered beginning with the carbonyl carbon as carbon-1. All other substituents are named and numbered as usual.

Common names for aldehydes are derived from the same Latin root as the corresponding carboxylic acids (see Table 13.1 in your textbook). Substituted aldehydes are named as derivatives of the straight-chain parent compound, using Greek letters to indicate the positions of substituents. The carbon atom next to the carbonyl carbon is referred to as the α-carbon.

Naming Ketones

To name ketones, first determine the parent compound. Replace the *–e* ending of the parent alkane with the *-one* suffix of the ketone family. The longest carbon chain is numbered to give the carbonyl carbon the lowest possible number. All other substituents are named and numbered as usual.

The common names for ketones are derived by naming the alkyl groups that are bonded to the carbonyl carbon. These are used as prefixes followed by the word *ketone*. Alkyl groups may be arranged alphabetically or by size.

Example 2: Naming Aldehydes and Ketones

Determine the IUPAC and common names of the following aldehydes and ketones:

1. $H-\overset{O}{\overset{\|}{C}}-H$
2. $CH_3CH_2-\overset{O}{\overset{\|}{C}}-H$
3. $CH_3CH_2-\overset{O}{\overset{\|}{C}}-CH_3$
4. $CH_3CH_2CH_2-\overset{O}{\overset{\|}{C}}-CH_3$

Solution:

1. This is an aldehyde with one carbon. Drop the *–e* ending of methane and add *–al*. The compound is methanal (Table 13.1 gives the common name: formaldehyde.)
2. This is an aldehyde with three carbon atoms. Drop the *–e* of propane and add *–al*. The compound is propanal (The common name is propionaldehyde.)
3. This ketone has a four carbon chain. The *–one* suffix is added to butane, giving butanone. No number is needed because this is the only possible location for the carbonyl in butanone. (To name a ketone by the common name system, name the groups on either side of the carbonyl group. A two carbon group is ethyl. A one carbon group is methyl. The compound is ethyl methyl ketone.)
4. This ketone has a five carbon chain. The carbonyl is located on the second carbon when it is numbered from the end closest to the carbonyl, giving the name 2-pentanone (To name according to the common system, name the groups on either side of the carbonyl, giving methyl propyl ketone.)

Example 3: Drawing Structural Formulas for Aldehydes and Ketones

Write structural formulas for each of the following:

1. 5-bromohexanal
2. 3-methylheptanal
3. 3-bromo-2-pentanone
4. cyclohexanone

Example 3 Solution:

1. The parent name is hexanal. This tells us that it is a six carbon atom chain and that it is an aldehyde, so the carbonyl carbon is C-1.

```
O
‖
C—C—C—C—C—C
1  2  3  4  5  6
```

Place the bromine atom on C-5, and add the remaining hydrogen atoms to give each carbon four bonds.

```
   O  H  H  H  Br H
   ‖  |  |  |  |  |
H—C—C—C—C—C—C—H
      |  |  |  |  |
      H  H  H  H  H
```

2. The parent name is heptanal; therefore, the parent chain is seven carbons long. It is an aldehyde as evident by the *–al* suffix. Draw the 7 carbons. The carbonyl is on C-1, and the methyl group is on C-3.

```
   O  H  CH3 H  H  H  H
   ‖  |  |   |  |  |  |
H—C—C—C—C—C—C—C—H
      |  |   |  |  |  |
      H  H   H  H  H  H
```

3. The parent name is pentanone. This tells us that the chain is 5 carbons long and the *–one* suffix lets us know it is a ketone. The carbonyl group is on C-2, and the bromine atom is on C-3.

```
   H  O  Br H  H
   |  ‖  |  |  |
H—C—C—C—C—C—H
   |     |  |  |
   H     H  H  H
```

4. The parent name of cyclohexane informs us that this is a 6 member cyclic structure which is a ketone as denoted by the *–one* suffix.

```
      H2
      C
    /   \\
 H2C     C=O
  |      |
 H2C     CH2
    \\   /
      C
      H2
```

13.3 Important Aldehydes and Ketones

LG 3

Methanal (formaldehyde) is a gas. It is available commercially as an aqueous solution (formalin) that is used to preserve tissue samples.

Ethanal (acetaldehyde) is produced from ethanol in the liver and is responsible for the symptoms of a hangover.

Propanone (acetone) is the simplest ketone. It is an important solvent because it can dissolve organic compounds and is also miscible (mixes) with water. It is found as a solvent in adhesives, paints, and nail polish remover.

Many complex members of the ketone family are important in the food industry as food additives. Others are useful as medicinals and agricultural chemicals.

13.4 Reactions Involving Aldehydes and Ketones

Preparation of Aldehydes and Ketones

In the laboratory, aldehydes and ketones are often prepared by the **oxidation** of the corresponding alcohol. Any aldehyde or ketone can be prepared if the correct alcohol is available. These reactions were covered in detail in Chapter 12. Oxidation of a primary alcohol produces an aldehyde; oxidation of a secondary alcohol yields a ketone. Tertiary alcohols do not undergo oxidation.

Example 4: Oxidation of Alcohols

Give the formula and the IUPAC name for the product of the oxidation of the following alcohols.

1. 1-butanol   2. 2-propanol

Solution:

1. The oxidation on an alcohol removes an H atom from the -OH group and an H atom from the carbinol carbon (that is, the carbon containing the -OH group.) In 1-butanol, the OH group is on C-1, and therefore, the alcohol is a primary alcohol, and the product will be butanal (butyraldehyde).
2. 2-Propanol is an alcohol with the -OH group on C-2. Oxidation will produce a ketone with three carbons, forming propanone.

Oxidation Reactions

Aldehydes are very easily oxidized further to form carboxylic acids. Potassium permanganate and potassium chromate are two oxidizing agents that work well for this reaction. The oxidation of an aldehyde to a carboxylic acid is depicted on the next page.

$$R-\overset{O}{\overset{\|}{C}}-H \xrightarrow{[O]} R-\overset{O}{\overset{\|}{C}}-OH$$

Aldehyde → Carboxylic acid

Ketones do not undergo further oxidation reactions because a carbon-hydrogen bond to the carbonyl carbon is necessary for the reaction to occur.

Aldehydes and ketones can be distinguished from one another based on their ability to undergo oxidation reactions. The **Tollens' test**, or Tollens' silver mirror test, is the most common test used to distinguish between aldehydes and ketones. Tollens' reagent consists of a basic solution of $Ag(NH_3)_2^+$. An aldehyde will undergo an oxidation-reduction reaction in which the silver ion (Ag^+) is reduced to silver metal (Ag^0) as the aldehyde is oxidized to a carboxylic acid anion. The silver metal precipitates from solution and coats the vessel, giving a smooth silver mirror. Because ketones cannot undergo further oxidation, they do not react with the Tollens' reagent.

Example 5: Tollens' Test for Aldehydes

Write the major products for each of the following reactions, using Tollens' reagent as the oxidizing agent:

1. $H_3C-\overset{O}{\overset{\|}{C}}-H + Ag(NH_3)_2^+ \longrightarrow ?$

2. $H_3CH_2C-\overset{O}{\overset{\|}{C}}-H + Ag(NH_3)_2^+ \longrightarrow ?$

3. $H_3C-\overset{O}{\overset{\|}{C}}-CH_3 + Ag(NH_3)_2^+ \longrightarrow ?$

Solution:

Treatment of an aldehyde with Tollens' reagent gives an oxidation-reduction reaction. The aldehyde is oxidized to the carboxylic acid anion, and the silver ion (Ag^+) is reduced to silver metal (Ag^0).

1. $H_3C-\overset{O}{\overset{\|}{C}}-H + Ag(NH_3)_2^+ \longrightarrow H_3C-\overset{O}{\overset{\|}{C}}-O^- + Ag^0$

2. $H_3CH_2C-\overset{O}{\overset{\|}{C}}-H + Ag(NH_3)_2^+ \longrightarrow H_3CH_2C-\overset{O}{\overset{\|}{C}}-O^- + Ag^0$

3. No reaction—Tollens' reagent can not oxidize ketones.

Another test used to distinguish between aldehydes and ketones is **Benedict's test.** In this test, Cu^{2+} is reduced to Cu^{+}. Cu^{2+} is soluble and gives a blue solution, while the Cu^{+} precipitates as the red solid, Cu_2O. The color change is observed for aldehydes but not for ketones (as ketones typically do not undergo further oxidation). Benedict's test has been used to determine the concentrations of glucose in urine, since glucose is an aldehyde. Glucose may be found in the urine in cases of uncontrolled diabetes. This reaction will be discussed in more detail in chapter 16.

Reduction Reactions

LG 6

Aldehydes and ketones are easily reduced to the corresponding alcohol by a large number of different reducing agents, designated [H]. This reaction was covered in Section 12.4 for the preparation of alcohols. The general reduction reaction is shown:

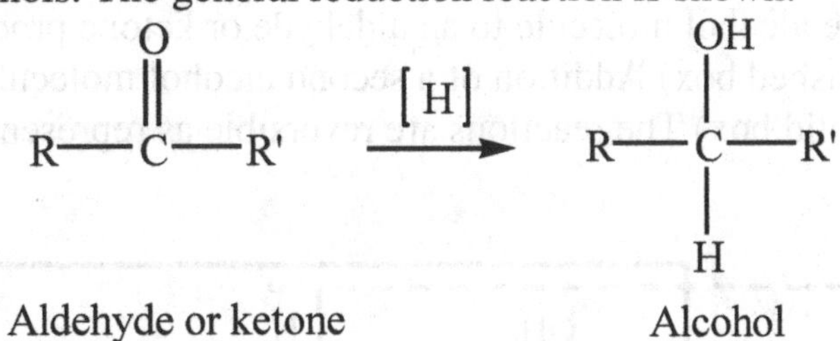

The classical reaction for aldehyde and ketone reduction is **hydrogenation**, in which the carbonyl compound is reacted with hydrogen gas. This reaction requires a metal catalyst, pressure and/or heat. The carbon-oxygen double bond (the carbonyl group) is reduced to a carbon-oxygen single bond. Aldehydes are reduced to primary alcohols and ketones are reduced to secondary alcohols.

$$R-\overset{\overset{\displaystyle O}{\|}}{C}-R' + H_2 \xrightarrow{Pt} R-\underset{\underset{\displaystyle H}{|}}{\overset{\overset{\displaystyle OH}{|}}{C}}-R'$$

Aldehyde or Ketone — Alcohol

Example 6: Predicting Oxidation and Reduction Reactions
Label the following as either oxidation or reduction reactions:

1. methanol to methanoic acid
2. methanol to formaldehyde
3. 2-pentanol to 2-pentanone
4. acetone to 2-propanol

Solution:

1. Methanol is an alcohol. When oxidized, it will first form an aldehyde, and upon further oxidation it will be converted to a carboxylic acid.
2. Again, methanol is an alcohol. The first step of oxidation (the removal of 2H) is to form an aldehyde. (The aldehyde can further oxidize to form the carboxylic acid as in part 1.)

Example 6 Continued

3. 2-Pentanol is a 2° alcohol. The process of converting a 2° alcohol to a ketone (2-pentanone) requires the removal of 2 hydrogen atoms. This is an oxidation process.
4. Acetone is a ketone. To create an alcohol (2-propanol) from a ketone requires the addition of H_2 across the C=O bond. This is a reduction reaction.

Addition Reactions

The most common reaction of the carbonyl group is **addition** across the carbon-oxygen double bond. The reaction often requires catalytic amounts of acid, represented as H^+ over the reaction arrow.

The addition of one alcohol molecule to an aldehyde or ketone produces a **hemiacetal.** (Reaction (1), in the dashed box) Addition of a second alcohol molecule yields an **acetal.** (Reaction (2), in the solid box) The reactions are reversible as represented by the two-way arrows.

$$R^1-\overset{O}{\overset{\|}{C}}-R^2 + H-OR^3 \underset{(1)}{\overset{H^+}{\rightleftharpoons}} R^1-\overset{OH}{\underset{R^2}{C}}-OR^3 + H-OR^3 \underset{(2)}{\overset{H^+}{\rightleftharpoons}} R^1-\overset{OR^4}{\underset{R^2}{C}}-OR^3 + H_2O$$

Aldehyde or Ketone (1) Hemiacetal (2) Acetal

This same reaction may occur in an intramolecular fashion (meaning within the same molecule). For this to occur, both an aldehyde or ketone functional group, as well as an alcohol functional group must be present in the same molecule.

$$H-\overset{O}{\overset{\|}{C}}\sim\sim\sim O-H \xrightarrow{\text{redrawn}} \text{(–O–H and –C(=O)–H on the same chain)} \longrightarrow O-\overset{H}{C}-OH$$

intramolecular (cyclic) hemiacetal

Hemiacetals are commonly found in the structures of carbohydrates. Linear sugar molecules have many hydroxyl groups and at least one carbonyl group that can undergo an intramolecular reaction in solution to give cyclic hemiacetals. This will be explained in greater detail in chapter 16.

Example 7: Addition Reaction of a Ketone and Alcohol.
Show the reaction of acetone with methyl alcohol to form a hemiacetal compound.

Example 7 Solution:
The alcohol will add across the C=O bond, breaking one of the bonds and leaving a C-O single bond. The H of the alcohol is added to the O of the carbonyl group. The CH_3O will add to the C of the carbonyl group.

$$H_3C-C(=O)-CH_3 + CH_3OH \xrightleftharpoons{H^+} CH_3-C(OH)(CH_3)-O-CH_3$$

Acetone — Hemiacetal

Keto-Enol Tautomerism

An aldehyde or a ketone may exist in an equilibrium mixture of two isomers called **tautomers**. One isomer is the **keto** form (it contains a carbonyl); the other is the **enol** form (that is, it has a double bond, *en*, and is an alcohol, *ol*). The two forms are shown below:

$$R-CH(R')-C(=O)-R'' \rightleftharpoons (R)(R')C=C(OH)(R'')$$

Keto form — Enol form

Typically, the keto form is more stable than the enol form.

Self Test for Chapter Thirteen

1. Rank the following compounds in order of increasing boiling point.
   a. 1-propanol, propane, propanone
   b. pentane, methoxyethane, propanal
2. Why are ketones polar compounds?
3. Draw the polar attraction between two butanone molecules.
4. Why are 2-propanone and 2-butanone incorrect IUPAC names?
5. Write structural formulas for the following:
   a. 2-bromobutanal
   b. 2-methyl-3-pentanone
6. Give the common name of the two-carbon aldehyde.

7. Give the IUPAC names of the following:

a. $CH_3-\overset{\overset{O}{||}}{C}-H$

b. $CH_3-\overset{\overset{O}{||}}{C}-CH_3$

c. $CH_3CH_2-\overset{\overset{O}{||}}{C}-H$

d. $CH_3CH_2-\overset{\overset{O}{||}}{C}-CH_2CH_3$

e. $CH_3CH_2\underset{\underset{Br}{|}}{C}HCH_2CH_2-\overset{\overset{O}{||}}{C}-H$

8. Draw the structure of formaldehyde. What is one use of formaldehyde?

9. Ethanal is responsible for the symptoms of a hangover. How and where is ethanol produced in the body?

10. Label the following as oxidation or reduction reactions:
    a. Conversion of methyl alcohol by the liver into formaldehyde.
    b. Conversion of acetaldehyde by the liver into acetic acid.
    c. Conversion of 3-pentanone into 3-pentanol.
    d. Conversion of ethylene glycol (1,2-ethanediol) to oxalic acid (ethanedioic acid).

11. Complete the following oxidation reactions:

a. $H-\overset{\overset{O}{||}}{C}-H \xrightarrow{[O]} ?$

b. $CH_3-\overset{\overset{O}{||}}{C}-H \xrightarrow{[O]} ?$

c. $CH_3-\overset{\overset{O}{||}}{C}-CH_3 \xrightarrow{[O]} ?$

d. $CH_3CH_2-\overset{\overset{O}{||}}{C}-H \xrightarrow{[O]} ?$

e. $CH_3-\underset{\underset{CH_3}{|}}{\overset{\overset{OH}{|}}{C}}-CH_3 \xrightarrow{[O]} ?$

12. Give the IUPAC names of the oxidation products for the following alcohols:
    a. CH_3CH_2OH
    b. 1-propanol
    c. 2-propanol
    d. 2-methyl-2-propanol
13. What class of alcohol will not undergo oxidation to form an aldehyde or ketone?
14. Show the major products for each of the following oxidation reactions:

$$\text{a.}\quad CH_3-\overset{\overset{\Large O}{||}}{C}-H \quad + \; Ag(NH_3)_2^+ \longrightarrow \; ?$$

$$\text{b.}\quad CH_3-\overset{\overset{\Large O}{||}}{C}-CH_3 \quad + \; Ag(NH_3)_2^+ \longrightarrow \; ?$$

$$\text{c.}\quad CH_3CH_2-\overset{\overset{\Large O}{||}}{C}-CH_3 \quad + \; Ag(NH_3)_2^+ \longrightarrow \; ?$$

15. Aldehydes are easily further oxidized to what compounds?
16. Draw the structure for the product of the oxidation of benzaldehyde?
17. Briefly describe two common laboratory tests used to distinguish between aldehydes and ketones.
18. Complete the following reaction:

$$RCHO + ROH \rightleftharpoons \; ?$$

19. What is the classification of the product of the following reaction:

$$R-\underset{\underset{\displaystyle H}{|}}{\overset{\overset{\displaystyle OH}{|}}{C}}-O-R \quad + \quad ROH \xrightarrow{\text{acid}} \; ?$$

20. Identify the missing starting material in the following reaction.

$$H-\overset{\overset{\Large O}{||}}{C}-CH_2-CH_3 \quad + \;\; ? \xrightarrow{H+} H-\underset{\underset{\displaystyle CH_2CH_3}{|}}{\overset{\overset{\displaystyle OH}{|}}{C}}-O-CH_3$$

21. Identify the missing starting material in the following reaction.

$$? \quad + \quad 2\,CH_3CH_2OH \xrightarrow{H+} \text{(cyclic: } H_3CH_2CO\text{, }OCH_2CH_3 \text{ on C; ring } H_2C, CH_2, H_2C, CH_2, \underset{H_2}{C}\text{)}$$

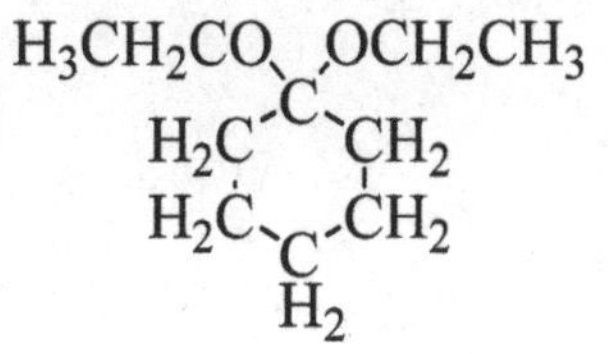

22. Draw the enol tautomer of the following compound:

```
          H     H     O
          |     |     ||
H3C — C — C — C — CH3
          |     |
          H    CH3
```

23. Is the keto or enol tautomer typically more stable?

14 *Carboxylic Acids and Carboxylic Acid Derivatives*

Learning Goals

1. Write structures and describe the physical properties of carboxylic acids.
2. Determine the common and IUPAC names of carboxylic acids.
3. Describe the biological, medical, or environmental significance of several carboxylic acids.
4. Write equations that show the synthesis of a carboxylic acid.
5. Write equations representing acid-base reactions of carboxylic acids.
6. Write equations representing the preparation of an ester.
7. Write structures and describe the physical properties of esters.
8. Determine the common and IUPAC names of esters.
9. Write equations representing the hydrolysis of an ester.
10. Define the term *saponification,* and describe how soap works in the emulsification of grease and oil.
11. Determine the common and IUPAC names of acid chlorides.
12. Determine the common and IUPAC names of acid anhydrides.
13. Write equations representing the synthesis of acid anhydrides.
14. Discuss the significance of thioesters and phosphoesters in biological systems.

Introduction

We frequently encounter carboxylic acids and their derivatives, the esters, in our daily lives. Some carboxylic acids contribute a tart or tangy flavor to the foods we eat. Other carboxylic acids, the fatty acids, form esters called triglycerides that make up our dietary fats. Several other esters are present in fruits and provide their unique flavors.

Carboxylic acids and esters such as those in our diet are the focus of this chapter.

14.1 Carboxylic Acids

Structure and Physical Properties

Carboxylic acids have the following general structure:

$$R-\overset{\overset{\displaystyle O}{\|}}{C}-OH$$

The –COOH functional group is called a **carboxyl group**. The term *carboxylic acid* tells us that the carboxyl group is derived from a carbonyl group and hydroxyl group. It further tells us that these molecules are acids.

Carboxylic acids are very polar compounds because the carboxyl group consists of two very polar functional groups, the carbonyl group and the hydroxyl group. The electronegative oxygen allows dipole-dipole attractions. Carboxylic acids can also hydrogen bond to one another and to molecules of a polar solvent, such as water. Because of the intermolecular hydrogen bonding and the strong dipole-dipole attractions, carboxylic acids boil at higher temperatures than aldehydes, ketones, or alcohols of comparable molar mass.

Although small carboxylic acids are water soluble, solubility falls off dramatically as the carbon chain length increases.

The lower molar mass carboxylic acids have sharp, sour tastes and unpleasant aromas. The longer chain carboxylic acids are called fatty acids and are components of many biologically important lipids (discussed in Chapter 17.)

Example 1: Recognizing Classes of Organic Compounds Containing Oxygen.
The following general formulas are used to represent different classes of organic compounds. Name each of the families.

1. ROH
2. RCHO
3. RCOR
4. ROR
5. RCOOH
6. RCOOR

Solution:
It is important that you can identify the class of a compound by looking at the structure. Chapters 12, 13 and 14 cover these classes of organic compounds.

1. alcohol (Chapter 12)
2. aldehyde (Chapter 13)
3. ketone (Chapter 13)
4. ether (Chapter 12)
5. carboxylic acid (Chapter 14)
6. ester (Chapter 14)

Example 2: Ranking Compounds According to Boiling Point
Which member of each of the following sets of compounds has the highest boiling point?

1. butane, 1-propanol, or ethanoic acid
2. propanal, propanone, or 1-propanol
3. ethanol, ethanal, or methanoic acid

Solution:
The boiling points of most alkanes, alcohols, aldehydes, and carboxylic acids that have similar molar masses obey the following relationship:

alkane < aldehyde or ketone < alcohol < carboxylic acid
(lowest) (highest)

Note: nomenclature of carboxylic acids has not yet been covered. However, they are easily recognized with *acid* in their name.

Example 2 continued

1. Butane is an alkane, 1-propanol is an alcohol, and ethanoic acid is a carboxylic acid. The three have similar molar masses. According to the ranking system, ethanoic acid would have the highest boiling point.
2. The *–al* ending indicates that propanal is an aldehyde. Propanone is a ketone, and 1-propanol is an alcohol. Each substance has three carbon atoms and they have similar molar masses. Alcohols have the highest boiling point of these three classes of compounds; therefore, 1-propanol has the highest boiling point.
3. Ethanol, with its *–ol* suffix, is an alcohol. Ethanal is an aldehyde, and methanoic acid is a carboxylic acid. The three molecules have comparable molar masses. When comparing these three classes of compounds, the methanoic acid has the highest boiling point.

Nomenclature

LG 2

In the IUPAC system, carboxylic acids are named by replacing the *-e* ending of the parent alkane with the suffix *-oic acid.* The suffix *–dioic* acid is used if there are two carboxyl groups. The parent chain is numbered so that the carboxyl carbon is carbon-1. Other groups are named and numbered in the usual way.

To name the acyl group of a carboxylic acid, replace the *-oic* acid suffix with *–yl*. For instance, the acyl group of acetic acid (ethanoic acid) is the acetyl group (ethanoyl group).

The carboxylic acid derivatives of cycloalkanes are named by adding the suffix *carboxylic acid* to the name of the cycloalkane or substituted cycloalkane. The carboxyl group is defined as being on carbon-1.

Common names of carboxylic acids are frequently used. Table 14.1 shows the IUPAC and common names of several carboxylic acids. Use flash cards to learn these names.

TABLE 14.1 Names of Common Carboxylic Acids

Common Name	IUPAC Name	Structure
formic acid	methanoic acid	$HCOOH$
acetic acid	ethanoic acid	CH_3COOH
propionic acid	propanoic acid	CH_3CH_2COOH
butyric acid	butanoic acid	$CH_3CH_2CH_2COOH$
valeric acid	pentanoic acid	$CH_3(CH_2)_3COOH$

Aromatic carboxylic acids are usually named as derivatives of benzoic acid. The *-oic acid* or *-ic acid* suffix is attached to the appropriate prefix. However, common names of substituted benzoic acids are frequently used.

Example 3: Naming Carboxylic Acids
Give the IUPAC names for the following carboxylic acids:

1. CH_3COOH 2. CH_3CH_2COOH 3. $CH_3CHBrCH_2COOH$

Example 3 Solution:

1. This carboxylic acid has two carbon atoms with no other substituents. The name of the parent alkane is ethane, giving **ethanoic acid**. This compound is much more often referred to by its common name of acetic acid.
2. A carboxylic acid with three carbon atoms gives the name **propanoic acid**. The common name is propionic acid.
3. This carboxylic acid has a substitution on the carbon chain. Begin by finding the longest carbon chain. In this case there are four carbons, giving the parent name of butane. Number the carbon atoms with the carboxyl group as C-1. To which carbon is the Br group attached? Remember, the Br is written after the number of the carbon to which it is attached. The compound is **3-bromobutanoic acid**. (Common name: β-bromobutyric acid)

Some Important Carboxylic Acids

LG 3

Many carboxylic acids found in nature are listed in Table 14.1 of the text. Fatty acids can be isolated from a variety of fats and oils. More complex carboxylic acids are also found in a variety of foodstuffs. Citric acid is found in citrus fruits and is added to foods to give a sharp taste (sour candies) or as a preservative and antioxidant. Bacteria in milk produce lactic acid as a product of fermentation of sugars. It contributes a tangy flavor to yogurt and buttermilk and acts as a food preservative. Lactic acid is also produced by muscles during strenuous exercise. Adipic acid adds tartness to soft drinks and helps to retard spoilage. Another important use of adipic acid is the synthesis of polymers such as nylon and polyurethane.

Acetylsalicylic acid, commonly known as aspirin is a carboxylic acid.

Reactions Involving Carboxylic Acids

LG 4

Preparation of Carboxylic Acids

Simple carboxylic acids can be made by **oxidation** of the appropriate primary alcohol or aldehyde. A variety of oxidizing agents, including oxygen, can be used. The general reaction is:

$$RCH_2OH \xrightarrow{[O]} R-\overset{\overset{\displaystyle O}{\|}}{C}-H \xrightarrow{[O]} R-\overset{\overset{\displaystyle O}{\|}}{C}-OH$$

Primary alcohol — Aldehyde — Carboxylic acid

LG 5

Acid-Base Reactions

The carboxylic acids behave as weak acids because they are proton donors. They are weak acids (typically less than 5% dissociation) that produce a carboxylate anion and a hydronium ion in water, as seen in the following example:

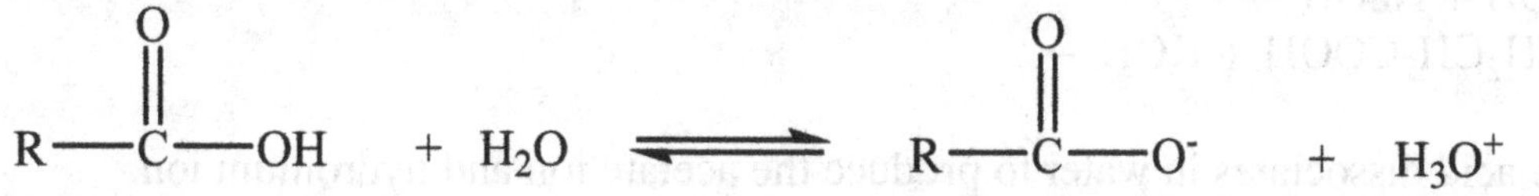

Carboxylic acid | Water | Carboxylate anion | Hydronium ion

When strong bases are added to a carboxylic acid, neutralization occurs:

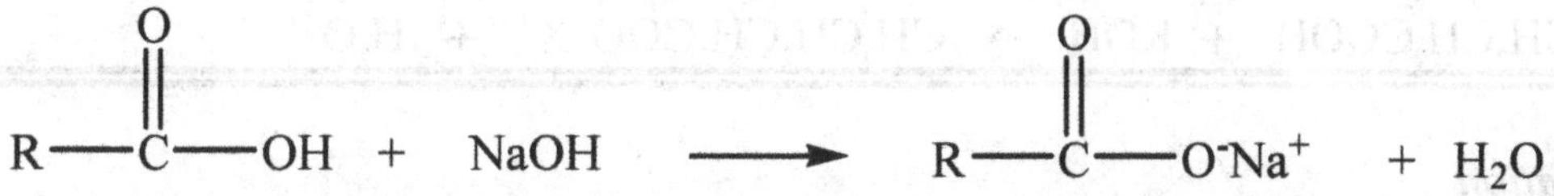

Carboxylic acid | Strong base | Carboxylic acid salt | Water

The salt of a carboxylic acid is named by replacing the *-ic acid* suffix with *-ate*. This name is preceded by the name of the appropriate cation, for instance, sodium.

The carboxylic acid salts formed are ionic substances, and hence, are quite soluble in water. The long-chain carboxylic acid salts (fatty acid salts) are called **soaps**.

Example 4: Drawing the Structure of Carboxylic Acid Salts

Draw the structural formula and line formula of sodium pentanoate.

Solution:

The parent name is pentane, which means there are five carbons in the chain. The *–ate* suffix indicates it is an anion of a carboxylic acid, giving:

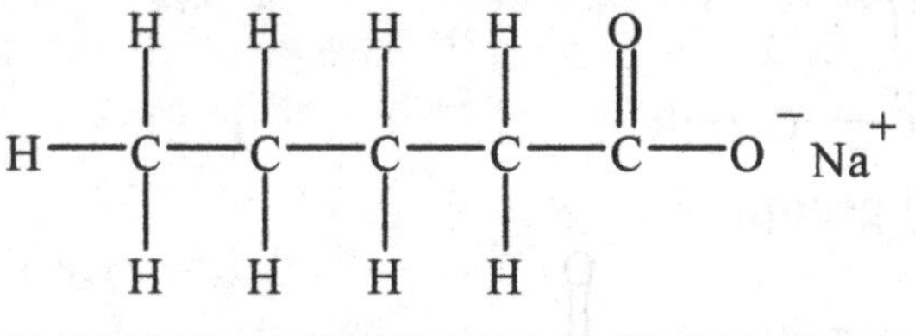

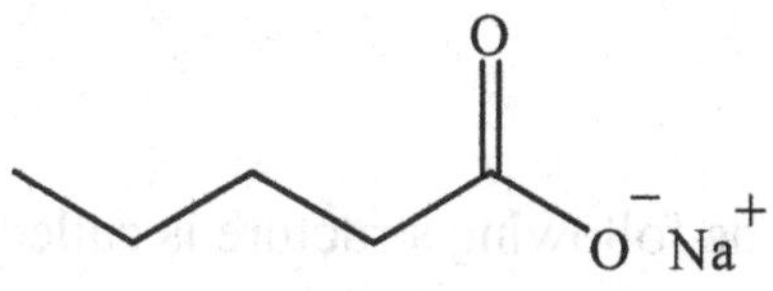

Example 5: Reactions of Carboxylic Acids

Give the products of the following reactions:

1. $CH_3COOH + H_2O \rightleftharpoons$
2. $HCOOH + NaOH \rightarrow$
3. $CH_3CH_2CH_2COOH + KOH \rightarrow$

Solution:

1. Acetic acid dissociates in water to produce the acetate ion and hydronium ion.
   $CH_3COOH + H_2O \rightleftharpoons CH_3COO^- + H_3O^+$
2. Formic acid reacts with a strong base to produce sodium formate and water.
   $HCOOH + NaOH \rightarrow HCOO^-Na^+ + H_2O$
3. KOH is also a strong base. Butyric acid reacts with a strong base to produce potassium butyrate and water.
   $CH_3CH_2CH_2COOH + KOH \rightarrow CH_3CH_2CH_2COO^-K^+ + H_2O$

Esterification

LG 6

In the esterification reaction, carboxylic acids react with alcohols to form esters and water according to the following general reaction:

$$R{-}\overset{\overset{\large O}{\|}}{C}{-}OH + R'{-}OH \underset{\text{heat}}{\overset{H^+}{\rightleftharpoons}} R{-}\overset{\overset{\large O}{\|}}{C}{-}OR' + H_2O$$

Carboxylic acid   Alcohol   Ester   Water

14.2 Esters

Structure and Physical Properties

LG 7

Esters have the following general structure:

$$R{-}\overset{\overset{\large O}{\|}}{C}{-}O{-}R$$

The following structure is called the **acyl group**:

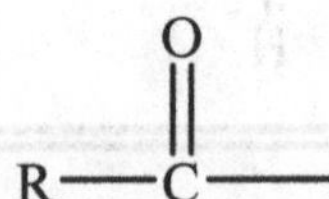

The acyl group is the functional group of the carboxylic acid derivatives, including the esters, acid chlorides, acid anhydrides, and amides. (Amides have the general form $RCONH_2$ and will be discussed in Chapter 15.)

Esters have pleasant aromas. Many are found in natural foodstuffs. The carbonyl group of the ester is polar and can participate in dipole-dipole attractions. However the carbonyl group is flanked by hydrocarbon chains. Because of these hydrophobic chains, the polarity of esters is comparable to that of aldehydes and ketones. In fact, esters boil at approximately the same temperatures of aldehydes or ketones of comparable molar mass.

Like aldehydes and ketones, esters can form hydrogen bonds with water molecules. As a result, smaller esters are somewhat soluble in water.

Nomenclature

LG 8

Esters are **carboxylic acid derivatives,** organic compounds derived from carboxylic acids. They are formed in the reaction of a carboxylic acid with an alcohol and both of these families are reflected in the naming of the ester. The first part of the name of an ester is the alkyl (or aryl) group of the alcohol. The second part of the name is derived from the carboxylic acid. The *alkyl* or *aryl* portion of the alcohol name is used as the prefix, and the -*ic acid* ending of the name of the carboxylic acid is replaced with *-ate*. See Example 6.

Example 6: Naming Esters

Give the common and IUPAC names of the following esters:

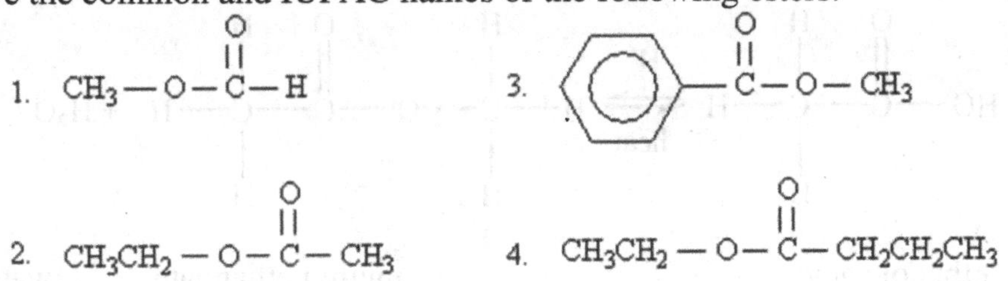

1. $CH_3-O-C(=O)-H$
2. $CH_3CH_2-O-C(=O)-CH_3$
3. $C_6H_5-C(=O)-O-CH_3$
4. $CH_3CH_2-O-C(=O)-CH_2CH_2CH_3$

Solution:

1. Begin by finding the oxygen which is between two carbons. See the arrow below. The side of the O containing the carboxyl group is named for the carboxylic acid. (In this case it is to the right of the arrow.) The name of this group is based on the one carbon carboxylic acid (formic acid) to give the name formate.

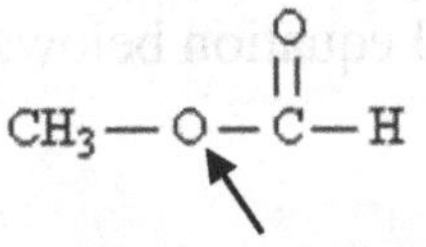

   $CH_3-O-C(=O)-H$ (arrow pointing to the O)

   On the other side of the arrow is the portion derived from the alcohol. A one carbon group is methyl. Combining the two names gives **methyl formate.** The IUPAC name is **methyl methanoate**.
2. Find the oxygen between the two carbons. The side containing the carboxyl group has two carbons, giving the common name acetate. The alcohol-derived side is a two carbon chain, giving the ethyl group. Combining these two names give **ethyl acetate**. The IUPAC name is **ethyl ethanoate**.
3. Again, we locate the oxygen between the two carbons, but in this example, the carboxyl group is on the left-hand side. You must be careful to realize that there is no set way to draw the structure. The carboxyl group is attached to a benzene ring, giving benzoate. The other side of the molecule contains a methyl group (one carbon), giving **methyl benzoate**.
4. The side with the carboxyl group has four carbons (butyrate), and the other side has two carbons. This gives the name **ethyl butyrate.** The IUPAC name is **ethyl butanoate**.

Reactions Involving Esters

Preparation of Esters

The conversion of a carboxylic acid to an ester requires heat and is catalyzed by a trace of acid (H+). This is considered a *condensation* reaction since a water molecule is removed during the reaction.

Example 7: Esterification
Write the reaction of methanol and ethanoic acid. Give the structure, and use IUPAC nomenclature to name of all reactants and products.

Solution:
To easily see the linkage, write the structure of the two reactants with their –OH groups facing each other. Remove an H_2O from the two OH groups, which leaves an O for the linkage between the groups.

$$\mathrm{H{-}CH_2{-}OH} + \mathrm{HO{-}C({=}O){-}CH_2{-}H} \underset{\text{heat}}{\overset{H^+}{\rightleftharpoons}} \mathrm{H{-}CH_2{-}O{-}C({=}O){-}CH_2{-}H} + H_2O$$

methanol — ethanoic acid — methyl ethanoate — water

Hydrolysis of Esters

Esters undergo **hydrolysis** reactions in water. This reaction requires heat and a small amount of acid (H+) or base (OH–) to catalyze the reaction. The acid catalyzed reaction is essentially the reverse reaction of esterification, as seen in the first general equation below.

$$\mathrm{R{-}C({=}O){-}OR'} + H_2O \underset{\text{Heat}}{\overset{H^+}{\rightleftharpoons}} \mathrm{R{-}C({=}O){-}OH} + \mathrm{R'{-}OH}$$

Ester — Water — Carboxylic acid — Alcohol

$$\mathrm{R{-}C({=}O){-}OR'} + NaOH \longrightarrow \mathrm{R{-}C({=}O){-}O^-Na^+} + \mathrm{R'{-}OH}$$

Ester — Strong base — Carboxylic acid salt — Alcohol

LG 10

The second reaction above is the base-catalyzed hydrolysis of an ester. The reaction is called **saponification**, which produces a carboxylic acid salt. The carboxylic acid is formed when the reaction mixture is neutralized with an acid. Saponification is used to hydrolyze fats and oils (which are esters) to the salts of long chain fatty acids – *soaps*.

Soaps function by forming micelles around grease and oil. The hydrophobic alkane tail of the soap dissolves the oils and grease, and the hydrophilic carboxylate end of the molecule dissolves in the water. This produces a micelle, a tiny sphere with the alkyl group tails of the soap molecules, grease, and oils in the center and the carboxylate ends of the soap on the outside of the sphere, dissolved in the water.

Condensation Polymers

Condensation polymers are formed by the polymerization of monomers in a reaction that forms a small molecule, such as water or an alcohol. Polyesters are formed by reacting a dicarboxylic acid and a dialcohol.

$$HO—R'—OH + HOOC—R—COOH \xrightarrow{H^+} HO—R'—O—\overset{\overset{O}{\|}}{C}—R—COOH + H_2O$$

Because each of the reactants has two reacting groups, the new molecule formed still has two reactive groups, —OH and —COOH. So the polymerization reaction continues, forming very long polyester molecules.

When formed as fibers, polyesters can be used to make fabrics. They can also be formed into films, such as Mylar, which is used as the base for recording tapes and photographic film.

Polyethylene terephthalate (PET), a polyester, is used to make plastic bottes, like those used for soft drinks. A downside of PET is that it cannot withstand high temperatures. Another polyester, polyethylene naphthalate (PEN), can be used at higher temperatures.

14.3 Acid Chlorides and Acid Anhydrides

Acid Chlorides

Acid chlorides are compounds which have a chlorine atom in place of the –OH group in a carboxylic acid. They are named by dropping the *-ic acid* ending of the common or the *-oic* acid ending of the IUPAC name of the carboxylic acid and replacing it with *-oyl chloride.*

Example 8: Naming Acid Chlorides
Name the following acid chlorides:

1. $CH_3CH_2—\overset{\overset{O}{\|}}{C}—Cl$ 2. $BrCH_2—\overset{\overset{O}{\|}}{C}—Cl$

Solution:

1. This acid chloride has a three carbon chain. The IUPAC name for the acid would be propanoic acid. Replace the *–oic acid* suffix with *–oyl chloride,* giving the name **propanoyl chloride.**
2. The parent chain is two carbons long, so it is based on the IUPAC name of ethanoic acid. It has a bromo group attached to C-2. The compound is **2-bromoethanoyl chloride.**

Acid chlorides are noxious, irritating chemicals that must be handled with care. They are slightly polar and boil at about the same temperature as the corresponding aldehyde or ketone. They cannot be dissolved in water because they react violently with it. Acid chlorides are used primarily in the synthesis of esters and amides.

Acid Anhydrides

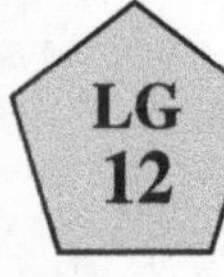

Acid anhydrides are compounds in which two carboxylic acids are joined together by the removal of a water molecule. They can be classified as *symmetrical* if both acyl groups are the same or *unsymmetrical* if there are two different acyl groups. They are named by replacing the *acid* ending of the carboxylic acid name with the term *anhydride*.

Example 9: Drawing Structures of Anhydrides.
Draw the structures of the following symmetrical acid anhydrides:

1. acetic anhydride 2. propanoic anhydride 3. 3-bromobutanoic anhydride

Solution:

1. Draw two acetic acid molecules with their OH groups facing each other. Remove an H_2O and join the two groups through the remaining O.

$$\mathrm{H{-}\overset{H}{\underset{H}{C}}{-}\overset{O}{\overset{\|}{C}}{-}OH\;\;HO{-}\overset{O}{\overset{\|}{C}}{-}\overset{H}{\underset{H}{C}}{-}H \;\Rightarrow\; H{-}\overset{H}{\underset{H}{C}}{-}\overset{O}{\overset{\|}{C}}{-}O{-}\overset{O}{\overset{\|}{C}}{-}\overset{H}{\underset{H}{C}}{-}H}$$

remove the H_2O

2. Propanoic acid is a three carbon acid. Joining two of the acids together by way of their OH groups (removing H_2O) gives:

$$\mathrm{CH_3CH_2-\overset{O}{\overset{\|}{C}}-O-\overset{O}{\overset{\|}{C}}-CH_2CH_3}$$

3. 3-Bromobutanoic acid has four carbons. A bromine atom is attached to C-3. Remember, the carboxyl carbon is C-1. Joining these two acids together will produce 3-bromobutanoic anhydride.

$$\mathrm{CH_3CHBrCH_2-\overset{O}{\overset{\|}{C}}-O-\overset{O}{\overset{\|}{C}}-CH_2CHBrCH_3}$$

LG
13

The anhydrides are not generally formed by the removal of water from two acids (as depicted in Example 9.) They are formed in a reaction between an acid chloride and a carboxylate ion.

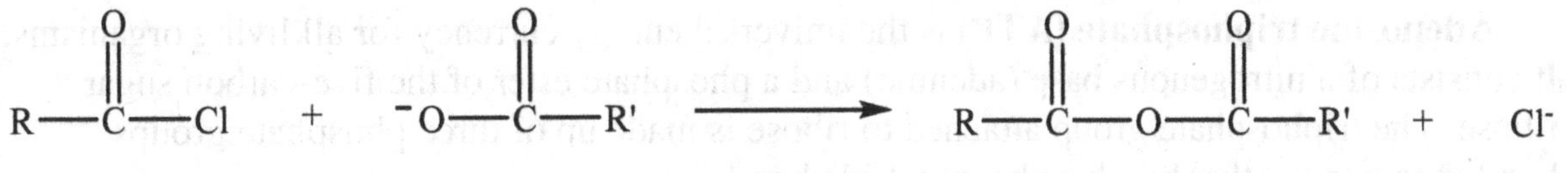

$$R-\overset{O}{\overset{\|}{C}}-Cl + {}^{-}O-\overset{O}{\overset{\|}{C}}-R' \longrightarrow R-\overset{O}{\overset{\|}{C}}-O-\overset{O}{\overset{\|}{C}}-R' + Cl^-$$

An acid anhydride can be hydrolyzed back to its corresponding carboxylic acids by water. This reaction is sped up by the addition of acid. See the reaction below.

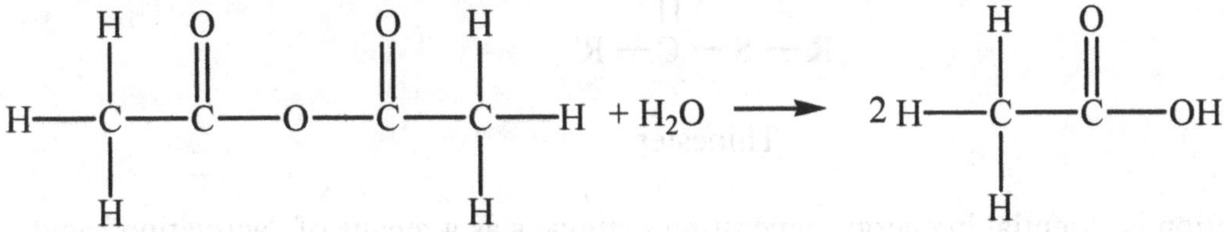

$$CH_3-\overset{O}{\overset{\|}{C}}-O-\overset{O}{\overset{\|}{C}}-CH_3 + H_2O \longrightarrow 2\,CH_3-\overset{O}{\overset{\|}{C}}-OH$$

Acyl transfer reactions can occur when acid anhydrides react with an alcohol to produce an ester and a carboxylic acid.

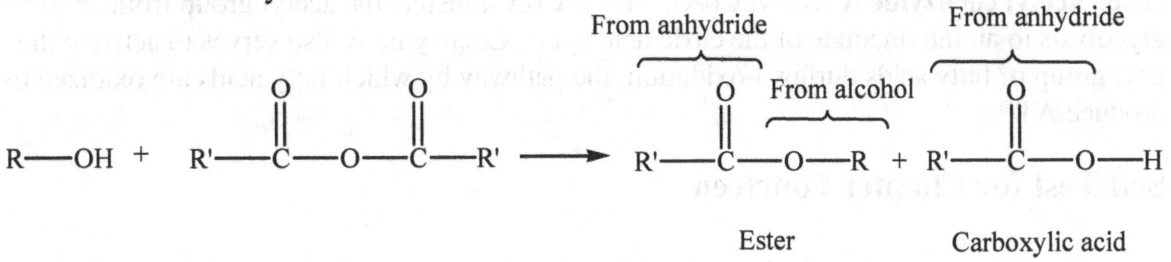

$$R-OH + R'-\overset{O}{\overset{\|}{C}}-O-\overset{O}{\overset{\|}{C}}-R' \longrightarrow \underbrace{R'-\overset{O}{\overset{\|}{C}}-O-R}_{\text{Ester}} + \underbrace{R'-\overset{O}{\overset{\|}{C}}-O-H}_{\text{Carboxylic acid}}$$

From anhydride | From alcohol | From anhydride

Ester | Carboxylic acid

14.4 Nature's High Energy Compounds: Phosphoesters and Thioesters

LG 14

An alcohol can react with phosphoric acid to produce a phosphate ester, or **phosphoester**, as seen here:

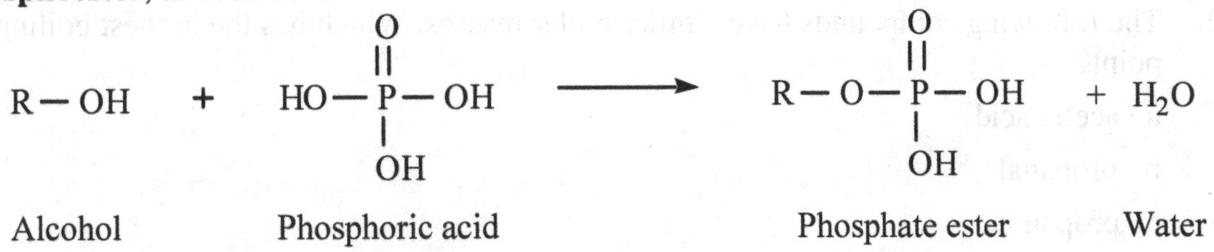

$$R-OH + HO-\overset{O}{\overset{\|}{P}}(OH)-OH \longrightarrow R-O-\overset{O}{\overset{\|}{P}}(OH)-OH + H_2O$$

Alcohol | Phosphoric acid | Phosphate ester | Water

Phosphoesters of monosaccharides are very important in energy harvesting reactions in the cell.

When two phosphate groups react with one another, a water molecule is lost. Because water is lost, the resulting bond is called a **phosphoanhydride** bond.

$$R-O-\overset{O}{\overset{\|}{P}}(OH)-OH + HO-\overset{O}{\overset{\|}{P}}(OH)-OH \longrightarrow R-O-\overset{O}{\overset{\|}{P}}(OH)-O-\overset{O}{\overset{\|}{P}}(OH)-OH + H_2O$$

Phosphate ester | Phosphoric acid | Phosphoanhydride bond

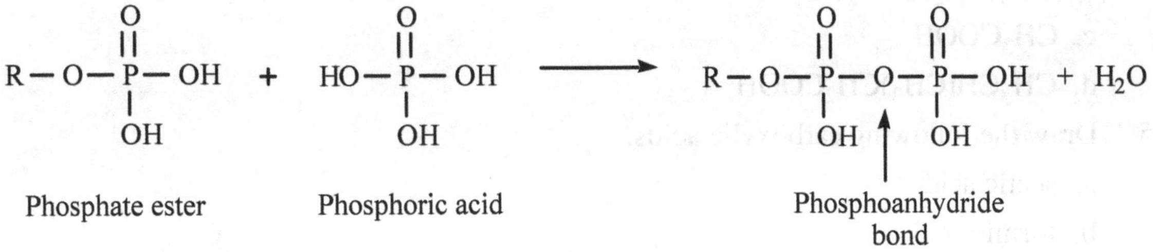

Adenosine triphosphate (ATP) is the universal energy currency for all living organisms. It consists of a nitrogenous base (adenine) and a phosphate ester of the five-carbon sugar ribose. The triphosphate group attached to ribose is made up of three phosphate groups bonded to one another by phosphoanhydride bonds.

Cellular enzymes can carry out a reaction between a carboxylic acid and a thiol to produce a **thioester.**

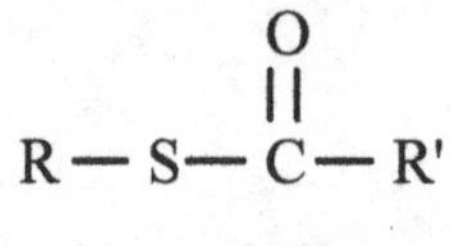

$$R-S-\overset{\overset{\large O}{||}}{C}-R'$$

Thioester

This reaction is essential in energy-generating pathways as a means of "activating" acyl groups for subsequent breakdown reactions. The complex thiol coenzyme A is the most important acyl group activator in the cell. The most common thioester is the acetyl ester, called **acetyl coenzyme A** (acetyl CoA). Acetyl CoA transfers the acetyl group from glycolysis to an intermediate of the citric acid cycle. Coenzyme A also serves to activate the acyl group of fatty acids during β-oxidation, the pathway by which fatty acids are oxidized to produce ATP.

Self Test for Chapter Fourteen

1. Give the general formula for each of the following classes of organic compounds:

   a. alcohol d. aldehyde

   b. ether e. carboxylic acid

   c. ester f. ketone

2. The following compounds have similar molar masses. Which has the highest boiling point?
   a. acetic acid
   b. propanal
   c. propanone
   d. 1-propanol
3. Would you expect palmitic acid, $CH_3(CH_2)_{14}COOH$, to be soluble in water?
4. Give the IUPAC names of the following:
   a. HCOOH
   b. $CH_3(CH_2)_4COOH$
   c. CH_3COOH
   d. $CH_3CH(CH_3)CH_2COOH$
5. Draw the following carboxylic acids:
   a. acetic acid
   b. formic acid

c. benzoic acid

d. capric acid

6. Provide the common names of each of the following carboxylic acids:

a. $CH_3(CH_2)_2\overset{O}{\overset{\|}{C}}-OH$ b. CH_3CH_2COOH c. $CH_3\underset{CH_3}{\underset{|}{CH}}-\overset{O}{\overset{\|}{C}}-OH$

7. How is lactic acid used in the food industry? Where is lactic acid formed in the body?
8. What two ions are formed when a soluble carboxylic acid reacts with a molecule of water?
9. For the following reactions, name the products that are formed:

a. butyric acid + NaOH

b. acetic acid + $Ca(OH)_2$

10. A primary alcohol is mixed with potassium permanganate solution. What reaction occurs, and what is the function of the $KMnO_4$?
11. Complete the following reactions:

a. $CH_3COOH + H_2O \rightleftharpoons$?

b. $CH_3COOH + NaOH$?

c. $CH_3COOH + CH_3OH$ (acid as a catalyst) ?

12. The long-chain carboxylic acid salts (fatty acid salts) are good _______________.
13. What carboxylic acid derivatives are mildly polar and often have pleasant aromas?
14. Give the common name of the following esters:

a. $H-\overset{O}{\overset{\|}{C}}-O-CH_2CH_3$

b. $CH_3CH_2-\overset{O}{\overset{\|}{C}}-O-CH_3$

c. $CH_3CH_2CH_2CH_2CH_2-O-\overset{O}{\overset{\|}{C}}-CH_3$

15. Give the IUPAC names of the following esters:

a. methyl acetate

b. $CH_3CH_2-\overset{O}{\overset{\|}{C}}-O-CH_2CH_3$

16. What two compounds must be reacted to form methyl propanoate?

17. Provide the IUPAC name for the product of the reaction between ethanoic acid and methanol in the presence of heat and a trace of acid.
18. Provide the missing reactant in the following equation:

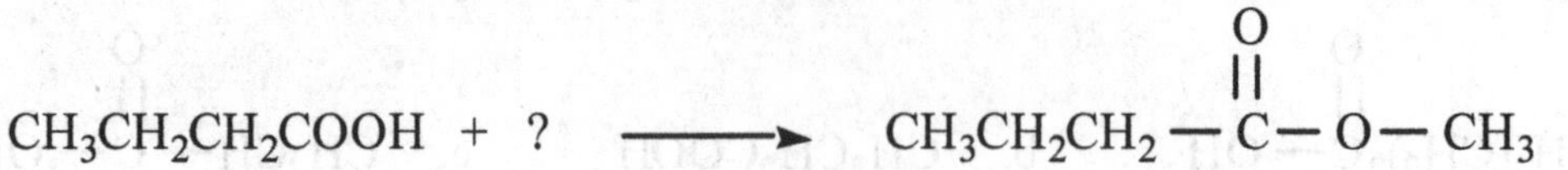

$$CH_3CH_2CH_2COOH + ? \longrightarrow CH_3CH_2CH_2-\overset{\overset{\displaystyle O}{||}}{C}-O-CH_3$$

19. What are the hydrolysis products of an ester?
20. Name the products of the hydrolysis of methyl acetate.
21. What polymer is produced by the condensation reaction between a dialcohol and a dicarboxylic acid?
22. What is the term for base-catalyzed hydrolysis of an ester?
23. What two functional groups in a soap molecule allow it to act as soap?
24. Provide the common and IUPAC name for the following molecule:

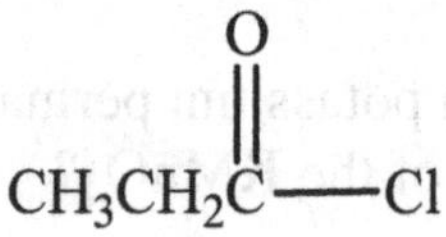

$$CH_3CH_2\overset{\overset{\displaystyle O}{||}}{C}-Cl$$

25. What are the structure and the name of the product of the following reaction?

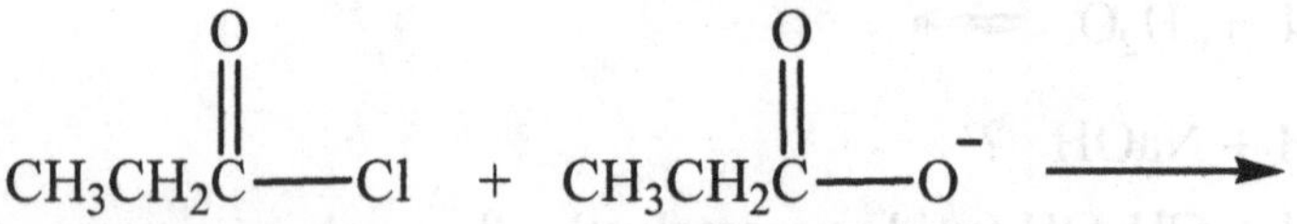

$$CH_3CH_2\overset{\overset{\displaystyle O}{||}}{C}-Cl + CH_3CH_2\overset{\overset{\displaystyle O}{||}}{C}-O^- \longrightarrow$$

26. Name the product of the hydrolysis of acetic anhydride.
27. What is the product of the reaction between an alcohol and phosphoric acid?
28. What is the term for the bond formed between two phosphate groups?
29. What molecule is the universal energy currency for all cells?
30. What is the term for the product of a reaction between a thiol and a carboxylic acid?
31. What is the name of the molecule that carries 2-carbon acetyl groups to the citric acid cycle?

15 *Amines and Amides*

Learning Goals

1. Classify amines as primary, secondary, or tertiary.
2. Describe the physical properties of amines.
3. Draw and name simple amines using the systematic and common nomenclature systems.
4. Write equations representing the synthesis of amines.
5. Write equations showing the basicity and neutralization of amines.
6. Describe the structure of quaternary ammonium salts, and discuss their use as antiseptics and disinfectants.
7. Discuss the biological significance of heterocyclic amines.
8. Describe the physical properties of amides.
9. Draw the structure and write the common and IUPAC names of amides.
10. Write equations representing the preparation of amides.
11. Write equations showing the hydrolysis of amides.
12. Draw the general structure of an amino acid.
13. Draw and discuss the structure of a peptide bond.
14. Describe the function of neurotransmitters.

Introduction

Oxycodone and hydrocodone are examples of pain medications that are amines. They are often combined with acetaminophen, an amide, in order to control severe pain. Amines and amides are important in pharmaceutical applications, and they are essential for normal physiological functions of the body.

In this chapter, you will learn about the structure, properties, and chemistry of amines and amides.

15.1 Amines

Structure and Physical Properties

Amines are organic molecules that contain the amino group, $-NH_2$, or substituted amino group. They may be aromatic or aliphatic and have the general formula $R-NH_2$. You can think of the amines as substituted ammonia molecules in which one or more of the hydrogens has been substituted by an organic group. Like ammonia, amines are pyramidal molecules. The nitrogen atom is attached to three groups and has a nonbonding pair of electrons.

Amines are classified by the number of hydrocarbon groups attached to the nitrogen. **Primary (1°) amines** have one R group; **secondary (2°) amines** have two R groups; and **tertiary (3°) amines** have three R groups attached to the nitrogen.

Example 1: Classification of Amines

Classify the following amines as either primary, secondary, or tertiary:

1. CH_3NH_2
2. CH_3NHCH_3
3. $CH_3N(CH_3)_2$

Solution:

1. If we draw the structural formula it is easy to see that the nitrogen has one carbon attached:

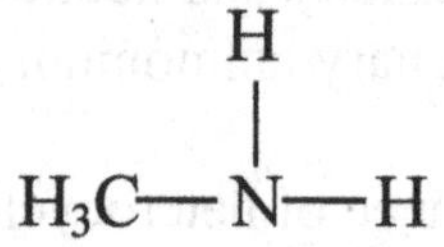

$$H_3C-\underset{}{\overset{H}{\overset{|}{N}}}-H$$

   The compound is a primary (1°) amine.

2. This compound can be drawn as

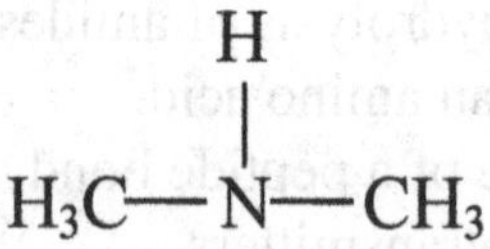

$$H_3C-\overset{H}{\overset{|}{N}}-CH_3$$

   There are two carbons connected directly to the nitrogen; therefore, the compound is a secondary (2°) amine.

3. The compound is a tertiary (3°) amine which is easily seen in the following structure with three carbons connected to the nitrogen atom.

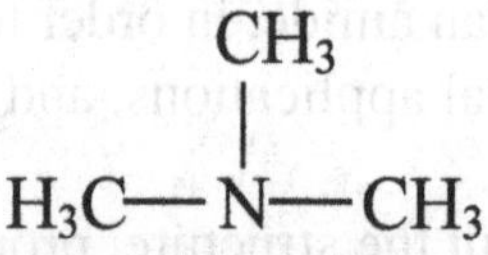

$$H_3C-\overset{CH_3}{\overset{|}{N}}-CH_3$$

Example 2: Classification of Amines

Write the general formulas for the following:

1. primary amine
2. secondary amine
3. tertiary amine

Solution:

1. RNH_2
2. R_2NH
3. R_3N

LG 2

The N–H bond is polar. In addition, the nitrogen atom contains an unshared pair of electrons. As a result, hydrogen bonding can occur between amine molecules or between amine molecules and water. This feature determines the physical properties of the amines, such as boiling point and water solubility. Keep in mind that tertiary amines do not have N-H bonds, and thus, do not exhibit hydrogen bonding between amine molecules. As a result, they have much lower boiling points than primary or secondary amines.

The –NH group is less polar than the –OH group. Therefore, the boiling points of amines are lower than those of comparable alcohols, but higher than those of comparable ethers or alkanes. The smaller amines are readily soluble in water, but as the size of the hydrocarbon groups increases, their solubility in water decreases. The following diagram depicts the hydrogen bonding between an amine molecule and water.

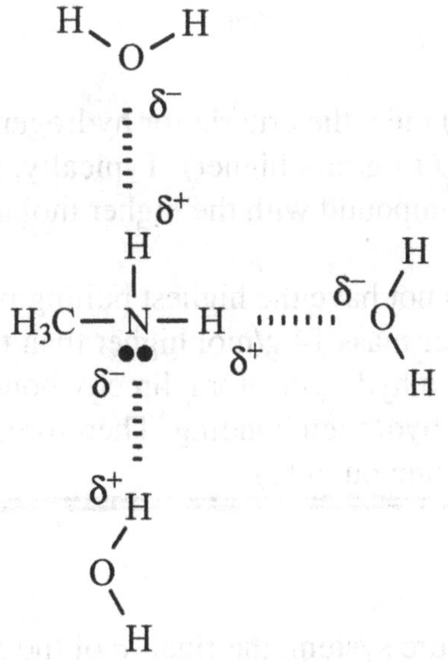

H   H
O
δ⁻
δ⁺
H   H
H₃C—N—H ...... δ⁻ O
δ⁻   δ⁺   H
δ⁺ H
O
H

Example 3: Ranking Compounds According to Boiling Point
Which member of each of the following sets of compounds (with similar molar masses) has the highest boiling point?

1. propane, dimethyl ether, or ethanamine
2. ethanol, dimethyl ether, or ethanamine

Solution:
The normal boiling points of alkanes, ethers, primary amines, and primary alcohols obey the following relationship:

alkane < ether < primary amine < primary alcohol

1. Although we have not yet learned the nomenclature of amines, the name ethanamine indicates that the compound is an amine. When comparing the boiling point of an amine to that of an alkane (propane) and ether (dimethyl ether) of comparable molar masses, the amine will have the highest boiling point.
2. Ethanol (an alcohol) will have the highest boiling point among these three organic compounds we have studied with comparable molar masses.

Example 4: Boiling Point Trends
Consider the following three amines and their corresponding boiling points.
1. Why is dimethylamine higher than methylamine?
2. Why does trimethylamine not have the highest boiling point of the three?

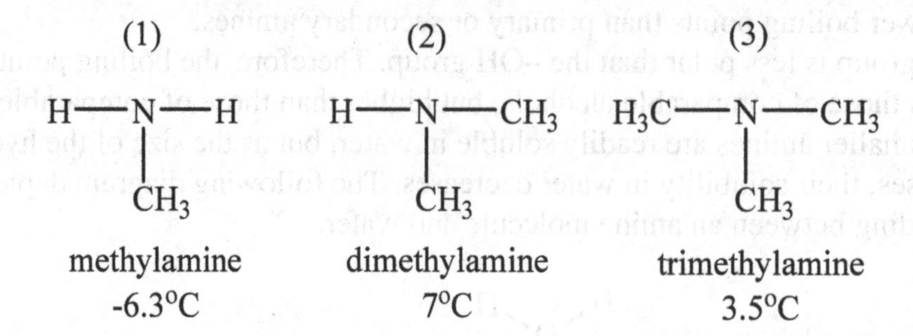

Solution:
1. Both compounds (1) and (2) meet the criteria for hydrogen bonding. However, compound (2) has a higher molar mass (14 g/mol higher). Typically, if the intermolecular attractions are of the same types, the compound with the higher molar mass will have the higher boiling point.
2. So, why does compound (3) not have the highest boiling point of the three compounds? After all, it does have a molar mass 14 g/mol higher than that of compound (2). The tertiary amine no longer has a hydrogen atom directly bonded to the nitrogen, so it no longer meets the criteria for hydrogen bonding. Therefore, it has a lower boiling point than the lower molar mass compound (2).

Nomenclature

In the systematic nomenclature system, the final *-e* of the name of the parent compound is dropped, and the suffix *-amine* is added. The parent chain is numbered to give the amine group the lowest possible number and other substituents are named and numbered as usual. For secondary or tertiary amines, the prefix *N*-alkyl is added to the name of the parent compound. Many aromatic amines have special names. An example of this is aniline, a benzene molecule bonded to an $-NH_2$ group.

Common names are also used, especially for the simple amines. The common names of the alkyl groups bonded to the amine nitrogen are followed by the suffix *-amine*. All alkyl groups are listed alphabetically as one continuous word followed by the ending *-amine*.

Example 5: Naming Amines Using the Systematic Name
Determine the systematic names of the following amine compounds:

1. CH_3NH_2

2. $CH_3-\underset{NH_2}{\overset{H}{C}}-CH_3$

3. $CH_3CH_2CH_2\underset{NH_2}{C}HCH_3$

4. $CH_3CH_2CH_2NHCH_3$

Example 5 Solution:

1. This amine only has one alkyl group on the nitrogen; that is, it is a primary amine. The alkyl group only has one carbon. Start with the name methane. Drop the *–e* suffix and add the ending *-amine* giving the name **methanamine.**
2. This primary amine has a parent chain with three carbon atoms, giving the parent name of propane. The amine group is attached to C-2 giving the name **2-propanamine.**
3. The parent chain of this primary amine has 5 carbons, giving the parent name of pentane. Again, the amine group is attached to C-2, giving the name **2-pentanamine.**
4. This is a secondary amine because it has two alkyl groups attached to the nitrogen. The parent chain is the longer chain, in this case having three carbon atoms. The parent name is propane with the amine group attached to the end carbon (or C-1). The smaller alkyl group is named with an *N* before the methyl group, giving the name ***N*-methyl-1-propanamine.**

Example 6: Naming Amines Using the Common Naming System
Determine the common names for each of the following amines.

1. CH_3NH_2
2. $(CH_3)_2NH$
3. $(CH_3CH_2)_2NH$
4. $CH_3-N(H)-C(H)(CH_3)-CH_2CH_3$

Solution:

1. This amine has a methyl group attached to the nitrogen, giving the name **methylamine.**
2. This secondary amine has two methyl groups and has the common name **dimethylamine**.
3. Similar to the compound in 2, this secondary amine has two ethyl groups attached to the nitrogen, giving the common name **diethylamine.**
4. This secondary amine has two different alkyl groups attached to the nitrogen. A methyl group and a four carbon alkyl group are attached at the second carbon of the butyl chain. The name of the four carbon alkyl group is sec-butyl. Name the groups in alphabetical order (m before s), giving the name **methyl-*sec*-butylamine.**

Medically important amines include pain relievers or pain blockers such as novocaine and demerol; decongestants such as ephedrine; and antibiotics such as sulfanilamide.

Reactions Involving Amines

Preparation of Amines

Amines are prepared by the reduction of amides. If two of the R groups of an amide are hydrogen atoms, the product will be a *primary amine*. If only one R group is a hydrogen atom and the other is an organic group, the product will be a *secondary amine*. If both R groups are organic groups, the product will be a *tertiary amine*.

$$\underset{\text{Amide}}{R-\overset{\overset{O}{\|}}{C}-\overset{\overset{R}{|}}{N}-R} \xrightarrow{[H]} \underset{\text{Amine}}{R-CH_2-\overset{\overset{R}{|}}{N}-R}$$

(R may represent a hydrogen atom or an organic group.)

Aromatic amines can be prepared by the reduction of nitro compounds.

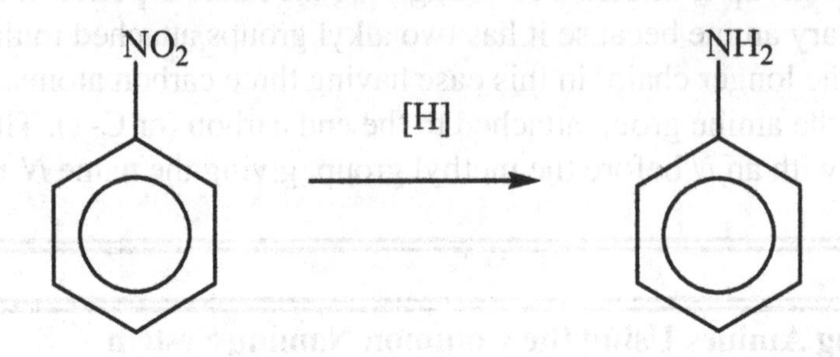

Basicity

LG 5

Amines have a nonbonding pair of electrons that can be shared with an electron-deficient group to form a new bond. For instance, an amine may react with a proton (H^+), producing a new N–H bond. The original unshared pair of electrons on the nitrogen atom has been shared with the electron-deficient proton. The product is an **alkylammonium ion**.

Neutralization

Amines react with most acids to form alkylammonium salts. The generalized reaction is represented here:

$$R-\underset{\underset{H}{|}}{\overset{\overset{H}{|}}{N}}: + HCl \longrightarrow R-\underset{\underset{H}{|}}{\overset{\overset{H}{|}}{N^+}}-H\ Cl^-$$

The product, an *alkylammonium salt*, is named by replacing the term *amine* with the term *ammonium* followed by the name of the anion.

Example 7: Base Properties of Amines
Complete the following reactions and name the products:

1. $CH_3CH_2NH_2 + HCl \rightarrow ?$
2. $(CH_3)_2NH + HI \rightarrow ?$
3. $CH_3NH_2 + H_2O \rightleftharpoons ?$

Example 7 Solution:

1. This is the neutralization reaction of the amine and the acid. The amine base accepts the proton to produce the ammonium salt: $CH_3CH_2NH_3^+ Cl^-$.
   To name the salt, replace the suffix amine with the suffix ammonium, giving **ethylammonium chloride.**
2. HI is also a strong acid and will neutralize the base, giving the salt $(CH_3)_2NH_2^+ I^-$, named **dimethylammonium iodide.**
3. This is the ionization of the weak base. The water donates the proton to the base, giving $CH_3NH_3^+ OH^-$, **methylammonium hydroxide.**

Several important drugs are amines. They are generally administered as alkylammonium salts because the salts are ionic and much more soluble in water and body fluids.

Quaternary Ammonium Salts

LG 6

Quaternary ammonium salts are ammonium salts having four organic groups bonded to the nitrogen. They have the general structure $R_4N^+ A^-$.

Quaternary ammonium salts with very long carbon chains are often used as disinfectants and antiseptics because they have detergent activity.

15.2 Heterocyclic Amines

Heterocyclic amines are cyclic amines in which one of the atoms of the backbone of the ring is nitrogen. They are found in many important cellular macromolecules, including DNA, RNA, and coenzymes. The indole and pyridine rings are found in many **alkaloids**. Alkaloids include cocaine, nicotine, quinine, morphine, heroin, and LSD (lysergic acid diethylamide). In small doses, cocaine is used as an **anesthetic** for the sinus and eyes. Morphne is a strong analgesic, a drug that acts as a painkiller

15.3 Amides

Amides are an important class of nitrogen-containing organic compounds with the functional group shown below:

$$R-\overset{\overset{\displaystyle O}{\|}}{C}-NH_2$$

Amides are carboxylic acid derivatives. The –OH of the carboxylic acid is replaced with an amine ($-NH_2$). The **amide bond** is the bond between the carbonyl carbon and the amine nitrogen.

Structure and Physical Properties

Most amides are solids at room temperature, with boiling points even higher than the corresponding carboxylic acid. The simpler amides are quite soluble in water. Both of these properties reflect the strong intermolecular hydrogen bonding. The water solubility decreases as the molar mass increases. Unlike amines, amides are not bases.

Nomenclature

Amides are named by removing the *-ic acid* ending of the common name or the *-oic acid* ending of the IUPAC name of the carboxylic acid and replacing it with *-amide*. If there are substituents on the nitrogen, they are placed as prefixes and are indicated by *N*-, followed by the name of the substituent. There are no spaces between the prefix and the amide name.

Example 8: Naming Amides

Determine the IUPAC names of the following amides:

1. $CH_3-\overset{\overset{\displaystyle O}{||}}{C}-NH_2$

2. $CH_3CH_2-\overset{\overset{\displaystyle O}{||}}{C}-\overset{\overset{\displaystyle H}{|}}{N}-CH_3$

3. $CH_3CH_2CH_2CH_2-\overset{\overset{\displaystyle O}{||}}{C}-\overset{\overset{\displaystyle H}{|}}{N}-CH_2CH_3$

4. $CH_3CH_2CONH_2$

Solution:

1. The amide has two carbons. The parent name is ethane, giving **ethanamide.**
2. The parent chain has three carbon atoms; therefore, it is based on the name propane. There is a methyl group attached to the nitrogen, giving the name ***N*-methylpropanamide.**
3. The parent chain has five carbon atoms. There is an ethyl group attached to the nitrogen, giving ***N*-ethylpentanamide**.
4. You need to be able to recognize an amide when it is written in this fashion. The C=O is written in line with the rest of the molecule. This amide has a three carbon parent chain with no substitutions on the nitrogen. The name is **propanamide.**

Medically important amides include barbiturates and acetaminophen.

Reactions Involving Amides

Preparation of Amides

Amides can be prepared in a reaction between a carboxylic acid derivative and an amine or ammonia. The carboxylic acid derivative can be either an acid chloride or an acid anhydride. The product of the reaction between an acid chloride and either ammonia or an amine produces an amide and either ammonium chloride or an alkylammonium chloride. Two molar equivalents of ammonia or amine are required for this reaction, which is termed an acyl group transfer because the **acyl group** of the acid chloride is transferred from the Cl atom of the acid chloride to the N atom of the amine. The product of a reaction between an acid chloride and ammonia is a **primary (1°) amide**.

$$H_3C-\overset{\overset{\displaystyle O}{||}}{C}-Cl \;+\; 2\,NH_3 \xrightarrow{\text{heat}} R-\overset{\overset{\displaystyle O}{||}}{C}-NH_2 \;+\; NH_4^+Cl^-$$

Acid chloride — Ammonia — Amide — Ammonium chloride

The product of a reaction between an acid chloride and a primary amine is a **secondary (2°) amide**, and the product of a reaction between an acid chloride and a secondary amine is a **tertiary (3°) amide**.

Amides may also be prepared by the reaction between an acid anhydride and ammonia or an amine.

$$R-\overset{\overset{\displaystyle O}{||}}{C}-O-\overset{\overset{\displaystyle O}{||}}{C}-R \;+\; 2\,NH_3 \longrightarrow R-\overset{\overset{\displaystyle O}{||}}{C}-NH_2 \;+\; R-\overset{\overset{\displaystyle O}{||}}{C}-O^-NH_4^+$$

Acid anhydride — Ammonia — Amide — Ammonium salt

When subjected to heat, the ammonium salt loses a water molecule and produces a second amide molecule.

Hydrolysis of Amides

LG 11

Hydrolysis of an amide results in breaking the amide bond to produce the carboxylic acid and an ammonium ion or an alkyl ammonium ion. This reaction requires heat and the presence of a strong acid or base.

$$R-\overset{\overset{\displaystyle O}{||}}{C}-NH-R' \;+\; H_3O^+ \longrightarrow R-\overset{\overset{\displaystyle O}{||}}{C}-OH \;+\; R'-NH_3^+$$

Amide — Strong Acid — Carboxylic acid — Alkyl ammonium ion

$$R-\overset{\overset{\displaystyle O}{||}}{C}-NH-R' \;+\; OH^- \longrightarrow R-\overset{\overset{\displaystyle O}{||}}{C}-O^- \;+\; R'-NH_2$$

Amide — Strong Base — Carboxylate ion — Amine

Example 9: Reactions Involving Amides
Complete the following reactions:

1. $$H_3C-\overset{\overset{\large O}{\|}}{C}-Cl + 2\,NH_3 \xrightarrow{\text{heat}} ?$$

2. $$CH_3CH_2-\overset{\overset{\large O}{\|}}{C}-NH-CH_3 + H_3O^+ \longrightarrow ?$$

Solution:

1. The reaction of an acid chloride with ammonia produces an amide. The products are

$$H_3C-\overset{\overset{\large O}{\|}}{C}-NH_2 + NH_4^+Cl^-$$

2. Amides react with a strong acid to produce the corresponding carboxylic acid and an ammonium ion. The products are:

$$CH_3CH_2-\overset{\overset{\large O}{\|}}{C}-OH + CH_3NH_3^+$$

15.4 A Preview of Amino Acids, Proteins, and Protein Synthesis

LG 12

Proteins are polymers of amino acids. Amino acids have an amino group and a carboxyl group and typically have the following general structure:

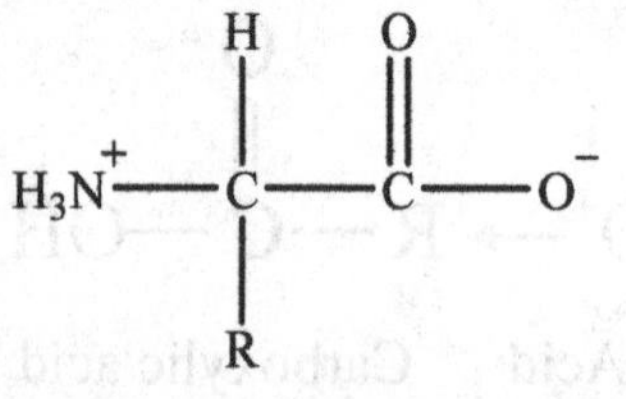

$$H_3\overset{+}{N}-\underset{\underset{R}{|}}{\overset{\overset{H}{|}}{C}}-\overset{\overset{O}{\|}}{C}-O^-$$

LG 13

The R group may be a hydrogen atom or an organic group. The primary structure of a protein consists of many amino acids bonded to one another by amide bonds called **peptide bonds**. Each peptide bond is formed by a reaction joining the amino group of one amino acid and the carboxyl group of another.

```
            peptide bond
                 |
          H  O   v  H  O
    +     |  ||     |  ||   -
  H3N  -  C -C  - N-C -C -O
          |       | |
          R1      H R2
       \________/ \________/
     Amino acid 1  Amino acid 2
```

During protein synthesis, the **aminoacyl group** of the amino acid is transferred from a special carrier molecule called **transfer RNA (tRNA).** When the aminoacyl group is covalently bonded to a tRNA, the resulting structure is:

```
          H     O
          |     ||
   +      |     ||
  H3N --- C --- C --- transfer RNA
          |
          |
          R
```

15.5 Neurotransmitters

Neurotransmitters are chemicals that carry messages, or signals, from a nerve cell to a target cell, which may be another nerve cell or a muscle cell. They may be *inhibitory* or *excitatory* and all are nitrogen-containing compounds.

Catecholamines

The catecholamines are synthesized from the amino acid tyrosine and include dopamine, norepinephrine, and epinephrine (adrenaline). Too little dopamine results in Parkinson's disease. Too much is associated with schizophrenia. Dopamine is also associated with addictive behavior. Norepinephrine and epinephrine are involved in the "fight or flight" response.

Serotonin

A deficiency of serotonin is associated with depression and eating disorders. Serotonin is involved in pain perception, regulation of body temperature, and sleep. Serotonin is synthesized from the amino acid tryptophan.

Histamine

Histamine contributes to allergy symptoms. Antihistamines block histamines and provide relief from allergies. Histamine is synthesized from the amino acid histidine.

γ-Aminobutyric Acid and Glycine

γ-Aminobutyric acid (GABA) and glycine are inhibitory neurotransmitters. It is believed that GABA is involved in control of aggressive behavior. GABA is synthesized from the amino acid glutamate; glycine is an amino acid.

Acetylcholine

Acetylcholine is a neurotransmitter that functions at the neuromuscular junction, carrying signals from the nerve to the muscle. Acetylcholine is synthesized in a reaction between choline and acetyl coenzyme A.

Nitric Oxide (NO) and Glutamate

Nitric oxide and glutamate function in a positive feedback loop that is thought to be involved in learning and the formation of memories. NO is synthesized from the amino acid arginine; glutamate is an amino acid.

Self Test for Chapter Fifteen

1. Amines may be viewed as compounds that are substituted products of which inorganic compound?
2. Identify the following as a primary, a secondary, or a tertiary amine or as a quaternary ammonium salt.
   a. $(CH_3)_3N$
   b. $(CH_3)_3N^+CH_2CH_3\ Cl^-$
   c. $CH_3CH_2NH_2$
   d. $CH_3CH_2\text{-}NH\text{-}CH_2CH_3$
3. Write the general formulas for the following:
   a. primary amine
   b. tertiary amine
4. Would you expect diethyl ether or diethylamine to have a higher boiling point? Why?
5. Provide the systematic names of the following amine compounds:
   a. $CH_3CH_2NH_2$
   b. $(CH_3)_2CHNH_2$
   c. $(CH_3)_3N$
6. What is the name of the amine of benzene?
7. Draw the structure of the amide that could be reduced to give diethylamine.
8. Name the product that results when methylamine reacts with HCl.
9. Complete the following reactions:
   a. $(CH_3)_2NH + H_2O \rightleftharpoons$?
   b. $(CH_3)_3N + HCl \rightarrow$?
10. Why are most drugs that are amines delivered as alkylammonium salts?
11. Why are quaternary ammonium salts effective as disinfectants and antiseptics?

12. In what biological molecules are the heterocyclic compounds, pyrimidines and purines, found?

13. What is the physical state of most amides at room temperature?

14. Determine the IUPAC names of the following amides:

    a. acetamide

    b. $CH_3(CH_2)_4\text{-}CONH\text{-}CH_2CH_2CH_3$

    c. $CH_3CH_2CH_2\text{-}CONH\text{-}CH_3$

15. Write an equation showing how acetamide could be prepared from ammonia and any other reagents of your choice.

16. Complete each of the following reactions:

    a. $CH_3CONH_2 + H_3O^+$

    b. $CH_3CH_2\text{-}CONH\text{-}CH_3 + OH^-$

17. The amide bond is the central feature in the structure of what class of biological molecules?

18. What functional groups of amino acids react to form the peptide bond?

19. Draw the peptide formed from the joining of the two amino acids shown below.

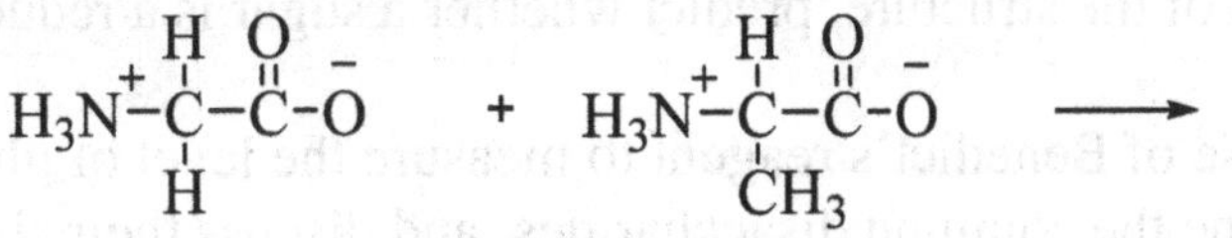

$$H_3\overset{+}{N}-\underset{H}{\overset{H}{C}}-\overset{O}{\overset{\|}{C}}-\overset{-}{O} \quad + \quad H_3\overset{+}{N}-\underset{CH_3}{\overset{H}{C}}-\overset{O}{\overset{\|}{C}}-\overset{-}{O} \quad \longrightarrow$$

20. What is the general term for a drug, such as morphine, that acts as a painkiller?

21. What is the function of a neurotransmitter?

22. What disease is caused by a deficiency of dopamine in the human body?

23. Nicotine is an agonist of which neurotransmitter?

24. Which neurotransmitter is involved in allergic reactions?

16 *Carbohydrates*

Learning Goals

1. Describe the learning strategies that will allow mastery of biochemistry.
2. Explain the difference between complex and simple carbohydrates, and know the amounts of each recommended in the daily diet.
3. Apply the systems of classifying and naming monosaccharides according to the functional group and number of carbons in the chain.
4. Determine whether a molecule has a chiral center.
5. Explain stereoisomerism.
6. Identify monosaccharides as either D- or L-.
7. Draw and name the common monosaccharides using structural formulas.
8. Given the linear structure of a monosaccharide, draw the Haworth projection of its α and β-cyclic forms and vice versa.
9. By inspection of the structure, predict whether a sugar is a reducing or a non-reducing sugar.
10. Discuss the use of Benedict's reagent to measure the level of glucose in urine.
11. Draw and name the common disaccharides, and discuss their significance in biological systems.
12. Describe the difference between galactosemia and lactose intolerance.
13. Discuss the structural, chemical, and biochemical properties of starch, glycogen, and cellulose.

Introduction

Although carbohydrates form a large part of our diet, they also are present on the surfaces of our cells. In our focus on carbohydrates in this chapter, we will study the simplest to the most complex forms. You will learn about their structure, biological function, reactions, and stereochemistry.

16.1 Strategies for Success in Biochemistry

The key concepts of general and organic chemistry will help you master biochemistry. You are encouraged to continue applying the components of the Study Cycle (covered in Sections 1.1 and 10.1) while learning this material. However, there are additional strategies that may help you learn biochemistry.

Biological molecules are just very large organic molecules. When studying reactions, notice that the reactions between molecules are essentially reactions between the functional groups that you have already studied.

Many biochemical terms have Latin or Greek roots. Learning the root will help you learn patterns in their use.

Enzyme names also often provide clues that will help you understand the reactions that they are catalyzing.

When you are learning about biochemical processes, it may be helpful if you focus on the big concepts before you begin to master the details.

You may find that understanding the significance of biochemistry to healthcare makes learning biochemistry more enjoyable and more relevant to your career goals.

16.2 Types of Carbohydrates

LG 2

Carbohydrates are produced in plants by photosynthesis. They are the main source of energy for both plants and animals. They are found in many natural sources, such as grains and cereals, breads, fruits, sugar cane, and sugar beets. A healthy diet should include both complex and simple carbohydrates. It is recommended that 45-65% of the diet should be carbohydrates and the World Health Organization recommends that no more than 5% of the daily caloric intake should be sucrose.

Carbohydrates may be categorized by size. **Monosaccharides** are composed of a single (mono-) sugar (saccharide) unit. A **disaccharide** consists of two monosaccharides. Intermediate in size are the **oligosaccharides**, polymers consisting of three to ten monosaccharide units. The largest and most complex carbohydrates are the **polysaccharides**; polymers consisting of greater than ten monosaccharide units. The largest and most complex carbohydrates are the polysaccharides. Oligosaccharides and polysaccharides are chains of monosaccharides held together by **glycosidic bonds** through "bridging" oxygen atoms.

Example 1: Classifying Carbohydrates

List each of the following as either a monosaccharide, disaccharide, or polysaccharide:

1. starch
2. glucose
3. ribose
4. maltose
5. glycogen
6. sucrose
7. cellulose
8. glyceraldehyde

Solution:

Monosaccharides are those carbohydrates that cannot be broken down into any simpler substance by hydrolysis. *Disaccharides* are composed of two monosaccharides joined together by a glycosidic bond. *Polysaccharides* are composed of many monosaccharides joined together by glycosidic bonds between each pair.

1. polysaccharide
2. monosaccharide
3. monosaccharide
4. disaccharide
5. polysaccharide
6. disaccharide
7. polysaccharide
8. monosaccharide

16.3 Monosaccharides

Nomenclature

LG 3

If a monosaccharide is a ketone, it is called a **ketose**. In a ketose, the carbonyl group is located on carbon-2. If it is an aldehyde, it is called an **aldose**. In an aldose, the carbonyl group is located on carbon-1. All carbohydrates contain a large number of hydroxyl groups and are, therefore, *polyhydroxyaldehydes* or *polyhydroxyketones*.

A second system of classification is based on the number of carbon atoms in the main skeleton. A **triose** has three carbons, a **tetrose** has four carbons, and so on. By combining the two systems, a classification can be obtained that provides information about both the structure and composition of a sugar. For instance, a sugar may be an aldotriose, aldohexose, ketotriose, ketotetrose, etc.

In addition, each carbohydrate also has a specific unique name, such as glucose and fructose. Finally, it is important to indicate which stereoisomer is present; thus, a D- or L-designation (explained in Section 16.4) is placed in front of the name.

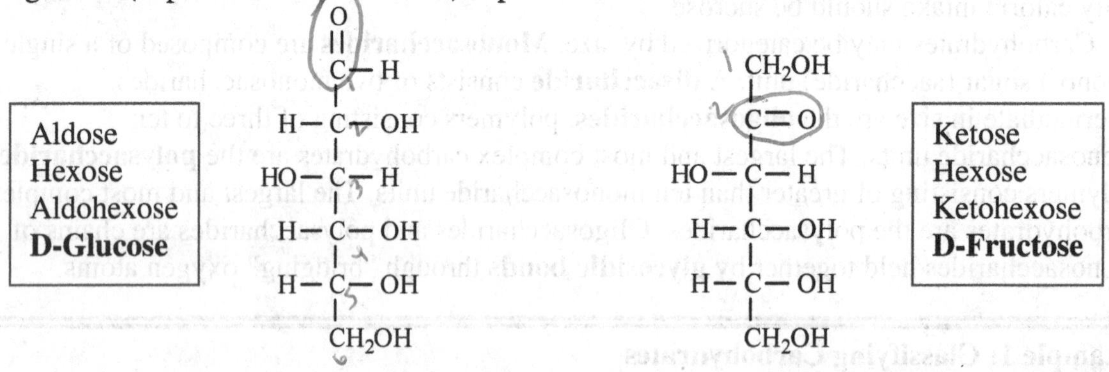

Example 2: Classifying Monosaccharides
Classify each of the following monosaccharides based on whether they contain an aldose or a ketose and the number of carbon atoms found in each:

1. ribose
2. glyceraldehyde
3. galactose
4. fructose

Solution:
Monosaccharides are classified as an aldose (which means they have an aldehyde structure) or as a ketose (ketone structure). Also, by using suffixes, we can indicate the number of carbon atoms found in each. For example, glucose is classified as an aldohexose, since it has six carbon atoms, and it also contains an aldehyde structure. At this point, we do not expect you to know the structure of the monosaccharide by the name. Find each monosaccharide in your textbook.

1. aldopentose
2. aldotriose
3. aldohexose
4. ketohexose

16.4 Stereoisomers and Stereochemistry

LG 4

LG 5

LG 6

Stereochemistry is the study of the spatial arrangement of atoms in a molecule. The prefixes D- and L- are used to distinguish **stereoisomers** in monosaccharides. In each member of a pair of stereoisomers (D- and L-), all of the atoms are bonded together using the same bonding pattern; they differ only in the arrangements of their atoms in space. D- and L-stereoisomers are **enantiomers**, nonsuperimposable mirror images of one another. In other words, a pair of enantiomers have the same molecular formula and the same bonding pattern, but if you pick one enantiomer up and set it on the other enantiomer, there is no way for every atom to line up.

A carbon atom that has *four different* groups bonded to it is called an asymmetric or **chiral carbon**. A molecule that has a chiral carbon (a **chiral molecule**) can exist as a pair of enantiomers.

Rotation of Plane-Polarized Light and the Relationship between Molecular Structure and Optical Activity

When considering a pair of enantiomers, one key difference is that each enantiomer will rotate **plane-polarized light** (which consists of light waves in only one plane) in different directions. A polarimeter is used to measure the direction and the extent to which plane-polarized light is rotated after it interacts with a molecule. Compounds that rotate light in a clockwise direction are termed **dextrorotatory** and are designated by a plus sign (+). Compounds that rotate light in a counterclockwise direction are called **levorotatory** and are indicated by a minus sign (–). So, (+)-glyceraldehyde will rotate plane-polarized light to the right, and (–)-glyceraldehyde will rotate plane-polarized light by the same amount to the left. The ability to rotate plane-polarized light is called *optical activity*.

Fischer Projection Formulas

The **Fischer Projection** is a two-dimensional way to represent a three-dimensional molecule. A chiral carbon is represented by the intersection of two lines. Horizontal lines represent bonds projecting out of the page and vertical lines represent bonds that project into the page. The most oxidized carbon is always represented at the "top" of the structure.

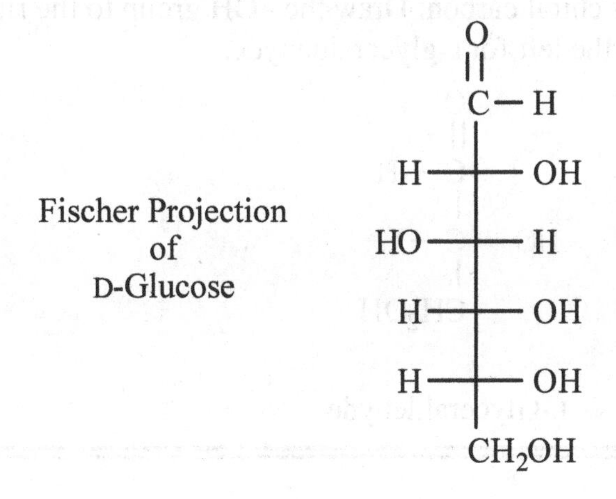

Racemic Mixtures

A mixture composed of equal amounts of two enantiomers is called **a racemic mixture**. Racemic mixtures do not rotate plane polarized light, and thus, are considered optically inactive, because the rotation of one enantiomer is canceled out by the equal but opposite rotation of the other enantiomer.

Diastereomers and Meso Compounds

Some compounds contain more than one chiral center. The term **diastereomers** is used to describe a pair of stereoisomers having two or more chiral centers and that are not enantiomers. They differ from one another both in chemical and physical properties, as well as in the direction of rotation of plane-polarized light.

In some cases, a molecule with chiral carbons may not be optically active. For that to be the case, there must be an internal plane of symmetry (i.e., the molecule can be superimposed on its mirror image). A **meso compound** has two or more chiral carbon atoms and an internal plane of symmetry that causes it to be optically inactive.

The D- and L- System of Nomenclature

By convention, it is the position of the -OH group on the chiral carbon farthest from the carbonyl group (most oxidized end of the molecule) that determines whether a monosaccharide is in the D- or L- configuration. Take a look at the Fischer Projection of D-glucose on the preceding page. The -OH group farthest from the carbonyl group is on the right. This is the D-configuration (D-glucose). If that -OH had been drawn on the left of the Fischer Projection, the molecule would have been L-glucose.

Example 3 demonstrates the D- and L- System using the example of the simplest monosaccharide, glyceraldehyde.

Example 3: Drawing D- and L- Stereoisomers
Draw and label the two different stereoisomers of glyceraldehyde.

Solution:
Glyceraldehyde is an aldehyde with three carbons. Put the carbonyl group at the top of the structure. C-2 is the only chiral carbon. Draw the –OH group to the right for D-glyceraldehyde. Draw it the left for L-glyceraldehyde.

```
      O                    O
      ||                   ||
      C—H                  C—H
      |                    |
  H—C—OH              HO—C—H
      |                    |
      CH2OH                CH2OH

D-Glyceraldehyde      L-Glyceraldehyde
```

Example 4: Determining D- or L- Stereoisomers

Determine the configuration (D- or L-) for each of the following:

```
   1.                2.                 3.                 4.
   O                 O                  O                  O
   ||                ||                 ||                 ||
   C—H               C—H                C—H                C—H
   |                 |                  |                  |
HO—C—H            H—C—OH             H—C—OH             H—C—H
   |                 |                  |                  |
   CH2OH         HO—C—H              H—C—OH             H—C—OH
                     |                  |                  |
                 HO—C—H              H—C—OH             H—C—OH
                     |                  |                  |
                  H—C—OH                CH2OH              CH2OH
                     |
                     CH2OH
```

Solution:

1. The chiral atom which determines the stereoisomer is C-2. The –OH is to the left; therefore, this is an L-isomer. The compound is named L-glyceraldehyde.
2. The chiral atom which determines the stereoisomer is C-5; that is, it the chiral carbon farthest from the carbonyl group. The –OH is to the right, which makes it a D-stereoisomer. The name of this sugar is D-galactose.
3. The chiral atom farthest from the C=O is C-4. The –OH group is drawn to the right, giving a D-stereoisomer. The compound is D-ribose.
4. C-4 is the chiral carbon which determines the D- or L-notation. In this case, the –OH lies to the right making a D-stereoisomer. The compound is D-2-deoxyribose.

Note that the (+) and (–) designations are a result of a physical measurement with a polarimeter, but the D- and L- designations are based solely on the configuration of the chiral carbon furthest from the aldehyde or ketone group. To determine whether a specific D-sugar is (+) or (–), it must be analyzed using a polarimeter. To determine whether a specific (+) sugar is D- or L- (if the chemical structure is unknown), further experimentation must be done.

As an example, given a sample of D-galactose, it is impossible to predict whether it will rotate plane-polarized light to the right or to the left until the sample is actually measured in a polarimeter. It is certain, however, that L-galactose will rotate plane-polarized light by the same amount as D-galactose, but *in the opposite direction.*

16.5 Biologically Important Monosaccharides

Glucose

Glucose (dextrose), specifically D-glucose, is the most important sugar in the human body and is the preferred energy source of many tissues, especially the brain. Thus, the blood glucose concentration must be carefully controlled to allow the body to function optimally.

LG 7

D-Glucose is an D-aldohexose and, like other sugars of five or more carbons, exists primarily in cyclic form under physiological conditions. This results from the reaction of the

LG 8

carbonyl group at C-1 of glucose with the hydroxyl group at C-5 to produce a six-membered ring. Recall that the reaction between an aldehyde and an alcohol yields a **hemiacetal** (see Section 13.4 in the text). In the case of glucose, where the carbonyl group of the aldehyde and the hydroxyl group of the alcohol are on the same molecule, the two ends of the molecule react together and the product is a cyclic *intramolecular* (meaning within the same molecule) *hemiacetal.*

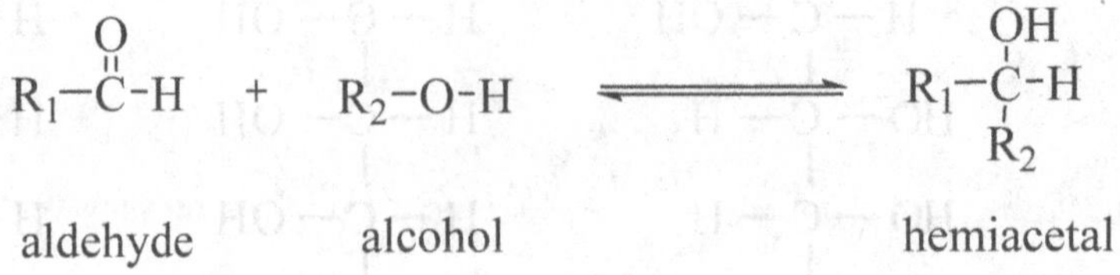

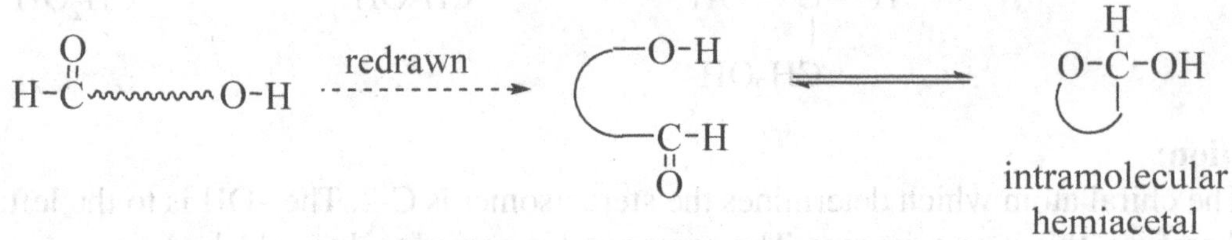

When the cyclic hemiacetal of glucose forms, a new asymmetric carbon is created, in this case at C-1. Two isomers result because there are two different ways the "H" and "OH" can be arranged on C-1. These isomers, called **anomers**, are designated as either α- or β- forms. In the α-isomer, the C-1 hydroxyl group is below the ring. In the β-isomer, it is above the ring.

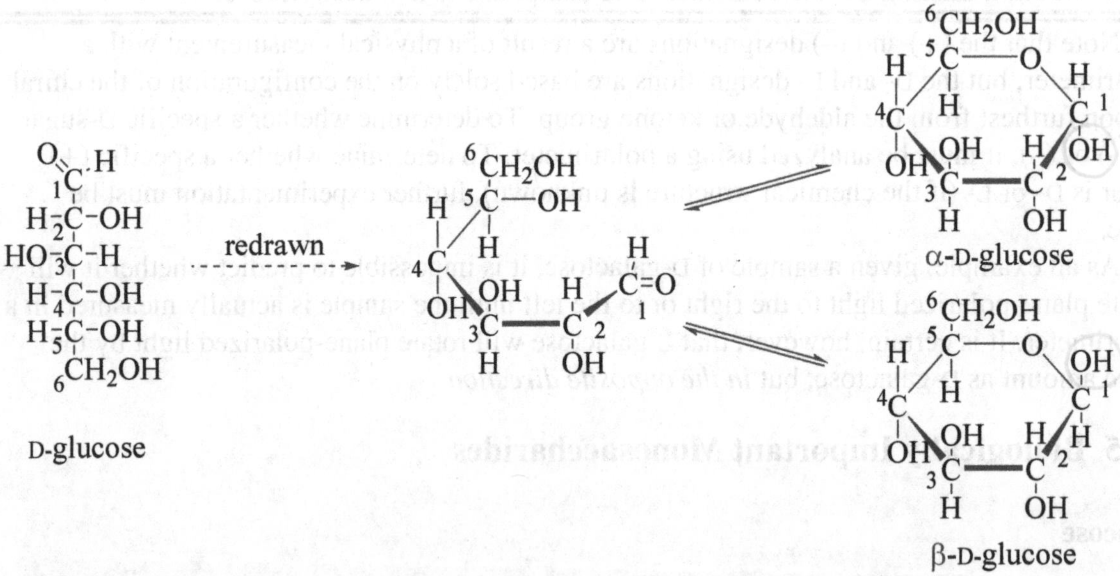

Haworth projections are used to depict the three-dimensional configuration of cyclic monosaccharide molecules. To draw a Haworth projection, begin with the structural formula of the linear form of the sugar (refer to Example 16.5 in the text). Chemical groups to the left of the carbon chain are placed above the ring in the Haworth projection. Chemical groups to the right of the carbon chain are drawn beneath the carbon ring in the Haworth projection.

Fructose

Fructose, the sweetest of all sugars, is found in honey and fruits. It is also called levulose and fruit sugar. Like glucose, fructose contains 6 carbons. Unlike glucose, it contains a ketone functional group and is classified as a ketohexose.

The most common ring structure formed by D-fructose consists of a five-membered ring. Like glucose, two anomers of the ring structure can form. Cyclization of D-fructose results in an –OH group being placed above (β-D-fructose), or below (α-D-fructose) the ring.

Galactose

Galactose is an aldohexose found most commonly as a component of the disaccharide lactose. Lactose, or milk sugar, is the most abundant sugar found in milk. Some sugars can be modified to serve different roles in the body. β-D-galactose and a modified form, *N*-acetyl-β-D-galactosamine, are found in the oligosaccharides on the surface of red blood cells. These oligosaccharides are referred to as the *blood group antigens*. They determine whether an individual has Type A, B, AB, or O blood.

Ribose and Deoxyribose, Five-Carbon Sugars

Ribose, an aldopentose, is a component of ribonucleic acids (RNA), a molecule which aids in the production of proteins, and of several coenzymes, compounds required for the action of some enzymes. The molecule that carries the genetic information in the cell is deoxyribonucleic acid (DNA). DNA contains a modified form of D-ribose in which the –OH group at C-2 has been replaced by a hydrogen. This sugar is called *2-deoxyribose*.

Reducing Sugars

The aldehyde group of aldoses is easily oxidized. Thus, **Benedict's reagent** can be used to detect the presence of aldoses, or **reducing sugars**. In this redox reaction, the aldehyde group of the sugar is oxidized to a carboxylate group, and Cu^{2+} of Benedict's reagent is reduced to Cu^{+}. This is accompanied by a visible change in the reaction as the blue Cu^{2+} is reduced to brick-red Cu_2O. The sugar is causing the metal ions to be reduced, and thus acts as a reducing agent, for that reason the sugar is called a **reducing sugar**.

Although ketones are not readily oxidized, ketoses, such as fructose, can rearrange to aldoses which then will react with Benedict's reagent. Under the basic conditions of the Benedict's reaction, the ketose can be converted to an enediol and then to an aldose, a reaction called the *enediol reaction*.

Benedict's reagent can be used to detect the presence of glucose in the urine, a condition which may arise as a result of uncontrolled *Type I insulin-dependent diabetes mellitus*. The color change in the reaction can indicate the presence and amount of glucose in the urine. Use of Benedict's reagent to test urine glucose levels has been replaced by chemical tests that provide more accurate results. The most common technology is a strip test in which a strip is impregnated with the enzyme glucose oxidase and reagents to produce a color change when the enzyme catalyzes the oxidation of glucose.

Frequently, physicians recommend that diabetics monitor *blood* glucose levels several times a day. This provides a more accurate indication of how well the diabetes is being controlled.

Example 5: Oxidation Reactions

Write an equation representing the oxidation of L-glyceraldehyde by Benedict's reagent.

Solution:

Benedict's reagent oxidizes the aldehyde to a carboxylate anion.

$$\begin{array}{c} O \\ \| \\ C-H \\ | \\ HO-C-H \\ | \\ CH_2OH \end{array} + 2\,Cu^{2+} + 5\,OH^- \longrightarrow \begin{array}{c} O \\ \| \\ C-O^- \\ | \\ HO-C-H \\ | \\ CH_2OH \end{array} + Cu_2O + 3\,H_2O$$

Example 6: Oxidation Reactions

Write an equation representing the oxidation of D-glucose by Benedict's reagent.

Solution:

Benedict's reagent oxidizes the aldehyde to a carboxylate anion.

$$\begin{array}{c} O \\ \| \\ C-H \\ | \\ H-C-OH \\ | \\ HO-C-H \\ | \\ H-C-OH \\ | \\ H-C-OH \\ | \\ CH_2OH \end{array} + 2\,Cu^{2+} + 5\,OH^- \longrightarrow \begin{array}{c} O \\ \| \\ C-O^- \\ | \\ H-C-OH \\ | \\ HO-C-H \\ | \\ H-C-OH \\ | \\ H-C-OH \\ | \\ CH_2OH \end{array} + Cu_2O + 3\,H_2O$$

16.6 Biologically Important Disaccharides

Recall that when a hemiacetal reacts with an alcohol the product is an acetal (see Section 13.4 in the text). This same reaction can occur between monosaccharides when the α- or β-hydroxyl group of a cyclic monosaccharide reacts with a hydroxyl group of another sugar to form an "oxygen bridge" (ether bond) and produce a disaccharide. This ether bond is referred to as a **glycosidic bond.** This reaction between two molecules of glucose is illustrated in Figure 16.9.

Maltose

Maltose is a disaccharide composed of two glucose molecules. Since the C-1 hydroxyl group of one α-D-glucose molecule is attached to C-4 of another glucose molecule (which could be α or β), the bond between the two monosaccharides is called an α(1→4) glycosidic bond. The "1→4" indicates that the functional group on C-1 of one glucose is bonding to the functional group on C-4 of the other glucose. The "α" indicates the way the two monosaccharides are linked. With an "α" linkage the O of the glycosidic bond is below the ring, while with a "β" glycosidic bond the O would be above the ring (similar to how α- and β- were determined for monosaccharides).

In Figure 16.9 in the text, the C-1 hemiacetal –OH of the glucose drawn on the left bonds with the C-4 alcohol –OH of the glucose molecule drawn on the right. This leaves the C-1 hemiacetal structure of the glucose on the right unreacted. The hemiacetal structure can break open forming a free aldehyde and this free aldehyde can act as a reducing agent with Benedict's or other reagents. Maltose is termed a reducing sugar. Any disaccharide with an unreacted hemiacetal –OH group at C-1 is a reducing sugar.

Lactose

Milk sugar, or **lactose**, is a dimer of β-D-galactose and either α- or β-D-glucose. In lactose, the C-1 hydroxyl group of β-D-galactose is bonded to the C-4 hydroxyl group of glucose. Thus the bond between these two monosaccharides is called a β(1→4) glycosidic bond. Notice in Figure 16.11 of the text that the O of the glycosidic bond is above the galactose ring and therefore this is referred to as a "β" linkage.

LG 12

Lactose, the principal sugar in milk, must be broken down to glucose and galactose before it can be used by the body as an energy source. Glucose can be used directly in energy-harvesting metabolic reactions. However, galactose must be converted into a phosphorylated form of glucose before it can enter the energy harvesting metabolic reactions in the cells. **Galactosemia** is a genetic disease caused by the absence of an enzyme needed for this conversion. Dulcitol, a reduced form of galactose, accumulates to toxic levels, causing severe mental retardation, cataracts, and early death. Galactosemic infants must be provided a diet that does not contain any milk or milk products.

The inability to digest lactose, called **lactose intolerance**, is caused by the absence of the enzyme lactase. The symptoms include intestinal cramping, diarrhea, and dehydration. Because some intestinal bacteria metabolize lactose and release organic acids and CO_2 gas, the individual suffers further discomfort. Avoiding milk and milk products will eliminate these unpleasant symptoms. Lactase is available in tablet form and can be ingested along with dairy products to avoid the symptoms of lactose intolerance.

Sucrose

Sucrose, common table sugar, is a disaccharide of α-D-glucose joined to β-D-fructose. Examine the structure of sucrose in Figure 16.12 in the text. The glycosidic linkage between C-1 anomeric carbon of α-D-glucose and C-2 anomeric carbon of β-D-fructose involves both of the carbons that were previously part of the hemiacetal. This is called an (α1→β2)-

glycosidic linkage. This glycosidic linkage involving the anomeric carbons of both monosaccharides results in no hemiacetal structures remaining in the sucrose molecule. Neither of the rings of the disaccharide can open to form a free aldehyde or ketone, and thus, sucrose is a **nonreducing sugar**.

16.7 Polysaccharides

LG 13

Starch

Most carbohydrates found in nature are polymers of glucose. **Homopolysaccharides** are large polymers composed of monomers of a single monosaccharide. **Heteropolysaccharides** are those made up of two or more different monosaccharides.

Plants use the energy of sunlight to produce glucose from CO_2 and H_2O. Most plants store glucose in the form of the polysaccharide **starch**. Starch is composed of the glucose polymers **amylose** (20%) and **amylopectin** (80%). Amylose is a linear polymer of up to 4,000 α-D-glucose molecules connected by α(1→4) glycosidic bonds. Amylose exists as a helix that coils at every sixth glucose unit.

During digestion, amylose is degraded by two enzymes. The enzyme α-amylase cleaves the glycosidic bonds of amylose chains at random along the chain, and β-amylase sequentially removes maltose molecules from the reducing end of the amylose chain.

Amylopectin is a highly branched form of amylose. The main chain consists of α(1→4) glycosidic bonds (like amylose), however, branches form off of this main chain by glucose molecules attaching to the C-6 hydroxyl groups by α(1→6) glycosidic bonds (see Figure 16.15). The branches tend to be 20–25 glucose molecules in length.

Glycogen

Glycogen is the principal glucose storage form of animals. It is stored in granules in liver and muscle cells. The structure of glycogen is like that of amylopectin, except that glycogen has more branches, and they are shorter.

Blood glucose levels must be carefully regulated in the body, and the liver controls this by the formation and degradation of glycogen. When blood glucose levels are high, liver cells take up glucose from the blood and convert it to glycogen. When the blood glucose levels are too low, the liver breaks down glycogen and releases the glucose into the blood stream.

Cellulose

Cellulose, a polymer of about 3,000 β-D-glucose units linked by β (1→4) glycosidic bonds, is the most abundant organic molecule in the world. Cellulose is an unbranched polymer that forms long fibrils. These long, straight, parallel chains of cellulose form a rigid cage that serves as a structural component of plant cell walls.

Humans cannot digest cellulose because we lack the enzyme *cellulase*, which hydrolyzes the β (1→4) glycosidic bond of cellulose. However, cellulose serves as a source of dietary fiber, a necessary component of a healthful diet.

Example 7

What are the products of hydrolysis of the following carbohydrates:

1. starch 2. maltose 3. sucrose 4. glycogen 5. Lactose

Solution:

1. Starch is hydrolyzed into many glucose molecules
2. The disaccharide maltose is hydrolyzed into two glucose molecules
3. The disaccharide sucrose is hydrolyzed into the monosaccharides glucose and fructose
4. Glycogen, a polysaccharide, is hydrolyzed into many glucose molecules
5. Lactose is hydrolyzed into the monosaccharides glucose and galactose

For each of the above reactions, acid conditions, or an enzyme are required.

Example 8: Digestion of Cellulose

Why are termites, cows, and goats able to digest cellulose, while humans are unable to do so?

Solution:

The termites, cows, and goats have microorganisms within their digestive systems that produce the enzyme *cellulase*, which allows them to break down the β (1→4) glycosidic bonds in cellulose. Humans do not have the cellulase enzyme needed for the hydrolysis of cellulose.

Self Test for Chapter Sixteen

1. What is the difference between a simple and a complex carbohydrate? Give an example of each.
2. What is the recommended percent of calories in our diet coming from carbohydrates?
3. Classify each of the following as a monosaccharide, disaccharide, or polysaccharide:

   a. deoxyribose d. starch

   b. galactose e. sucrose

   c. glucose f. fructose
4. What is the difference between a triose and a pentose?
5. Classify each of the following as an aldose or a ketose. Also, include in the classification the number of carbon atoms in each.

   a. glucose d. ribose

   b. fructose e. deoxyribose

   c. galactose f. glyceraldehyde
6. How would you determine which carbons of a monosaccharide are chiral?
7. Consider a pair of stereoisomers. What is the same between the two structures? What is different between the two structures?

8. What is the term for a pair of stereoisomers that are nonsuperimposable mirror images of one another?
9. Determine the D- or L- configurations for each of the following sugars:

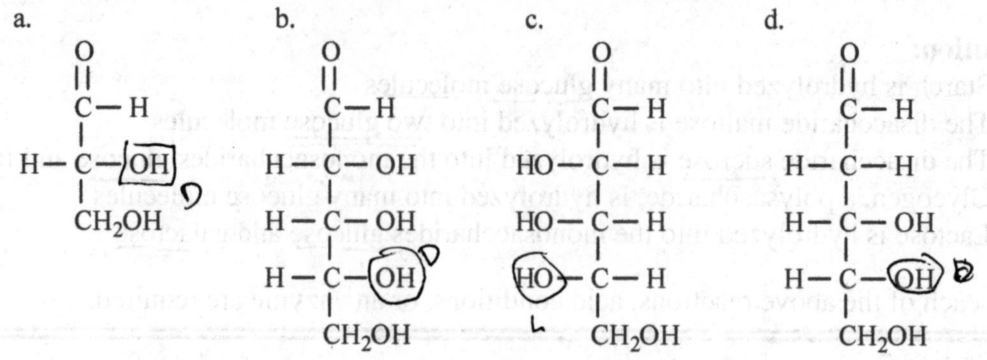

10. Which enantiomeric form of glucose is found in humans?
11. What monosaccharide is found in RNA?
12. Which monosaccharide is the sweetest of all carbohydrates? Is this monosaccharide an aldose or a ketose?
13. What compound is formed in the reaction between an alcohol and an aldehyde?
14. Which form of cyclic D-glucose has the –OH at carbon-1 below the ring?
15. What is the term for carbohydrates that can reduce metal ions?
16. How can Benedict's reagent be used to determine the presence or amount of glucose in the urine? Why is it important to detect the presence of glucose in the urine?
17. Name the bond that holds two monosaccharides together in a disaccharide.
18. How does an $\alpha(1\rightarrow6)$ glycosidic bond differ from a $\beta(1\rightarrow4)$ glycosidic bond?
19. Compare and contrast maltose, lactose and sucrose.
20. What is the term for the condition that results from an inability to digest lactose?
21. What is the name of the human genetic disorder in which galactose cannot be converted into a form that can be used in cellular metabolic reactions?
22. Which disaccharide will not react with Benedict's solution? Why is no reaction observed?
23. What are the roles of the two enzymes involved in the degradation of amylose?
24. How does amylose differ from amylopectin?
25. Name two ways amylopectin differs from glycogen?
26. How is glycogen used to regulate blood glucose levels?
27. Which polysaccharide serves as a structural element in plant cell walls?
28. Why are humans unable to digest cellulose?
29. What are the products of hydrolysis of the following carbohydrates?

a. glucose c. sucrose
b. lactose d. starch

17 *Lipids and Their Functions in Biochemical Systems*

Learning Goals

1. Discuss the physical and chemical properties and biological function of each of the families of lipids.
2. Write the structures of saturated and unsaturated fatty acids.
3. Compare and contrast the structure and properties of saturated and unsaturated fatty acids.
4. Describe the functions of prostaglandins.
5. Discuss the mechanism by which aspirin reduces pain.
6. Write equations representing the reactions that fatty acids and glycerides undergo.
7. Draw the structure of a phospholipid, and discuss its amphipathic nature.
8. Discuss the general classes of sphingolipids and their functions.
9. Draw the structure of the steroid nucleus, and discuss the functions of steroid hormones.
10. Describe the function of lipoproteins in triglyceride and cholesterol transport in the body.
11. Draw the structure of the cell membrane, and discuss its functions.

Introduction

Lipids are a diverse collection of organic molecules with both potential hazards and essential roles in our bodies. In this chapter, we will study the various lipid molecules including those that are a dietary source of energy and a storage form of energy, those that serve as structural components of cell members, hormones, and vitamins.

17.1 Biological Functions of Lipids

LG 1

Lipids are grouped together on the basis of their solubility in nonpolar solvents. The four groups of lipids that will be considered in this chapter are fatty acids, glycerides, nonglyceride lipids, and complex lipids.

Lipids are involved in a variety of biological processes. They are structural components of cell membranes. They are an energy source and an energy storage form in the body, stored in adipocytes (fat cells). Some lipids, like the steroids, are hormones. Others are vitamins, required for processes such as blood clotting, proper bone development, and vision. Additionally, dietary fat is needed for vitamin absorption. Lipids also have functions in terms of protection and insulation for the body.

Example 1: Lipid Properties
List two properties that all lipids have in common.

Example 1 Solution:
Most are insoluble in water but very soluble in nonpolar solvents.

Example 2: Lipid Functions
List four important functions of the class of compounds called the lipids.

Solution:
1. Cell membranes are composed of lipids, including phospholipids and steroids. The cell membrane creates a barrier between the cell and its environment and provides a means for the controlled passage of materials into and out of the cell.
2. Most of the available stored energy in animals is in the form of lipids known as triglycerides.
3. Several of the lipids have hormone like properties.
4. Others are lipid-soluble vitamins.

17.2 Fatty Acids

Structure and Properties

Fatty acids are long-chain monocarboxylic acids. The general formula for a **saturated fatty acid** is

$$CH_3(CH_2)_nCOOH$$

where n is an even number between 10 and 22. In a saturated fatty acid, each carbon in the chain is bonded to the maximum number of hydrogen atoms.

An **unsaturated fatty acid** is one that contains at least one carbon-to-carbon double bond. Oleic acid is an eighteen-carbon unsaturated fatty acid, as seen in this example:

$$CH_3(CH_2)_7CH = CH(CH_2)_7COOH$$

As we observed with alkanes, the melting points of saturated fatty acids increase with increasing carbon number due to London dispersion forces. This general trend is also seen for unsaturated fatty acids, except that the melting point also decreases markedly as the number of carbon-to-carbon double bonds increases.

Example 3: Properties of Fatty Acids
List the properties that most fatty acids have in common.

Solution:
1. They are long-chain monocarboxylic acids.
2. They often contain an even number of carbon atoms.
3. They form continuous chains of carbon atoms with no branches.

Example 3 Continued

4. They are either saturated, with only single bonds in the acyl group, or unsaturated, with one or more double bonds in the acyl group.

Example 4: Saturated and Unsaturated Fatty Acids

What physical property clearly distinguishes a saturated and unsaturated fatty acid with the same number of carbon atoms?

Solution:

The melting points of saturated fatty acids are much higher. The greater the number of double bonds in the unsaturated fatty acid, the lower its melting point.

Example 5: Saturated and Unsaturated Fatty Acids

Is the fatty acid $C_{13}H_{25}COOH$ saturated or unsaturated?

Solution:

It is unsaturated, since all saturated fatty acids (similar to alkanes) will fit the general formula of $C_nH_{2n+1}COOH$. If it were saturated, its formula would have to be $C_{13}H_{27}COOH$; thus, the $C_{13}H_{25}COOH$ has two fewer hydrogen atoms than the saturated one. This means that the $C_{13}H_{25}COOH$ contains one double bond.

Omega-3 Fatty Acids

Several omega-3 fatty acids, which derive their name from the location of the double bond on the third carbon from the end, are a recommended part of a healthy diet. In particular, the American Heart Association recommends inclusion of eicosapentaenoic acid (EPA), docosahexaenoic acid (DHA), and α-linolenic acid (ALA) in the diet. It is thought that these omega-3 fatty acids may reduce the risk of cardiovascular disease.

Eicosanoids: Prostaglandins, Leukotrienes, and Thromboxanes

LG 4

Linolenic acid and linoleic acid are **essential fatty acids**. They must be obtained in the diet because they cannot be synthesized by the body. Linoleic acid is required for the biosynthesis of **arachidonic acid**, the precursor of a class of 20-carbon, hormone-like molecules known as **eicosanoids**. The eicosanoids include *prostaglandins, leukotrienes,* and *thromboxanes*.

The **prostaglandins** are extremely potent biological molecules with hormone-like activity. They are carboxylic acids with a five-carbon ring. All are composed of the basic C_{20} carbon skeleton of prostanoic acid and are grouped under the designations of A, B, E, F, and so on, which indicate the basic carbon skeletal arrangement. Each prostaglandin also has a number designation that indicates the number of carbon-carbon double bonds in the compound. Figure 17.3 in the text provides examples of prostaglandin structure and nomenclature.

Prostaglandins are made in all tissues and exert biological effects on the cells that produce them and on neighboring cells. Their functions are briefly summarized below.

1. *Blood clotting:* Thromboxane A_2, produced by blood platelets, enhances blood clotting. PGI_2 (prostacyclin), produced by the cells lining the blood vessels, inhibits the clotting process. Working together, these molecules ensure that blood clots form only when necessary.
2. *The inflammatory response*: Prostaglandins promote the pain and fever associated with the inflammatory response.
3. *Reproductive system:* PGE_2 stimulates smooth muscle contraction, especially uterine contractions. Dysmenorrhea (painful menstruation) is the result of an excess of two prostaglandins. Drugs that inhibit prostaglandin synthesis provide relief from symptoms.
4. *Gastrointestinal tract:* Prostaglandins inhibit the secretion of stomach acid and increase the secretion of a protective mucous layer into the stomach. Aspirin inhibits prostaglandin synthesis; thus, prolonged use may contribute to stomach ulcers.
5. *Kidneys*: Prostaglandins increase the excretion of water and electrolytes.
6. *Respiratory tract:* Leukotrienes promote the constriction of the bronchi associated with asthma. Other prostaglandins have the opposite effect, bronchodilation.

Aspirin relieves pain by inhibiting prostaglandin synthesis. It works by inhibiting the enzyme cyclooxygenase, which catalyzes the first step in prostaglandin synthesis. The acetyl group from aspirin is covalently bound to the enzyme, inhibiting its activity.

17.3 Glycerides

Glycerides are lipids that contain the alcohol glycerol. They may be subdivided into two classes: **neutral glycerides** and **phosphoglycerides**.

Neutral Glycerides

The esterification of glycerol with one, two, or three fatty acids produces a **mono-**, **di-**, or **triglyceride**. These are also referred to as *mono-*, *di-*, or *triacylglycerols*. Although mono- and diglycerides are present in nature, the most common is the triglyceride (general structure shown below), the major storage form of lipids found in fat cells.

```
      H       O
      |       ||
  H — C — O — C — R
      |
      |       O
      |       ||
  H — C — O — C — R
      |
      |       O
      |       ||
  H — C — O — C — R
      |
      H
```

Neutral glycerides do not dissociate into charged species, because bonding throughout the molecule is covalent and nonpolar. Consequently, they readily stack with one another and are easily stored in the body's fat cells (adipocytes). In fact, their major function in the body is energy storage. If more nutrients are consumed than are needed for daily metabolic processes, the excess is converted to triglycerides and stored in adipocytes, which form *adipose tissue*. If energy is needed, the triglycerides are broken down, and their stored energy is released for use by the body.

Chemical Reactions of Fatty Acids and Glycerides

The reactions of fatty acids are similar to those of short-chain carboxylic acids (discussed in Chapter 14). The major reactions of fatty acids include esterification and addition at the double bond.

The major reactions of glycerides are acid hydrolysis and saponification.

Esterification

Esterification is the reaction of a fatty acid with an alcohol to form an ester and water.

Example 6: Esterification
Complete the following esterification reaction:

```
         H
         |
    H————C————OH      C17H35COOH
         |                            H+, heat
    H————C————OH   +  C17H35COOH   ——————————→
         |
    H————C————OH      C17H35COOH
         |
         H
```

Solution:
A triglyceride is produced by the reaction of three fatty acids and glycerol. Line up the carboxyl group of each fatty acid with the hydroxyl groups (–OH) of the glycerol molecule. Then remove a water molecule between each to form an ester structure at each of the three positions. The resulting triglyceride will be formed, along with three molecules of water:

```
         H           O
         |           ||
    H————C————O——————C————C17H35
         |           O
         |           ||
    H————C————O——————C————C17H35
         |           O
         |           ||
    H————C————O——————C————C17H35
         |
         H
```

Notice that there are no hydroxyl or carboxyl groups present; only three ester groups are now holding the glycerol to the three fatty acids.

Reaction at the Double Bond (Unsaturated Fatty Acids)

Hydrogenation of an unsaturated fatty acid is the addition of hydrogen to a double bond. This is an example of an addition reaction. Hydrogenation is used in the food industry to convert polyunsaturated vegetable oils into solid fats.

Example 7: Hydrogenation
Show how linoleic acid can be converted in the laboratory into stearic acid.

Solution:

$$CH_3(CH_2)_4CH{=}CHCH_2CH{=}CH(CH_2)_7COOH \ + \ 2\ H_2 \xrightarrow{\text{heat, pressure, Ni}} CH_3(CH_2)_{16}COOH$$

Stearic acid

Acid Hydrolysis

Hydrolysis is the addition of water to a fatty acid ester to produce a carboxylic acid (fatty acid in this case) and an alcohol.

Example 8: Acid Hydrolysis
Give the complete acid hydrolysis of the following triglyceride:

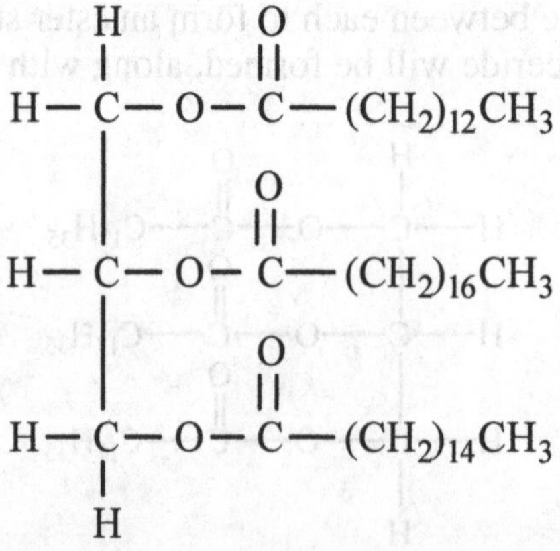

$$\begin{array}{l} \quad\ \ H \qquad\quad O \\ \quad\ \ | \qquad\quad \| \\ H-C-O-C-(CH_2)_{12}CH_3 \\ \quad\ \ | \qquad\quad O \\ \quad\ \ | \qquad\quad \| \\ H-C-O-C-(CH_2)_{16}CH_3 \\ \quad\ \ | \qquad\quad O \\ \quad\ \ | \qquad\quad \| \\ H-C-O-C-(CH_2)_{14}CH_3 \\ \quad\ \ | \\ \quad\ \ H \end{array}$$

1-Myristoyl -2-stearoyl-3-palmitoylglycerol

Example 8 Solution:
The hydrolysis of a triglyceride requires three water molecules and a catalyst (generally acid). The products formed are glycerol and one molecule of each of the fatty acids used to prepare the triglyceride. The name of the above triglyceride tells us that glycerol plus myristic acid, stearic acid, and palmitic acid were used to make it.

$$
\begin{array}{l}
H{-}C{-}O{-}\overset{O}{\overset{\|}{C}}{-}(CH_2)_{12}CH_3 \\
H{-}C{-}O{-}\overset{O}{\overset{\|}{C}}{-}(CH_2)_{16}CH_3 \\
H{-}C{-}O{-}\overset{O}{\overset{\|}{C}}{-}(CH_2)_{14}CH_3
\end{array}
+ 3\,H_2O \xrightarrow{\text{catalyst}}
\begin{array}{l}
H{-}C{-}OH \\
H{-}C{-}OH \\
H{-}C{-}OH
\end{array}
\begin{array}{l}
+\ HO{-}\overset{O}{\overset{\|}{C}}{-}(CH_2)_{12}CH_3 \\
+\ HO{-}\overset{O}{\overset{\|}{C}}{-}(CH_2)_{16}CH_3 \\
+\ HO{-}\overset{O}{\overset{\|}{C}}{-}(CH_2)_{14}CH_3
\end{array}
$$

Example 9: Hydrolysis Reactions
Write the hydrolysis products for each of the following:

1. Tristearoylglycerol + $3H_2O$
2. 1-Lauroyl-2-stearoyl-3-palmitoylglycerol + $3H_2O$
3. phosphatidylcholine + $4H_2O$ (See phospholipids in the next section.)

Solution:
1. glycerol + three molecules of stearic acid
2. glycerol + lauric acid + stearic acid + palmitic acid
3. glycerol + 2 fatty acids + H_3PO_4 + choline

Saponification

Saponification is the base-catalyzed hydrolysis of an ester. In saponification, the products are a carboxylic acid salt (fatty acid salt, in this case) and an alcohol. The long-chain carboxylic acid salt or fatty acid salt that is the product of this reaction is soap.

$$R{-}\overset{O}{\overset{\|}{C}}{-}OR \quad + \quad NaOH \longrightarrow R{-}\overset{O}{\overset{\|}{C}}{-}O^-Na^+ \quad + \quad R{-}OH$$

Recall from Section 14.2, that soaps have long hydrophobic tails that dissolve oil and dirt, and hydrophilic head groups which allow the soap micelles (containing the oil and dirt) to be rinsed away.

Phosphoglycerides

Phospholipids contain a phosphoryl group (PO_4^{3-}) and are phosphate esters. The presence of the phosphoryl group produces an amphipathic molecule with a polar head (the phosphoryl group) and a nonpolar tail (the hydrocarbon chain of the fatty acid).

Phosphoglycerides or phosphoacylglycerols contain acyl groups derived from long-chain fatty acids esterified at C-1 and C-2 of glycerol-3-phosphate. The simplest, **phosphatidate**, contains a free phosphoryl group. More complex phosphoglycerides are formed when the phosphoryl group is bonded to another hydrophilic molecule.

Phosphatidylcholine (lecithin), phosphatidylserine, and *phosphatidylethanolamine* (*cephalin*) are commonly found in cell membranes. The structures of these three phospholipids are shown in Figure 17.7 of the text. Each possesses a polar "head" and a nonpolar "tail." The ionic "head" is hydrophilic and interacts with water molecules, and the nonpolar "tail" is hydrophobic and interacts with nonpolar molecules. This bipolar nature is essential to the structure and properties of biological membranes.

In addition to being a component of cell membranes, lecithin is also used an emulsifying agent in various foods, such as ice cream. An **emulsifying agent** aids in the suspension of triglycerides in water.

Example 10: Recognizing Hydrophilic vs. Hydrophobic Regions of Phospholipids

Draw a phosphatidylcholine that contains the R groups from linolenic acid (position 1) and linoleic acid (position 2), and then label the hydrophobic and hydrophilic ends.

Solution:

The general structure of a phosphatidylcholine is shown on the left. The R groups of linolenic and linoleic acids are esterified at positions 1 and 2 on the glycerol molecule, respectively, in the structure on the right.

General Structure

```
   H     O                                      H     O
   |     ||                                     |     ||
H—C—O—C—R1                                   H—C—O—C—C17H29
   |     O                                      |     O
   |     ||                                     |     ||
H—C—O—C—R2                                   H—C—O—C—C17H31
   |     O         CH3                          |     O         CH3
   |     ||        |                            |     ||        |
H—C—O—P—CH2CH2—N+—CH3                        H—C—O—P—CH2CH2—N+—CH3
   |     |         |                            |     |         |
   H     O-        CH3                          H     O-        CH3

                                                ↑                 ↑
                                           Hydrophobic       Hydrophilic
```

17.4 Nonglyceride Lipids

Sphingolipids

LG 8

Sphingolipids are lipids that are not derived from glycerol. They are phospholipids derived from the amino alcohol sphingosine.

Sphingomyelin, a sphingolipid, is abundant in the myelin sheath that surrounds and insulates cells of the central nervous system and is essential to proper nerve transmission.

Glycosphingolipids, or *glycolipids*, include the cerebrosides, sulfatides, and gangliosides. They are built on a ceramide backbone structure, a fatty acid amide of sphingosine. The cerebroside galactocerebroside is found almost exclusively in brain cells. The *gangliosides* are found in most tissues of the body, although they were first isolated from nervous tissue.

Steroids

Steroids are an important family of lipids derived from cholesterol. The steroids include the bile salts that aid in the emulsification and digestion of lipids, and the sex hormones testosterone and estrone. They are members of a large, diverse collection of lipids called isoprenoids, all of which are derived from the five-carbon isoprene unit, seen here:

$$H_2C{=}\underset{}{\overset{CH_3}{\overset{|}{C}}}{-}CH{=}CH_2$$

All steroids are structured around the steroid nucleus (steroid carbon skeleton), which consists of four fused rings (shown below). Two fused rings share one or more common bonds as part of their ring backbones.

Cholesterol is found in most cell membranes, where it functions to regulate the fluidity of the membrane. A high serum cholesterol concentration is associated with heart disease, especially **atherosclerosis** or hardening of the arteries. Cholesterol and other substances coat the inside of the arteries. This causes the arteries to become narrower, and more pressure is needed to cause blood to flow through them. This results in elevated blood pressure (hypertension), which is also linked to heart disease.

Bile salts are derivatives of cholesterol which serve to emulsify dietary fats and allow them to be more readily digested by lipid digesting enzymes (lipases) in the small intestine.

Many steroids play roles in the reproductive cycle. *Progesterone*, the most important hormone associated with pregnancy, is synthesized from cholesterol. Produced in the ovaries and placenta, it prepares the uterine lining to accept the fertilized egg. Progesterone is also involved in fetal development and suppression of further ovulation during pregnancy.

Progesterone is the precursor of *testosterone*, a male sex hormone, and estrone, a female sex hormone. Both these hormones are involved in the development of secondary sexual characteristics.

Development of birth control agents has involved application of steroid chemistry. One of the first effective synthetic birth control agents was 19-norprogesterone. Unfortunately, it had to be taken by injection. Norlutin (17-α-ethynyl-19-nortestosterone) is equally effective and can be taken orally. More common now are "combination" oral contraceptives that include both progesterone and an estrogen. These compounds all act by inducing a false pregnancy, which prevents ovulation.

Cortisone is involved in carbohydrate metabolism and is an important drug used to treat rheumatoid arthritis, asthma, gastrointestinal, and skin disorders. Care must be taken in the use of cortisone because of the possible side effects.

Aldosterone is produced by the adrenal cortex. In the kidney, it causes sodium ions and water to be returned to the blood when sodium levels are too low.

Waxes

The chemical composition of **waxes** is highly variable. All are insoluble in water and are solid at room temperature. Paraffin wax is a mixture of solid straight-chain hydrocarbons. The natural waxes are composed of a long-chain fatty acid esterified to a long-chain alcohol.

Examples of naturally occurring waxes include beeswax; lanolin, used in skin creams; carnauba wax, used in automobile polish; and whale oil, once used as a fuel and for candles.

17.5 Complex Lipids

Complex lipids are lipids that are bonded to other types of molecules. **Plasma lipoproteins** are complex lipids that transport other lipids through the blood stream. Because lipids are only slightly soluble in water, their movement from organ to organ requires such a transport system. Lipoprotein particles are spheres that consist of a core of hydrophobic lipids surrounded by an outer layer or shell of amphipathic proteins, phospholipids, and cholesterol.

There are four major classes of human plasma lipoproteins. **Chylomicrons** carry triglycerides from the intestine to other tissues. **Very low-density lipoproteins (VLDL)** carry triglycerides synthesized in the liver to other tissues for storage. **Low-density lipoproteins (LDL)** carry cholesterol from the liver to peripheral tissues and help regulate cholesterol levels. **High-density lipoproteins (HDL)** transport cholesterol from peripheral tissues to the liver.

The path of lipid transport begins with dietary fat, which is emulsified in the small intestine. Triglycerides are hydrolyzed by lipase, releasing fatty acids and monoglycerides that are absorbed by intestinal cells and reassembled into triglycerides. Chylomicrons are produced, which eventually enter the bloodstream for transport to cells throughout the body. Triglycerides and cholesterol synthesized in the liver are transported in lipoproteins. Triglycerides are carried in VLDL particles, and cholesterol is transported in LDL particles.

LDL particles bind to specific protein receptors (LDL receptors) on cell membranes. The LDL-receptor complex is taken into the cell by a process called *receptor-mediated endocytosis*. In this process, the membrane is pulled into the cell, bringing the entire LDL particle into the cytoplasm in a membrane-bound sac called an endosome. Lysosomes, membrane-bound sacs containing digestive enzymes, fuse with the endosome, and the LDL particles are digested by the lysosomal enzymes. Cholesterol is released into the cell cytoplasm, where it inhibits its own biosynthesis, activates an enzyme for its own storage, and inhibits the synthesis of LDL receptors.

A genetic defect in the gene for the LDL receptor causes an accumulation of LDL cholesterol in the plasma, resulting in atherosclerosis in the sufferer. While high concentrations of LDL in the blood are associated with hardening of the arteries, high levels of HDL in the blood appear to reduce the incidence of atherosclerosis.

17.6 The Structure of Biological Membranes

Biological membranes are *lipid bilayers* consisting of phospholipids and cholesterol. The hydrophobic, hydrocarbon tails stack in the center of the bilayer, and the ionic head groups are exposed on the surfaces.

Fluid Mosaic Structure of Biological Membranes

Membranes are fluid, having the consistency of olive oil. The degree of fluidity is determined by the amounts of saturated and unsaturated fatty acids in the membrane phospholipids. More unsaturated fatty acid tails produce a more fluid membrane. Floating in the sea of phospholipids are many proteins that are critical to normal cellular function. When viewed by electron microscopy, these proteins look like a mosaic. Because of the fluid consistency of the membrane and the presence of numerous proteins, our concept of membrane structure is called the **fluid mosaic model**.

Peripheral proteins are found only on the surfaces of the membrane. **Transmembrane proteins** are embedded within the membrane and extend completely through it.

Self Test for Chapter Seventeen

1. Lipids are a diverse class of biological compounds. What is the unifying feature of all lipids?
2. Why are most lipids insoluble in water?
3. What is the simplest type of lipid?
4. List two properties of all fatty acids.
5. Saturated fatty acids containing ten or more carbon atoms are found in what physical state at room temperature?

6. How might an 18-carbon saturated and unsaturated fatty acid differ in both structure and physical properties?

7. List the three different kinds of eicosanoids.

8. What molecule is the precursor to prostaglandins?

9. Prostaglandins produced in the kidneys cause what effect on renal blood vessels?

10. How does aspirin reduce pain?

11. What is the main energy storage compound in animals?

12. What are the properties of neutral glycerides?

13. What products are formed in the reaction of a fatty acid with an alcohol?

14. What starting materials are necessary to prepare a diglyceride?

15. What products are released by the complete hydrolysis of a triglyceride?

16. Hydrogenation is used in the food industry to convert polyunsaturated vegetable oils into what?

17. List two ions found in hard water.

18. How can soap be formed from a triglyceride?

19. What two regions characterize phosphoglycerides?

20. How does a sphingolipid differ from a glyceride?

21. What role does sphingomyelin play in the central nervous system?

22. What structure is found in all steroids?

23. Name foods in your diet that have a high concentration of cholesterol.

24. Which lipid is found in egg yolks and soybeans and is used as an emulsifying agent in ice cream?

25. Which steroid is responsible for both the successful initiation and completion of pregnancy?

26. Which steroid is used to suppress the inflammatory response in the treatment of rheumatoid arthritis?

27. What is the purpose of the orally ingested synthetic steroid hormone called norlutin?

28. What products are formed by hydrolysis of a natural wax?

29. List the four major classes of human plasma lipoproteins.

30. Evidence indicates that high levels of which plasma lipoprotein help reduce the incidence of atherosclerosis?

31. **Which type of lipoprotein is bound to plasma cholesterol and transports it from the peripheral tissues to the liver?**

32. **Patients whose diets are high in saturated fat tend to have high levels of what substance in their blood?**

33. **In the lipid bilayer of cell membranes, which part of the phospholipid is found in the center of the bilayer?**

34. **What substance dissolved in the hydrophobic region of a biological membrane helps to regulate membrane fluidity?**

18 *Protein Structure and Function*

Learning Goals

1. List the functions of proteins.
2. Draw the general structure of an amino acid, and classify amino acids based on their R groups.
3. Describe the primary structure of proteins, and draw the structure of the peptide bond.
4. Draw the structures of small peptides and name them.
5. Describe the types of secondary structure of a protein.
6. Discuss the forces that maintain secondary structure.
7. Describe the structure and function of fibrous proteins.
8. Describe the tertiary and quaternary structure of a protein.
9. List the R group interactions that maintain protein conformation, and draw examples of each.
10. List examples of proteins that require prosthetic groups, and explain the way in which they function.
11. Discuss the importance of the three-dimensional structure of a protein to its function.
12. Describe the roles of hemoglobin and myoglobin.
13. Describe how extremes of pH and temperature cause denaturation of proteins.
14. Explain the difference between essential and nonessential amino acids.

Introduction

In this chapter, we will study the proteins, their building blocks the amino acids, their multiple levels of structure, and their many functions in the body.

18.1 Biological Functions of Proteins

Proteins are polymers made up of α-amino acids. Proteins are very important for our diet because they provide the body with nitrogen and sulfur, however, proteins have a large variety of other extremely important functions in the body.

LG 1

Enzymes are proteins that serve as biological catalysts. **Defense proteins** include **antibodies** (immunoglobulins) that are specific protein molecules produced in the immune system in response to foreign **antigens**. **Transport proteins** carry materials from one place to another in the body. They include soluble proteins, such as hemoglobin, which transports oxygen in the body. **Regulatory proteins** control cell function and include hormones such as insulin and glucagon. **Structural proteins** provide mechanical support and outer covering to animals. **Movement proteins** are necessary for all forms of movement. Our muscles contract and expand through the interaction of actin and myosin proteins. **Nutrient proteins** are sources of amino acids for embryos and infants.

Example 1: Cellular Functions of Proteins
Provide a summary of the cellular functions of proteins.

Solution:

1. Most *enzymes* are proteins. Enzymes catalyze almost all the chemical reactions that occur in living cells. Reactions that would take days in the laboratory occur in a matter of seconds or minutes in the presence of enzymes.
2. *Antibodies* are proteins that help stop infections by binding specifically to an antigen and then causing its destruction or removal from the body.
3. *Transport proteins* carry materials from one place to another in the body.
4. *Regulatory proteins* control many aspects of cell function, including metabolism and reproduction.
5. *Structural proteins* provide mechanical support to large animals and provide them with their outer coverings.
6. *Proteins of motion* allow a single-celled organism or higher organism to have different types of motility.
7. *Nutrient proteins* are sources of amino acids for infants and embryos.

18.2 Protein Building Blocks: the α-Amino Acids

Structure of Amino Acids

LG 2

The general structure of an **α-amino acid** is shown below:

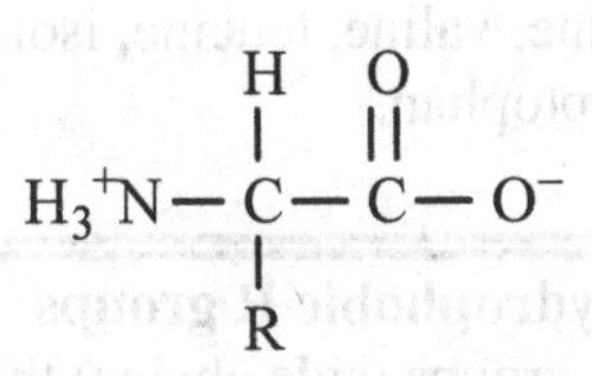

The α-carbon is attached to a carboxylate group ($-CO_2^-$), a protonated amino group ($-NH_3^+$), a hydrogen, and an R group. The carboxylate group and protonated amino groups are necessary for the covalent binding of amino acids to one another to form a protein. The R groups cause the proteins to fold into precise, three-dimensional configurations that determine their ultimate function. Amino acid molecules are neutral overall, however contain equal numbers of positive and negative changes. Molecules with this feature are referred to as *zwitterions*.

Stereoisomers of Amino Acids

The α-carbon of all the α-amino acids, except glycine, is attached to four different groups and is therefore chiral. Thus, the α-amino acids exist as D- or L- stereoisomers, just as we saw for the monosaccharides in Chapter 16. The difference is that the configuration of all the naturally occurring α-amino acids isolated from proteins is L-, while that of the naturally occurring monosaccharides is D-. See Figure 18.2 in the textbook for a comparison of D- and L- glyceraldehyde with D- and L- alanine.

Example 2: Comparison of Glyceraldehyde and Alanine
Draw and compare the structures of glyceraldehyde and alanine isolated from natural sources.

Solution:
In living organisms, we find that glyceraldehyde is in the D-isomer form, but the amino acids exist in the L-isomer form.

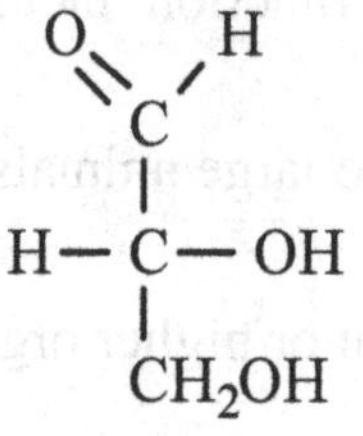

```
 O    H                 O    O⁻
  \\ /                   \\ /
   C                      C
   |                      |
H— C— OH          H3+N— C— H
   |                      |
 CH2OH                   CH3

D-Glyceraldehyde        L-Alanine
```

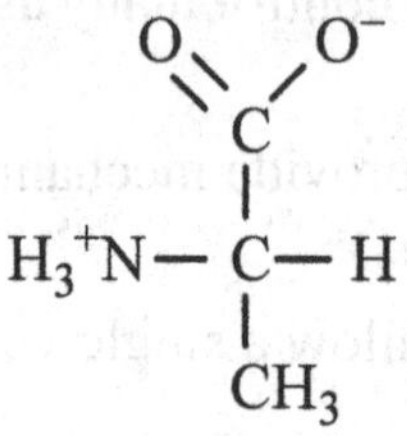

Classes of Amino Acids

The amino acids are grouped according to the polarity of their side chains. The side chains of some amino acids are nonpolar. These are **hydrophobic** (water-fearing) **amino acids**, and they are generally found buried in the interior of proteins. The hydrophobic, nonpolar amino acids include alanine, valine, leucine, isoleucine, proline, glycine, methionine, phenylalanine, and tryptophan.

Example 3: Amino Acids with Hydrophobic R groups
List the amino acids that contain R groups (side chains) that are hydrophobic. Where are these usually found in a protein molecule?

Solution:
Alanine, valine, leucine, isoleucine, proline, glycine, methionine, phenylalanine and tryptophan contain hydrophobic R groups. These amino acids are generally found buried in the interior of proteins, where they can associate with one another and remain isolated from interaction with polar water molecules.

The side chains of the remaining amino acids are polar, and therefore, these are **hydrophilic** (water-loving) **amino acids**; they are often found on the surfaces of proteins. These polar amino acids can be subdivided into three classes:

1. Polar, neutral amino acids - including serine, threonine, tyrosine, cysteine, asparagine, and glutamine
2. Negatively charged (acidic) amino acids – including aspartate and glutamate
3. Positively charged (basic) amino acids – including lysine, arginine, and histidine

The amino acids are often referred to using an abbreviation of three letters. Most of these abbreviations are the first three letters of their names. (For example, the abbreviation for glutamate is Glu, but glutamine is Gln.)

Example 4: Amino Acids with Hydrophilic R groups
List and categorize the amino acids that have hydrophilic side chains.

Solution:
1. Arginine, aspartate, cysteine, glutamate, histidine, lysine, asparagine, glutamine, serine, threonine, and tyrosine have side chains that are hydrophilic.
2. These polar amino acids are divided into three classes:
   a. Polar, neutral amino acids: These have side chains that have a high affinity for water, but they are not ionic. Serine, threonine, tyrosine, cysteine, asparagine, and glutamine fall into this category.
   b. Acidic amino acids are those that have R groups that can ionize to form the negatively charged carboxylate ion ($-CO_2^-$ ion). Aspartate and glutamate are in this category.
   c. Basic amino acids are those with R groups that can ionize to form the positively charged amine ion ($-NH_3^+$). These amino acids act like bases in water, since the side chains react with water, picking up a proton and releasing a hydroxide ion. Lysine, arginine, and histidine are in this category.

18.3 The Peptide Bond

Proteins are polymers of L–amino acids joined by **peptide bonds**. This is the covalent amide bond formed between the α-carboxylate group of one amino acid and the α-amino group of another amino acid. The reaction to form a peptide bond is a condensation reaction since a molecule is lost in the formation of the peptide bond.

The amino acid of a protein chain which has a free $\alpha\text{–}NH_3^+$ group is known as the amino terminal, or simply the **N-terminal amino acid** residue or **N-terminus**, and the amino acid with a free $\alpha\text{–}CO_2^-$ group is known as the carboxyl, or **C-terminal amino acid** residue, or **C-terminus**. By convention, the N-terminal amino acid is always drawn on the left, and the C-terminal amino acid on the right when depicting a series of covalently linked amino acids.

The names of peptides are derived from the C-terminal amino acid, which receives its entire name. For all other amino acids, the ending *-ine* is changed to *-yl*. Thus, the dipeptide tryptophanyl-leucine has leucine as the C-terminal amino acid and tryptophan as the N-terminal amino acid.

To draw a peptide chain, begin with the backbone of the peptide (N-C-C-N-C-C-N-C-C-…etc.). Next add oxygens to the carboxyl carbons, hydrogens to the amino nitrogens, and hydrogens to the α-carbons. Finally add the side chains specific to each amino acid. Example 18.1 in the text describes the procedure for drawing the structure of a peptide chain.

Example 5: Tripeptide Formation

Draw and name the tripeptide made from alanine + glycine + serine.

Solution:

1. Begin by drawing the backbone of the tripeptide and add the appropriate oxygens and hydrogens. Make sure that the N-terminal amino acid is on the left and the C-terminal amino acid is on the right, and that the amino acids are joined correctly by amide bonds.

$$H_3\overset{+}{N}-\underset{R}{\overset{H}{C}}-\overset{O}{\overset{\|}{C}}-\underset{H}{N}-\underset{R}{\overset{H}{C}}-\overset{O}{\overset{\|}{C}}-\underset{H}{N}-\underset{R}{\overset{H}{C}}-\overset{O}{\overset{\|}{C}}-O^-$$

N-terminal amino acid — C-terminal amino acid

2. Next, fill in the correct R– groups to correspond with the correct amino acids.

$$H_3\overset{+}{N}-\underset{CH_3}{\overset{H}{C}}-\overset{O}{\overset{\|}{C}}-\underset{H}{N}-\underset{H}{\overset{H}{C}}-\overset{O}{\overset{\|}{C}}-\underset{H}{N}-\underset{\underset{OH}{CH_2}}{\overset{H}{C}}-\overset{O}{\overset{\|}{C}}-O^-$$

3. The name of the tripeptide is alanyl-glycyl-serine.

The lone pair of electrons of the nitrogen atom of the peptide bond interacts with the carbon and oxygen of the carbonyl group. Thus, a resonance structure is formed that gives the peptide bond partially double bond character. As a result, the peptide bond is planar, and the two adjacent α-carbons lie *trans* to it. The hydrogen of the amide nitrogen is also *trans* to the oxygen of the carbonyl group

18.4 Primary Structure of Proteins

LG 3

The **primary structure** of a protein is the linear sequence of amino acids. This is dictated by the genetic information in the DNA. The order and interactions of the side chains (R groups) on the amino acids of the peptide determine the three-dimensional structure the protein will ultimately have, and thus, it's biological function.

LG 5

18.5 Secondary Structure of Proteins

LG 6

Secondary structure of a protein results from folding of the chain of covalently linked amino acids into regularly repeating structures. The folding pattern is maintained by numerous hydrogen bonds between the amide nitrogen and the carbonyl oxygen of the peptide chain background.

α-Helix

The most common type of secondary structure is a right-handed coiled, helical conformation known as the **α-helix**. Each carbonyl oxygen in the helix is hydrogen bonded to an amide hydrogen four amino acids away in the chain, producing an array of hydrogen bonds that are parallel to the long axis of the helix. There are 3.6 amino acids in each turn of the helix.

The **α-keratins** are α-helical proteins. These insoluble **fibrous proteins** form the covering (hair, wool, and fur) of most land animals. The individual α-helices of the keratins coil together in a bundle, producing a three-stranded protofibril that is part of an array known as a microfibril. These "molecular pigtails" possess great mechanical strength.

β-Pleated Sheet

The second common secondary structure in proteins resembles the pleated folds of drapery, and is known as **β-pleated sheet.** In this secondary structure, the polypeptide chain is nearly completely extended, with all the carbonyl oxygens and amide hydrogens involved in hydrogen bonds. The polypeptide chains in a β-pleated sheet may be *parallel* or *antiparallel.* In the parallel structure, the N-termini are aligned head to head, and in the antiparallel structure, the N-terminus of one chain is aligned with the C-terminus of a second chain (head to tail). Silk fibroin is an example of a protein whose structure is an antiparallel β-pleated sheet.

18.6 Tertiary Structure of Proteins

Soluble cellular proteins are usually **globular**. This globular, three-dimensional structure is called the **tertiary structure**. The peptide chain, with its regions of secondary structure, further folds on itself to achieve the tertiary structure.

The forces that maintain the tertiary structure of a protein include the following:

1. *van der Waals attractions* between the R groups of nonpolar amino acid residues
2. *hydrogen bonds* between the polar R groups of the polar amino acid residues
3. *ionic bonds* between the R groups of oppositely charged amino acid residues
4. *covalent bonds* between the thiol-containing amino acid residues; two cysteines can be oxidized to a dimeric amino acid, cystine, that can be a cross-link between different proteins or hold two segments within a protein together.

The tertiary structure of the protein determines its biological function; therefore, the weak interactions that hold the protein in its correct three-dimensional shape are extremely important. See Figure 18.9 in the textbook for examples of the forces that maintain tertiary protein structure.

18.7 Quaternary Structure of Proteins

LG 8 LG 9 LG 10

The active form of some proteins is an aggregate of two or more smaller globular proteins. This is the **quaternary protein structure**. The attractions that hold two or more peptides together are the same as those that maintain tertiary structure: hydrogen bonds, ionic bridges, disulfide bridges, and van der Waals forces. Some proteins must be bound to a nonprotein **prosthetic group** in order to be functional. Hemoglobin is an example of a protein with quaternary structure that is bound to an iron containing heme group. **Glycoproteins** are proteins with covalently bonded sugar groups.

18.8 An Overview of Protein Structure and Function

Primary structure is the linear sequence of amino acids covalently joined by peptide bonds. **Secondary structure** is the result of folding of the peptide chain into an α-helix or a β-pleated sheet. It is maintained by hydrogen bonding. **Tertiary structure** is the further folding of a peptide to produce a globular structure. It is maintained by a variety of noncovalent interactions between the R groups of amino acids, including hydrogen bonding, ionic bonding, and van der Waals forces, as well as by covalent bonds (disulfide bridges). **Quaternary structure** is the association of two or more peptides to form the functional protein. It is maintained by the same R group interactions that are responsible for tertiary structure.

Example 6: Protein Structure
Summarize the four types of protein structure and their relationship to one another.

Solution:

1. The primary structure of a protein is the linear sequence of amino acids bonded to one another through amide (peptide) bonds.

$$H_3\overset{+}{N}-\underset{R}{\overset{H}{C}}-\overset{O}{\overset{\|}{C}}-\underset{H}{N}-\underset{R'}{\overset{H}{C}}-\overset{O}{\overset{\|}{C}}-\underset{H}{N}-\underset{R''}{\overset{H}{C}}-\overset{O}{\overset{\|}{C}}-O^-$$

amino acid 1   amino acid 2   amino acid 3

2. The secondary structure involves the hydrogen bonding that can occur between the carbonyl oxygen and the amide hydrogen of the peptide bonds. α-Helix and the β-pleated sheet structures are the most common kinds.
3. The tertiary structure involves the gross overall folding of the entire protein molecule. Both noncovalent interactions (hydrogen bonding, ionic bonding, and van der Waals forces) between the side chains (R groups) and covalent –S–S– bridges play a role in determining the overall tertiary structure.

Example 6 Continued

4. The quaternary structure involves the aggregation of two or more peptide chains with respect to one another. The quaternary structure is maintained by R group interactions.

LG 11

The final structure of a protein ultimately defines its biological function. It is the precise shape of a protein which allows transport proteins to correctly recognize and transport particular molecules into the cell; regulatory proteins to bind to receptors on the cell surface; and enzymes to bind their reactants; to give just a few examples.

18.9 Myoglobin and Hemoglobin

Myoglobin and Oxygen Storage

LG 12

Myoglobin is the oxygen storage protein of skeletal muscle. Myoglobin is bound to a **heme group** which contains an Fe^{2+} ion that serves as the site for oxygen binding. Myoglobin has a greater affinity for oxygen than hemoglobin does, and therefore it serves as an efficient molecule to receive and store oxygen in the muscle.

Hemoglobin and Oxygen Transport

Hemoglobin (Hb) is the blood protein responsible for oxygen transport from the lungs to other tissues. It is composed of four peptide subunits, two α subunits and two β subunits, each of which contains a heme group. Due to the presence of four heme groups, each hemoglobin protein can bind four molecules of oxygen.

Example 7: Function of Hemoglobin
Explain the general structure and function of hemoglobin.

Solution:
Hemoglobin is the oxygen transport protein of the blood. It is composed of four separate peptide subunits. There are two identical α chains and two identical β chains. In addition, each subunit of hemoglobin contains a heme group. A hemoglobin molecule therefore has the ability to bind and carry four molecules of oxygen.

Oxygenation of hemoglobin in the lungs is favored by the high partial pressure of oxygen (pO_2) in the air we breathe and the low pO_2 in the blood. Thus, oxygen diffuses from the region of high pO_2 to the region of low pO_2. When the blood reaches actively metabolizing tissues, this situation is reversed and oxygen diffuses into tissues from the blood.

Oxygen Transport from Mother to Fetus

A fetus receives oxygen from its mother by simple diffusion across the placenta. The fetus makes *fetal hemoglobin*, which has a greater affinity for oxygen than the hemoglobin of the mother. Thus, the fetus is assured of a constant and adequate supply of oxygen.

Sickle Cell Anemia

Sickle cell anemia is a human genetic disease that first appeared in tropical west and central Africa. It afflicts about 0.4% of African Americans. These individuals produce mutant hemoglobin known as sickle cell hemoglobin (Hb S). Sickle cell anemia receives its name from the sickled appearance of the red blood cells that form in this condition. Sickle cell hemoglobin differs from normal hemoglobin by a single amino acid that has been replaced in each of the β-chains. A valine has replaced the amino acid glutamic acid in each of the two chains. This substitution provides a basis for the binding of hemoglobin S molecules to one another. When oxyhemoglobin S molecules release their oxygen, they bind to one another as long fibers. These fibers radically alter the shape of the red blood cell, resulting in the sickling effect. The sickled cells are unable to pass through the small capillaries, thereby causing severe medical problems.

In individuals with sickle cell anemia, both copies of the β-globin gene produce the mutant protein. Individuals with the *sickle cell trait* have one mutant and one normal β-globin gene. Typically, they do not suffer the symptoms of the disease; however, they may pass the mutant gene on to their children.

Example 8: Sickle Cell Anemia
Explain what causes sickle cell anemia.

Solution:
Sickle cell anemia is a human genetic disease caused by a mutation in the gene that encodes the β-subunit of hemoglobin. This mutation results in the synthesis of sickle cell hemoglobin (Hb S). There is only a single amino acid difference between normal and sickle cell hemoglobin. That single change causes deoxyhemoglobin to polymerize, causing the red blood cells to have a sickled appearance. This can result in damage to many organs and death at an early age.

18.10 Proteins in the Blood

The blood plasma of healthy individuals contains between 60 and 80 g/L of protein. These proteins are separated into five classes (α through γ) depending on the charge on each type of protein.

About 55% of blood protein is albumin, a molecule which transports many important metabolites through the blood. The α_1- and α_2-globulins comprise about 13% of the blood proteins. A variety of blood proteins fall into this category including α_1-antitrypsin, which inactivates an enzyme that can damage the lungs, and α_1-antichymotrypsin, another inhibitor found in the bloodstream. Another 13% of the blood plasma proteins are β-globulins including transferrin, an iron transport protein. The remaining proteins in the blood include fibrogen and the γ-globulins (antibodies).

18.11 Denaturation of Proteins

When a protein solution is exposed to extremes of temperatures and pH levels, the protein's secondary, tertiary, and quaternary structure is disrupted and the characteristic three-dimensional shape is lost. This is referred to as **denaturation**.

Temperature

When temperatures are too high, the bonds within a protein molecule begin to vibrate violently. The weak interactions that hold the three-dimensional structure of the protein are disrupted, and the protein molecules are denatured. These molecules then **coagulate** as they clump together and precipitate out of solution.

pH

If the pH of a solution of proteins becomes too high or too low, the characteristic electric charge on the surface of the protein is changed. This results in a disruption and the salt bridges and hydrogen bonds and thus changes the three-dimensional structure of the protein. This causes the proteins to aggregate with one another, and coagulation occurs.

Organic Solvents

Polar organic solvents, such as rubbing alcohol (2-propanol) can denature proteins by disrupting hydrogen bonds and interfering with hydrophobic interactions in the interior of the protein. As a result, this disrupts the conformation of the protein.

Detergents

Since detergents have both a hydrophobic and hydrophilic region, they can disrupt hydrophobic interactions within a protein molecule.

Heavy Metals

Heavy metals can form bonds with negatively charged side chain groups or bind to sulfhydryl groups of a protein. This can cause a loss of conformation of the protein. Mercury (Hg^{2+}) and lead (Pb^{2+}) are two examples of heavy metals which can denature proteins in this way.

Mechanical Stress

Since the interactions that maintain the three-dimension structures are proteins are relatively weak, quick actions such as stirring, whipping, or shaking can disrupt these interactions. This can be observed when whipping egg whites into a stiff meringue.

18.12 Dietary Protein and Protein Digestion

Proteins are the third major type of energy source in the diet. They are hydrolyzed to amino acids that may be oxidized to provide energy or used directly in protein synthesis. Amino acids are the precursors of a large, diverse group of nitrogen compounds that includes some hormones and the heme group.

Protein digestion begins in the stomach. The enzyme pepsin begins the degradation of dietary protein to amino acids. Further digestion occurs in the small intestine. These hydrolysis reactions are carried out by enzymes such as trypsin and chymotrypsin.

Essential amino acids are those that cannot be synthesized by the body and are required in the diet. **Nonessential amino acids** are those amino acids that can be synthesized by the body and need not be included in the diet.

Proteins are also classified as **complete** or **incomplete**. Animal protein is generally complete, providing all of the essential and nonessential amino acids in approximately the correct amounts for biosynthesis. Vegetable protein is generally incomplete, lacking a sufficient amount of one or more essential amino acids. Only a few vegetable sources provide complete protein; among these are soy, quinoa, and buckwheat. However, the correct mixture of vegetable protein sources can provide all the necessary amino acids.

Example 9: Dietary Sources of Protein
What are some common dietary sources of proteins?

Solution:
Meat is an excellent source of protein. Also, soy, dried beans, peas, and nuts are good sources.

Example 10: Vegetable Proteins
How can vegetable proteins be mixed to provide all of the essential amino acids?

Solution:
A complete protein provides all of the essential amino acids needed by the human body. Proteins derived from animal sources are generally complete. In contrast, proteins derived from vegetable sources are generally incomplete because they lack a sufficient amount of one or more amino acids. Soy protein one of the few known sources of complete vegetable protein. By mixing many vegetable sources, an adequate amount of each essential amino acid may be obtained.

Self Test for Chapter Eighteen

1. What elements found in dietary proteins are not found in fats and carbohydrates?
2. What is the function of transport proteins?
3. What is the function of structural proteins? Name one example of a structural protein.

4. Amino acids in water at pH = 7 exist as zwitterions, what does this mean?
5. What is meant by the term *hydrophobic*?
6. Which amino acids have negatively charged side chains at physiological pH?
7. Which amino acids have positively charged side chains at physiological pH?
8. An amino acid contains the following R group. Would you classify this amino acid as an amino acid with a hydrophobic R group, polar, neutral R group, negatively charged R group, or positively charged R group?

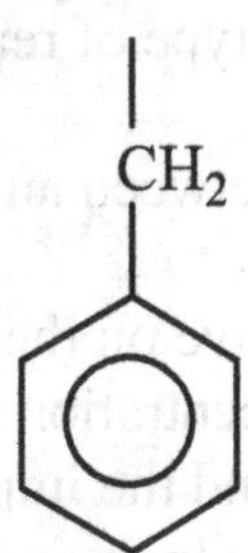

9. Draw the structure of serine as it appears in a living cell.
10. What is the term for the linkage between two amino acids in a peptide?
11. Draw the structure of alanyl-valyl-glycyl-cysteine.
12. What level of protein structure is defined by the sequence of amino acids in the protein?
13. What interactions maintain the secondary structure of a protein?
14. What kind of secondary structure characterizes the fibrous proteins of muscles?
15. Name the interactions that maintain the tertiary structure of a protein.
16. What is different between the tertiary and quaternary structure of a protein?
17. How does the role of myoglobin and hemoglobin differ in the body?
18. What is the prosthetic group found in myoglobin and hemoglobin that allow these proteins to perform their function?
19. Why is the specific three-dimensional structure of a protein so important?
20. What can occur to a protein if the temperature were to rise too high?
21. What happens to the overall charge on a protein when the pH drops too low?
22. What is produced by the controlled hydrolysis of proteins?
23. What enzyme begins the digestion of protein in the stomach?
24. What is the term used to describe amino acids that must be acquired from the diet?

19 *Enzymes*

Learning Goals

1. Classify enzymes according to the type of reaction catalyzed and the type of specificity.
2. Give examples of the correlation between an enzyme's common name and its function.
3. Describe the effect that enzymes have on the activation energy of a reaction.
4. Explain the effect of substrate concentration on enzyme-catalyzed reactions.
5. Discuss the role of the active site and the importance of enzyme specificity.
6. Describe the difference between the lock-and-key model and the induced fit model of enzyme-substrate complex formation.
7. Discuss the roles of cofactors and coenzymes in enzyme activity.
8. Explain how pH and temperature affect the rate of an enzyme-catalyzed reaction.
9. Describe the mechanisms used by cells to regulate enzyme activity.
10. Discuss the mechanisms by which certain chemicals inhibit enzyme activity.
11. Discuss the role of the enzyme chymotrypsin and other serine proteases.
12. Provide examples of medical uses of enzymes.

Introduction

In this chapter, we are going to study a group of proteins that do the majority of the work for the cell, the enzymes. In addition to learning about the ways in which they catalyze biochemical reactions, you will earn about how they are named and classified.

19.1 Nomenclature and Classification

Classification of Enzymes

LG 1

Enzymes are proteins that function as catalysts in cell processes, greatly speeding up chemical reactions. They are essential for the thousands of metabolic reactions that allow life to exist. Enzymes may be classified according to the type of reaction that they catalyze. These six classes are as follows:

1. **Oxidoreductases** – catalyze oxidation-reduction (redox) reactions
2. **Transferases** - catalyze the transfer of functional groups from one molecule to another
3. **Hydrolases** – catalyze hydrolysis reactions
4. **Lyases** – catalyze the addition or removal of a group to or from a double bond
5. **Isomerases** – catalyze the rearrangement of functional groups within a molecule
6. **Ligases** – catalyze a reaction which makes or breaks of C-C, C-S, C-O, or C-N bond

Example 1: Characteristics of Enzymes
Describe some characteristics of enzymes.

Solution:

1. Life depends upon the simultaneous occurrence of hundreds of chemical reactions that must take place rapidly under mild conditions. Enzymes allow these critical reactions to occur under the mild conditions required for life.
2. The enzymes are proteins that speed up biochemical reactions by lowering the energy of activation and increasing the rate of the reaction.
3. An enzyme is very specific. It generally recognizes only one, or occasionally a few, molecules (substrates) upon which it will work its magic.
4. Enzyme-catalyzed reactions often occur from 1 to 100 million times faster than the uncatalyzed reactions.

Example 2: The Functions of an Oxidoreductase
Explain completely the functions of oxidoreductases.

Solution:

These enzymes catalyze oxidation-reduction reactions. They are responsible for the removal of protons and electrons from a substrate to cause its oxidation, or the addition of protons and electrons to a substrate to cause its reduction. Lactate dehydrogenase is a good example. This enzyme transiently binds the coenzyme NAD^+, which accepts a hydride anion ($H:^-$) from the substrate lactate. The product is pyruvate, as shown below:

$$\underset{\text{lactate (reduced form)}}{HO-\overset{\overset{\displaystyle COO^-}{|}}{\underset{\underset{\displaystyle CH_3}{|}}{C}}-H} \; + \underset{\text{lactate dehydrogenase (oxidized form)}}{\text{enzyme-NAD}^+} \longrightarrow \underset{\text{pyruvate (oxidized form)}}{\overset{\overset{\displaystyle COO^-}{|}}{\underset{\underset{\displaystyle CH_3}{|}}{C}}{=}O} \; + \underset{\text{lactate dehydrogenase (reduced form)}}{\text{enzyme-NADH}}$$

Example 3: Enzyme Classification
List the six types of enzymes classified according to the type of reaction that they catalyze. Briefly describe the function of each class.

Example 3 Solution:

1. *Oxidoreductases*: these enzymes catalyze electron transfers from one molecule to another.
2. *Transferases*: these enzymes catalyze the transfer of functional groups from one molecule to another.
3. *Isomerases*: these enzymes catalyze the rearrangement of functional groups within a molecule to convert the substrate into a different isomeric form.
4. *Hydrolases*: these enzymes catalyze hydrolysis reactions.
5. *Lyases*: these enzymes catalyze the addition of a group to a double bond or the removal of a group to form a double bond.
6. *Ligases*: these enzymes catalyze the condensation or joining of two molecules.

Example 4: Enzyme Classification

To which class does each of the following enzymes belong?

1. pyruvate kinase
2. lipase (hydrolysis of triglycerides)
3. triose isomerase
4. lactate dehydrogenase

Solution:

1. transferase
2. hydrolase
3. isomerase
4. Oxidoreductase

Nomenclature of Enzymes

LG 2

Enzymes are remarkably specific, usually recognizing and binding to only a single type of **substrate**, or reactant, and facilitating its conversion to a **product**. An enzyme's name often tells us the substrate of the reaction and the nature of the reaction. For instance, the enzyme sucrase hydrolyzes the disaccharide sucrose and the enzyme lactate dehydrogenase removes hydrogen ($H:^-$) from lactate ions. Some enzymes have historical names that do not reveal the nature of the substrate or of the reaction. The substrate and reaction catalyzed by such enzymes as catalase and trypsin must simply be memorized.

Example 5: Enzyme Nomenclature

Answer the following questions based on what you know about how the common names of enzymes are derived.

a. What substrate is acted on by urease?
b. What substrate is acted on by lactase?
c. What does the enzyme dehydrogenase do?
d. What does the enzyme decarboxylase do?

Solution:

The common name of an enzyme is derived from the name of the substrate with which the enzyme interacts, and/or the type of reaction that it catalyzes.

Example 5 continued

a. Urea is the substrate acted on by the enzyme urease.
b. Lactose is the substrate of lactase.
c. Dehydrogenase is an enzyme that removes hydrogen from a substrate.
d. Decarboxylase is an enzyme that removes a carboxyl group from a substrate.

19.2 The Effect of Enzymes on the Activation Energy of a Reaction

Every chemical reaction is characterized by an equilibrium constant. **Enzymes do not alter that equilibrium constant.** They do, however, provide a lower energy path for the conversion of reactant to product and, in this way, speed up the reaction. Thus, enzymes increase the rate of a chemical reaction by lowering the activation energy and therefore increasing the rate at which the reaction reaches equilibrium.

19.3 Effect of Substrate Concentration on Enzyme-Catalyzed Reactions

For an enzyme-catalyzed reaction, the rate of the reaction increases as the substrate concentration increases. However, at a certain substrate concentration, the reaction rate reaches a maximum. At its maximum rate, the active sites of all the enzyme molecules are occupied by substrate molecules. The reaction rate is then limited by the speed with which the substrate is converted to product and product is released.

19.4 The Enzyme-Substrate Complex

Enzyme catalyzed reactions occur in four general steps. The first step in an enzyme-catalyzed reaction is the binding of the substrate by the enzyme to form the **enzyme-substrate complex**. The groove or pocket in the enzyme that binds to the substrate is the **active site**. The active site of an enzyme is small compared to the overall size of the enzyme. The substrate is held within the active site by weak, noncovalent interactions. These include hydrogen bonding, van der Waals forces, dipole-dipole attractions, and electrostatic attractions. The conformation (shape and charge distribution) of the active site is complementary to the conformation of the substrate. Thus, the conformation of the active site determines the specificity of the enzyme. Only those substrates that fit into the active site can bind the enzyme.

Originally it was thought that the substrate simply snapped into place in the active site, like a piece of a jigsaw puzzle. This view was called the **lock-and-key model**. Our current model describes enzymes as flexible molecules and an active site that is not a rigid pocket into which the substrate fits precisely. The active site is thought to be a flexible pocket that approximates the shape of the substrate. When the substrate binds, the active site molds itself around the substrate. This is called the **induced-fit model**.

Example 6: Enzyme-Substrate Complex
Describe the formation of the enzyme-substrate complex.

Solution:
The first step in an enzyme-catalyzed reaction involves the enzyme binding to the substrate to form the enzyme-substrate complex. The portion of the enzyme that is in contact with the substrate is called the active site.

19.5 Specificity of the Enzyme-Substrate Complex

Enzymes show a high degree of specificity. For instance, the enzyme urease catalyzes the decomposition of urea to carbon dioxide and ammonia, but it will not accept the related molecule methylurea.

Enzyme specificities differ, and often enzymes are classified into the following four groups based on this property:

1. **Absolute specificity** - an enzyme that catalyzes the reaction of only one substrate into product is absolutely specific.
2. **Group specificity** - an enzyme that catalyzes reactions involving similar molecules containing the same functional group is group specific.
3. **Linkage specificity** - in this type of specificity, an enzyme always catalyzes the formation or breakage of only one kind of bond.
4. **Stereochemical specificity** - in this type of specificity, an enzyme is capable of catalyzing a reaction involving only one enantiomer.

Example 7: Enzyme Specificity
The enzyme trypsin selectively hydrolyzes peptide bonds. What type of specificity is this?

Solution
This represents linkage specificity. Trypsin will only catalyze the breakage of a specific bond, the peptide bond.

19.6 The Transition State and Product Formation

The overall process of an enzymatic reaction can be summarized by the following set of four reversible reactions:

$$E + S \underset{}{\overset{\text{Step I}}{\rightleftharpoons}} ES \overset{\text{Step II}}{\rightleftharpoons} ES^* \overset{\text{Step III}}{\rightleftharpoons} EP \overset{\text{Step IV}}{\rightleftharpoons} E + P$$

Enzyme + Substrate	Enzyme substrate complex	Transition state	Enzyme product complex	Enzyme + Product

Step one, formation of the enzyme-substrate complex was discussed above. In the second step, the shape of the substrate is altered, due to its interaction with the enzyme, into an intermediate form having features of both the substrate and the final product. This intermediate form is referred to as a **transition state.** This favors the conversion of the substrate into the product (Step three), which is subsequently released (Step four). There are several ways in which an enzyme could cause a reaction to proceed more quickly. In some cases, the enzyme exerts "stress" on a bond, thereby facilitating bond breakage. An enzyme may simply bring reactants close to one another and in the proper orientation for reaction to occur. Alternatively, the active site of an enzyme may modify the pH of the microenvironment surrounding the substrate by serving as a donor or acceptor of H^+. Remember that the enzyme is unchanged by the chemical reaction. If it transiently loses or gains a proton, it will be converted back to its original form by either regaining or losing a proton, respectively.

Example 8: Enzymes and Activation Energy
Summarize the three ways in which an enzyme may lower the activation energy of a reaction.

Solution:

1. The enzyme might exert a stress on a bond in a substrate and therefore facilitate bond breakage.
2. An enzyme may facilitate a reaction by bringing two reactant molecules close together and in the proper orientation, so that a reaction easily occurs.
3. The active site of an enzyme may so modify the pH in the microenvironment of the substrate that a reaction will occur quickly.

19.7 Cofactors and Coenzymes

LG 7

Like other proteins, some enzymes require an additional nonprotein *prosthetic group* in order to function. The protein portion of such an enzyme is called the **apoenzyme,** and the nonprotein prosthetic group is called the **cofactor**. Together, they form the active enzyme called the **holoenzyme**. Usually, cofactors are metal ions, however, they may also be organic compounds or organometallic compounds. These cofactors bind to the enzyme and help maintain the correct shape of the active site. The enzyme is only active when the cofactor is bound.

Other enzymes require the transient binding of a **coenzyme**. Coenzymes are organic molecules that generally serve as carriers of electrons or chemical groups. They take part in a reaction by either donating or accepting chemical groups. Most coenzymes contain modified **vitamins** as part of their structure (see Table 19.1 in the textbook).

Nicotinamide adenine dinucleotide (NAD+) is a coenzyme that accepts hydride ions (a hydrogen atom with two electrons) from the substrate that is oxidized. The portion of NAD+ derived from the vitamin niacin is reduced to produce NADH.

19.8 Environmental Effects

Effect of pH

As we learned in the previous chapter, if the pH of a solution becomes too acidic or too basic, a protein is denatured. The pH at which an enzyme functions best is called the **pH optimum**. Enzyme function decreases as the pH rises above or falls below the pH optimum, and at extremes of pH, they are denatured and cease to function.

Most cellular enzymes function optimally at a pH near 7. However, *pepsin*, a proteolytic enzyme found in the stomach where the pH is very low, has a pH optimum of 2. Another proteolytic enzyme, *trypsin*, functions under the conditions of higher pH found in the intestine and has a pH optimum around 8.5.

Effect of Temperature

The temperature at which an enzyme functions best is called the **temperature optimum**. Enzymes are rapidly denatured if the temperature of the solution rises much above 37°C, but they remain stable at much lower temperatures. For this reason, enzymes used for clinical assays are stored in refrigerators or freezers. Since heating enzymes destroys their activity by denaturing the protein, cells can't survive extremes of temperatures. Thus, heat is an effective means of sterilizing medical instruments and solutions.

Some bacteria and yeast can survive very high temperatures, living in active volcanoes or in hot springs where the temperature is near the boiling point of water. The proteins of these bacteria have a structure that is stable at these extraordinary temperatures.

19.9 Regulation of Enzyme Activity

Often enzyme activity is regulated by the cell as a means of controlling metabolic processes, to conserve energy. One mechanism of regulation, used by bacteria, is to produce the enzyme only when the substrate is present.

Allosteric Enzymes

A more complex level of enzyme regulation involves enzymes that have more than a single binding site. These are called **allosteric enzymes**. One site, the active site, binds a substrate and catalyzes a biochemical reaction. The second site binds an *effector molecule.* Effector binding alters the shape of the active site. In **negative allosterism**, effector binding converts the active site to an inactive configuration. In **positive allosterism**, effector binding converts the active site to an active configuration.

Feedback Inhibition

Feedback inhibition usually involves a biosynthetic pathway of many enzymatic steps. If the product of this multistep pathway builds up to high levels, it inhibits the entire pathway by serving as a negative allosteric effector of an enzyme early in the pathway.

Proenzymes

Proenzymes are enzymes produced in an inactive form. They are converted to the active form when they reach the site of their activity. This is a protective mechanism. Most of the proteolytic digestive enzymes are produced in an inactive form so that they do not destroy the cells that make them. An example of this strategy is seen with the enzyme pepsin, a proteolytic enzyme that acts in the stomach. The cells that produce pepsin actually produce an inactive proenzyme, called *pepsinogen.*

Protein Modifications

In **protein modification**, adding or removing a covalently bound group either activates or inactivates an enzyme. The most common type of protein modification is the addition (or removal) of a phosphoryl group to (or from) a free –OH on the R group of a serine, tyrosine, or threonine in the protein chain of the enzyme. The covalent modification of an enzyme is catalyzed by other enzymes, such as *protein kinases* and *phosphatases*. Protein modification is reversible, allowing an enzyme to be turned on or off quickly in response to environmental or physiological conditions.

19.10 Inhibition of Enzyme Activity

Enzyme inhibitors are chemicals that bind to enzymes and either eliminate or drastically reduce their catalytic ability. They are classified on the basis of whether the inhibition is reversible or irreversible, competitive or noncompetitive.

Irreversible Inhibition

Irreversible enzyme inhibitors, bind very tightly, sometimes even covalently, to the enzyme. This binding irreversibly blocks substrate binding or enzyme catalysis.

Reversible, Competitive Inhibitors

Generally, these inhibitors are **structural analogs**, molecules that "look like" the structure of the natural substrate for an enzyme because of similarities in shape and charge distribution. Because of this resemblance, the inhibitor can occupy the enzyme active site, but no reaction can occur. Since it is only bound by weak interactions, the inhibitor is easily removed from the active site, providing an opportunity for the substrate to bind. This is **competitive inhibition** because the degree of inhibition depends on the relative concentrations of substrate and inhibitor and their relative affinities for the active site.

19.11 Proteolytic Enzymes

Proteolytic enzymes are enzymes that catalyze the hydrolysis of the peptide bonds that maintain the primary protein structure.

The **pancreatic serine proteases** *trypsin*, *chymotrypsin*, and *elastase* are produced in the pancreas and all digest protein in the small intestine. They also have similar primary and tertiary structures and share the same mechanism of action. They evolved by divergent evolution of a single gene that was probably duplicated. Each copy of the gene underwent mutation, producing many different proteolytic enzymes with different specificities. Chymotrypsin cleaves peptide bonds on the carbonyl side of aromatic amino acids. This specificity results from the hydrophobic pocket, a portion of the active site into which the flat aromatic side chains of these amino acids can fit. Trypsin cleaves peptide bonds on the carbonyl side of basic amino acids, and elastase cleaves peptide bonds on the carbonyl side of the amino acids glycine and alanine.

19.12 Use of Enzymes in Medicine

LG 12

The serum concentration of some enzymes is used in disease diagnosis. An enzyme assay is a test that is performed to measure the activity or concentration of an enzyme, expressed in international units.

Heart attacks can cause some of the cells in the heart muscle to die; when this occurs, the enzymes in the cell are released into the bloodstream (along with the other contents of the cell). Three cardiac enzymes in particular are often measured to assess *acute myocardial infarction*, *AMI* (blockage of blood supply to the heart). These include myoglobin, creatine kinase-MB, and cardiac troponin I.

Certain enzymes can be used to treat blood clots. The enzyme *streptokinase* catalyzes production of the enzyme plasmin which can degrade a fibrin clot into subunits. Additionally, injection of the proteolytic enzyme *tissue-type plasminogen activator (TPA)* can improve blood circulation to the heart.

Liver disease is indicated by elevated levels of alanine aminotransferase/serum glutamate-pyruvate transaminase (ALT/SGPT) and aspartate aminotransferase/serum glutamate-oxaloacetate transaminase (AST/SGOT) in blood serum. Elevated blood serum concentrations of amylase and lipase are indications of pancreatitis, an inflammation of the pancreas.

In the clinical laboratory, enzymes are valuable analytical reagents. For example, enzymes are used in the BUN test (Blood Urea Nitrogen test). Direct measurement of urea levels is difficult due to the complexity of blood. So the enzyme urease is added, which converts each molecule of urea into two molecules of ammonia (the indicator of urea). Ammonia concentration is easily measured.

The enzyme *glucocerebrosidase* is used to treat Gaucher's disease. Those who suffer this genetic disorder are incapable of producing active glucocerebrosidase and are thus unable to carry out the degradation of the glycolipid called glucocerebroside. Glucocerebroside accumulates in the macrophages of the liver, spleen, and bone marrow. If untreated, these abnormal macrophages will displace normal bone marrow cells, causing severe anemia. Glucocerebrosidase, administered intravenously, degrades glucocerebroside to glucose and ceramide, which are further metabolized.

Self Test for Chapter Nineteen

1. What do kinase enzymes do?
2. Which enzyme catalyzes the hydrolysis of triglycerides?

3. What does the enzyme lactate dehydrogenase do?
4. What is the substrate for each of the following enzymes:
   a. succinate dehydrogenase
   b. sucrase
   c. glycogen phosphorylase
5. How does an enzyme speed up a biological chemical reaction?
6. What effect does an enzyme have on the equilibrium constant of a reaction?
7. What is the term for the reactant in an enzyme-catalyzed reaction?
8. In enzyme-catalyzed reactions, what happens as the concentration of the substrate is increased?
9. At the maximum rate of an enzyme-catalyzed reaction, what may be said about all the enzymes and substrate molecules?
10. What is the first step in an enzyme-catalyzed reaction?
11. What is the difference between the lock-and-key model and the induced-fit model of enzyme-substrate complex formation?
12. What are the final three steps (after formation of the enzyme-substrate complex) in a enzyme-catalyzed reaction?
13. List the different classes of enzyme specificity.
14. What is the term used to describe an enzyme that catalyzes the reaction of only one substrate?
15. What are apoenzymes and cofactors?
16. What molecule is commonly used as a coenzyme of an enzyme that is a dehydrogenase?
17. Why are the water-soluble vitamins important?
18. What is the term for the pH at which an enzyme has the greatest activity?
19. At extreme pH values, an enzyme may lose the normal three-dimensional shape of the active site. What is the term used to describe this condition?
20. Most enzymes are rapidly destroyed if the temperature is much higher than what value?
21. List two classes of enzyme inhibitors.
22. How do the sulfa drugs work as inhibitors?
23. What process regulates the activity of allosteric enzymes?
24. In feedback inhibition, which compound causes the first reaction of a pathway to be shut off?
25. What is the term for an enzyme that is first produced in an inactive form?
26. What is the function of the pancreatic enzymes trypsin, chymotrypsin, and elastase?
27. Why are measurements done on the amount of enzymes such as myoglobin, creatine kinase-MB, and cardiac troponin I in the blood when acute myocardial infarction is suspected?

20 *Introduction to Molecular Genetics*

Learning Goals

1. Draw the general structure of DNA and RNA nucleotides.
2. Describe the structure of DNA, and compare it with RNA.
3. Explain DNA replication.
4. List three classes of RNA molecules, and describe their functions.
5. Explain the process of transcription.
6. List and explain the three types of post-transcriptional modifications of eukaryotic mRNA.
7. Describe the essential elements of the genetic code and develop a "feel" for its elegance.
8. Describe the process of translation.
9. Define mutation and understand how mutations can cause cancer and cell death.
10. Describe the tools used in the study of DNA and in genetic engineering.
11. Describe the process of polymerase chain reaction, and discuss potential uses of the process.
12. Discuss strategies for genome analysis and DNA sequencing.

Introduction

DNA makes up the genes in our cells, and the expression of those genes ultimately determines who and what we are. In this chapter, you will learn about the structure of DNA and the processes that translate the genetic information of genes into the structure of proteins.

20.1 The Structure of the Nucleotide

Chemical Composition of DNA and RNA

Deoxyribonucleic acid (DNA) carries all the genetic information in the cell. DNA is found in the nucleus of the cell and is wound into structures called chromosomes. Through a series of events, the information in the DNA is translated into the structure of a protein.

There are two important types of nucleic acids, DNA and **ribonucleic acid (RNA)**. DNA carries all the genetic information for an organism, while RNA interprets the genetic information of DNA into proteins. The chemical composition of DNA and RNA is summarized in the following table.

	DNA	RNA
Sugar	2'-Deoxyribose	Ribose
Purine nitrogenous bases	Adenine (A)	Adenine (A)
	Guanine (G)	Guanine (G)
Pyrimidine nitrogenous bases	Cytosine (C)	Cytosine (C)
	Thymine (T)	Uracil (U)

In addition, both DNA and RNA contain phosphoryl groups.

Example 1: Chemical Composition of DNA and RNA
Explain the chemical composition of DNA and RNA.

Solution:
DNA includes the sugar 2'-deoxyribose, phosphoryl groups, and the following four heterocyclic bases: adenine, guanine, cytosine, and thymine. RNA includes the sugar ribose, phosphoryl groups, and the following four heterocyclic bases: adenine, guanine, cytosine, and uracil.

Nucleosides

Nucleosides are made up of a sugar (ribose or 2'-deoxyribose) and a nitrogenous base. Because this large structure contains two cyclic molecules, the sugar and the base, the ring atoms of the sugar are designated with a prime to distinguish them from atoms in the nitrogenous base. The covalent bond that forms to join the sugar and the base is called a β-N-glycosidic bond. See Figure 20.2 in the textbook for examples of nucleosides.

Nucleotide Structure

Nucleotides are composed of a nitrogenous base, a five-carbon sugar, and at least one phosphoryl group. A nucleotide containing the sugar ribose is referred to as a **ribonucleotide**. These are the monomers from which RNA molecules are formed. A nucleotide containing the sugar 2'-deoxyribose is called a **deoxyribonucleotide**. These are the monomers from which DNA is made. The covalent bond between the sugar and the phosphoryl group is a phosphoester bond.

20.2 The Structure of DNA and RNA

LG 2

A strand of DNA is a polymer of 2'-deoxyribonucleotide units bonded to one another by 3'-5' phosphodiester bonds. The backbone of the polymer is called the *sugar-phosphate backbone* because it is composed of alternating deoxyribose and phosphoryl groups in phosphodiester linkage. A nitrogenous base is bonded to each sugar by a β-*N*-glycosidic linkage.

Example 2: Structure of DNA
Summarize the primary structure of DNA.

Example 2 Solution:
The primary structure of DNA is the linear sequence of its 2'-deoxyribonucleotides. The nucleotides are linked by 3'-5' phosphodiester bonds. Each single strand of DNA is characterized by a backbone of alternating deoxyribose and phosphoryl groups in phosphodiester linkages. Each deoxyribose is also linked to one of the nitrogenous bases by a β-*N*-glycosidic bond.

DNA Structure: The Double Helix

DNA is a **double helix** of two strands of DNA wound around one another held together by hydrogen bonds. The sugar-phosphate backbone spirals around the outside of the helix, and the nitrogenous bases extend into the center at right angles to the axis of the helix. Adenine forms two hydrogen bonds with thymine, and cytosine forms three hydrogen bonds with guanine. The hydrogen-bonded bases are called **base pairs**. The hydrogen bonds between the base pairs help hold the two strands together and maintain the double helix structure. The two strands of DNA are c**omplementary**, because the sequence of bases on one strand automatically determines the sequence of bases on the opposite strand. There are ten base pairs in each turn of the helix. The two strands of the double helix are **antiparallel** to one another; they proceed in opposite directions.

Example 3: Double Helix of DNA
Summarize the double helix structure of DNA.

Solution:
DNA consists of two strands wound around each other. Each strand has a helical conformation, and the resultant structure is called a double helix. The major properties of the DNA double helix are as follows:

1. The sugar phosphate backbone winds around the outside of the bases like the handrails of a spiral staircase.
2. Each purine base is hydrogen bonded to a pyrimidine base in the interior of the double helix. Adenine is always paired with thymine, and guanine is always paired with cytosine. Each base pair lies at nearly right angles to the long axis of the helix, like the stairs of the spiral staircase. Due to the base pairing of A to T and G to C, the two strands are complementary to one another. Therefore, the sequence of bases on one strand automatically determines the sequence of the other strand.
3. The two strands of the DNA double helix are antiparallel. One strand advances in the 5' → 3' direction, and the second strand advances in the 3' → 5' direction.
4. The double helix of DNA completes one full turn every ten nucleotides.

Chromosomes

Chromosomes are pieces of DNA that carry the genetic instructions, or genes, of an organism. The complete set of genetic information in all the chromosomes of an organism is called the **genome**.

The genome of **prokaryotes** is circular and supercoiled, and consists of a single chromosome. Chromosomes in **eukaryotes** are much more complex. The first level of eukaryotic chromosome structure is the **nucleosome**, which consists of a strand of DNA wrapped around a small disk made up of histone proteins. At this level the DNA looks like beads along a string. This string of beads then coils into a larger structure called the *condensed fiber*. This complex of DNA and protein is termed *chromatin* and makes up the eukaryotic chromosomes.

RNA Structure

RNA molecules are single stranded. The sugar-phosphate backbone of RNA consists of ribonucleotides linked by 3'-5' phosphodiester bonds. The sugar in RNA is ribose rather than 2'-deoxyribose, and uracil (U) replaces thymine.

20.3 DNA Replication

LG 2

DNA must be replicated before cell division so that each daughter cell inherits a copy of each gene. The mechanism of faithful replication of DNA involves an enzyme which "reads" each parental strand and catalyzes the polymerization of a complementary daughter strand. Thus, each daughter DNA molecule consists of one parental strand and one newly synthesized daughter strand. This mode of DNA replication is called **semiconservative replication.**

Bacterial DNA Replication

Bacterial chromosomes are circular DNA molecules. Replication begins at a **replication origin** and proceeds bidirectionally around the circular chromosome. The site at which the new DNA is being synthesized is called the **replication fork.**

The first step in DNA replication is the separation of the two strands of DNA (breaking of the hydrogen bonds between the base pairs) by the enzyme *helicase*. *Topoisomerase* relieves the stress of the supercoiling, and *single-stranded binding protein* binds to each strand. *Primase* catalyzes the synthesis of a short RNA primer to which the enzyme **DNA polymerase III** will add DNA nucleotides to begin synthesis of the daughter strand.

Because the two strands of DNA are antiparallel, and replication can proceed in only the 5'→3' direction, the precise mechanism for replication on the two strands is different. The **leading strand** is replicated continuously from a single RNA primer; and the **lagging strand** is replicated discontinuously from many RNA primers. Each of these primers is removed and replaced with DNA. The small fragments produced in this way are joined together by *DNA ligase*.

Since even a small error in replicating DNA could have disastrous consequences for the cell, DNA polymerase III is also able to proofread the new DNA strand. If a mistake is found, the incorrect nucleotide is removed and replaced with the correct one.

Eukaryotic DNA Replication

In eukaryotes, DNA replication begins at many replication origins and proceeds bidirectionally along each chromosome.

Example 4: DNA Replication
Summarize DNA replication

Solution:
DNA replication occurs each time a cell divides; thus, all the genetic information is passed from one generation to the next. As a result of specific base pairing, the sequence of bases along each strand of DNA automatically specifies the sequence of bases in the complementary strand. DNA replication begins at the replication origin, and each parental strand is copied. The enzyme DNA polymerase III not only catalyzes the replication of new DNA, it also proofreads the new DNA strand to be sure there are no mistakes.

20.4 Information Flow in Biological Systems

The genetic information in the DNA must be expressed to produce the proteins that actually carry out the work of the cell. Gene expression involves two steps. First, DNA is transcribed to produce a variety of RNA molecules. This process is called **transcription**. Then the RNA molecules participate in **translation**, a process which uses this genetic information to produce proteins. This unidirectional expression of the genetic information is called the **central dogma** of molecular biology and can be summarized as follows:

$$\text{DNA} \rightarrow \text{RNA} \rightarrow \text{PROTEIN}$$

Example 5: Terms Pertaining to Information Flow
Explain the following terms: central dogma, transcription, and translation.

Solution:
The *central dogma* of molecular biology states that, in cells, the flow of genetic information contained in DNA is a one-way street that leads from DNA to RNA to protein synthesis. The process by which a single strand of DNA serves as a template for the synthesis of an RNA molecule is called *transcription.* In this process, part of the information of DNA is copied into a strand of RNA. The process by which the message in that RNA is converted into protein is called *translation.*

Classes of RNA Molecules

There are three classes of RNA molecules found in the cell, and they are classified by their function in gene expression.

Messenger RNA (mRNA) carries the genetic information for a protein from DNA to the ribosomes. It is a complementary RNA copy of a gene on the DNA.

Ribosomal RNA (rRNA) is a structural and functional component of the ribosomes, which are "platforms" on which protein synthesis occurs.

Transfer RNA (tRNA) translates the genetic code of the mRNA into the primary sequence of amino acids in the protein. In addition to the primary structure, tRNA molecules have a cloverleaf-shaped secondary structure resulting from base pair hydrogen bonding (traditional A–U and G–C, as well as other unusual base pairings), and a tertiary structure.

The sequence CCA is found at the 3'-end of the tRNA. The 3'-adenosine of this sequence can be covalently attached to an amino acid. Three nucleotides at the base of the cloverleaf structure form the **anticodon**. This triplet of bases forms hydrogen bonds to a **codon** (complementary sequence of bases) in a mRNA molecule on the surface of a ribosome during protein synthesis. This interaction assures that the correct amino acid is brought to the site of protein synthesis at the appropriate location in the growing peptide chain.

> **Example 6: Classification of RNA**
> Describe the different classes of RNA molecules.
>
> **Solution:**
> An RNA molecule is classified by its function.
>
> 1. *Messenger RNA (mRNA)* carries the genetic information for a protein.
> 2. *Ribosomal RNA (rRNA)* is a structural and functional component of the ribosomes.
> 3. *Transfer RNA (tRNA)* is responsible for translating the genetic code of the mRNA.

Transcription

Transcription is catalyzed by the enzyme **RNA polymerase**. First, RNA polymerase binds to a specific nucleotide sequence, the **promoter**, at the beginning of a gene. It then separates the two strands of DNA so that it can "read" the sequence of the DNA. Chain elongation begins as the RNA polymerase reads the DNA template strand and catalyzes the polymerization of a complementary RNA copy. The final stage of transcription is termination. The RNA polymerase finds a termination sequence at the end of the gene and releases the newly formed RNA molecule.

Post-Transcriptional Processing of RNA

In eukaryotes, transcription produces a **primary transcript** that undergoes extensive **post-transcriptional modification** before it is exported from the nucleus for translation. The primary transcripts undergo three post-translational modifications:

1. Addition of a **cap structure** to the 5' end of the RNA. This facilitates efficient translation.
2. Addition of a **poly(A) tail** to the 3' end of the RNA. This protects the mRNA from enzymatic degradation.
3. **RNA splicing**. Bacterial genes are continuous, and all the nucleotide sequences of the gene are found in the mRNA. Eukaryotic genes are discontinuous; there are extra DNA sequences within the genes. The initial mRNA, or primary transcript, carries both the protein-coding sequences **(exons)** and these extra sequences, which are termed intervening sequences or **introns**. The introns are removed by the process of RNA splicing. Splicing is carried out with the assistance of ribonucleoprotein particles called *spliceosomes.* They aid in the recognition of specific sequences at the splice boundaries, stabilize the splicing complex, and catalyze the splicing events.

20.5 The Genetic Code

LG 7

The **genetic code** is a triplet code. Each code word (codon) consists of three nucleotides. The three-letter genetic code contains 64 codons, but there are only 20 amino acids. Thus, there are 44 more codons than are required. Three of the codons (UAA, UAG, and UGA) are translation termination signals, leaving 41 additional codons. The genetic code is said to be **degenerate**, meaning that several triplet codons may code for the same amino acid. Methionine and tryptophan are the only amino acids encoded by only a single codon. All others are encoded by at least two codons, and serine and leucine each have six codons. As a result of this, the genetic code is quite resistant to mutation. For those amino acids that have multiple codons, the first two bases define the amino acid, and the third position is variable. A mutation (the change of a nucleotide sequence) in the third position, therefore, often has no effect upon the amino acid that is incorporated into a protein.

20.6 Protein Synthesis

Protein synthesis, or **translation,** occurs on ribosomes. Ribosomes are complexes of ribosomal RNA and proteins consisting of two subunits, a small and a large ribosomal subunit. Many ribosomes simultaneously translate each mRNA. These structures are called polyribosomes or **polysomes**.

The Role of Transfer RNA

The molecule that decodes the information on the mRNA molecule into the primary structure of a protein is transfer RNA (tRNA). In order to do this, the tRNA must be covalently linked to one specific amino acid. The enzyme that binds the specific tRNA to its specific amino acid is an **aminoacyl tRNA synthetase**. The product is called an **aminoacyl tRNA**. The tRNA recognizes the appropriate codon on the mRNA by codon-anticodon hydrogen bonding.

The Process of Translation

Initiation

Initiation factors mediate the formation of a translation initiation complex composed of an mRNA molecule, the small and large ribosomal subunits, and the initiator tRNA. The initiator tRNA specifically recognizes the initiation codon, AUG, on the mRNA. The ribosome has two sites for binding tRNA molecules. The first site is called the **peptidyl tRNA binding site (P-site)**. It holds the initiator tRNA in the initiation complex and then carries the tRNA bound to the growing peptide during the remainder of protein synthesis. The second site, called the **aminoacyl tRNA binding site (A-site),** holds the aminoacyl tRNA carrying the next amino acid to be added to the peptide chain.

Chain Elongation

Chain elongation occurs in three steps that are repeated until protein synthesis is complete. This stage requires the breakdown of GTP to GDP and P_i as well as the involvement of several **elongation factors.**

1. An aminoacyl tRNA molecule binds to the empty A-site.
2. Peptidyl transferase, a catalytic region of the 28S rRNA, catalyzes the formation of the peptide bond, and the peptide chain is shifted to the tRNA that occupies the A-site.
3. The uncharged tRNA molecule is discharged, and the ribosome changes positions (**translocation**), so that the next codon on the mRNA occupies the A-site, shifting the new peptidyl tRNA from the A-site to the P-site.

Termination

When a stop, or **termination codon** (UAA, UAG, or UGA), is encountered, translation is terminated. A **release factor** binds the empty A-site and causes release of the newly formed peptide chain and the ribosomal subunits.

The newly synthesized protein folds into its characteristic three-dimensional shape, and if it has quaternary structure, it may associate with other protein subunits. Some proteins are further modified following protein synthesis by the addition of carbohydrate or lipid molecules.

20.7 Mutations, Ultraviolet Light, and DNA Repair

The Nature of Mutations

Any change in the nucleotide sequence of a DNA molecule is called a **mutation**. Mutations can arise from errors during DNA replication or may be the result of chemicals, called **mutagens**, which damage the DNA.

Mutations are classified by the kind of change that occurs in the DNA. The substitution of a single nucleotide for another is called a **point mutation**. Loss of one or more nucleotides is a **deletion mutation**. Addition of one or more nucleotides to a DNA sequence is an **insertion mutation**.

The Results of Mutations

Some mutations are **silent mutations**; that is, they cause no change in the organism. This occurs because in many cases more than one codon codes for the same amino acid. Often, however, the result of a mutation has a negative effect on the health of the organism. This would result if the mutation coded for a different amino acid leading to a nonfunctional or improperly functioning protein. The effect of a mutation depends on how it alters the genetic code for a protein. There are approximately 4000 human genetic disorders that are known to be caused by mutations, including sickle cell anemia, hemophilia, cystic fibrosis, and color-blindness.

Mutagens and Carcinogens

Often mutagens are also **carcinogens**, cancer-causing chemicals. Most cancers result from mutations in a single normal cell. These mutations result in the loss of normal growth control, causing the abnormal cell to proliferate. If that growth is not controlled or destroyed, it may result in the death of the individual.

Ultraviolet Light Damage and DNA Repair

Ultraviolet (UV) light causes damage to DNA by inducing the formation of **pyrimidine dimers.** Such dimers consist of adjacent pairs of pyrimidine bases covalently bonded to one another. This interferes with normal hydrogen bonding between these pyrimidines and the complementary bases on the opposite strand. As a result, this region of DNA cannot be replicated or transcribed and the cell dies.

Bacteria have four different mechanisms to repair ultraviolet light damage. However, mutations can still occur when the repair system makes an error and causes a change in the nucleotide sequence of the DNA. Ultraviolet lights (germicidal lamps) are used in hospitals to kill bacteria in the air and on environmental surfaces, such as in a vacant operating room.

Human cells which have had excessive exposure to UV light (as during sun tanning) can also form many pyrimidine dimers and this has been correlated with an increased incidence of skin cancer.

Consequences of Defects in DNA Repair

The human repair system for pyrimidine dimers requires at least five enzymes. The first enzyme involved is a *repair endonuclease*. Without this enzyme pyrimidine dimers cannot be repaired. A mutation in the gene for this enzyme, or any of the enzymes of the repair system, results in the genetic skin disorder *xeroderma pigmentosum*. Individuals with this disorder are extremely sensitive to the UV rays of sunlight and usually develop multiple skin cancers before the age of twenty.

20.8 Recombinant DNA

Tools Used in the Study of DNA

LG 10

Many of the techniques and tools used to study DNA and to clone genes were developed or discovered during basic studies on bacterial DNA replication and gene expression. These include many enzymes that catalyze reactions of DNA molecules, cloning vectors, and hybridization techniques.

Restriction Enzymes

Restriction enzymes (or *restriction endonucleases)* are bacterial enzymes that "cut" the sugar-phosphate backbone of DNA molecules at specific nucleotide sequences. Often these enzymes leave short single-stranded stretches at the ends of the DNA molecules. These are called "sticky ends" because they can reassociate with one another by hydrogen bonding.

This property of the DNA fragments generated by restriction enzymes is very important to gene cloning.

These enzymes are used to digest large DNA molecules into smaller fragments of specific size. DNA from an individual will produce a reproducible set of DNA fragments when digested with a particular restriction enzyme. This is essential for the study or cloning of DNA from any source.

DNA Cloning Vectors

A **cloning vector** is a piece of DNA having its own replication origin so that it can be replicated inside a host cell. Often the bacterium *Escherichia coli* serves as the host cell in which the vector carrying the cloned DNA is replicated in abundance.

There are two major kinds of cloning vectors: *phage vectors* and *plasmid vectors*. Phage vectors are specially modified bacterial viruses and plasmids are extra pieces of circular DNA found in most kinds of bacteria. Both can carry a cloned DNA fragment into a bacterial cell and carry out replication that allows the amplification of the vector and the cloned DNA fragment.

Genetic Engineering

To clone a gene, vector DNA and target DNA (the source of the gene to be cloned) are digested with the same restriction enzyme. They are then mixed together under conditions that allow the sticky ends of the target and vector DNA to hybridize with one another. **Hybridization** is a technique used to identify the presence of a gene on a particular DNA fragment. It is based on the fact that complementary DNA sequences will hydrogen bond, or hybridize, to one another. DNA ligase is added to covalently join the ends of the DNA molecules.

Finally, these recombinant DNA molecules are introduced into a bacterial cell. This can be done by transformation, a process in which bacterial cells are specially treated to favor the entry of DNA into the cells. Antibiotic selection and hybridization can be used to detect those clones carrying the target gene.

Genetic engineers have had to overcome many obstacles to clone eukaryotic genes of special interest medically. For instance, if the researcher desires to produce the protein product of a eukaryotic gene, he or she cannot simply clone cellular DNA because there are introns within eukaryotic genes. Molecular biologists found that a DNA copy of a eukaryotic mRNA could be made using the enzyme reverse transcriptase of a class of viruses called retroviruses. This DNA copy of the mRNA carries all the protein-coding sequences of a gene, but none of the intron sequences. Thus bacteria are able to transcribe and translate the cloned DNA and produce valuable products for use in medicine and other applications.

Great progress has been made by applying genetic engineering to genetic disorders, but most of these products represent only a treatment, not a cure. The ultimate dream for the future is to be able to replace mutated genes.

20.9 Polymerase Chain Reaction

LG 11

Polymerase chain reaction (PCR) is a sensitive technique for the amplification of specific DNA sequences. A primer, which is a short piece of single-stranded DNA that will hybridize to a region of the gene of interest, is required. The DNA, primer, four DNA nucleotides, and a heat stable DNA polymerase (Taq polymerase) are mixed and placed in a thermocycler at 94-96°C to separate the two strands of DNA. The temperature is dropped to 50-56°C to allow the primers to hybridize to the template DNA. Finally, the temperature is raised to 72°C to allow Taq polymerase to synthesize a daughter strand. These three steps are repeated many times, doubling the amount of target DNA with each cycle.

20.10 The Human Genome Project

The goals of the Human Genome Project (HGP) were to identify and map all of the genes of the human genome and to determine the complete DNA sequence of each of the chromosomes. The initial goal of the HGP was to complete this sequence by the year 2005. Technological advances that developed as a result of the project allowed its completion by 2003.

Genetic Strategies for Genome Analysis

To sequence the human genome, genomic libraries were generated which were a set of clones representing the entire genome. Then the DNA sequences of the clones were determined. To map the sequences along each chromosome, chromosome walking was used. This provides information on both the DNA sequence and the DNA sequences next to it on the chromosome.

DNA Sequencing

DNA sequencing involves reactions in which DNA polymerase copies specific DNA sequences. Nucleotide analogues that cause chain termination (dideoxynucleotides) are incorporated randomly into the growing DNA chain. This generates a family of DNA fragments that differ in size by one nucleotide. DNA sequencing gels separate these fragments and provide DNA sequence data.

Modern technology uses dideoxynucleotides which are labeled with fluorescent dyes (a different color for each nucleotide). This allows all the reactions to be done in a single reaction mixture and the products to be separated on a single lane of a gel. A computer will then "read" the gel based on the color of each DNA band.

With the abundance of DNA information now available, the field of **bioinformatics** brings together the disciplines of computer science, mathematics, statistics, DNA technology, and engineering to devise methods and software tools for organizing, understanding, analyzing, and applying the knowledge we gain from these DNA sequences.

Self Test for Chapter Twenty

1. Which molecule in living things is the carrier of genetic information?
2. In which type of nucleic acid is thymine usually found?
3. Give the correct abbreviations for the following compounds:
   a. adenine + deoxyribose + triphosphate unit
   b. cytosine + ribose + triphosphate unit
   c. thymine + deoxyribose + monophosphate unit
4. What type of bonding helps to hold the two strands of DNA together as a double helix?
5. Which base in one strand of DNA is always hydrogen-bonded to adenine in the other strand?
6. One strand of the DNA molecule has a 5' → 3' orientation. What is the orientation of the opposite strand?
7. As a result of specific base pairing in the DNA molecule, what is the relationship of the sequence of bases on one strand to the sequence of bases on the other strand?
8. What are three differences between the structure of DNA and RNA?
9. What is the first event to occur during replication of DNA?
10. DNA replication in *E. coli* begins at a unique sequence on the circular chromosome. What is the term for this site?
11. What is the term for the site at which the new deoxyribonucleotides are added to the growing daughter strand?
12. What are the two major functions of the enzyme DNA polymerase?
13. State the central dogma of molecular biology.
14. What is the function of mRNA?
15. Which form of RNA carries the needed amino acids to the ribosomes for use in protein synthesis?
16. What sequence on DNA indicates the site at which the RNA polymerase should begin transcription?
17. Why is a poly(A) tail added to the 3' end of RNA after transcription?
18. In eukaryotic cells, the initial mRNA (primary transcript) carries the sequences for a protein but also contains noncoding sequences. What is the term for these noncoding sequences?
19. What is the term for the coding sequences that remain after mRNA splicing?
20. The three-letter genetic code contains 64 words. What is the term for these genetic words?
21. Why is the genetic code said to be degenerate?
22. What role do ribosomes play in translation?
23. What is a point mutation?

24. What is the term used to describe a mutation that results in no change in the organism?
25. What is the name of the DNA damage that is caused by ultraviolet light?
26. How can mutations result in cell death?
27. What disease is caused by the accumulation of mutations that cause uncontrolled cell growth and division?
28. What is the term for an enzyme that recognizes a specific DNA nucleotide sequence and "cuts" the sugar-phosphate backbone at that site?
29. List two types of cloning vectors.
30. What is the polymerase chain reaction used for?
31. Briefly describe the steps in the polymerase chain reaction.
32. Why are dideoxynucleotides used in gene sequencing?

21 Carbohydrate Metabolism

Learning Goals

1. Discuss the importance of ATP in cellular energy transfer processes.
2. Describe the three stages of catabolism of dietary proteins, carbohydrates, and lipids.
3. Discuss glycolysis in terms of its two major segments.
4. Looking at an equation representing any of the chemical reactions that occur in glycolysis, describe the kind of reaction that is occurring and the significance of that reaction to the pathway.
5. Describe the mechanism of regulation of the rate of glycolysis. Discuss particular examples of that regulation.
6. Discuss the practical and metabolic roles of fermentation reactions.
7. List several products of the pentose phosphate pathway that are required for biosynthesis.
8. Compare glycolysis and gluconeogenesis.
9. Summarize the regulation of blood glucose levels by glycogenesis and glycogenolysis.

Introduction

Glycolysis and fermentation are two pathways used by many organisms to provide energy for cellular work. In this chapter, we will also study pathways for the synthesis and breakdown of glycogen, a major energy storage form.

21.1 ATP: The Cellular Energy Currency

Cells need a constant supply of energy to maintain essential life processes such as *active transport*, *biosynthesis*, and *mechanical work*. A supply of energy-rich food molecules, carbohydrates, protein, and fats, is required to provide this needed cellular energy. Carbohydrates are the most readily used energy source, and glycolysis is the pathway for the first stages of carbohydrate degradation.

Catabolism is the set of metabolic pathways that breakdown complex macromolecules into simpler ones and, in the process, harvest part of their potential energy for use by the cell. One of those energy-requiring functions is **anabolism**, or biosynthesis. The energy released in catabolism is stored as *chemical bond energy*. The molecule used for the storage of chemical energy, and often called the universal energy currency, is **adenosine triphosphate (ATP)**.

ATP is a **nucleotide** composed of the nitrogenous base adenine bonded in *N*-glycosidic linkage to the sugar ribose. Ribose is bonded to one (AMP), two (ADP), or three (ATP) phosphoryl groups. The molecule is a high-energy compound because the phosphoanhydride bonds holding the terminal phosphoryl groups are high-energy bonds. This means that a large amount of energy is released when these bonds are broken. Such high-energy bonds are indicated as squiggles (~).

ATP has higher energy content than the compounds to which it donates energy, but, as must be the case, it contains less energy than the compounds that are involved in forming it. The secret to the success of ATP as the energy currency is that both the reactions that produce ATP and breakdown ATP in order to provide energy for cellular work are energetically favored reactions. ATP is an ideal "go-between," shuttling energy from exothermic reactions to endothermic reactions.

Example 1: ATP Function
What is the function of ATP?

Solution:
Adenosine triphosphate (ATP) is a "go-between" molecule that can store or release chemical energy. The secret to the function of ATP as a go-between lies in its chemical structure. The molecule is a high-energy compound because of the phosphoanhydride bonds holding the terminal phosphoryl groups. When these bonds are broken, they release a large amount of energy that can be used for cellular work. Then other pathways allow the resynthesis of ATP by providing the energy needed for the reaction between ADP + P_i which produces ATP.

Example 2: Breaking the Phosphoanhydride Bond of ATP
Show how breaking the phosphoanhydride bond in ATP is coupled with the synthesis of glucose-6-phosphate. Use the equation below to help you answer this question.

$$\textbf{ATP + glucose} \rightarrow \textbf{glucose-6-phosphate + ADP + 4 kcal/mol}$$

Solution:
In the first step of glycolysis, there is a transfer of a phosphoryl group ($-PO_3^{2-}$) from ATP to the C-6 hydroxyl group of a glucose molecule. This reaction is catalyzed by the enzyme hexokinase. This reaction can be "dissected" to reveal the role of ATP as a source of energy. First, the phosphoanhydride bond of ATP breaks to form ADP and a phosphoryl group. This reaction releases 7 kcal/mole of energy:

$$\textbf{ATP + } H_2O \rightarrow \textbf{ADP + phosphoryl group + 7 kcal/mole}$$

Next, the synthesis of glucose-6-phosphate from glucose and the phosphoryl group requires 3.0 kcal/mole:

$$\textbf{3.0 kcal/mole + glucose + phosphoryl group} \rightarrow \textbf{glucose-6-phosphate + } H_2O$$

Thus, by the coupling of a reaction needing energy with breaking the phosphoanhydride bond of ATP to release energy, the cell can undergo hundreds of reactions.

21.2 Overview of Catabolic Processes

The first stage of catabolism is the hydrolysis of dietary macromolecules into small subunits. Large polymeric molecules are degraded into their constituent subunits, which are taken into the cells. Polysaccharides are hydrolyzed to monosaccharides. This is catalyzed by several different enzymes and begins in the mouth with salivary amylase and continues with pancreatic amylase in the small intestine. Proteins are hydrolyzed to oligopeptides and amino acids. This process begins with pepsin in the stomach and continues in the small intestine. Fats are hydrolyzed into fatty acids, glycerol, and monoglycerides beginning in the small intestine by pancreatic lipase. All these molecules are then absorbed by the cells lining the intestine.

The second stage of catabolism is the conversion of monomers into a form that can enter one of the catabolic pathways and be degraded to yield energy. The final stage of catabolism is the complete oxidation of nutrients and the production of ATP.

21.3 Glycolysis

An Overview

LG 3

Glycolysis is the first stage of carbohydrate catabolism. It is an anaerobic process, requiring no oxygen, and it is carried out by enzymes in the cytoplasm of the cell. The degradation of glucose by glycolysis yields chemical energy in the form of ATP. Four ATP molecules are produced by **substrate-level phosphorylation,** a reaction in which a high-energy phosphoryl group from one of the substrates is transferred to ADP to form ATP. However, the net yield of ATP is only 2, because 2 ATP are used in the early stages of glycolysis. Chemical energy in the form of reduced **nicotinamide adenine dinucleotide,** NADH, is also produced. Under aerobic conditions, the electrons are donated to an electron transport system for the generation of ATP by **oxidative phosphorylation**; but under anaerobic conditions, NADH is used as a source of electrons in fermentation reactions. The final product of glycolysis is two pyruvate molecules. Under anaerobic conditions, pyruvate is used as a substrate in fermentation reactions.

Example 3: Carbohydrate Metabolism
Describe the first stage of carbohydrate metabolism.
Solution:
The pathway called glycolysis is the beginning stage for the use of glucose as a source of chemical energy. This pathway is anaerobic, which means that no oxygen is required. The glycolysis pathway releases only about 2% of the potential energy of the glucose molecule. In the pathway, glucose is finally transformed after many steps into two molecules of pyruvate.

Example 4: NADH
Explain how the NADH produced by glycolysis is regenerated into NAD^+.
Solution:
1. If the cell is functioning under aerobic conditions, the NADH will be reoxidized, and pyruvate will be completely oxidized by aerobic respiration.
2. Under anaerobic conditions, cells of different types employ a variety of fermentation reactions to accomplish the conversion of NADH to NAD^+.

Biological Effects of Genetic Disorders of Glycolysis

Each step of glycolysis is controlled by specific enzymes. When genetic defects affect these enzymes various symptoms can occur. One symptom is muscle myopathy, muscle damage caused by the inability to extract energy from food molecules. Another is hemolytic anemia, which results from the lysis of red blood cells that cannot produce sufficient ATP.

Deficiency of phosphofructokinase can result in Tarui's disease. Deficiency in phosphoglycerate kinase may have symptoms ranging from metal challenge and seizures to myopathy and hemolytic anemia. Deficiency of phosphoglycerate mutase results in exercise intolerance, muscle pain, and myoglobinuria.

Reactions of Glycolysis

Glycolysis is divided into two major segments. The first segment consists of five reactions that involve the investment of ATP energy. In the final five reactions, energy is harvested to produce a net gain of ATP.

The structures of the intermediates of glycolysis can be found in Section 21.3 of the text. The following is a brief description of those reactions.

Reaction 1
Glucose is phosphorylated by the enzyme *hexokinase*. A phosphoryl group from ATP is transferred to C-6 of glucose, and the product is glucose-6-phosphate.

Reaction 2
The enzyme *phosphoglucose isomerase* catalyzes the isomerization of glucose-6-phosphate to produce fructose-6-phosphate.

Reaction 3
The enzyme *phosphofructokinase* transfers a phosphoryl group from ATP to the C-1 hydroxyl of fructose-6-phosphate, producing fructose-1,6-bisphosphate.

Reaction 4
Aldolase splits the fructose-1,6-bisphosphate into glyceraldehyde-3-phosphate (G3P) and dihydroxyacetone phosphate (DHAP).

Reaction 5
Only G3P can be used in glycolysis. In this step, the DHAP is isomerized by the enzyme *triose phosphate isomerase* to produce a second molecule of G3P.

Reaction 6
Glyceraldehyde-3-phosphate dehydrogenase catalyzes the oxidation of the aldehyde group of G3P to a carboxyl group. The coenzyme in this reaction is NAD^+, which is reduced to NADH. Next, an inorganic phosphate group is transferred to the carboxylate group to produce 1,3-bisphosphoglycerate.

Reaction 7
Phosphoglycerate kinase catalyzes the transfer of a phosphoryl group from 1,3-bisphosphoglycerate to ADP. This is the first substrate-level phosphorylation of glycolysis to produce ATP.

Reaction 8
Phosphoglycerate mutase catalyzes the isomerization of 3-phosphoglycerate to 2-phosphoglycerate.

Reaction 9
Enolase catalyzes the dehydration of 2-phosphoglycerate, producing the energy-rich product phosphoenolpyruvate.

Reaction 10
Pyruvate kinase catalyzes the last substrate-level phosphorylation, in which a phosphoryl group from phosphoenolpyruvate is transferred to ADP to produce ATP. The final product of glycolysis is pyruvate.

Reactions 6 through 10 occur twice per glucose molecule; thus, the final products of glycolysis are 2 NADH, 2 pyruvate, and 4 ATP. However, since 2 ATP/glucose were invested early in glycolysis, the net yield is 2 ATP/glucose molecule.

Example 5: Steps in Glycolysis
List the steps in glycolysis in which a phosphoanhydride bond of an ATP molecule is broken in order to provide energy for specific reactions.

Solution:
1. Reaction 1 is a coupled reaction in which the enzyme *hexokinase* catalyzes the transfer of a phosphoryl group from ATP to glucose. The product is glucose-6-phosphate.
2. Reaction 3 is a coupled reaction in which the enzyme *phosphofructokinase* catalyzes the transfer of a phosphoryl group from ATP to fructose-6-phosphate. The product is fructose-1,6-bisphosphate.

Example 6: Steps in Glycolysis
List the two steps of the anaerobic glycolysis pathway that result in the production of ATP molecules.

Solution:
1. Reaction 7: This reaction is the first step of the pathway in which energy is harvested in the form of ATP. One of the phosphoryl groups of 1,3-bisphosphoglycerate is transferred to an ADP molecule in the first substrate-level phosphorylation of glycolysis. This reaction is catalyzed by the enzyme *phosphoglycerate kinase*.
2. Reaction 10: The final substrate-level phosphorylation in glycolysis involves the transfer of a phosphoryl group from phosphoenolpyruvate to ADP to form ATP. This reaction is catalyzed by the enzyme *pyruvate kinase*.

Entry of Fructose into Glycolysis

Fructose can enter glycolysis by different pathways, depending on the tissue. In muscle, the enzyme hexokinase converts fructose into fructose 6-phosphate which can directly enter glycolysis. In the liver, fructose is converted by fructokinase into fructose-1-phosphate which in turn is converted into DHAP and glyceraldehyde. The glyceraldehyde is converted into G3P, which, along with DHAP, enter glycolysis.

Regulation of Glycolysis

LG 5

Energy-harvesting pathways are responsive to the energy needs of the cell. Reactions of the pathway speed up when there is a demand for ATP. They slow down when there is abundant ATP to meet the energy requirements of the cell.

One of the major mechanisms for the control of the rate of the glycolytic pathway is the use of allosteric enzymes (introduced in Chapter 19). The allosteric effectors that indicate the energy needs of the cell include molecules such as ATP. If the ATP concentration is high, the cell must have a sufficient energy reserve. ADP and AMP indicate that the cell is in need of ATP.

Examples can be seen in the first, third, and final steps of glycolysis. Hexokinase is allosterically inhibited by the product of the reaction it catalyzes, glucose-6-phosphate. ATP is an allosteric inhibitor of the enzyme phosphofructokinase, while AMP, ADP, and citrate are allosteric activators. Pyruvate kinase is activated by fructose-1,6-bisphosphate.

21.4 Fermentation

In order for glycolysis to continue to degrade glucose and produce ATP, the NADH must be reoxidized, and the pyruvate must be utilized and removed. Under anaerobic conditions, these two requirements are met through **fermentation** reactions. Fermentations are catabolic reactions with no net oxidation.

Lactate Fermentation

Some cells, such as muscle cells and dairy bacteria, utilize lactate fermentation. Under anaerobic conditions, the enzyme *lactate dehydrogenase* reduces pyruvate to lactate. NADH is the reducing agent for this reaction, and thus NAD^+ is regenerated, while the pyruvate is used up.

This reaction occurs in muscle cells during strenuous exercise, when the body cannot provide sufficient oxygen for aerobic respiration to provide enough ATP. Lactate eventually builds up in the muscle to the extent that glycolysis can no longer proceed, and the muscle cells can no longer function. This point of exhaustion is called the **anaerobic threshold**.

Lactate fermentation is used by bacteria that produce yogurt and some cheeses. Lactate is one of the molecules that gives these dairy products their tangy flavor.

Alcohol Fermentation

Under anaerobic conditions, yeast cells ferment sugars produced by fruit and grains to produce ethanol. First, *pyruvate decarboxylase* removes CO_2 (decarboxylates) from the pyruvate, producing CO_2 and ethanal (acetaldehyde). Then, *alcohol dehydrogenase* catalyzes the reduction of ethanal to ethanol, reoxidizing NADH in the process. Eventually, the stable fermentation end product, ethanol, builds up to a concentration that kills the yeast cells. It is characteristic of fermentations that the end product eventually builds up to levels that inhibit the cells carrying out the reaction.

Example 7: Fermentation Pathways
Summarize the lactate and ethanol fermentation pathways.

Solution:

1. *Lactate fermentation*: This pathway occurs when you have exercised beyond the capacity of your lungs and circulatory system to deliver enough oxygen to the working muscles. The aerobic energy-generating pathways will no longer be able to supply enough ATP, but the muscles still demand energy. Under anaerobic conditions, pyruvate can be reduced by lactate dehydrogenase to form lactate. Simultaneously, NADH is oxidized to NAD^+, which is required for the continued anaerobic functioning of the glycolysis pathway.
2. *Alcohol fermentation:* Under anaerobic conditions in yeast (usually not humans!), the pyruvate formed by glycolysis is converted into ethanol and CO_2 in two enzyme-catalyzed steps. First, pyruvate is converted into ethanal as CO_2 is released. Then the ethanal is reduced by NADH. This produces ethanol and NAD^+ that allows anaerobic glycolysis to continue.

21.5 The Pentose Phosphate Pathway

The **pentose phosphate pathway** is an alternative pathway for glucose degradation that is particularly abundant in the liver and adipose tissue. It can be divided into three stages, each of which provides a unique biosynthetic requirement for the cell.

In the reactions of the first stage, glucose-6-phosphate is oxidized and decarboxylated to produce ribulose-5-phosphate, NADPH, and CO_2. NADPH is the reducing agent required for many biosynthetic pathways. The second stage involves isomerization reactions that convert ribulose-5-phosphate into other pentose phosphates. Among these is ribose-5-phosphate, which is required for nucleotide biosynthesis. In the final stage, a complex series of reactions occurs that involve carbon-carbon bond breakage and formation. The products are two molecules of fructose-6-phosphate and one molecule of glyceraldehyde-3-phosphate. Among the intermediates of the third stage is erythrose-4-phosphate, a precursor in the biosynthesis of the aromatic amino acids (such as phenylalanine, tyrosine, and tryptophan).

21.6 Gluconeogenesis: The Synthesis of Glucose

LG 8

During starvation and following extended exercise, the body must make glucose. **Gluconeogenesis** is the process by which glucose is produced from noncarbohydrate precursors. Lactate, all amino acids, except leucine and lysine, and glycerol from fats all can serve as precursors for glucose biosynthesis. Gluconeogenesis, shown in Figure 21.9 in the text, appears to be the reverse of glycolysis. Certainly most of the intermediates of the two pathways are identical. However, steps 1, 3, and 10 of glycolysis are irreversible and must be bypassed by other enzymes. Glucose-6-phosphate is dephosphorylated by the enzyme glucose-6-phosphatase, found in the liver but not in muscle. Similarly, the phosphorylation of fructose-6-phosphate by phosphofructokinase is irreversible. Fructose bisphosphatase catalyzes the removal of the phosphoryl group from fructose-1,6-bisphosphate. Finally, step 10 of glycolysis is bypassed by two enzymes. Pyruvate carboxylase carboxylates pyruvate by the addition of atmospheric CO_2, producing oxaloacetate. Then phosphoenolpyruvate carboxykinase removes the CO_2 and adds a phosphoryl group, producing phosphoenolpyruvate. The phosphoryl group donor is **guanosine triphosphate, GTP**.

It is extremely important that both the glycolysis and gluconeogenesis pathways be regulated. In step 3 of glycolysis, the enzyme phosphofructokinase is stimulated by AMP, ADP and phosphate, indicating that the cell needs energy and glycolysis proceeds. However, when ATP (and thus energy) is plentiful, phosphofructokinase is inhibited and fructose-1,6-bisphosphatase is stimulated, shutting down glycolysis and initiating gluconeogenesis.

In the Cori Cycle, lactate produced by working muscle is transported to the liver by the bloodstream. In the liver, the lactate is converted to pyruvate, which may be used to produce glucose by gluconeogenesis. The glucose produced in the liver may be degraded for energy or stored as glycogen.

21.7 Glycogen Synthesis and Degradation

The liver helps regulate the blood glucose level. One of the ways this is accomplished is through the uptake and storage of excess glucose as glycogen (**glycogenesis**). Alternatively, liver cells may degrade glycogen (**glycogenolysis**) and release glucose into the blood.

The Structure of Glycogen

Glycogen is a highly branched glucose polymer. The primary chain is linked by $\alpha(1\rightarrow4)$ glycosidic bonds. The branches are linked to the primary chain by $\alpha(1\rightarrow6)$ glycosidic bonds. Glycogen is stored in the cytoplasm of liver and muscle cells as **glycogen granules**. These granules are complexes of glycogen and the enzymes that carry out glycogenesis and glycogenolysis.

Glycogenolysis: Glycogen Degradation

Glycogenolysis is controlled by two hormones, **glucagon** and *epinephrine*. Glucagon is released from the pancreas in response to low blood glucose, and epinephrine is released from the adrenal glands in response to a threat or a stress. Both hormones regulate the activity of glycogen phosphorylase and glycogen synthase. Glycogen phosphorylase, involved in glycogenolysis, is activated; glycogen synthase, involved in glycogenesis, is inactivated. The steps in glycogenolysis are summarized below:

Step 1: The enzyme *glycogen phosphorylase* catalyzes phosphorolysis of a glucose unit at one end of glycogen, producing glucose-1-phosphate.

Step 2: The enzyme *$\alpha(1\rightarrow6)$ glycosidase*, sometimes called the debranching enzyme, hydrolyzes the $\alpha(1\rightarrow6)$ glycosidic bond at a branch point, thereby removing the branches. Branches are further degraded by glycogen phosphorylase to produce glucose-1-phosphate.

Step 3: *Phosphoglucomutase* converts glucose-1-phosphate to glucose-6-phosphate, which can be degraded by glycolysis or dephosphorylated for transport into the bloodstream.

Example 8: Importance of Glycogenolysis
Explain why glycogenolysis is important.

Solution:
Glucose is the sole source of energy for mammalian red blood cells and the major source of energy for the brain. Neither of these can store glucose; thus, a constant supply must be available as blood glucose. This is provided by dietary glucose and by the production of glucose either by gluconeogenesis or by glycogenolysis, the degradation of glycogen molecules. Glycogen is a long, branched-chain polymer of glucose that is stored in the liver and skeletal muscles. Breakdown of glycogen in the liver mobilizes the glucose when hormonal signals register a need for increased levels of blood glucose.

Glycogenesis: Glycogen Synthesis

The hormone **insulin**, produced by the pancreas in response to high blood glucose levels, stimulates glycogenesis. It accelerates the uptake of glucose by most cells of the body. In the liver, insulin promotes glycogenesis by inhibiting glycogen phosphorylase, thus inhibiting glycogenolysis, and stimulating glycogen synthase and glucokinase, two enzymes involved in glycogen synthesis. The following is a summary of the steps of glycogenesis:

Step 1: *Glucokinase* phosphorylates glucose, using ATP as a phosphoryl group donor and forming glucose-6-phosphate.

Step 2: *Phosphoglucomutase* isomerizes glucose-6-phosphate to glucose-1-phosphate.

Step 3: *Pyrophosphorylase* catalyzes bond formation between the C-1 phosphoryl group of glucose and the α-phosphoryl group of uridine triphosphate (UTP) to produce UDP-glucose. A pyrophosphate group (PP_i) is released in the reaction.

Step 4: *Glycogen synthase* breaks the phosphoester linkage of UDP-glucose and forms an $\alpha(1\rightarrow4)$ glycosidic bond between the glucose and the growing glycogen chain, releasing UDP.

Step 5: Finally, the branches are added. The *branching enzyme* removes sections of the linear α(1→4)-linked glycogen and reattaches them elsewhere in the chain by α(1→6) glycosidic linkages.

Compatibility of Glycogenesis and Glycogenolysis

Glycogenesis and glycogenolysis are regulated by hormonal controls. When blood glucose levels are too high (**hyperglycemia**), insulin stimulates glucose uptake. It also stimulates glucokinase and glycogen synthase activity and inhibits the first enzyme in glycogen degradation, glycogen phosphorylase. The net effect is the removal of glucose from the blood and its conversion into glycogen in the liver.

Glucagon is produced in response to low blood glucose levels (**hypoglycemia**). Glycogenolysis is stimulated because the activity of glycogen phosphorylase, which catalyzes the first stage of glycogenolysis, is accelerated. Additionally, glucagon inhibits the activity of glycogen synthase.

Self Test for Chapter Twenty-One

1. Of the several high-energy compounds produced in cells, which one is the principal energy storage compound?
2. How does ATP store and release energy?
3. Would a biosynthetic reaction that requires ATP as an energy source be considered a catabolic or anabolic reaction?
4. Which aspect of metabolism involves the degradation of absorbed food molecules to produce energy for the body?
5. Of the three classes of food molecules, which one is the most readily used by the human body?
6. What is the end product of the first stage of catabolism of carbohydrates? Proteins? Lipids?
7. What type of chemical reaction is occurring when food molecules are digested in the first stage of catabolism?
8. What is the name of the enzyme found in saliva that begins the hydrolysis of starch?
9. What enzyme in the stomach starts the breakdown of proteins?
10. Where does the digestion of triglycerides begin?
11. What occurs during the second and third stages of catabolism?
12. Why is glycolysis thought to be the most ancient metabolic pathway, having existed for at least 3.5 billion years?
13. What is different between the first segment of glycolysis (steps 1-5) and the second segment of glycolysis (steps 6-10)?
14. Describe glycolysis in terms of its oxygen requirements.
15. Where in the cell does glycolysis occur?
16. What is the net yield of ATP produced by anaerobic glycolysis?

17. Provide the missing product in the following reaction representing the first step of glycolysis:

$$\text{ATP} + \text{water} + \text{glucose} \rightarrow \text{ADP} + 4\ \text{kcal/mole} + ?$$

18. In step 3 of glycolysis, fructose-6-phosphate is converted into fructose-1,6-bisphosphate. What other molecule is needed for this reaction? What type of enzyme catalyzes this reaction?
19. During glycolysis, 1,3-bisphosphoglycerate is converted into 3-phosphoglycerate. Simultaneously, an ATP is produced. What is the general term for this kind of reaction?
20. In step 6 of glycolysis, glyceraldehyde-3-phosphate, phosphate, and NAD^+ are converted into 1,3-bisphosphoglycerate and NADH. How is this step important to the glycolysis pathway?
21. Name one molecule that inhibits glycolysis. Explain why this inhibits glycolysis.
22. Why do fermentation reactions occur?
23. What is the final fermentation end product in yeast cells?
24. What is the final fermentation end product produced in muscle cells under anaerobic conditions?
25. What role does the pentose phosphate pathway serve?
26. In what ways does gluconeogenesis differ from glycolysis?
27. Where does the Cori Cycle occur? What are the starting material and products of this cycle?
28. How does the body respond to low blood glucose levels?

22 *Aerobic Respiration and Energy Production*

Learning Goals

1. Name the regions of the mitochondria and the function of each region.
2. Describe the reaction that results in the conversion of pyruvate to acetyl CoA, describing the location of the reaction and the components of the pyruvate dehydrogenase complex.
3. Summarize the reactions of aerobic respiration.
4. Looking at an equation representing any of the chemical reactions that occur in the citric acid cycle, describe the kind of reaction that is occurring and the significance of that reaction to the pathway.
5. Explain the mechanisms for the control of the citric acid cycle.
6. Describe the process of oxidative phosphorylation.
7. Describe the conversion of amino acids to molecules that can enter the citric acid cycle.
8. Explain the importance of the urea cycle, and describe its essential steps.
9. Discuss the cause and effect of hyperammonemia.
10. Summarize the role of the citric acid cycle in catabolism and anabolism.

Introduction

Many mutations of mitochondrial DNA result in genetic disorders of energy metabolism. As a result, research is underway that has the potential to relieve the suffering of those with mitochondrial disorders.

In this chapter, we will study some of the biochemical reactions that take place in the mitochondria. For example, the final oxidations of carbohydrates, lipids, and proteins occur in the mitochondria and lead to the production of ATP.

22.1 Mitochondria

The energy-harvesting reactions that produce the greatest energy yield are aerobic reactions that occur in the **mitochondria**. Mitochondria are bounded by an **outer mitochondrial membrane** and an **inner mitochondrial membrane**. The region between the two membranes is known as the **intermembrane space,** and the region enclosed by the inner membrane is known as the **matrix space**. The enzymes of the citric acid cycle and those for the oxidation of fatty acids and amino acids are located in the matrix space.

Structure and Function

LG 1

The outer membrane is freely permeable to substances of molar mass less than 10,000 g/mol. Thus, metabolites to be oxidized via the citric acid cycle easily enter the intermembrane space through channel proteins.

The inner membrane is a highly folded, continuous structure. This gives the inner membrane a large surface area. The individual folds are known as **cristae**. The inner mitochondrial membrane is virtually impermeable to most substances and contains three types of proteins. *Transport proteins* allow the transport of metabolites across the inner mitochondrial membrane into the matrix. *Electron transport system proteins* are involved in electron transfers, for which O_2 serves as the terminal electron acceptor. The third protein is a very large multiprotein complex known as ATP synthase, which is responsible for phosphorylation of ADP to yield ATP.

Example 1: Mitochondria Function
What is the function of the mitochondria?
Solution:
The oxidative reactions of metabolism are responsible for most cellular ATP production. These reactions occur in metabolic pathways located in the mitochondria. This cytoplasmic organelle is the power plant of the cell, producing most of the ATP for cellular processes. The mitochondria are responsible for the final oxidation of the acetyl group of acetyl CoA from glycolysis, fatty acid degradation, and amino acid catabolism, and for the production of ATP.

Example 2: Mitochondria Structure
Describe the structure of the mitochondria.
Solution:
Mitochondria are football-shaped organelles that are roughly the size of bacteria. This organelle has both an outer and an inner membrane. The region between the two membranes is called the intermembrane space, and the region enclosed by the inner membrane is known as the matrix space. The outer membrane is freely permeable to substances of less than 10,000 g/mol. This is because of the presence of a large number of small pores in the membrane.

The inner membrane is a continuous structure that is highly folded. These folds of the inner membrane are called cristae. This membrane is virtually impermeable to most substances. The inner membrane contains transport proteins to allow the movement of metabolites across the membrane. Protein complexes for electron transport and ATP synthesis are found in the inner membrane. The enzymes of the citric acid cycle and fatty acid oxidation are found in the matrix space.

Origin of the Mitochondria

Mitochondria are roughly the size of a bacterium, and are thought to have evolved from free-living bacteria that were captured by eukaryotic cells. Some of the evidence for this is that they have their own DNA and protein-synthesizing system. In addition, they are self-replicating; growing and dividing to produce new mitochondria.

22.2 Conversion of Pyruvate to Acetyl CoA

Under aerobic conditions, cells use oxygen and completely oxidize glucose to CO_2 in a metabolic pathway called the **citric acid cycle. Acetyl CoA** is the molecule that carries two-carbon fragments (acetyl groups) produced from pyruvate into the citric acid cycle.

The **coenzyme A** portion of acetyl CoA is derived from ATP and the vitamin pantothenic acid. The acetyl groups are linked to the thiol group of coenzyme A by a high-energy thioester bond. Acetyl CoA is an "activated" form of the acetyl group.

The reaction that converts pyruvate to acetyl CoA is carried out by the **pyruvate dehydrogenase complex**. Pyruvate is decarboxylated and oxidized and acetyl CoA is formed. The pyruvate dehydrogenase complex is a composed of three enzymes and five coenzymes. Four of the coenzymes are made from the vitamins thiamine, riboflavin, niacin, and pantothenic acid.

Acetyl CoA plays a central role in cellular metabolism. It is the product of the degradation of glucose, fatty acids, and some amino acids. It carries acetyl groups to the citric acid cycle for complete oxidation and the ultimate production of large amounts of ATP. It is also used for *anabolic* or biosynthetic reactions to produce cholesterol and fatty acids. Thus, acetyl CoA is the intermediate through which all energy sources (fats, proteins, and carbohydrates) are interconvertible.

22.3 An Overview of Aerobic Respiration

Aerobic respiration is the oxygen-requiring degradation of food molecules and production of ATP. Acetyl groups derived from the breakdown of sugars, amino acids, or lipids are completely oxidized to CO_2 in the reactions of the citric acid cycle. The electrons harvested in these oxidation reactions are used to reduce 3 NAD^+ and 1 FAD, producing 3 NADH and 1 $FADH_2$.

These electrons are then passed through an electron transport system that simultaneously pumps protons (H^+) into the H^+ reservoir in the mitochondrial intermembrane space. The energy of the H^+ reservoir is used by ATP synthase to produce ATP. The entire process is called *oxidative phosphorylation* because the energy of electrons from the oxidation of substrates is used to phosphorylate ADP and produce ATP.

22.4 The Citric Acid Cycle (The Krebs Cycle)

Biological Effects of Disorders of the Citric Acid Cycle

Mutations leading to deficiencies in the enzymes of the citric acid cycle are often very serious. Mutations in the fumarase gene cause encephalopathy, which results in a variety of neurological symptoms and many children with this deficiency die in infancy or in childhood. A deficiency of α-ketoglutarate dehydrogenase leads to lactic acidosis, encephalopathy and hypotonia. Death typically results in the first 2-3 years of life. Mutations of the succinate dehydrogenase gene can lead to Leigh disease, which results in loss of motor skills and later, death.

Reactions of the Citric Acid Cycle

The following is a summary of the reactions of the **citric acid cycle**.

1. The acetyl group of acetyl CoA is transferred to oxaloacetate by the enzyme *citrate synthase*, forming citrate.
2. *Aconitase* catalyzes the dehydration of citrate, producing *cis*-aconitate. This same enzyme then adds a water molecule to the *cis*-aconitate, converting it to isocitrate. Overall, the result of these two steps is the isomerization of citrate to isocitrate.
3. *Isocitrate dehydrogenase* catalyzes the first oxidative reaction of the citric acid cycle. This is a complex reaction in which three things happen. First, the hydroxyl group of isocitrate is oxidized to a ketone; then carbon dioxide is released; finally NAD^+ is reduced to NADH. The product of this oxidative decarboxylation reaction is α-ketoglutarate.
4. The *α-ketoglutarate dehydrogenase* complex, an enzyme complex very similar to the pyruvate dehydrogenase complex, mediates the next reaction. Once again, three chemical events occur. First, α-ketoglutarate loses a carboxylate group as CO_2; next, NAD^+ is reduced to NADH; and finally, coenzyme A combines with the product to form succinyl CoA. The bond between succinate and coenzyme A is a high-energy thioester linkage.
5. Succinyl CoA is converted to succinate by the enzyme *succinyl CoA synthase*, which removes the CoA group and uses the energy of the thioester bond to add an inorganic phosphate group to GDP, producing GTP. *Dinucleotide diphosphokinase* then shifts the phosphoryl group from GTP to ADP, producing ATP.
6. *Succinate dehydrogenase* then catalyzes the oxidation of succinate to fumarate. The oxidizing agent flavin adenine dinucleotide (FAD) is reduced in this step to $FADH_2$.
7. *Fumarase* catalyzes the addition of H_2O to the double bond of fumarate, producing malate.
8. Finally, *malate dehydrogenase* reduces NAD^+ to NADH and oxidizes malate to oxaloacetate. Since the citric acid cycle began with the addition of an acetyl group to oxaloacetate, we have come full circle.

Example 3: Citric Acid Cycle
Write the specific reactions in the citric acid cycle in which the coenzymes NAD^+ and FAD act as oxidizing agents.

Solution:

1. The first oxidative step, Step 3, is catalyzed by isocitrate dehydrogenase:
   isocitrate + NAD^+ → α-ketoglutarate + CO_2 + NADH
2. The next step involves the oxidation of α-ketoglutarate into succinate:
   α-ketoglutarate + NAD^+ → succinate + CO_2 + NADH

Example 3 Continued

3. In Step 6, succinate is oxidized to fumarate:
   succinate + FAD $\rightarrow$ fumarate + $FADH_2$

4. Then in the final step of the citric acid cycle, malate is oxidized to produce oxaloacetate:
   malate + NAD^+ $\rightarrow$ oxaloacetate + NADH

22.5 Control of the Citric Acid Cycle

LG 5

The citric acid cycle is regulated so that cellular energy demand determines the rate of cellular energy production. This regulation is possible because some of the enzymes of the citric acid cycle are allosteric enzymes. Binding of an effector, such as ATP, ADP, or NADH, alters the shape of the active site. Effector binding may turn the enzyme on (positive allosterism) or it may inhibit the enzyme (negative allosterism). Four allosteric enzymes involved in the citric acid cycle are listed below.

- The pyruvate dehydrogenase complex is inhibited by high concentrations of ATP, acetyl CoA, and NADH.
- Citrate synthase is an allosteric enzyme inhibited by ATP.
- Isocitrate dehydrogenase is an allosteric enzyme that is stimulated by ADP binding and is also inhibited by high levels of NADH and ATP.
- The α-ketoglutarate dehydrogenase complex is inhibited by high levels of NADH, succinyl CoA, and ATP.

22.6 Oxidative Phosphorylation

Oxidative phosphorylation is a series of reactions that couples the oxidation of NADH and $FADH_2$ to the phosphorylation of ADP to produce ATP.

Electron Transport Systems and the Hydrogen Ion Gradient

Embedded within the mitochondrial inner membrane is a series of electron carriers called the **electron transport system**. Prominent among these electron carriers are the cytochromes. These molecules are arranged within the membrane so that they pass electrons from one to the next. At three sites in the electron transport system, protons can be pumped into the intermembrane space to create a region of high proton concentration. NADH donates its electrons at the beginning of the electron transport system; $FADH_2$ donates its electrons at a later point. As a result, $3H^+$ are pumped into the intermembrane space for each NADH oxidized, but only $2H^+$ for each $FADH_2$ oxidized.

As the electron transport system continues to function, a high concentration of hydrogen ions accumulates in the intermembrane space, creating a concentration gradient across the inner membrane. The result is a high concentration of protons in the intermembrane space and a low concentration in the matrix.

The last component needed for oxidative phosphorylation is a multiprotein complex called **ATP synthase**, or **F_0F_1 complex**. The F_0 portion provides a channel in the membrane through which protons may pass. The F_1 portion phosphorylates ADP to produce ATP, using the energy of the proton gradient.

ATP Synthase and the Production of ATP

NADH carries electrons to the first carrier of the electron transport system, *NADH dehydrogenase*. There it is oxidized to NAD^+, donating a pair of hydrogen atoms, and returns to the site of the citric acid cycle to be reduced again. The pair of electrons is passed to the next electron carrier, but the protons are pumped into the intermembrane compartment. The electrons are passed sequentially through the electron transport system and, at two additional points, protons from the matrix are pumped into the intermembrane compartment. $FADH_2$ donates its electrons to a carrier later in the electron transport system, so only four protons are pumped into the intermembrane space. In aerobic organisms, the **terminal electron acceptor** is molecular oxygen, O_2, and the product is water.

ATP synthase harvests the energy of this gradient and uses it to produce ATP. As the protons pass into the matrix through F_0, some of their energy is used by the enzymatic portion of ATP synthase (F_1) to catalyze the phosphorylation of ADP to ATP.

Example 4: ATP Synthesis
What are the two different ways that ATP can be synthesized by the cell? Compare the location and energy yields of these two methods of ATP synthesis.

Solution:
The two ways that ATP can be synthesized by the cell are substrate-level phosphorylation and oxidative phosphorylation.

1. Substrate-level phosphorylation occurs in the cytoplasm of the cell and produces only a few ATP molecules per glucose molecule.
2. Oxidative phosphorylation occurs in the mitochondria and produces a large number of ATP molecules per glucose molecule.

Example 5: Electron Transport System

1. Explain how the electron transport system converts NADH back into NAD^+, so that it may be used again by the citric acid cycle.
2. Write an equation to represent the last step of the electron transport system.

Solution:

1. The NADH carries a hydride anion, which was originally from glucose or other food molecules, to the first carrier of the electron transport system. There it is oxidized to NAD^+ as the first carrier, NADH dehydrogenase is reduced. NAD^+ returns to be used again in the citric acid cycle.

2. $2H^+ \; + \; 2e^- + 1/2\, O_2 \; \rightarrow \; H_2O$

Summary of the Energy Yield

Oxidative phosphorylation yields 3 ATP per NADH molecule and 2 ATP per $FADH_2$ molecule. However, NADH produced from glycolysis in the cytoplasm only yield two ATP per NADH. With these conversions in mind, the total energy yield of a molecule of glucose can be calculated:

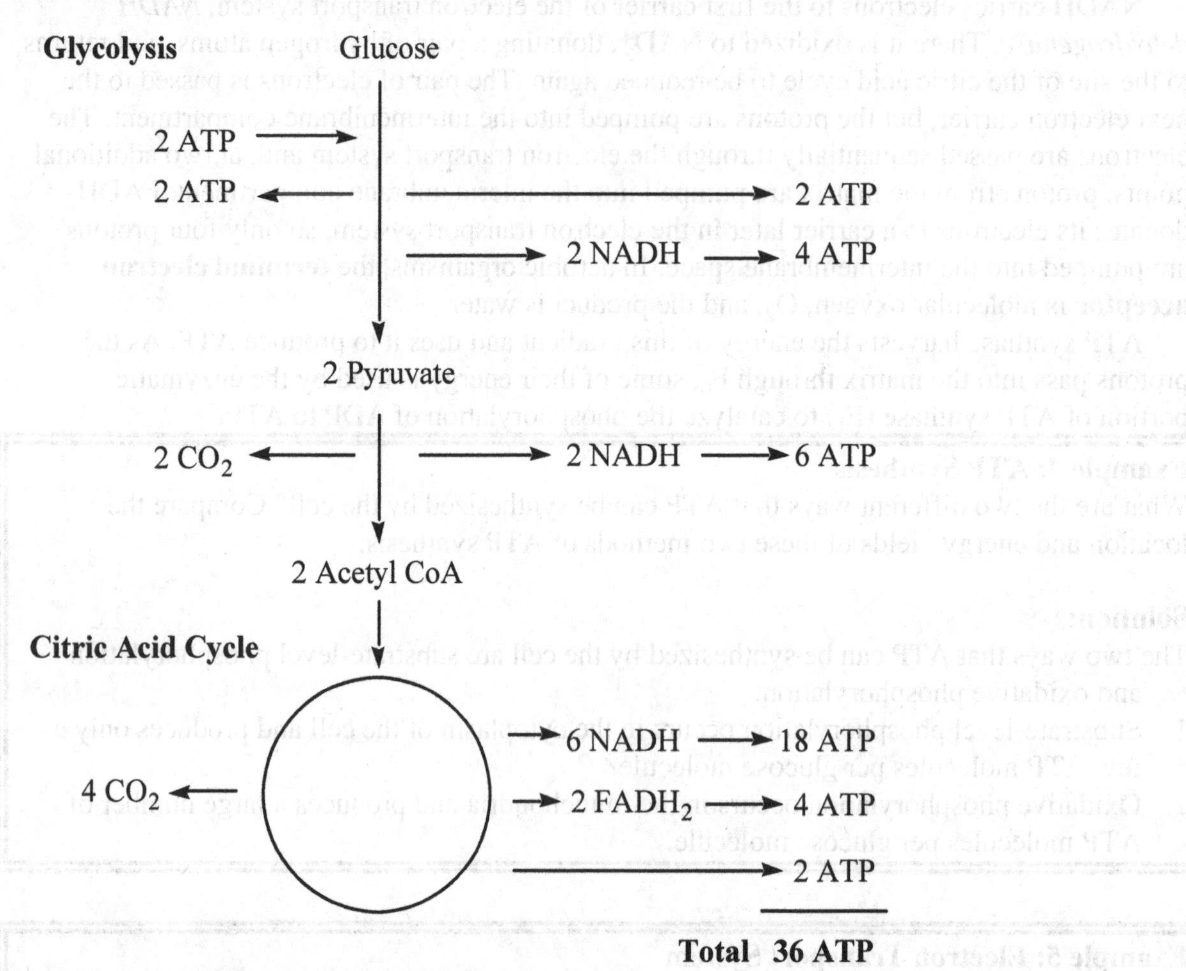

22.7 The Degradation of Amino Acids

LG 7

Carbohydrates are not the only source of energy. Dietary protein is digested to amino acids, which also may be used as an energy source. Most of the amino acids used for energy come from dietary protein. Only under starvation conditions, when stored glycogen and lipids have been greatly depleted, does the body begin to use its own muscle protein for energy.

Amino acid degradation occurs mainly in the liver and takes place in two stages: (1) the removal of the α-amino group and (2) the degradation of the remaining carbon skeleton. In land mammals, the amino group is excreted in the urine as urea. The carbon skeletons of amino acids can be converted into a variety of compounds.

Example 6: Protein as a Source of Energy
When and how is body protein used for energy?

Solution:
Proteins are used for energy when there are extra amino acids which are not needed for biosynthesis, and under starvation conditions. Proteins are used for energy by first removing the α-amino group and then degrading the carbon skeleton.

Removal of α-Amino Groups: Transamination

A **transaminase** catalyzes the transfer of the α-amino group from an α-amino acid to an α-keto acid. This general reaction is summarized below.

$$\underset{\substack{\text{Donor}\\\text{amino}\\\text{acid}}}{H-\overset{\overset{+}{N}H_3}{\underset{R}{C}}-COO^-} + \underset{\substack{\text{Acceptor}\\\text{keto acid}}}{\overset{O}{\underset{R}{C}}-COO^-} \xrightleftharpoons{\text{transaminase}} \underset{\substack{\text{Carbon}\\\text{skeleton of}\\\text{amino acid}}}{\overset{O}{\underset{R}{C}}-COO^-} + \underset{\text{New amino acid}}{H-\overset{\overset{+}{N}H_3}{\underset{R}{C}}-COO^-}$$

The α-amino groups of many amino acids are transferred to α-ketoglutarate to produce the amino acid glutamate. This glutamate family of transaminases is especially important because the α-keto acid corresponding to glutamate is the citric acid cycle intermediate, α-ketoglutarate. The *glutamate transaminases* thus provide a direct link between amino acid degradation and the citric acid cycle. *Alanine transaminase* and *aspartate transaminase* are also important members of this family.

All transaminases require the prosthetic group **pyridoxal phosphate**, a coenzyme derived from vitamin B_6 (pyridoxine). During transamination, the α-amino group is first transferred to pyridoxal phosphate and then from pyridoxal phosphate to the α-keto acid.

Example 7: Transamination of Amino Acids
Explain how the *transamination* of amino acids provides a direct linkage between amino acid degradation and the citric acid cycle.

Solution:
Transaminases catalyze the transfer of the α-amino group from an α-amino acid to an α-keto acid. The α-amino groups of many of the amino acids are transferred to α-ketoglutarate to produce the amino acid glutamate. The glutamate family of transaminases is especially important because the α-keto acid corresponding to glutamate is the citric acid cycle intermediate, α-ketoglutarate. Thus, the glutamate transaminases, as well as others, provide a direct link to the citric acid cycle.

Removal of α-Amino Groups: Oxidative Deamination

Ammonium ion is now liberated from the glutamate by **oxidative deamination,** a reaction that is catalyzed by *glutamate dehydrogenase*:

$$\begin{array}{c} COO^- \\ | \\ CH-NH_3^+ \\ | \\ CH_2 \\ | \\ CH_2 \\ | \\ COO^- \end{array} + NAD^+ + H_2O \rightleftharpoons NH_4^+ + \begin{array}{c} COO^- \\ | \\ C{=}O \\ | \\ CH_2 \\ | \\ CH_2 \\ | \\ COO^- \end{array} + NADH$$

Glutamate α-Ketoglutarate

The Fate of Amino Acid Carbon Skeletons

The carbon skeletons produced by these and other deamination reactions enter the energy-harvesting pathways at many steps, as seen in Figure 22.8 in the text.

22.8 The Urea Cycle

Oxidative deamination of glutamate and deamination of other amino acids produce considerable quantities of ammonium ion. This must be incorporated into a biological molecule and removed from the body so that it does not reach toxic levels. It is critically important to the survival of the organism to be able to excrete ammonium ions, regardless of the energy required. In humans and most terrestrial vertebrates, the means of ammonium ion removal is the urea cycle, which occurs in the liver. This cycle keeps excess ammonium ion out of the blood and allows the excretion of the excess in the form of urea.

Reactions of the Urea Cycle

The reactions of the urea cycle, summarized below, are shown in Figure 22.12 of the text.

Step 1. CO_2 and NH_4^+ react to form carbamoyl phosphate. This reaction requires ATP and H_2O and is catalyzed by the enzyme carbamoyl phosphate synthase.

Step 2. Carbamoyl phosphate condenses with the amino acid ornithine to produce the amino acid citrulline. This is catalyzed by the enzyme ornithine transcarbamoylase.

Step 3. Citrulline now condenses with aspartate to produce argininosuccinate. This reaction requires energy released by breaking a phosphoanhydride bond of ATP and is catalyzed by the enzyme argininosuccinate synthase.

Step 4. The enzyme argininosyccinate lyase catalyzes the cleavage of argininosuccinate to produce the amino acid arginine and the citric acid cycle intermediate fumarate.

Step 5. Finally, arginine is hydrolyzed, producing urea, which will be excreted, and ornithine, the original reactant in the cycle. This is catalyzed by arginase. Note that one of the amino groups of urea is the ammonium ion, and the second is derived from the amino acid aspartate.

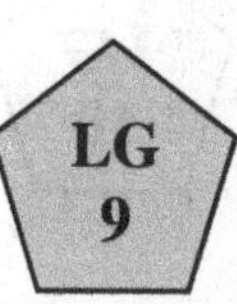

A deficiency of urea cycle enzymes causes an elevation of the concentration of NH_4^+, a condition known as **hyperammonemia**. If there is a complete deficiency of one of the enzymes of the urea cycle, the result is death in early infancy. A partial deficiency results in less severe symptoms, such as mental challenge, convulsions, and vomiting. This can be treated with a low-protein diet.

Example 8: Urea Cycle
Give a summary of the reactions of the urea cycle.

Solution:

1. *Step 1*: The first step involves the reaction of the waste product CO_2 with ammonium ions from the amino acids to form carbamoyl phosphate.

$$CO_2 + NH_4^+ + 2\,ATP \longrightarrow H_2N{-}\overset{\displaystyle O}{\overset{\|}{C}}{-}O{-}\overset{\displaystyle O}{\overset{\|}{\underset{\underset{\displaystyle O^-}{|}}{P}}}{-}O^- + 2\,ADP + P_i + 3\,H^+$$

2. *Step 2*: The carbamoyl phosphate then condenses with the amino acid ornithine to produce citrulline (another amino acid).
3. *Step 3*: Citrulline condenses with aspartate to produce argininosuccinate.
4. *Step 4*: The argininosuccinate is cleaved to produce arginine and fumarate.
5. *Step 5*: Finally, arginine is hydrolyzed to generate urea and ornithine. The ornithine can then return to be used again in Step 2.

Example 9: Enzyme Deficiency
Describe the genetic disorder that results from the deficiency of an enzyme in the urea cycle.

Solution:

A deficiency of a urea cycle enzyme causes an elevation of the concentration of ammonium ion, a condition known as *hyperammonemia*. If there is a complete deficiency of one of these enzymes, the result is death in early infancy. If, however, there is a partial deficiency of one of the cycle's enzymes, the result can be retardation, convulsions, and vomiting. If the disorder is caught early, a diet low in protein may lead to less severe clinical symptoms.

22.9 Overview of Anabolism: The Citric Acid Cycle as a Source of Biosynthetic Intermediates

LG 10

The citric acid cycle plays a key role in biosynthetic reactions, as well as ATP production. Glycolysis and the citric acid cycle provide precursors for the biosynthesis of many other molecules in the body.

Since amino acids can be converted into citric acid cycle intermediates, those same intermediates can be used for the synthesis of amino acids. For example, aspartate can be synthesized through the transamination reaction and asparagine can be synthesized through the amination of aspartate. Additionally, α-ketoglutarate can serve as a precursor for the synthesis of glutamate, glutamine, proline, and arginine.

Glycolysis and the citric acid cycle also produce precursors for the synthesis of lipids, the nitrogenous bases found in DNA and the precursors for heme, the prosthetic group required for hemoglobin, myoglobin, cytochrome c, and catalase.

Metabolic pathways that function in both **anabolism** and **catabolism** are called **amphibolic pathways**. Cellular energy demand sometimes exceeds the supply of citric acid intermediates. Mammalian cells can produce more oxaloacetate by the carboxylation of pyruvate, a reaction that is also important in gluconeogenesis. This reaction is called an **anaplerotic reaction**, meaning "to fill up or replenish."

Self Test for Chapter Twenty-Two

1. The oxidative reactions of metabolism provide most cellular energy. In what specific part of the cell do these reactions occur?
2. The number of mitochondria in a eukaryotic cell varies widely. What need is reflected by this variation?
3. To what compound is the acetyl group transferred in the first step of the citric acid cycle?
4. Why would a deficiency in the B vitamins result in a deficiency in the amount of acetyl CoA our bodies could produce?
5. Where in the mitochondrion do the reactions of the citric acid cycle occur?
6. Malate dehydrogenase reduces NAD^+ to NADH and oxidizes malate into what product in the final step of the citric acid cycle? Why is this reaction important in the citric acid cycle?
7. How many molecules of CO_2 are produced by the complete oxidation of an acetyl group by the citric acid cycle?
8. How many molecules of NADH are formed by the complete oxidation of an acetyl group by the citric acid cycle?
9. ATP is a negative effector for several of the enzymes of the citric acid cycle. Why is it a negative effector? What does this signal about the cell?
10. What is the term for the series of electron and proton carriers embedded within the mitochondrial inner membrane?

11. When NADH dehydrogenase oxidizes NADH in the mitochondrial matrix, it can pass the electrons to the next carrier. What happens to the protons?
12. NADH carries electrons, originally from glucose, to the first carrier of the electron transport system. What compound is formed by the oxidation of NADH at this site?
13. After NADH is oxidized to NAD+ in the first step of the electron transport system, what happens to the NAD+?
14. Spanning the inner mitochondrial membrane is a protein complex (F_0) that serves as a channel protein. What specific particles pass through this channel?
15. Protruding into the mitochondrial matrix is a spherical protein complex (F_1). What is the enzymatic activity of F_1?
16. In the last step of the electron transport system, the electrons have too little energy to accomplish any more work. In aerobic organisms, what is the terminal electron acceptor?
17. Why does $FADH_2$ produce less ATP than NADH?
18. How many ATP are produced for each NADH generated in the mitochondria?
19. How many ATP are produced for each $FADH_2$ generated in the mitochondria?
20. What is the source of most of the amino acids that are used by the human body?
21. Where in the human body does the degradation of amino acids primarily occur?
22. What is the first step of amino acid catabolism?
23. The α-amino groups of many amino acids can be transferred to an α-ketoacid. What are the products of this reaction?
24. One of the most important transaminases is aspartate transaminase, which catalyzes the transfer of the α-amino group of aspartate to α-ketoglutarate. What are the two new compounds produced in this reaction?
25. The glutamate transaminases provide a direct link between amino acid degradation and which other process?
26. Where in the human body does the conversion of ammonium ions to urea occur?
27. Provide the missing reactant in the following reaction of the urea cycle:

$$CO_2 + NH_4^+ + ? + H_2O \rightarrow 2ADP + P_i + 3H^+ + \text{carbamoyl phosphate}$$

28. What causes hyperammonemia?
29. What is the term used to describe metabolic pathways that function in both anabolism and catabolism?
30. How does the citric acid cycle play a role in both catabolic and anabolic metabolism?

23 *Fatty Acid Metabolism*

Learning Goals

1. Summarize the digestion and storage of lipids.
2. Describe the degradation of fatty acids by β-oxidation.
3. Explain the role of acetyl CoA in fatty acid metabolism.
4. Understand the role of ketone body production in β-oxidation.
5. Compare β-oxidation of fatty acids and fatty acid biosynthesis.
6. Describe the regulation of lipid metabolism in relation to the liver, adipose tissue, muscle tissue, and the brain.
7. Summarize the antagonistic effects of glucagon and insulin.

Introduction

In this chapter, we focus on the metabolism of triglycerides. The catabolism of triglycerides results in fatty acids that are degraded to acetyl CoA, which is oxidized by the citric acid cycle. Acetyl CoA is the key molecule of lipid catabolism. On the other hand, acetyl CoA is also the precursor for the biosynthesis of fatty acids, cholesterol, and steroid hormones.

23.1 Lipid Metabolism in Animals

Digestion and Absorption of Dietary Fats

LG 1

Fats are hydrophobic molecules and must be extensively processed before they can be digested and absorbed. The enzymes that hydrolyze lipids are called **lipases**. The most effective lipid digestion occurs in the small intestine, where **bile** (composed of lecithin, cholesterol, protein, bile salts, inorganic ions, and bile pigments) causes formation of small lipid micelles. **Micelles** are aggregations of amphipathic molecules having a polar region facing the aqueous exterior and an internal nonpolar region that dissolves lipid. Bile is produced in the liver and stored in the gall bladder. The presence of lipids in the small intestine stimulates secretion of bile into the duodenum, the first part of the small intestine.

Triglycerides are the major lipids in the micelles. A protein called **colipase** binds to the surface of the micelles and facilitates hydrolysis of the triglycerides by pancreatic lipases. The fatty acids and monoglycerides produced by hydrolysis are absorbed by cells of the intestinal epithelium.

Example 1: Dietary Fats
Explain the digestion and absorption of dietary fats.

Example 1 Solution:
Most dietary fats arrive in the duodenum in the form of fat globules. The globules stimulate the secretion of bile from the gall bladder. The bile salts emulsify the fat globules into tiny droplets. The protein colipase binds to the surface of the lipid droplets and helps pancreatic lipases adhere to the surface and hydrolyze the triglycerides into monoglycerides plus free fatty acids. These products are absorbed through the membranes of the intestinal epithelial cells.

In intestinal cells the monoglycerides and fatty acids are reassembled into triglycerides and combined with protein, producing **chylomicrons** (introduced in Chapter 17). These chylomicrons eventually enter the blood, and the lipids are once again hydrolyzed to products that can be absorbed by cells of the body.

Example 2: Lipid Digestion
Explain what happens to lipid digestion products after absorption.

Solution:
The monoglycerides and fatty acids are reassembled into triglycerides, which are combined with protein to produce the class of plasma lipoproteins called chylomicrons. These are secreted into lymphatic vessels and eventually arrive in the bloodstream. Here triglycerides are again hydrolyzed, and the products are absorbed by the cells of the body.

Lipid Storage

Fatty acids are stored as triglycerides within **adipocytes** (fat cells) that compose **adipose tissue**. When the body demands energy, these triglycerides can be hydrolyzed, and the fatty acids are oxidized to generate ATP. Other cells, such as cardiac muscle, also contain small amounts of fat droplets. These droplets are surrounded by mitochondria and when energy is needed by the cell the triglycerides can be hydrolyzed and then oxidized to produce ATP.

Example 3: Lipid Storage
Explain lipid storage in the human body.

Solution:
Fatty acids are stored as triglycerides within fat droplets in the cytoplasm of adipocytes. These cells contain a large fat droplet, which accounts for nearly the entire volume of the cell.

23.2 Fatty Acid Degradation

Overview of Fatty Acid Degradation

Fatty acids are degraded in the mitochondrial matrix by a pathway called **β-oxidation,** which consists of a set of five reactions. Each trip through the β-oxidation pathway releases acetyl CoA and returns a fatty acyl CoA molecule that contains two fewer carbons. Additionally, one $FADH_2$ and one NADH are produced for each cycle of β-oxidation. The $FADH_2$ and NADH are oxidized by the electron transport system resulting in the production of two and three molecules of ATP, respectively, through the process of oxidative phosphorylation. Acetyl CoA can enter the citric acid cycle, resulting in production of twelve more ATP. The remaining fatty acyl CoA cycles through steps 2-5 of β-oxidation until the fatty acid is completely converted to acetyl CoA.

Example 4: Fatty Acid Degradation
Give an overview of fatty acid degradation.

Solution:
The β-oxidation cycle for fatty acid degradation occurs in the mitochondrial matrix. The fatty acids are degraded by the enzymes of the β-oxidation cycle until the fatty acids are all converted into acetyl CoA.

The Reactions of β-Oxidation

LG 2

Special transport mechanisms bring fatty acid molecules into the mitochondrial matrix, where β-oxidation occurs. The reactions of β-oxidation are summarized below.

1. *Activation*: A fatty acyl CoA molecule is formed between coenzyme A and the fatty acid. This reaction requires ATP, which is cleaved to AMP and pyrophosphate. The bond between the fatty acid and coenzyme A is a high-energy thioester bond.
2. *Oxidation*: A pair of hydrogen atoms are removed from the fatty acid producing a carbon-carbon double bond. In the process, FAD is reduced to $FADH_2$.
3. *Hydration*: The double bond produced in Step 2 undergoes a hydration reaction, and the β-carbon is hydroxylated.
4. *Oxidation*: The hydroxyl group of the β-carbon is now dehydrogenated, and NAD^+ is reduced to form NADH.
5. *Thiolysis*: A molecule of coenzyme A attacks the β-carbon, releasing acetyl CoA and a fatty acyl CoA that is two carbons shorter than the original fatty acid.

The fatty acyl CoA is further oxidized by cycling through Steps 2 through 5 until the fatty acid carbon chain is completely degraded to acetyl CoA. Each acetyl CoA enters the citric acid cycle where it is completely oxidized. Complete oxidation of fatty acids yields a significant amount of ATP. For example, palmitic acid, a sixteen-carbon fatty acid produces 129 ATP molecules. The balance sheet for ATP production in the complete oxidation of palmitic acid is found in Figure 23.8 of the text.

Example 5: ATP Production
Compare ATP production by β-oxidation of a fatty acid like palmitic acid to that produced by oxidation of an equivalent amount of glucose.

Example 5 Solution:
The complete oxidation of the C16-fatty acid palmitic acid produces 129 ATP molecules. The complete oxidation of glucose produces 36 ATP. Oxidation of the palmitic acid harvests 3.5 times more energy than results from the complete oxidation of the same amount of glucose.

23.3 Ketone Bodies

Since each acetyl CoA produced in β-oxidation must enter the citric acid cycle to be fully oxidized, there must be a sufficient supply of citric acid cycle intermediates to allow this to occur. If glycolysis and β-oxidation are occurring at the same rate, there will be a steady supply of the citric acid cycle intermediates. When there are not enough citric acid cycle intermediates, specifically oxaloacetate, acetyl CoA is converted to **ketone bodies**, such as β-hydroxybutyrate, acetoacetate, and acetone.

Example 6: Ketone Body Production
Explain why ketone bodies are produced in the human body.

Solution:
The β-oxidation of fatty acids produces a steady supply of acetyl CoA molecules. If glycolysis and the β-oxidation cycle are functioning at the same rate, there will be a steady supply of pyruvate, which can be converted into oxaloacetate. However, if the supply of oxaloacetate is too low to allow all of the acetyl CoA to enter the citric acid cycle, the acetyl CoA molecules are converted into ketone bodies. The ketone bodies are β-hydroxybutyrate, acetoacetate, and acetone.

Ketosis

Ketosis is the abnormal elevation of blood ketone body concentration. This may occur due to starvation, a diet that is extremely low in carbohydrates, or uncontrolled **diabetes mellitus**. In diabetes, this can lead to **ketoacidosis** because the ketone acids are relatively strong acids and dissociate to release H^+. This causes the pH of the blood to become acidic.

Ketogenesis

The production of ketone bodies is called ketogenesis. The first step in ketogenesis is the fusion of two molecules of acetyl CoA to produce acetoacetyl CoA. This reaction is catalyzed by the enzyme thiolase. Acetoacetyl CoA reacts with a third acetyl CoA to yield β-hydroxy-β-methylglutaryl CoA (HMG-CoA). HMG-CoA is cleaved to yield acetoacetate and acetyl CoA. While some acetoacetate spontaneously loses carbon dioxide, producing acetone, most of it undergoes NADH-dependent reduction to produce β-hydroxybutyrate.

Acetoacetate and β-hydroxybutyrate, produced in the liver, are circulated to other tissues through the blood. There they may be reconverted to acetyl CoA and used to produce ATP. Interestingly, the heart muscle uses the oxidation of ketone bodies, rather than glucose, for most of its energy. Other tissues can rely on ketone body oxidation if glucose becomes unavailable or limited.

Example 7: Ketogenesis
Explain ketogenesis.

Solution:
The pathway for the production of ketone bodies begins with the reversal of the last step of the β-oxidation cycle. When oxaloacetate levels are low, the enzyme that mediates the last reaction of β-oxidation, thiolase, now mediates the fusion of two acetyl CoA molecules to produce acetoacetyl CoA. The acetoacetyl CoA reacts with another molecule of acetyl CoA to produce HMG-CoA. The HMG-CoA is cleaved to yield *acetoacetate* and acetyl CoA. Some of the acetoacetate spontaneously loses carbon dioxide to produce *acetone*. However, most of the acetoacetate is reduced to form *β-hydroxybutyrate*.

23.4 Fatty Acid Synthesis

LG 5

All organisms possess the ability to synthesize fatty acids. Fatty acid synthesis appears to be simply the reverse of β-oxidation. In fact, the fatty acid chain is constructed by the sequential addition of two carbon acetyl groups, as seen in Figure 23.11 of the text. However, there are several differences between the two processes, including intracellular location, acyl group carriers, enzymes involved, and the electron carriers used. Fatty acid biosynthesis occurs in the cytoplasm. The acyl group carrier is **acyl carrier protein (ACP)**. The portion of ACP that binds the fatty acyl group is the **phosphopantetheine** group also found in coenzyme A. A multienzyme complex, *fatty acid synthase*, carries out fatty acid biosynthesis; NADPH is the reducing agent.

Example 8: Fatty Acid Synthesis
Compare fatty acid synthesis with fatty acid degradation.

Solution:
On first examination of fatty acid synthesis, it appears that it is simply the reverse of β-oxidation. Although fatty acid synthesis and breakdown are similar, there are several major differences:

1. The enzymes responsible for fatty acid biosynthesis are located in the cytoplasm of the cells of vertebrates, whereas those responsible for the degradation of fatty acids are found in the mitochondria.
2. The activated intermediates of fatty acid synthesis are bound to a thiol group of acyl carrier protein. Thus, the thioester intermediates of fatty acid synthesis are not derivatives of coenzyme A.
3. The seven steps of fatty acid biosynthesis are carried out by the multienzyme complex known as fatty acid synthase. The enzymes responsible for fatty acid degradation are not associated with one another as a complex.
4. NADH and $FADH_2$ are produced by fatty acid oxidation, whereas NADPH is the reducing agent for fatty acid biosynthesis.

23.5 The Regulation of Lipid Metabolism

Fatty acid metabolism occur at different levels in different organs.

The Liver

The liver plays a central role in lipid metabolism. When excess fuel is available, the liver synthesizes fatty acids, which are transported to adipose tissue and stored. During starvation, the liver converts fatty acids to ketone bodies that are exported to other organs.

Example 9: Lipid Metabolism
Explain the central role of the liver in lipid metabolism.

Solution:
When excess fuel is available, the liver synthesizes fatty acids. These are esterified to produce triglycerides that are transported to the adipose tissues by very low density lipoprotein complexes. This transport is very active when more calories are eaten than are burned. During fasting or starvation conditions, however, the liver converts fatty acids to acetoacetate and other ketone bodies. These ketone bodies are transported to other organs for energy usage.

Adipose Tissue

Adipose tissue is the major storage depot of fatty acids. Triglycerides produced in the liver are transported through the blood in very low density lipoprotein complexes. The triglycerides are hydrolyzed, and the fatty acids are absorbed by adipose tissue. In adipocytes, triglycerides are resynthesized from the absorbed fatty acids and glycerol-3-phosphate, which is produced from intermediates of glycolysis.

Muscle Tissue

The energy demand of resting muscle is generally supplied by the β-oxidation of fatty acids. The heart muscle actually prefers ketone bodies over glucose. Working muscle obtains energy by degradation of glycogen.

The Brain

Normally, the brain uses glucose as its sole source of metabolic energy. However, under starvation conditions, the brain converts to the use of ketone bodies produced by the liver. This is possible because ketone bodies can cross the *blood-brain barrier,* whereas fatty acids transported by lipoproteins cannot.

Example 10: Brain Production of ATP
Which compounds can be used by the brain for production of ATP?

Solution:
Under normal conditions, the brain uses glucose as its sole source of metabolic energy. If conditions of starvation occur, the glycogen stores drop sharply, and blood glucose concentrations drop. The ketone bodies, acetoacetate and β-hydroxybutyrate, derived from fatty acid degradation, then provide an alternative source of metabolic energy.

23.6 The Effects of Insulin and Glucagon on Cellular Metabolism

LG 7

Insulin is a polypeptide hormone secreted by the β-cells of the islets of Langerhans in the pancreas in response to an increase in the blood glucose level. Insulin lowers the concentration of blood glucose by stimulating its storage as glycogen or triglycerides. In general, insulin activates biosynthetic processes and inhibits catabolism.

Insulin acts only on those *target cells* that possess a specific insulin receptor protein in their membranes. The major target cells for insulin are liver, adipose, and muscle cells. Insulin lowers blood glucose levels by altering cellular carbohydrate, protein, and lipid metabolism. Insulin affects carbohydrate metabolism by stimulating glycogen synthesis, and inhibiting glycogenolysis and gluconeogenesis. With protein metabolism, it stimulates the transport and uptake of amino acids, as well as the incorporation of amino acids into proteins. Insulin affects metabolism by stimulating the uptake of glucose by adipose cells and the synthesis and storage of triglycerides.

A second hormone, **glucagon**, is secreted by the α-cells of the islets of Langerhans in response to decreased blood glucose levels. The effects of glucagon are generally the opposite of the effects of insulin.

Maintenance of adequate blood glucose levels is extremely important and insulin and glucagon have antagonistic effects to allow blood glucose level regulation to occur. The effects of insulin and glucagon on a variety of metabolic reactions are summarized in the following table:

Actions	Insulin	Glucagon
Cellular glucose transport	Increased	No effect
Glycogen synthesis	Increased	Decreased
Glycogenolysis in liver	Decreased	Increased
Gluconeogenesis	Decreased	Increased
Amino acid uptake and protein synthesis	Increased	No effect
Inhibition of amino acid release and protein degradation	Decreased	No effect
Lipogenesis	Increased	No effect
Lipolysis	Decreased	Increased
Ketogenesis	Decreased	Increased

Self Test for Chapter Twenty-Three

1. In what form do we find dietary fat when it arrives in the duodenum for digestion?
2. Name the two major bile salts. (Refer to the textbook.)
3. What is the role of bile and bile salts in the digestion of lipids?
4. What is a micelle?
5. Which protein molecule binds to the surface of lipid droplets and helps pancreatic lipases to adhere to the surface and hydrolyze the ester linkages?
6. Monoglycerides and fatty acids are reassembled into triglycerides and combined with protein in the cells of the small intestine. What class of lipoproteins is formed in this process?
7. What tissue stores most of the body's triglyceride molecules?
8. The metabolism of fatty acids and lipids revolves around which specific compound?
9. How many cycles of β-oxidation would have to occur to fully convert a 12-carbon fatty acid into acetyl CoA?
10. How many NADH and $FADH_2$ are produced in each round of β-oxidation?
11. What compound is produced in the first step of β-oxidation of fatty acids?
12. What compound is released in the final step of β-oxidation of fatty acids?
13. The complete oxidation of a molecule of palmitic acid (a 16-carbon fatty acid), using the β-oxidation cycle and the other oxidation pathways, produces how many molecules of ATP?
14. What is ketosis?
15. List three conditions that can produce ketosis.
16. Why does a limited amount of oxaloacetate lead to the production of ketone bodies?
17. In patients with uncontrolled diabetes, the very high concentration of ketone acids in the blood leads to what condition?
18. What happens to ketone bodies once they are formed?
19. Which muscle derives more of its energy from the metabolic use of ketone bodies than it does from glucose?
20. What molecule provides the carbons necessary to synthesize fatty acids?
21. Why is acetyl CoA central to the metabolism of fatty acids and lipids?
22. What is the name of the multienzyme complex used for biosynthesis of fatty acids?
23. As a general rule, NADH is produced by catabolic reactions, but what acts as the reducing agent for biosynthetic reactions?
24. What compounds are preferentially oxidized to supply the energy demand of the heart muscle?
25. What lipoprotein complex transports triglycerides produced in the liver through the bloodstream?

26. What compound is required for the synthesis of triglycerides in adipose tissue? What pathway must it use for its supply of this precursor?
27. What compounds are oxidized to supply the energy demand of resting muscles?
28. What are the major target cells for insulin?
29. What effect does insulin have on glycogen synthesis?
30. What effect does glucagon have on lipolysis?

APPENDIX A

Chapter 1

Chemistry: Methods and Measurement
Solutions to the Odd-Numbered Questions and Problems

In-Chapter Questions and Problems

1.1 The Study Cycle may be incorporated into your learning strategy. It involves previewing material before class, attending class, reviewing after class, studying the material, and assessing your learning. You may also choose to utilize intense study sessions that begin with a goal and then progress to studying with focus for up to an hour. This may be followed by a 15-min break. Then, your study session can conclude with a review of what you studied prior to your break.

1.3 Homogeneous mixture. The saline solution has a uniform composition.

1.5
a. Physical property
b. Chemical property
c. Physical property

1.7
a. Extensive property. Length depends on the quantity.
b. Intensive property. Color is independent of the quantity.

1.9
a. Three (all non-zero digits)
b. Three (all non-zero digits)
c. Four (zeros between non-zero digits are significant)
d. Two (trailing zero is significant due to decimal)
e. Three (leading zeros are insignificant)
f. Two (trailing zero is insignificant without decimal)

1.11
a. 2.4×10^{-3}
b. 1.80×10^{-2}
c. 2.24×10^{2}
d. 6.73×10^{5}
e. 7.2420×10^{1}
f. 8.3×10^{-1}

1.13
a. 61.404 rounds to 61.4
b. 6.1714 rounds to 6.17
c. 0.066494 rounds to 0.0665 (or, 6.65×10^{-2})
d. 63.669 rounds to 63.7
e. 8.7715 rounds to 8.77

1.15 a. 8.09 (3 significant figures)
b. 5.9 (2 significant figures)
c. 20.19 (4 significant figures)

1.17 Both numbers are converted to standard form so that they can be added together.

$$\begin{array}{r} 0.000823 \\ \underline{+\,0.000061} \\ 0.000884 \end{array}$$

(The position of the decimal indicates three significant figures.)
This answer has the proper number of significant figures and is converted to the final answer, which in scientific notation is 8.84×10^{-4}.

1.19 a. 51 (2 significant figures)
b. 8.0×10^{1} (2 significant figures)
c. 5.80×10^{1} (3 significant figures)
d. 8.6×10^{-1} (2 significant figures)
e. 4.63×10^{8} (3 significant figures)
f. 2.0×10^{-3} (2 significant figures)

1.21 $595 \text{ cal} \times \frac{4.18 \text{ J}}{1 \text{ cal}} = 2490 \text{ J}$ (3 significant figures) or 2.49×10^{3} J

End-of-Chapter Questions and Problems

1.23 If you attempt to solve the Questions within the chapter and the Questions and Problems at-the-end of the chapter, be sure that you do not look at your notes or other references. Once you solve them on your own, you can look at the answers to the odd-numbered problems in the Answer Appendix in the back of your textbook or detailed in the Student Study Guide/Solutions Manual. You can also solve the Self Test problems that are in the Student Study Guide/Solutions Manual.

1.25 The Study Cycle encourages you to begin by previewing the material. Perhaps you are able to designate 10-15 minutes before each day of class. Then, when you attend class, you will be able to ask thoughtful questions and take more meaningful notes. You are also encouraged schedule 10-15 minutes to review your notes as soon as possible after class. The next aspect of the Study Cycle involves scheduling three to five short intense study sessions each day. These study sessions should each last approximately 1 h. The final phase of the Study Cycle is assessment. It may be helpful if you schedule 1 h each week to determine what you have and haven't learned. One way that you can do this is by attempting to solve the problems at the end of the chapter.

1.27 Chemistry is the study of matter, its chemical and physical properties, the chemical and physical changes it undergoes, and the energy changes that accompany those processes.

When wood is burned, energy is released while its matter undergoes chemical and physical changes.

1.29 Experiments are necessary to test hypotheses and theories. Experimental results that are derived from measured or observed data may or may not support the hypotheses and theories.

1.31 To estimate the total cost of the gasoline required, you need the gasoline's cost per gallon, the average miles per gallon of gasoline according to the means of transportation that will be used, and the distance between New York City and Washington, D.C.

1.33 According to the model, methane has one carbon atom and four hydrogen atoms per molecule. The model also provides the special relationship of the atoms, and the location and number of bonds that hold the unit together.

1.35 A hypothesis is essentially an "educated guess". A theory is a hypothesis supported by extensive experimentation; it can explain and predict new facts.

1.37 Many examples exist; one could involve the time it takes to get to work: hypothesis – taking the "back roads" is faster than choosing the crowded interstate highway. The experiment would involve driving each path several times, starting at the same time, timing each trip and calculating an average time for each path.

1.39 The statement "stem-cell research has the potential to provide replacement 'parts' for the human body" is, at the time this book was written, a theory. Pay particular attention to the word "potential" in the statement. The meaning and the scientific status of this statement would be quite different if the words "has the potential" were replaced by the word "can."

1.41 Freeze the contents of two beakers, one containing pure water and the other containing salt dissolved in water. Slowly warm each beaker and measure the temperature of each as they convert from the solid state to the liquid state. This temperature is the melting point, hence, the freezing point. Note that the melting and freezing points of a solution are the same temperature. If you record the freezing point of the beaker with pure water, this will be the $\text{temperature}_{\text{initial}}$. Then, if you record the freezing point of the beaker with the salt dissolved in the water, this is the $\text{temperature}_{\text{final}}$. The freezing point change can be calculated using the following equation.

$$[\text{temperature}_{\text{final}} - \text{temperature}_{\text{initial}}] = \text{freezing temperature difference}$$

1.43 The gaseous state, the liquid state, and the solid state are the three states of matter.

1.45 A pure substance has constant composition with only a single substance, whereas a mixture is composed of two or more substances.

1.47 Mixtures are composed of two or more substances. A homogeneous mixture has uniform composition, and a heterogeneous mixture has non-uniform composition.

1.49 Physical properties are characteristics a substance that can be observed without the substance undergoing a change in chemical composition. Chemical properties can be observed only through chemical reactions that result in a change in chemical composition of the substance.

1.51 a. Gas
b. Solid
c. Solid

1.53 a. Heterogeneous
b. Homogeneous
c. Homogeneous
d. Homogeneous

1.55 The drawing below represents a heterogeneous mixture of two compounds. As you can see in the drawing, the compounds are nonuniformly distributed.

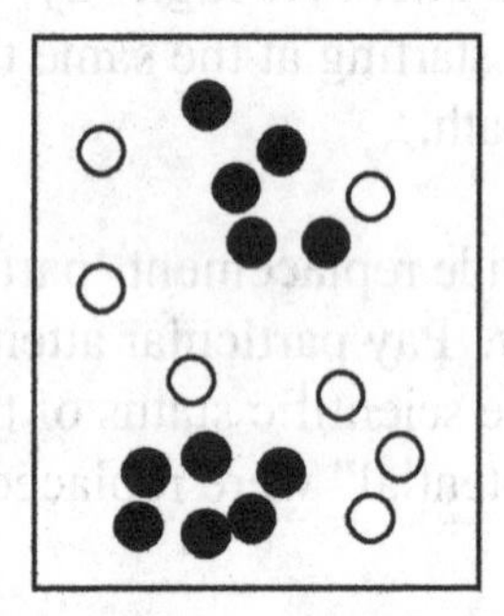

1.57 a. Chemical reaction
b. Physical change
c. Physical change

1.59 a. Physical property
b. Chemical property

1.61 a. Mixture
b. Pure substance
c. Mixture
d. Pure substance

1.63 The particles in the diagram are widely separated and fill the container. Therefore, the state of matter represented in the diagram is the gaseous state. Since there are two different compounds, the diagram represents a mixture. The molecules are well mixed and therefore the mixture is homogeneous.

1.65 An intensive property is a property of matter that is independent of the quantity of the substance. A substance's boiling point is an example of an intensive property.

1.67 a. Surface area and mass
b. Color and shape

1.69 Mass describes the quantity of matter in an object.

1.71 Length is the distance between two points.

1.73 A liter (L) is the volume occupied by 1000 g of water at 4 °C.

1.75 The shortest length is mm because the prefix m means 10^{-3}. The longest length is km because the prefix k means 10^{3}. Therefore, the order should be: mm < m < km

1.77 The temperature reading is made by determining the thermometer mark nearest the end of the red bar. That value is 23.9. Then, one additional digit is estimated by subdividing the marks into ten equal divisions. Therefore, the temperature reading to the correct number of significant figures is 23.92.

1.79 a. Precision is a measure of the agreement of replicate results.
b. Accuracy is the degree of agreement between the true value and measured value.

1.81 a. 3
b. 3
c. 3
d. 4
e. 4
f. 3

1.83 a. 3.87×10^{-3}
b. 5.20×10^{-2}
c. 2.62×10^{-3}
d. 24.3
e. 2.40×10^{2}
f. 2.41

1.85 a. $(23)\,(657) = 1.5 \times 10^{4}$

b. $0.00521 + 0.236 = 2.41 \times 10^{-1}$

c. $\dfrac{18.3}{3.0576} = 5.99$

d. $1157.23 - 17.812 = 1139.42$

e. $\frac{(1.987)(298)}{0.0821} = 7.21 \times 10^3$

1.87 a. 1.23×10^1
b. 5.69×10^{-2}
c. -1.527×10^3
d. 7.89×10^{-7}
e. 9.2×10^7
f. 5.280×10^{-3}
g. 1.279×10^0
h. -5.3177×10^2

1.89 Accuracy is the degree of agreement between the true value and the measured value. Precision is a measure of the agreement of replicate measurements. These measurements have high levels of precision and accuracy.

1.91 Unlike the English system, the metric system is systematic. Since it is composed of a set of units that are related to each other decimally, it is simpler to convert one unit to another.

1.93 a. The prefix kilo has the abbreviation k and means 10^3.
b. The prefix centi has the abbreviation c and means 10^{-2}.
c. The prefix micro has the abbreviation μ and means 10^{-6}.

1.95 Since 1 foot (ft) equals 12 inches (in), the conversion factors are $\frac{1 \text{ ft}}{12 \text{ in}}$ and $\frac{12 \text{ in}}{1 \text{ ft}}$.

1.97 a. $2.0 \text{ lb} \times \frac{16 \text{ oz}}{1 \text{ lb}} = 32 \text{ oz}$

b. $2.0 \text{ lb} \times \frac{1 \text{ t}}{2000 \text{ lb}} = 1.0 \times 10^{-3} \text{ t}$

c. $2.0 \text{ lb} \times \frac{454 \text{ g}}{1 \text{ lb}} = 9.1 \times 10^2 \text{ g}$

d. $2.0 \text{ lb} \times \frac{454 \text{ g}}{1 \text{ lb}} \times \frac{10^3 \text{ mg}}{1 \text{ g}} = 9.1 \times 10^5 \text{ mg}$

e. $2.0 \text{ lb} \times \frac{454 \text{ g}}{1 \text{ lb}} \times \frac{1 \text{ dag}}{10^1 \text{ g}} = 9.1 \times 10^1 \text{ dag}$

1.99 $1.50 \times 10^4 \text{ μg} \times \frac{10^{-6} \text{ g}}{1 \text{ μg}} \times \frac{1 \text{ mg}}{10^{-3} \text{ g}} = 1.50 \times 10^1 \text{ mg}$ or 15.0 mg

1.101 a. $3.0 \text{ g} \times \frac{1 \text{ lb}}{454 \text{ g}} = 6.6 \times 10^{-3} \text{ lb}$

b. $3.0 \text{ g} \times \frac{1 \text{ lb}}{454 \text{ g}} \times \frac{16 \text{ oz}}{1 \text{ lb}} = 1.1 \times 10^{-1} \text{ oz}$

c. $3.0 \text{ g} \times \frac{1 \text{ kg}}{10^{3} \text{ g}} = 3.0 \times 10^{-3} \text{ kg}$

d. $3.0 \text{ g} \times \frac{1 \text{ cg}}{10^{-2} \text{ g}} = 3.0 \times 10^{2} \text{ cg}$

e. $3.0 \text{ g} \times \frac{1 \text{ mg}}{10^{-3} \text{ g}} = 3.0 \times 10^{3} \text{ mg}$

1.103 In order to convert ft^2 to m^2, we must use a common unit. In the English system, one ft is equal to 12 inches. The meter is a common unit of length in the metric system. One meter is equal to 3.94×10^{1} inches. Note that the answer has 3 significant figures.

$$144 \text{ ft}^2 \times \left(\frac{12 \text{ inch}}{1 \text{ ft}}\right)^2 \times \left(\frac{2.54 \text{ cm}}{1 \text{ inch}}\right)^2 \times \left(\frac{10^{-2} \text{ m}}{1 \text{ cm}}\right)^2 = 13.4 \text{ m}^2$$

1.105 $9 \text{ pt} \times \frac{1 \text{ qt}}{2 \text{ pt}} \times \frac{0.946 \text{ L}}{1 \text{ qt}} = 4.257 \text{ L} \approx 4 \text{ L}$ (1 significant figure)

1.107 If $T_{°C} = \frac{T_{°F} - 32}{1.8}$, and $T_{°F} = 1.8(T_{°C}) + 32$

and $T_{°F} = 1.8(T_{°C}) + 32$

$$T_{°F} = [(1.8)(38.5)] + 32 = 101°F$$

1.109 In order to compare the magnitude of the two lengths, we must convert them to a common unit.

$$5.0 \text{ cm} \times \frac{1 \text{ in}}{2.54 \text{ cm}} = 2.0 \text{ in}$$

Therefore, 5 cm is shorter than 5 in.

1.111 In order to compare the magnitude of the two masses, we must convert them to a common unit.

$$5.0 \text{ mg x } \frac{10^{-3} \mu g}{1 \text{ mg}} = 5 \text{ x } 10^{-3} \ \mu\text{g}$$

Therefore, 5.0 μg is smaller than 5.0 mg.

1.113 a. In order to calculate the circumference of the property, the boundary measurements that are in feet should be converted to kilometers. This is because the lengths all must have the same units. Using the unit relationships provided in the textbook, feet can first be converted to yards. Then, yards can be converted to meters. Then, meters can be converted to kilometers. To complete these conversions, conversions factors are used that relate the units. The boundary measurements that are in meters also need to be converted to kilometers. Finally, after all of the lengths have the same units (km), they can be added together to determine the circumference of the property.

b. First, each boundary measurement is converted to kilometers.

$$435 \text{ ft x } \frac{1 \text{ yd}}{3 \text{ ft}} \text{ x } \frac{0.914 \text{ m}}{1 \text{ yd}} \text{ x } \frac{1 \text{ km}}{10^3 \text{ m}} = 0.133 \text{ km (3 significant figures)}$$

$$515 \text{ ft x } \frac{1 \text{ yd}}{3 \text{ ft}} \text{ x } \frac{0.914 \text{ m}}{1 \text{ yd}} \text{ x } \frac{1 \text{ km}}{10^3 \text{ m}} = 0.157 \text{ km (3 significant figures)}$$

$$85 \text{ m x } \frac{1 \text{ km}}{10^3 \text{ m}} = 0.085 \text{ km (2 significant figures)}$$

$$95 \text{ m x } \frac{1 \text{ km}}{10^3 \text{ m}} = 0.095 \text{ km (2 significant figures)}$$

Then, the circumference is calculated by adding the measurements in km together.

0.133 km + 0.157 km + 0.85 km + 0.095 km = 0.470 km (3 significant figures)

1.115 The three major temperature scales are Celsius, Fahrenheit, and Kelvin.

1.117 Energy may be categorized as either kinetic energy, the energy of motion, or potential energy, the energy of position.

1.119 a. Extensive property. Mass depends on the quantity.

b. Extensive property. Volume depends on the quantity.

c. Intensive property. Density is independent of the quantity.

1.121 a. $T_{°C} = \frac{T_{°F} - 32}{1.8} = \frac{50.0 - 32}{1.8} = 10.0°C$

b. $T_K = T_{°C} + 273.15 = 10.0 + 273.15 = 283.2\ K$

1.123 a. $T_K = T_{°C} + 273.15 = 20.0 + 273.15 = 293.2\ K$

b. If $T_{°C} = \frac{T_{°F} - 32}{1.8}$, then $1.8(T_{°C}) = T_{°F} - 32$

and $T_{°F} = 1.8(T_{°C}) + 32$

$T_{°F} = [(1.8)(20.0)] + 32 = 68.0°F$

1.125 First, kilocalories are converted to calories. Then, the relationship 1 cal = 4.18 J can be used to create a conversion factor to convert from calories to joules.

$6\ kcal\ x\ \frac{10^3\ cal}{1\ kcal}\ x\ \frac{4.18\ J}{1\ cal} = 3\ x\ 10^4\ J$ (1 significant figure)

1.127 $d = \frac{m}{V} = \frac{3.00\ x\ 10^2\ g}{50.0\ mL} = 6.00\ g/mL$

1.129 $1.50\ x\ 10^2\ mL\ x\ \frac{7.20\ g}{1\ mL} = 1.08\ x\ 10^3\ g$

1.131 $8.00\ x\ 10^2\ g\ x\ \frac{1\ L}{1.29\ g} = 6.20\ x\ 10^2\ L$

1.133 $7\ g\ salt\ x\ \frac{2\ times}{1\ day}\ x\ \frac{365\ days}{1\ year} = 5110\ g\ salt$

1.135 $d_{lead} = \frac{m}{V} = \frac{5.0\ x\ 10^1\ g}{6.36\ cm^3} = 7.9\ g/cm^3$

$d_{uranium} = \frac{m}{V} = \frac{75\ g}{3.97\ cm^3} = 19\ g/cm^3$

$$d_{\text{platinum}} = \frac{m}{V} = \frac{2140 \text{ g}}{1.00 \times 10^2 \text{ cm}^3} = 21.4 \text{ g/cm}^3$$

Lead has the lowest density, and platinum has the greatest density.

1.137 $d = \frac{m}{V}$

Thus $V = \frac{m}{d} = \frac{10.0 \text{ g}}{0.791 \text{ g/mL}}$

$V = 12.6$ mL

1.139 $d = \frac{m}{V}$

Thus $m = (d)(V) = (0.791 \text{ g/mL})(50.0\text{mL})$

$m = 39.6$ g

1.141 Specific gravity of urine = $\frac{\text{density of urine}}{\text{density of water}}$

Density of urine = (specific gravity of urine)(density of water) = 1.04 x 0.993 g/mL

Density of urine = 1.03 g/mL

1.143 Specific gravity of alcohol = $\frac{\text{density of alcohol}}{\text{density of water at 4°C}}$

Specific gravity of alcohol = $\frac{0.789 \text{ g/mL}}{1.00 \text{ g/mL}} = 0.789$

Chapter 2

The Structure of the Atom and the Periodic Table

Solutions to the Odd-Numbered Questions and Problems

In-Chapter Questions and Problems

2.1 a. Bromine, atomic number = 35, therefore 35 protons and 35 electrons. Mass number - atomic number = 79 – 35 = 44, therefore 44 neutrons.
b. Bromine, atomic number = 35, therefore 35 protons and 35 electrons. Mass number - atomic number = 81 – 35 = 46, therefore 46 neutrons.
c. Iron, atomic number = 26, therefore 26 protons and 26 electrons. Mass number - atomic number = 56 – 26 = 30, therefore 30 neutrons.

2.3 Electron density is the probability that an electron will be found in a particular region of an atomic orbital.

2.5 The periodic table shows the symbols for these elements.
a. Na, metal
b. Ra, metal
c. Mn, metal
d. Mg, metal

2.7 The periodic table shows the information needed to solve this problem.
a. Zr (zirconium)
b. 22.99 amu
c. Cr (chromium)
d. At (astatine)

2.9 The periodic table shows the information needed to solve this problem.
a. Helium, atomic number = 2, mass = 4.009 amu
b. Fluorine, atomic number = 9, mass = 19.00 amu
c. Manganese, atomic number = 25, mass = 54.94 amu

2.11 a. Oxygen (O) has the atomic number 8. Therefore, O^{2-} has 8 protons. A neutral oxygen atom has 8 electrons and must gain 2 more electrons to form the O^{2-} anion. As a result, O^{2-} has 10 electrons.
b. Magnesium (Mg) has the atomic number 12. Therefore, Mg^{2+} has 12 protons. A neutral magnesium atom has 12 electrons and must lose 2 electrons to form the Mg^{2+} cation. As a result, Mg^{2+} has 10 electrons.
c. Iron (Fe) has the atomic number 26. Therefore, Fe^{3+} has 26 protons. A neutral iron atom has 26 electrons and must lose 3 electrons to form the Fe^{3+} cation. As a result, Fe^{3+} has 23 electrons.

2.13 Metallic elements tend to form cations. Nonmetals tend to form anions. Isoelectronic ions and atoms have the same number of electrons.

a. K^+ and Ar are isoelectronic.
b. Sr^{2+} and Kr are isoelectronic.
c. S^{2-} and Ar are isoelectronic.
d. Mg^{2+} and Ne are isoelectronic.
e. P^{3-} and Ar are isoelectronic.
f. Be^{2+} and He are isoelectronic

2.15
a. K^+ has 18 electrons. $1s^22s^22p^63s^23p^6$ [Ar]
b. Ca^{2+} has 18 electrons. $1s^22s^22p^63s^23p^6$ [Ar]
c. Se^{2-} has 36 electrons. $1s^22s^2\ 2p^6\ 3s^23p^64s^23d^{10}4p^6$ [Kr]
d. Br^- has 36 electrons. $1s^22s^22p^63s^23p^64s^23d^{10}4p^6$ [Kr]

2.17
a. (smallest) F, N, Be (largest)
Atomic size decreases as we go across the periodic table within a period.
b. (lowest) Be, N, F (highest)
Ionization energy generally increases as we go across a period.
c. (lowest) Be, N, F (highest)
Electron affinity generally increases as we go across a period.

End-of-Chapter Questions and Problems

2.19 The atomic mass is the weighted average of the masses of isotopes of an element in amu. The mass number is equal to the sum of the number of protons and neutrons in an atom.

2.21
a. neutrons
b. protons
c. protons, neutrons
d. nucleus, negative

2.23 Isotopes of an element have different numbers of neutrons. They have similar chemical behavior.

2.25 a. True b. True c. True

2.27
a. Atomic number 56, hence, 56 protons and 56 electrons.
Mass number – atomic number = 136 – 56 = 80 neutrons.
b. Atomic number 84, hence, 84 protons and 84 electrons
Mass number – atomic number = 209 – 84 = 125 neutrons.
c. Atomic number 48, hence, 48 protons and 48 electrons.
Mass number – atomic number = 113 – 48 = 65 neutrons.

2.29 Atomic number = Number of protons = 9, Fluorine (F) has an atomic number of 9.
Mass number = Number of protons + Number of neutrons = 9 + 10 = 19
Therefore, the symbol is $^{19}_{9}F$

2.31 a. From the periodic table, all isotopes of Rn have 86 protons. Isotopes differ in the number of neutrons.
b. Rn has 86 protons and 134 neutrons (220 – 86 = 134 neutrons)

2.33 a. Selenium-80 has the atomic number 34. Therefore, selenium-80 has 34 protons.
b. Selenium-80 has a mass number of 80 and 34 protons.
Number of neutrons = Mass number - Number of protons
Therefore, selenium-80 has 46 neutrons.

2.35 a. Atomic number = Number of protons = 1, Hydrogen (H) has an atomic number of 1.
Mass number = Number of protons + Number of neutrons = 1 + 0 = 1
Therefore, the symbol is $^{1}_{1}H$
b. Atomic number = Number of protons = 6, Carbon (C) has an atomic number of 6.
Mass number = Number of protons + Number of neutrons = 6 + 8 = 14
Therefore, the symbol is $^{14}_{6}C$

2.37 Step 1. Convert each percentage to a decimal fraction.

$$69.09\% \text{ Copper-63} \times \frac{1}{100\%} = 0.6909 \text{ Copper-63}$$

$$30.91\% \text{ Copper-65} \times \frac{1}{100\%} = 0.3091 \text{ Copper-65}$$

Step 2. Multiply the decimal fraction of each isotope by the mass of that isotope to determine the isotopic contribution to the atomic mass.

Contribution to atomic mass by copper-63 = fraction of all Cu atoms that are copper-63 X mass of a copper-63 atom

= 0.6909 X 62.93 amu = 43.48 amu

Contribution to atomic mass by copper-65 = fraction of all Cu atoms that are copper-65 X mass of a copper-65 atom

= 0.3091 X 64.9278 amu = 20.07 amu

Step 3. The weighted average is:

Atomic mass of naturally occurring Cu	=	contribution of copper-63	+	contribution of copper-65		
	=	43.48 amu	+	20.07 amu	=	63.55 amu

2.39 The major postulates of Dalton's atomic theory include the following:

- All matter consists of tiny particles called atoms.
- Atoms cannot be created, divided, destroyed, or converted to any other type of atom.
- All atoms of a particular element have identical properties.
- Atoms of different elements have different properties.
- Atoms combine in simple whole-number ratios.
- Chemical change involves joining, separating, or rearranging atoms.

2.41 Our understanding of the nucleus is based on the gold foil experiment performed by Geiger and interpreted by Rutherford. In this experiment, Geiger bombarded a piece of gold foil with alpha particles, and observed that some alpha particles passed straight through the foil, others were deflected, and some simply bounced back. This led Rutherford to propose that the atom consisted of a small, dense nucleus (alpha particles bounced back) surrounded by a cloud of electrons (some alpha particles were deflected). The size of the nucleus is small when compared to the volume of the atom (alpha particles were able to pass through the foil).

2.43 a. Dalton developed the Law of Multiple Proportions; determined the relative atomic weights of the elements known at that time; developed the first scientific atomic theory.

b. Crookes developed the cathode ray tube and discovered "cathode rays"; he characterized electron properties.

c. Chadwick demonstrated the existence of the neutron in 1932. He accomplished this with a series of experiments using small nuclei as projectiles to study the nucleus.

d. Goldstein identified the existence of positive charge in the atom.

2.45 A cathode ray is the negatively charged particle formed in a cathode ray tube. It was characterized as an electron, with a mass of nearly zero and a charge of 1–.

2.47 Electrons surround the nucleus in a diffuse region. Electrons are negatively charged, and electrons are very low in mass in contrast to protons and neutrons.

2.49 Prior to Rutherford's experiment, it was believed that protons and electrons were uniformly distributed throughout the atom.

2.51 Spectroscopy is the measurement of intensity and energy of electromagnetic radiation.

2.53 Electromagnetic radiation, or light, travels in waves from its source. Each wavelength of light has its own characteristic energy.

2.55 False. Each wavelength of light travels with the same velocity, 3.0×10^8 m/s.

2.57 Infrared radiation has greater energy than microwave radiation. Energy is inversely proportional to the wavelength. Since infrared radiation has shorter wavelengths than microwave radiation, it has more energy than microwave radiation.

2.59 When electrical energy is applied to a sample of hydrogen gas, the electrons in lower orbits are "excited" to higher orbits. As they fall back down into lower orbits, they release an amount of energy equal to the amount of energy they absorbed to jump to the higher orbit. This energy may be released in the form of light and the wavelength is proportional to the energy difference. This produces a line spectrum that is characteristic of hydrogen.

2.61 According to Bohr, Planck, and others, electrons exist only in certain allowed regions, quantum levels, and outside of the nucleus.

2.63 The most important points of Bohr's theory are:
- Electrons are found in orbits at discrete distances from the nucleus.
- The orbits are quantized - they are of discrete energies.
- Electrons can only be found in these orbits, never in between (they are able to jump instantaneously from orbit to orbit).
- Electrons can undergo transitions - if an electron absorbs energy, it will jump to a higher orbit; when the electron falls back to a lower orbit, it releases energy.

2.65 Bohr's atomic model was the first to successfully account for electronic properties of atoms, specifically the interaction of atoms and light (spectroscopy). Although significantly modified, it still remains a useful model to predict bonding in simple systems.

2.67 The periodic table shows the information needed to solve this problem.
a. Sodium (Na) has an atomic number of 11 and an atomic mass of 22.99.
b. Potassium (K) has an atomic number of 19 and an atomic mass of 39.10.
c. Magnesium (Mg) has an atomic number of 12 and an atomic mass of 24.31.
d. Boron (B) has an atomic number of 5 and an atomic mass of 10.81.

2.69 Group IA (or 1) is known collectively as the alkali metals and consists of lithium (Li), sodium (Na), potassium (K), rubidium (Rb), cesium (Cs), and francium (Fr).

2.71 Group VIIA (or 17) is known collectively as the halogens and consists of fluorine, chlorine, bromine, iodine, and astatine.

2.73 a. The metals are: Na, Ni, Al.
b. The representative metals are: Na, Al.
c. The element that is inert is: Ar.

2.75 According to periodic law, the physical and chemical properties of the elements are periodic functions of their atomic numbers. This is demonstrated in the graph for Group I elements. As atomic number increases, the melting point decreases.

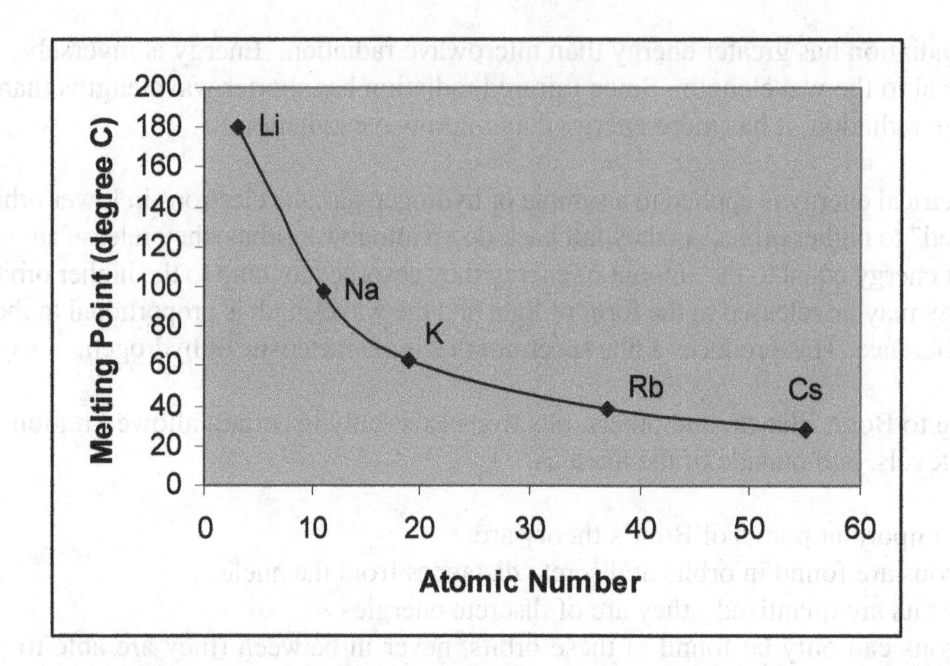

2.77 A principal energy level is designated $n = 1, 2, 3$, and so forth. It is similar to Bohr's orbits in concept. A sublevel is a part of a principal energy level and is designated *s, p, d,* and *f*.

2.79 The *s* orbital represents the probability of finding an electron in a region of space surrounding the nucleus. A diagram of an *s* orbital is found in Figure 2.11.

2.81 $2n^2 = 2(1)^2 = 2\ e^-$ for $n = 1$
$2n^2 = 2(2)^2 = 8\ e^-$ for $n = 2$
$2n^2 = 2(3)^2 = 18\ e^-$ for $n = 3$

2.83 According to the Pauli exclusion principle, each orbital can hold up to two electrons with the electrons spinning in opposite directions (paired). Therefore, since the *d* sublevel has five orbitals, it can contain a maximum of ten electrons.

$$5 \text{ orbitals} \times \frac{2 \text{ electrons}}{1 \text{ orbital}} = 10 \text{ electrons}$$

2.85 a. Al $1s^2 2s^2 2p^6 3s^2 3p^1$
b. Na $1s^2 2s^2 2p^6 3s^1$
c. Sc $1s^2 2s^2 2p^6 3s^2 3p^6 4s^2 3d^1$

2.87 a. B (5 e⁻) $1s^2 2s^2 2p^1$
The orbital diagram is:

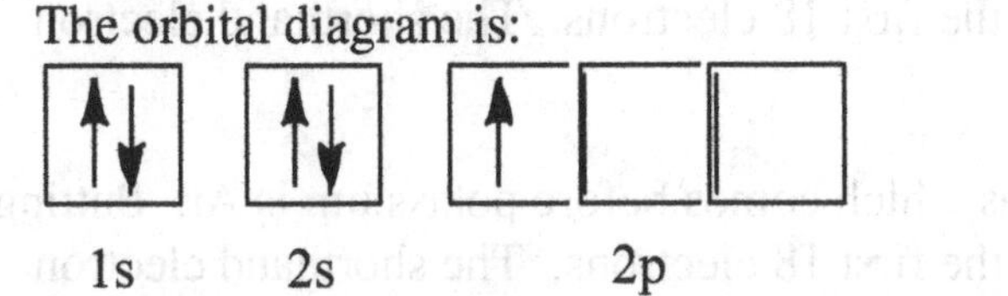

b. S (16 e⁻) $1s^2 2s^2 2p^6 3s^2 3p^4$
The orbital diagram is:

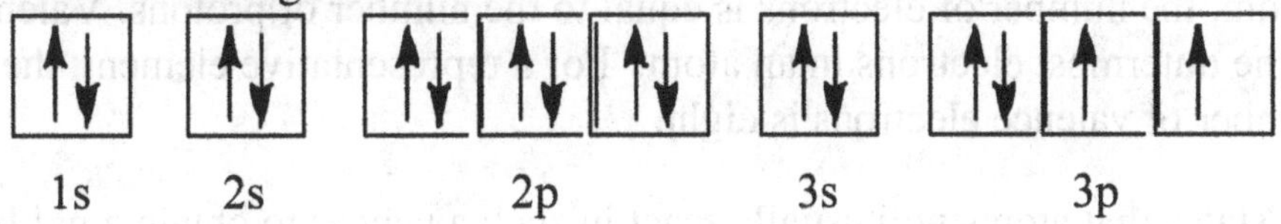

c. Ar (18 e⁻) $1s^2 2s^2 2p^6 3s^2 3p^6$
The orbital diagram is:

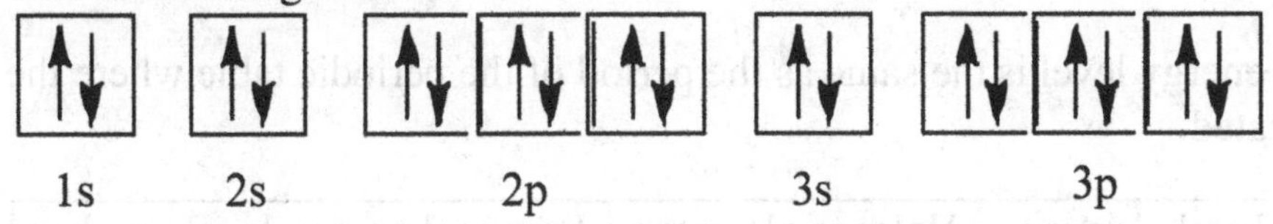

2.89 a. Not possible; $n = 1$ level can have only *s*-level orbitals.
b. Possible; the electron configuration that is shown represents the carbon atom.
c. Not possible; it cannot have two identical orbitals ($2s^2$).
d. Not possible; it cannot have three electrons in an s orbital ($2s^3$).

2.91 Diagram A is incorrect. Since it contains five electrons, the corrected diagram should show that the 1s and 2s orbitals completely fill. This leaves one electron in the 2p orbital.

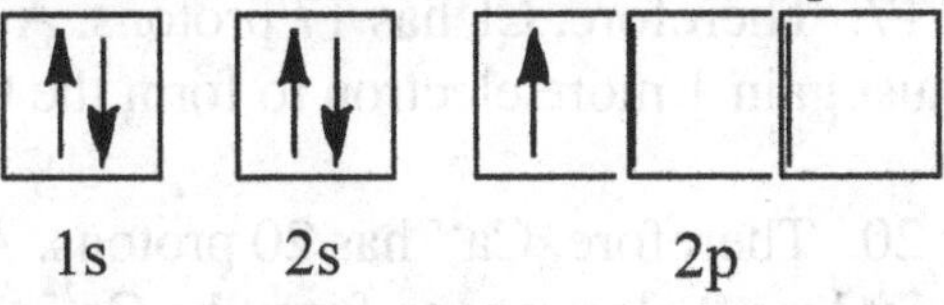

Diagram B is correct.
Diagram C is incorrect. According to Hund's rule, each orbital of equal energy should be half-filled before any become completely filled. The corrected diagram shows one electron in each of the three 2p orbitals.

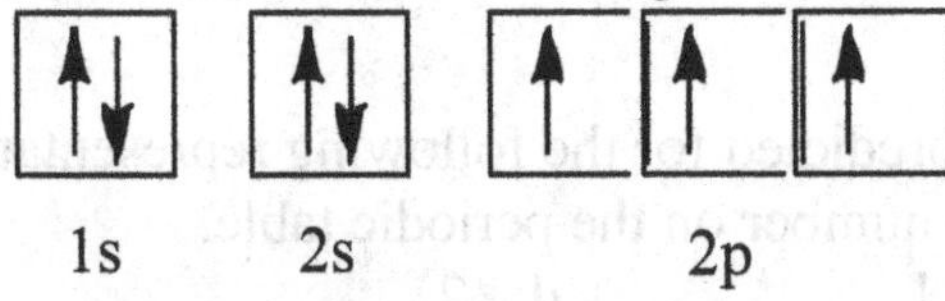

2.93 a. Zr has 40 electrons. The noble gas which comes before zirconium is Kr. Putting [Kr] in the configuration accounts for the first 36 electrons. The shorthand electron configuration is: $[Kr]5s^2 4d^2$

b. Br has 35 electrons. The noble gas which comes before bromine is Ar. Putting [Ar] in the configuration accounts for the first 18 electrons. The shorthand electron configuration is: $[Ar]4s^2 3d^{10} 4p^5$

c. K has 19 electrons. The noble gas which comes before potassium is Ar. Putting [Ar] in the configuration accounts for the first 18 electrons. The shorthand electron configuration is: $[Ar]4s^1$

2.95 The total number of electrons in an atom would include all of the electrons in the atom. In a neutral atom, the number of electrons is equal to the number of protons. Valence electrons are the outermost electrons in an atom. For a representative element, the maximum number of valence electrons is eight.

2.97 The octet rule states that atoms will usually react in such a way as to obtain a noble gas configuration.

2.99 Metals tend to lose electrons to become positively charged cations.

2.101 The principal energy level is the same as the period of the periodic table where the element is located.

	Atom	Total electrons	Valence electrons	Principal energy level number
a.	H	1	1	1
b.	Na	11	1	3
c.	B	5	3	2
d.	F	9	7	2
e.	Ne	10	8	2
f.	He	2	2	1

2.103 a. Chlorine (Cl) has the atomic number 17. Therefore, Cl^- has 17 protons. A neutral chlorine atom has 17 electrons and must gain 1 more electron to form the Cl^- anion. As a result, Cl^- has 18 electrons.

b. Calcium (Ca) has the atomic number 20. Therefore, Ca^{2+} has 20 protons. A neutral calcium atom has 20 electrons and must lose 2 electrons to form the Ca^{2+} cation. As a result, Ca^{2+} has 18 electrons.

c. Iron (Fe) has the atomic number 26. Therefore, Fe^{2+} has 26 protons. A neutral iron atom has 26 electrons and must lose 2 electrons to form the Fe^{2+} cation. As a result, Fe^{2+} has 24 electrons.

2.105 The number of valence electrons can be predicted for the following representative (Group A) elements by using their group number on the periodic table.

a. 2 b. 1 c. 1 d. 2

2.107 a. Li^+ (Li loses 1 electron to attain outermost octet)

b. Ca^{2+} (Ca loses 2 electrons to attain outermost octet)

c. S^{2-} (S gains 2 electrons to attain outermost octet)

2.109 a. O^{2-} and Ne are isoelectronic because they both have 10 electrons.
b. S^{2-} and Cl^- are isoelectronic because they both have 18 electrons.

2.111
a. I^- (54 e−) $1s^22s^22p^63s^23p^64s^23d^{10}4p^65s^24d^{10}5p^6$ [Xe]
b. Ba^{2+} (54 e−) $1s^22s^22p^63s^23p^64s^23d^{10}4p^65s^24d^{10}5p^6$ [Xe]
c. Se^{2-} (36 e−) $1s^22s^22p^63s^23p^64s^23d^{10}4p^6$ [Kr]
d. Al^{3+} (10 e−) $1s^22s^22p^6$ [Ne]

2.113 Atomic size decreases from left to right across a period in the periodic table.

2.115 Ionization energy is the energy required to remove an electron from an isolated atom.

2.117 Na + ionization energy → Na^+ + e^-

2.119
a. (Smallest) F, O, N (Largest)
Atomic size decreases as we go across the periodic table within a period.
b. (Smallest) Li, K, Cs (Largest)
Atomic size increases as we go down a group.
c. (Smallest) Cl, Br, I (Largest)
Atomic size increases as we go down a group.
d. (Smallest) Be, Mg, Ra (Largest)
Atomic size increases as we go down a group.

2.121
a. (Smallest) N, O, F (Largest)
Ionization energy generally increases as we go across a period.
b. (Smallest) Cs, K, Li (Largest)
Ionization energy generally decreases as we go down a group.

2.123
a. (Largest) Li, Na, K (Smallest)
Electron affinity generally decreases as we go down a group
b. (Largest) Te, Sn, Sr (Smallest)
Electron affinity generally increases as we go across the periodic table within a period.

2.125
a. A positive ion is always smaller than its parent atom because the positive charge of the nucleus is shared among fewer electrons in the ion. As a result, each electron is pulled closer to the nucleus and the volume of the ion decreases.
b. The fluoride ion has a completed octet of electrons and an electron configuration resembling its nearest noble gas. The electron affinity of fluorine is very high; therefore, it is energetically favorable for the fluorine atom to gain the electron.

2.127 Cl^- is larger than Ar. Both have the same number of electrons; however, Ar has one more proton in its nucleus than Cl^-, providing a larger nuclear charge/electron. Therefore, argon's electrons are pulled closer to the nucleus, making argon smaller than the chloride ion.

Chapter 3
Structure and Properties of Ionic and Covalent Compounds
Solutions to the Odd-Numbered Questions and Problems

In-Chapter Questions and Problems

3.1 The rules for writing the names ionic compounds from the formula are used to name these ionic compounds. The name of the cation appears first, followed by the name of the anion.

a. Potassium cyanide. Cyanide is the name for the polyatomic anion CN^-.
b. Magnesium sulfide. Sulfide is the name for the S^{2-} ion.
c. Magnesium acetate. Acetate is the name for the polyatomic anion CH_3COO^-.

3.3 The distance of separation and bond strength are inversely related.

a. The bonded nuclei are closer together when a double bond exists, in comparison to a single bond.
b. The bond strength increases as the bond order increases. Therefore, a double bond is stronger than a single bond.

3.5 a.

Three groups and one lone pair of electrons surround the phosphorus atom; the molecular geometry is trigonal pyramidal (similar to the structure of ammonia).

b.

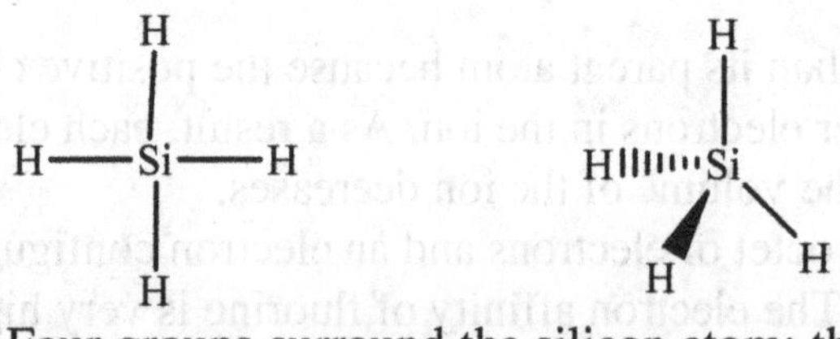

H
H—Si—H
H

H
H‖‖‖Si
H H

Four groups surround the silicon atom; the molecular geometry is tetrahedral (similar to the structure of methane).

3.7 a. Oxygen is more electronegative than sulfur; the bond is polar. The electrons are pulled toward the oxygen atom.

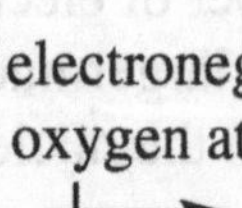

S—O

b. Nitrogen is more electronegative than carbon; the bond is polar. The electrons are pulled toward the nitrogen atom.

$$\overset{+\!\!\longrightarrow}{C\equiv N}$$

c. There is no electronegativity difference between two identical atoms; the bond is nonpolar.

d. Chlorine is more electronegative than iodine; the bond is polar. The electrons are pulled toward the chlorine atom.

$$\overset{+\!\!\longrightarrow}{I—Cl}$$

3.9 a. S=C=S

There are two identical groups around the central atom. The molecule has linear geometry and the polarity of the groups cancel. The molecule is nonpolar.

b. NF_3

Three groups and a lone pair of electrons surround the central atom. Due to the effect of the lone pair, the molecule is polar.

c. H—Cl:

The H-Cl bond is polar due to the electronegativity difference between hydrogen and chlorine. Since H-Cl is the only bond in the molecule, the molecule is polar.

d. $SiCl_4$

Four groups, all equivalent, surround the central atom. The structure is tetrahedral and the molecule is nonpolar.

3.11 a. The Lewis dot structures:

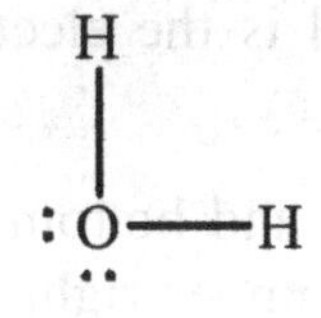

H₂C=CH₂

H_2O is polar covalent and has the higher melting and boiling points.
C_2H_4 is nonpolar covalent.

b. The Lewis dot structures:

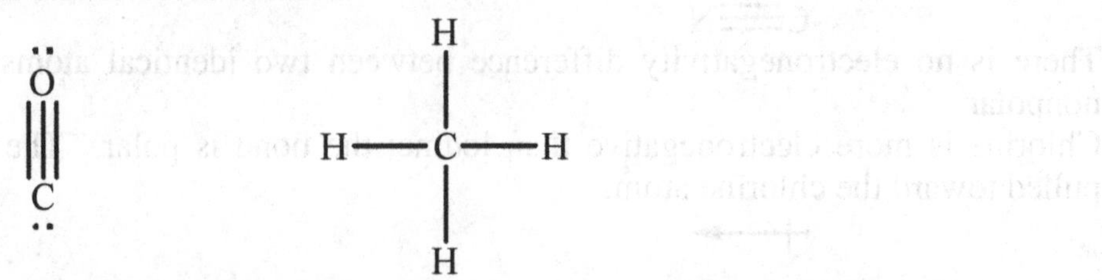

CO is polar covalent and has the higher melting and boiling points.
CH_4 is nonpolar covalent.

c. The Lewis dot structures:

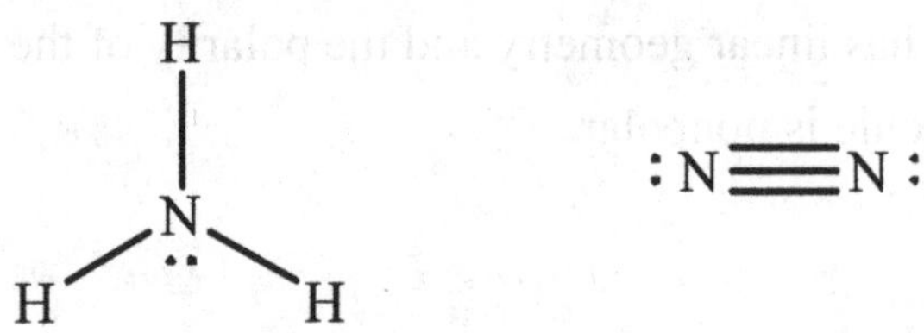

NH_3 is polar covalent and has the higher melting and boiling points.
N_2 is nonpolar covalent.

d. The Lewis dot structures:

:C̤̈l—C̤̈l: :C̤̈l—Ï̤:

ICl is polar covalent and has the higher melting and boiling points.
Cl_2 is nonpolar covalent.

End-of-Chapter Questions and Problems

3.13 a. H·   b. :He   c. ·Ṡi·   d. ·N̈·

3.15 a. Li^+ b. Mg^{2+} c. [:C̤̈l:]$^-$ d. [:P̤̈:]$^{3-}$

3.17 Covalent bonding involves a sharing of electrons between atoms to complete the octet of electrons for each atom participating in the bond. Ionic bonding involves a transfer of one or more electrons from one atom to another. An ionic bond is the electrostatic force between the resulting anion and a cation.

3.19 Electronegativity values increase as we proceed left to right and bottom to top of the table. The most electronegative elements are located in the upper right corner of the periodic table.

3.21 a. Ionic. The electronegativity difference between Mg (1.2) and F (4.0) is 2.8.
b. Polar covalent. The electronegativity difference between C (2.5) and O (3.5) is 1.0.
c. Nonpolar covalent. The electronegativity between H (2.1) and S (2.5) is 0.4.
d. Polar covalent. The electronegativity difference between N (3.0) and O (3.5) is 0.5.

3.23 a.

$$Li\cdot + :\ddot{\underset{..}{Br}}\cdot \longrightarrow Li^{+} + [:\ddot{\underset{..}{Br}}:]^{-}$$

b.

$$\cdot Mg\cdot + 2:\ddot{\underset{..}{Cl}}\cdot \longrightarrow Mg^{2+} + 2[:\ddot{\underset{..}{Cl}}:]^{-}$$

c.

$$\cdot\dot{\underset{..}{P}}\cdot + 3H\cdot \longrightarrow PH_3$$

3.25 He has two valence electrons (electron configuration $1s^2$) and a complete N = 1 level. It has a stable electron configuration, with no tendency to gain or lose electrons, and satisfies the octet rule (2 e^- for period 1). Hence, it is nonreactive.
:He

3.27 MgS

$$\cdot Mg\cdot + \cdot\ddot{\underset{..}{S}}\cdot \longrightarrow Mg^{2+} + [:\ddot{\underset{..}{S}}:]^{2-}$$

3.29 a. Sodium ion. The cation has the same name as the element.
b. Copper (I) ion (or cuprous ion). The stock system is used to indicate the cation's charge. The systematic name is in parenthesis.
c. Magnesium ion. The cation has the same name as the element.

3.31 a. Bicarbonate ion. This is the polyatomic cation name for HCO_3^-.
b. Hydronium ion. This is the polyatomic cation name for H_3O^+.
c. Carbonate ion. This is the polyatomic cation name for CO_3^{2-}.

3.33 The formula for these monoatomic ions can be determined using the elemental symbol from the periodic table and the charge that would be present on these ions.
a. K+
b. Ni^{2+}

3.35 The formula for these polyatomic anions corresponds to their names.
a. SO_4^{2-}
b. NO_3^-

3.37 a. Al_2O_3 (two 3^+ cancels out three 2^-)
b. Li_2S (two 1^+ cancels out one 2^-)

3.39 The rules for writing the names ionic compounds from the formula are used to name these ionic compounds. The name of the cation appears first, followed by the name of the anion.
a. Magnesium chloride
b. Aluminum chloride
c. Copper(II) nitrate

3.41 a. NaCl (one 1^+ cancels one 1^-)
b. $MgBr_2$ (one 2^+ cancels two 1^-)

3.43 a. AgCN (one 1^+ cancels out 1^-)
b. NH_4Cl (one 1^+ cancels out 1^-)

3.45 a. CuO (one 2^+ cancels out one 2^-)
b. Fe_2O_3 (two 3^+ cancels three 2^-)

3.47 Using the names provided, the cation appears first, followed by the anion. Nitrate is the name of the polyatomic anion represented by NO_3^-.
a. $NaNO_3$
b. $Mg(NO_3)_2$

3.49 Using the names provided, the cation appears first, followed by the anion. Ammonium is the name of the polyatomic cation represented by NH_4^+.
a. NH_4I
b. $(NH_4)_2SO_4$ Sulfate is the name of the polyatomic anion represented by SO_4^{2-}.

3.51 The rules for writing the names covalent compounds from the formula are used to name these covalent compounds.
a. Nitrogen dioxide
b. Selenium trioxide
c. Sulfur trioxide

3.53 a. SiO_2 The subscripts can be determined from the prefix *-di* in the name.
b. SO_2 The subscripts can be determined from the prefix *-di* in the name.

3.55 Ionic solid-state compounds exist in regular, repeating, three-dimensional structures: the crystal lattice. The crystal lattice is made up of positive and negative ions. Solid state covalent compounds are made up of molecules, which may be arranged in a regular crystalline pattern or in an irregular (amorphous) structure.

3.57 The boiling points of ionic solids are generally much higher than those of covalent solids.

3.59 KCl would be expected to exist as a solid at room temperature; it is an ionic compound, and ionic compounds are characterized by high melting points.

3.61 Water will have a higher boiling point. Water is a polar molecule with strong intermolecular attractive forces, whereas carbon tetrachloride is a nonpolar molecule with weak intermolecular attractive forces. More energy, hence, a higher temperature is required to overcome the attractive forces among the water molecules.

3.63 Yes, $MgCl_2$ in water forms an electrolytic solution. $MgCl_2$ is an ionic solid that dissociates in water, and the resulting solution is capable of conducting a current of electricity.

3.65 The least electronegative atom will usually be the central atom. Therefore, it is often the element farthest to the left and/or lowest in the periodic table. The central atom is often the element in the compound for which there is only one atom.

3.67 For polyatomic cations, subtract one electron for each unit of positive charge.

3.69 (lowest bond energy) single bond, double bond, triple bond (highest bond energy)

3.71 C_5H_{12} will have more isomers than C_4H_{10}. As the size of the hydrocarbon increases, the number of possible isomers increases.

3.73 Resonance can occur when more than one valid Lewis structure can be written for a molecule. Each individual structure that can be drawn is a resonance form. The true nature of the structure for the molecule is the resonance hybrid, which consists of the "average" of the resonance forms.

3.75 In a trigonal planar molecule, the bond angel is 120°.

3.77 True. A molecule containing only nonpolar bonds must be a nonpolar molecule.

3.79 a. NCl_3

```
        ..
       :Cl:
       /
 ..   /
:Cl——N:
 ..   \
       \
       :Cl:
        ..
```

One nitrogen atom has 5 valence electrons, and three chlorine atoms contribute 21 valence electrons (3 x 7 valence electrons/atom of chlorine).
This produces a total of 26 valence electrons.
This structure satisfies the octet rule for N and Cl.

b. CH_3OH

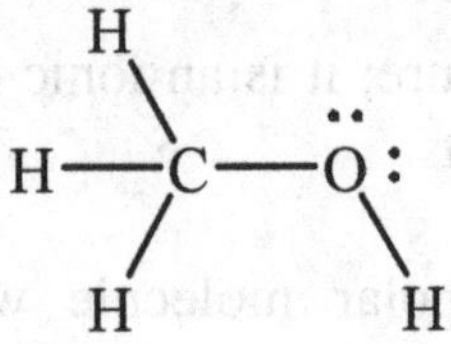

```
      H
        \   ..
  H ——— C ——— O:
        /      \
      H         H
```

One carbon atom has 4 valence electrons, one oxygen atom has 6 valence electrons, and four hydrogen atoms contribute 4 valence electrons (one for each hydrogen atom). This produces a total of 14 valence electrons.
This structure satisfies the octet rule for C and O.

c. CS_2

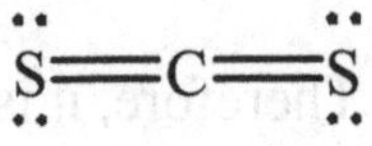

```
 ..        ..
 S ══ C ══ S
 ..        ..
```

One carbon atom contributes 4 valence electrons, and two sulfur atoms contribute 12 valence electrons (six from each sulfur atom) for a total of 16 valence electrons.
This structure satisfies the octet rule for C and S.

d. CH_2Cl_2

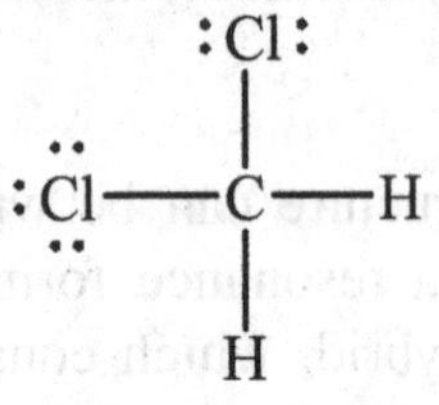

```
          :Cl:
           |
  ..       |
 :Cl ——— C ——— H
  ..       |
           |
           H
```

One carbon atom contributes 4 valence electrons, two hydrogen atoms contribute 2 valence electrons, and two chlorine atoms contribute 14 valence electrons (seven from each chlorine atom) for a total of 20 valence electrons. This structure satisfies the octet rule for C and Cl.

3.81 C_2H_4O

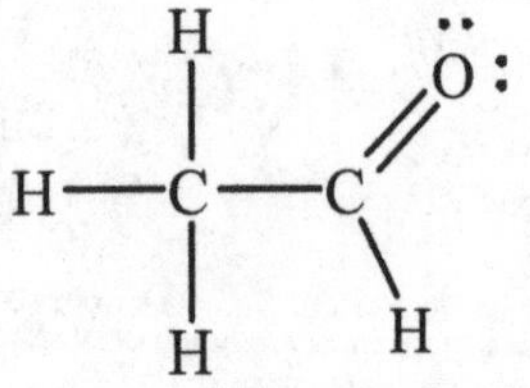

```
       H            ..
       |           O:
       |          //
  H —— C ——— C
       |          \
       |           \
       H            H
```

Both carbon atoms have 4 valence electrons ($2\ C \times 4\ e^- = 8\ e^-$), the oxygen atom has 6 valence electrons, and four hydrogen atoms contribute 4 valence electrons (one for each hydrogen atom).This produces a total of 18 valence electrons.
This structure satisfies the octet rule for C and O.

3.83

$$\begin{array}{c} \text{H}_3\text{C}-\text{C}(=\ddot{\text{O}}\text{:})-\text{CH}_3 \end{array}$$

(Lewis structure: H—C(H)(H)—C(=Ö:)—C(H)(H)—H)

3.85 Step 1. Draw a skeletal structure of the molecule.

N O

Step 2. Determine the number of valence electrons on each atom and add them to arrive at the total for the compound.

1	O atom	x	6 valence electrons	=	6 electrons
1	N atom	x	5 valence electrons	=	5 electrons
1	positive charge		(less 1 electron)	=	-1 electron
					10 electrons

Step 3. Distribute the electrons around the skeletal structure.

$$\left[:\text{O}{\equiv}\text{N}:\right]^{+}$$

Step 4. Confirm that the Lewis structure satisfies the octet rule for N and O. All 10 electrons are used in the process.

3.87 Step 1. Draw a skeletal structure of the molecule.

O H

Step 2. Determine the number of valence electrons on each atom and add them to arrive at the total for the compound.

1	O atom	x	6 valence electrons	=	6 electrons
1	H atom	x	1 valence electron	=	1 electron
1	negative charge			=	1 electron
					8 electrons

Step 3. Distribute the electrons around the skeletal structure.

$$\left[:\ddot{\underset{\cdot\cdot}{\text{O}}}-\text{H}\right]^{-}$$

Step 4. Confirm that the Lewis structure satisfies the octet rule for O. Appropriately, two electrons are around H. All 8 electrons are used in the process.

3.89 Acetate is the name for the polyatomic anion CH_3COO^-.

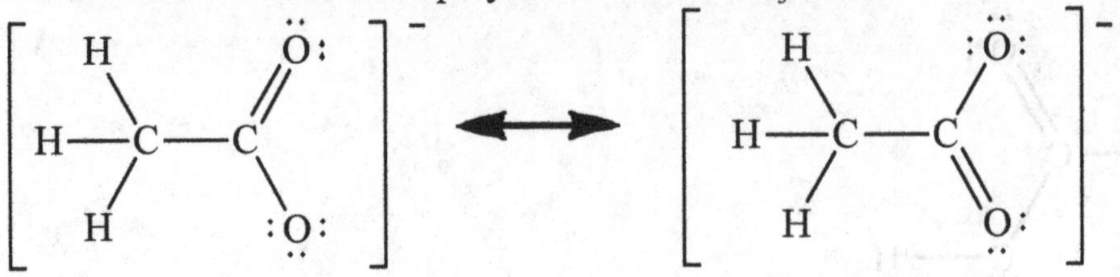

$$\left[\begin{array}{c} H \\ H-C-C(=\ddot{O}:)-\ddot{O}: \\ H \end{array} \right]^- \longleftrightarrow \left[\begin{array}{c} H \\ H-C-C(-\ddot{O}:)=\ddot{O}: \\ H \end{array} \right]^-$$

3.91 a. There should only be 10 electrons in this Lewis structure. The correct structure is:

$:C\equiv O:$

b. There should only be 8 electrons in this Lewis structure. The correct structure is:

$H-\ddot{O}-H$

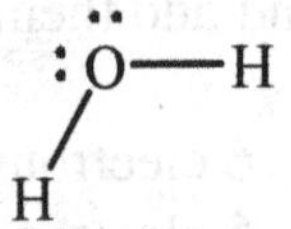

b. There should only be 16 electrons in this Lewis structure. The correct structure is:

$\ddot{O}=C=\ddot{O}$

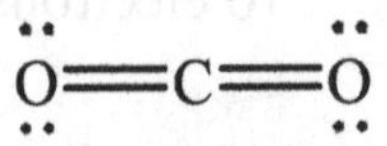

3.93 Step 1. Draw a skeletal structure of the molecule.

Cl Be Cl

Step 2. Determine the number of valence electrons on each atom and add them to arrive at the total for the compound.

2	Cl atoms	x	7 valence electrons	=	14 electrons
1	Be atom	x	2 valence electrons	=	2 electrons
					16 electrons

Step 3. Distribute the electrons around the skeletal structure.

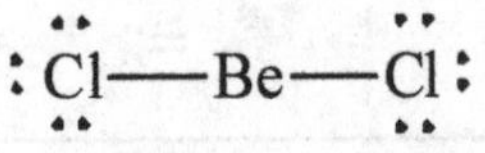

$:\ddot{Cl}-Be-\ddot{Cl}:$ All 16 electrons are used in the process.

3.95 Step 1. Draw a skeletal structure of the molecule.

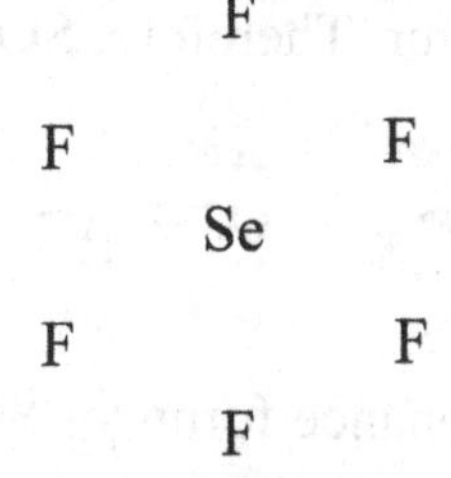

Step 2. Determine the number of valence electrons on each atom and add them to arrive at the total for the compound.

1	Se atom	x	6 valence electrons	=	6 electrons
6	F atoms	x	7 valence electrons	=	42 electrons
					48 electrons

Step 3. Distribute the electrons around the skeletal structure.

:F: :F: :F: Se :F: :F: :F: and the final structure

3.97 a. SO_2 has two resonance structures. Either Lewis structure can be used to determine the geometry using VSEPR. Looking at the central atom, it can be determined that there are two bonded atoms and one nonbonding electron pair around sulfur. The bond angle should be <120°. Therefore, the geometry is bent.

[:O—S=O ⟷ O=S—O:]

b. SO_3 has three resonance structures. Using the Lewis structures of SO_3, it can be seen that sulfur has three bonded atoms and zero nonbonding electron pairs. The bond angle should be 120°. Therefore, the geometry is trigonal planar.

3.99 a. The Lewis structures of the resonance forms SO_2, show that there is one pair of nonbonding electrons on the sulfur. Therefore, SO_2 is polar.

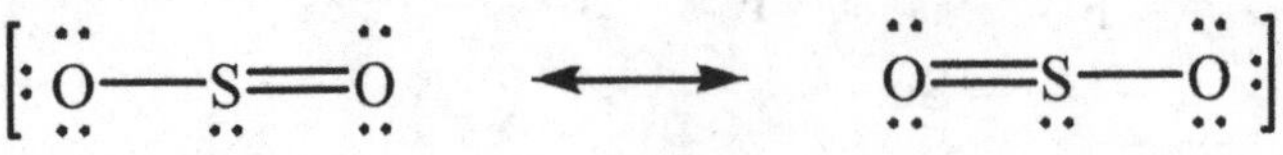

b. The Lewis structures of the resonance forms of SO_3 show that the bond polarities are arranged symmetrically and all three bonds are equivalent in the resonance hybrid. Therefore, SO_3 is nonpolar.

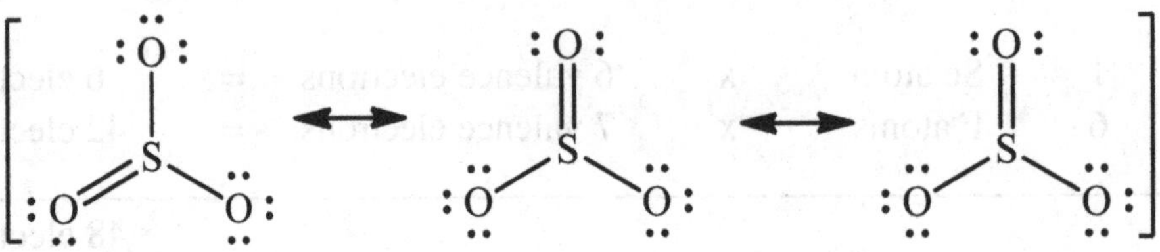

3.101 Compounds a. and c. have polar bonds, but are nonpolar compounds. This can be seen by looking at their Lewis structures. Compound b. is a polar compound because there is one pair of nonbonding electrons on the nitrogen.

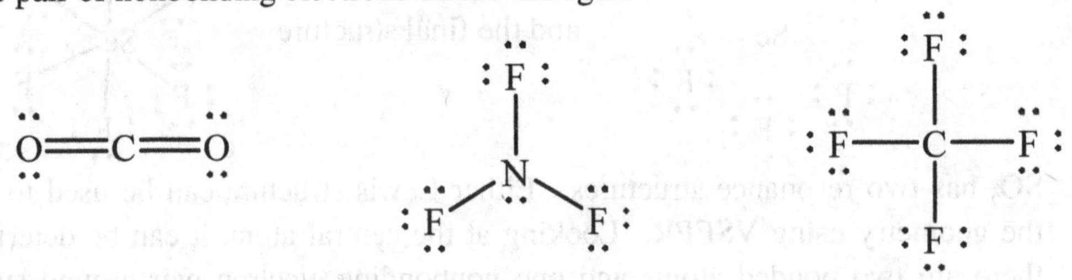

a. b. c.

3.103 A molecule containing no polar bonds *must* be nonpolar. A molecule containing polar bonds may or may not itself be polar. It depends upon the number and arrangement of the bonds. For example:

- If a molecule contains only one bond, and that bond is polar, the molecule must be polar.
- If the molecule contains more than one polar bond, the molecule will be nonpolar if the arrangement of the bonds causes their effects to cancel. If not, the molecule will be polar.
- If lone pairs of electrons are present, their effect must be considered as well.
- If a molecule contains no polar bonds, it cannot be a polar molecule.

3.105 a.

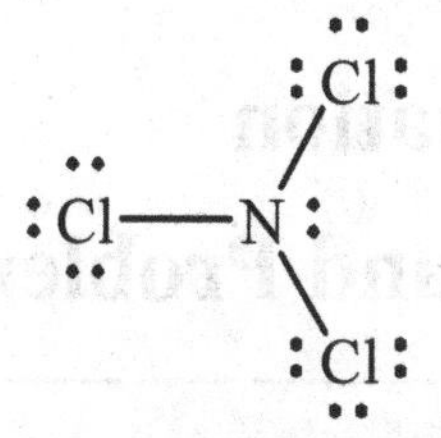

trigonal pyramidal
polar
water soluble

b.

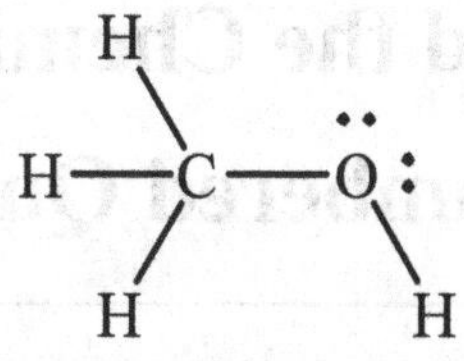

tetrahedral around C
bent around O
polar
water soluble

c.

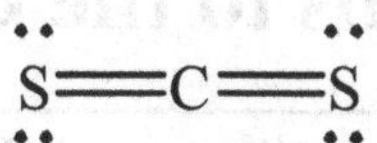

linear
nonpolar
not water soluble

d.

H
H—C—Cl:
:Cl:

tetrahedral
polar
water soluble

3.107 Polar compounds have strong intermolecular attractive forces. Higher temperatures are needed to overcome these forces and convert the solid to a liquid; hence, we predict higher melting points for polar compounds when compared to nonpolar compounds of similar molar mass.

3.109 Yes. Many ionic compounds are water soluble.

3.111 a. NH_3 has a higher boiling point than N_2 because NH_3 is a polar molecule and N_2 is nonpolar. Polar molecules have stronger intermolecular forces than nonpolar molecules.

b. Both CS_2 and CF_4 are nonpolar molecules. The molecular mass of CS_2 is 76.14 amu, and the molecular mass of CF_4 is 88.00 amu. The larger the mass of the molecule, the higher the boiling point. Therefore, CF_4 is predicted to have a higher boiling point.

c. NaCl has a higher boiling point than Cl_2 because NaCl is an ionic compound and Cl_2 is a nonpolar molecule. Ionic compounds have stronger intermolecular forces than nonpolar molecules.

Chapter 4

Calculations and the Chemical Equation

Solutions to the Odd-Numbered Questions and Problems

In-Chapter Questions and Problems

4.1 $$\text{Mass Hg} = 1.00 \times 10^{12} \text{ atoms Hg} \times \frac{1 \text{ mol Hg}}{6.022 \times 10^{23} \text{ atoms Hg}} \times \frac{200.6 \text{ g Hg}}{1 \text{ mol Hg}} = 3.33 \times 10^{-10} \text{g Hg}$$

4.3 $C_8H_{10}N_4O_2$

$$8 \text{ atoms C} \times \frac{12.01 \text{ amu C}}{1 \text{ atom C}} = 96.08 \text{ amu C}$$

$$10 \text{ atoms H} \times \frac{1.008 \text{ amu H}}{1 \text{ atom H}} = 10.08 \text{ amu H}$$

$$4 \text{ atoms N} \times \frac{14.01 \text{ amu N}}{1 \text{ atom N}} = 56.04 \text{ amu N}$$

$$2 \text{ atoms O} \times \frac{16.00 \text{ amu O}}{1 \text{ atom O}} = 32.00 \text{ amu O}$$

$$194.20 \text{ amu } C_8H_{10}N_4O_2$$

The formula mass of caffeine is 194.20 amu.

The molar mass of caffeine is 194.20 g/mol.

4.5
a. DR
b. SR
c. DR
d. D

4.7 The balanced equation is: $Na_2S(aq) + CuCl_2(aq) \rightarrow 2NaCl(?) + CuS(?)$
Solubility rules predict that NaCl is soluble and CuS is insoluble. Therefore, the black precipitate is CuS. The ionic equation is:

$$2Na^+(aq) + S^{2-}(aq) + Cu^{2+}(aq) + 2Cl^-(aq) \rightarrow 2Na^+(aq) + 2Cl^-(aq) + CuS(s)$$

The net ionic equation is: $Cu^{2+}(aq) + S^{2-}(aq) \rightarrow CuS(s)$

4.9 The balanced equation is: $O_3(g) + NO(g) \rightarrow O_2(g) + NO_2(g)$

a. The strategy will be to first convert 50.0g O_3 to moles of O_3. If we can calculate the number of moles of O_3 used, we can calculate the number of moles of NO_2 produced and then the mass of NO_2 produced. The formula mass for O_3 is 48.00g/mol. The formula mass for NO_2 is 46.01 g/mol.

$$50.0 \text{ g } O_3 \times \frac{1 \text{ mol } O_3}{48.00 \text{ g } O_3} \times \frac{1 \text{ mol } NO_2}{1 \text{ mol } O_3} \times \frac{46.01 \text{ g } NO_2}{1 \text{ mol } NO_2} = 47.9 \text{ g } NO_2$$

b. In part a, we found that the theoretical yield of the reaction was 47.9 g NO_2. Since only 25.0 g were obtained, the percent yield would be the fractional yield multiplied by 100%, or:

$$\frac{25.0 \text{ g } NO_2}{47.9 \text{ g } NO_2} \times 100 = 52.2\%$$

End-of-Chapter Questions and Problems

4.11 a. The average mass of Hg is 200.6 amu.
b. The average mass of Kr is 83.80 amu.
c. The average mass of Mg is 24.31 amu.

4.13 a. The average molar mass of Si (silicon) is 28.09 g/mol.
b. The average molar mass of Ag (silver) is 107.9 g/mol.
c. The average molar mass of As (arsenic) is 74.92 g/mol.

4.15 The mass of Avogadro's number of argon atoms is the same as the average molar mass, 39.95 g.

4.17 $$\text{C atoms} = 1.0 \times 10^{-4} \text{ mol C} \times \frac{6.022 \times 10^{23} \text{ atoms C}}{1 \text{ mol C}} = 6.0 \times 10^{19} \text{ carbon atoms}$$

4.19 $$\text{mol As} = 1.0 \times 10^{2} \text{ atoms As} \times \frac{1 \text{ mol As}}{6.022 \times 10^{23} \text{ atoms As}} = 1.7 \times 10^{-22} \text{ mol As}$$

4.21

$$2.00 \text{mol Ne} \times \frac{2.18 \text{ g Ne}}{1 \text{ mol Ne}} = 40.36 \text{ g Ne}$$

4.23 $$\frac{1.66 \times 10^{-24} \text{ g He}}{1 \text{ amu}} \times \frac{4.00 \text{ amu}}{1 \text{ atom He}} \times \frac{6.022 \times 10^{23} \text{ atoms He}}{1 \text{ mol He}} = 4.00 \frac{\text{g He}}{\text{mol He}}$$

4.25 a. $$20.0 \text{ g He} \times \frac{1 \text{ mol He}}{4.00 \text{ g He}} = 5.00 \text{ mol He}$$

b. $$0.040 \text{ kg Na} \times \frac{10^{3} \text{g Na}}{1 \text{ kg Na}} \times \frac{1 \text{ mol Na}}{22.99 \text{ g Na}} = 1.7 \text{ mol Na}$$

c. $$3.0 \text{ g } Cl_2 \times \frac{1 \text{ mol } Cl_2}{70.90 \text{ g } Cl_2} = 4.2 \times 10^{-2} \text{ mol } Cl_2$$

4.27 The mass of 1.00 mol of silver is 107.87 g. Therefore, the mass of 15.0 mol of silver is:

$$15.0 \text{ mol Ag} \times \frac{107.87 \text{ g Ag}}{1 \text{ mol Ag}} = 1.62 \times 10^3 \text{ g Ag}$$

4.29 $$\text{Ag atoms} = 15.0 \text{ g Ag} \times \frac{1 \text{ mol Ag}}{107.9 \text{ g Ag}} \times \frac{6.022 \times 10^{23} \text{ atoms Ag}}{1 \text{ mol Ag}} = 8.37 \times 10^{22} \text{ Ag atoms}$$

4.31 A molecule is a single unit made up of atoms joined by covalent bonds. An ion pair is composed of positive and negatively charged ions joined by electrostatic attraction, the ionic bond. The ion pairs, unlike the molecule, do not form single units; the electrostatic charge is directed to other ions in a crystal lattice.

4.33 a. $$1 \text{ atom Na} \times \frac{22.99 \text{ amu Na}}{1 \text{ atom Na}} = 22.99 \text{ amu Na}$$

$$1 \text{ atom Cl} \times \frac{35.45 \text{ amu Cl}}{1 \text{ atom Cl}} = \underline{35.45 \text{ amu Cl}}$$

58.44 amu NaCl (formula mass)

The mass of a single unit of NaCl is 58.44 amu/formula unit. Therefore, the mass of a mole of NaCl formula units is 58.44 g/mol.

b. $$2 \text{ atoms Na} \times \frac{22.99 \text{ amu Na}}{1 \text{ atom Na}} = 45.98 \text{ amu Na}$$

$$1 \text{ atom S} \times \frac{32.06 \text{ amu S}}{1 \text{ atom S}} = 32.06 \text{ amu S}$$

$$4 \text{ atoms O} \times \frac{16.00 \text{ amu O}}{1 \text{ atom O}} = \underline{64.00 \text{ amu O}}$$

142.04 amu Na_2SO_4 (formula mass)

The mass of a single unit of Na_2SO_4 is 142.04 amu/formula unit. Therefore, the mass of a mole of Na_2SO_4 formula units is 142.04 g/mol.

c. $$3 \text{ atoms Fe} \times \frac{55.85 \text{ amu Fe}}{1 \text{ atom Fe}} = 167.55 \text{ amu Fe}$$

$$2 \text{ atoms P} \times \frac{33.97 \text{ amu P}}{1 \text{ atom P}} = 61.94 \text{ amu P}$$

$$8 \text{ atoms O} \times \frac{16.00 \text{ amu O}}{1 \text{ atom O}} = \underline{128.00 \text{ amu O}}$$

357.49 amu $Fe_3(PO_4)_2$ (formula mass)

The mass of a single unit of $Fe_3(PO_4)_2$ is 357.49 amu/formula unit. Therefore, the mass of a mole of $Fe_3(PO_4)_2$ formula units is 357.49 g/mol.

4.35 $$2 \text{ atoms O} \times \frac{16.00 \text{ amu O}}{1 \text{ atom O}} = 32.00 \text{ amu (formula mass)}$$

The average mass of a single unit of O_2 is 32.00 amu/formula unit. Therefore the mass of a mole of O_2 units is 32.00 g/mol.

4.37

$$1 \text{ atom Cu} \times \frac{63.55 \text{ amu Cu}}{1 \text{ atom Cu}} = 63.55 \text{ amu C}$$

$$1 \text{ atom S} \times \frac{32.07 \text{ amu S}}{1 \text{ atom S}} = 32.07 \text{ amu S}$$

$$9 \text{ atom O} \times \frac{16.0) \text{ amu O}}{1 \text{ atom O}} = 144.00 \text{ amu O}$$

$$10 \text{ atom H} \times \frac{1.008 \text{ amu H}}{1 \text{ atom H}} = \underline{10.08 \text{ amu H}}$$

249.70 amu (formula mass)

The molar mass of $CuSO_4 \bullet 5H_2O$ is 249.70 g/mol.

4.39 a. The molar mass of NaCl is 58.44 g/mol.

$$15.0 \text{ g NaCl} \times \frac{1 \text{ mol NaCl}}{58.44 \text{ g NaCl}} = 0.257 \text{ mol NaCl}$$

b. The molar mass of Na_2SO_4 is 142.04 g/mol.

$$15.0 \text{ g } Na_2O_4 \times \frac{1 \text{ mol } Na_2SO_4}{142.04 \text{ g } Na_2SO_4} = 0.106 \text{ mol } Na_2SO_4$$

4.41 a. The molar mass of H_2O is 18.02 g/mol.

$$1.000 \text{ mol } H_2O \times \frac{18.02 \text{ g } H_2O}{1 \text{ mol } H_2O} = 18.02 \text{ g } H_2O$$

b. The molar mass of NaCl is 58.44 g/mol.

$$2.000 \text{ mol NaCl} \times \frac{58.44 \text{ g NaCl}}{1 \text{ mol NaCl}} = 116.9 \text{ g NaCl}$$

c. The molar mass of He is 4.00 g/mol.

$$10.0 \text{ mol He} \times \frac{4.00 \text{ g He}}{1 \text{ mol He}} = 40.0 \text{ g He}$$

d. The molar mass of H_2 = 2.016 g/mol.

$$1.00 \times 10^2 \text{ mol } H_2 \times \frac{2.016 \text{ g } H_2}{1 \text{ mol } H_2} = 2.02 \times 10^2 \text{ g } H_2$$

4.43 a. The molar mass of CH_4 is 16.04 g/mol.

$$0.100 \text{ mol } CH_4 \times \frac{16.04 \text{ g } CH_4}{1 \text{ mol } CH_4} = 1.60 \text{ g } CH_4$$

b. The molar mass of $CaCO_3$ is 100.09 g/mol.

$$0.100 \text{ mol } CaCO_3 \times \frac{100.09 \text{ g } CaCO_3}{1 \text{ mol } CaCO_3} = 10.0 \text{ g } CaCO_3$$

c. The molar mass of NaOH is 40.00 g/mol.

$$0.100 \text{ mol NaOH} \times \frac{40.00 \text{ g NaOH}}{1 \text{ mol NaOH}} = 4.00 \text{ g NaOH}$$

d. The molar mass of H_2SO_4 is 98.08 g/mol.

$$0.100 \text{ mol } H_2SO_4 \times \frac{98.08 \text{ g } H_2SO_4}{1 \text{ mol } H_2SO_4} = 9.81 \text{ g } H_2SO_4$$

4.45 a. The molar mass of KBr is 119.01 g/mol.

$$50.0 \text{ g KBr} \times \frac{119.01 \text{ g KBr}}{1 \text{ mol KBr}} = 0.420 \text{ mol KBr}$$

b. The molar mass of $MgSO_4$ is 120.37 g/mol.

$$50.0\text{g } MgSO_4 \times \frac{120.37 \text{ g } MgSO_4}{1 \text{ mol } MgSO_4} = 0.415 \text{ mol } MgSO_4$$

c. The molar mass of CS_2 is 76.13 g/mol.

$$50.0 \text{ g } CS_2 \times \frac{1 \text{ mol } CS_2}{76.13 \text{ g } CS_2} = 6.57 \times 10^{-1} \text{ mol } CS_2$$

d. The molar mass of $Al_2(CO_3)_3$ is 233.99 g/mol.

$$50.0 \text{ g } Al_2(CO_3)_3 \times \frac{1 \text{ mol } Al_2(CO_3)_3}{233.99 \text{ g } Al_2(CO_3)_3} = 2.14 \times 10^{-1} \text{ mol } Al_2(CO_3)_3$$

4.47 The ultimate basis for a balanced chemical equation is the law of conservation of mass. No mass may be gained or lost in a chemical reaction, and the chemical equation must reflect this fact.

4.49 A reactant is the starting material for a chemical reaction. Reactants are found on the left side of the reaction arrow.

4.51 The symbol Δ over the reaction arrow means that heat is necessary for the reaction to occur.

4.53 a. D
b. DR
c. C
d. SR

4.55 The subscript tells us the number of atoms or ions contained in one unit of the compound.

4.57 If we change the subscript, we change the identity of the compound.

4.59 a. $2C_2H_6(g) + 7O_2(g) \rightarrow 4CO_2(g) + 6H_2O(g)$

b. $6K_2O(s) + P_4O_{10}(s) \rightarrow 4K_3PO_4(s)$

c. $MgBr_2(aq) + H_2SO_4(aq) \rightarrow 2HBr(g) + MgSO_4(aq)$

d. $C_2H_5OH(l) + 3O_2(g) \rightarrow 2CO_2(g) + 3H_2O(g)$

4.61 a. $Ca(s) + F_2(g) \rightarrow CaF_2(s)$

b. $2Mg(s) + O_2(g) \rightarrow 2MgO(s)$

c. $3H_2(g) + N_2(g) \rightarrow 2NH_3(g)$

4.63 a. $2C_4H_{10}(g) + 13O_2(g) \rightarrow 10H_2O(g) + 8CO_2(g)$

b. $Au_2S_3(s) + 3H_2(g) \rightarrow 2Au(s) + 3H_2S(g)$

c. $Al(OH)_3(s) + 3HCl(aq) \rightarrow AlCl_3(aq) + 3H_2O(l)$

d. $(NH_4)_2Cr_2O_7(s) \rightarrow Cr_2O_3(s) + N_2(g) + 4H_2O(g)$

4.65 a. $N_2(g) + 3H_2(g) \rightarrow 2NH_3(g)$

b. $HCl(aq) + NaOH(aq) \rightarrow NaCl(aq) + H_2O(l)$

c. $C_6H_{12}O_6(s) + 6O_2(g) \rightarrow 6H_2O(l) + 6CO_2(g)$

d. $Na_2CO_3(s) \rightarrow Na_2O(s) + CO_2(g)$

4.67 a. Na_2SO_4 will not form a precipitate because most sulfates are soluble.
b. $BaSO_4$ will form a precipitate. Although many sulfates are soluble, this compound is an exception because it contains Ba^{2+}.
c. $BaCO_3$ will form a precipitate because most carbonates are insoluble.
d. K_2CO_3 will not form a precipitate. Although carbonates are insoluble, this compound is an exception because K^+ is an alkali metal ion.

4.69 Yes, a lead iodide precipitate will form (see Table 4.1, solubility rules).

$Pb(NO_3)_2\ (aq) + 2\ KI\ (aq) \rightarrow PbI_2\ (s) + 2K(NO_3)\ (aq)$

4.71 The balanced equation for the reaction is:

$(NH_4)_2CO_3(aq) + CaCl_2(aq) \rightarrow CaCO_3(?) + 2NH_4Cl(?)$

The solubility rules can be used to predict that NH_4Cl is soluble and $CaCO_3$ will be the precipitate formed. The ionic equation is:

$2NH_4^+(aq) + CO_3^{2-}(aq) + Ca^{2+}(aq) + 2Cl^-(aq) \rightarrow CaCO_3(s) + 2NH_4^+(aq) + 2Cl^-(aq)$

The net ionic equation is: $Ca^{2+}(aq) + CO_3^{2-}(aq) \rightarrow CaCO_3(s)$

4.73 An ionic equation shows all reactants and products as free ions unless they are precipitates. Ions that appear on both sides of the equation do not appear in the net ionic equation. The net ionic equation only shows the chemical species that actually undergo change.

4.75 The balanced equation is: $NaBr(aq) + AgNO_3(aq) \rightarrow NaNO_3(?) + AgBr(?)$

Solubility rules predict that $NaNO_3$ is soluble and AgBr is insoluble.
Therefore, the ionic equation is:

$Na^+(aq) + Br^-(aq) + Ag^+(aq) + NO_3^-(aq) \rightarrow Na^+(aq) + NO_3^-(aq) + AgBr(s)$

The net ionic equation is: $Ag^+(aq) + Br^-(aq) \rightarrow AgBr(s)$

4.77 An acid loses a hydrogen cation during an acid-base reaction.

4.79 HCN is the acid, and KOH is the base in this acid-base reaction:

$HCN(aq) + KOH(aq) \rightarrow KCN(aq) + H_2O(l)$

4.81 Oxidation. Cr(s) loses three electrons in order to form $Cr^{3+}(aq)$.

4.83 Mn(s) is being oxidized. $Sn^{2+}(aq)$ is being reduced.

4.85 The coefficients represent the relative number of moles of product(s) and reactant(s).

4.87 The molar mass of B_2H_6 is 27.67 g/mol.

$$3.00 \text{ mol } O_2 \text{ x } \frac{1 \text{ mol } B_2H_6}{3 \text{ mol } O_2} \text{ x } \frac{27.67 \text{ g } B_2H_6}{1 \text{ mol } B_2H_6} = 27.7 \text{ g } B_2H_6 \text{ (three significant figures)}$$

4.89 The molar mass of Cr_2O_3 is 151.99 g/mol.

$$50.0 \text{ g } Cr_2O_3 \text{ x } \frac{1 \text{ mol } Cr_2O_3}{151.99 \text{ g } Cr_2O_3} \text{ x } \frac{2 \text{ mol } CrCl_3}{1 \text{ mol } Cr_2O_3} = 0.658 \text{ mol } CrCl_3$$

4.91 a. $N_2(g) + 3H_2(g) \rightarrow 2NH_3(g)$

b. Three moles of H_2 will react with one mole of N_2, according to the coefficients in the balanced equation.

c. One mole of N_2 will produce two moles of the product NH_3, according to the coefficients in the balanced equation.

d. $$140.0 \text{ g } N_2 \text{ x } \frac{1 \text{ mol } N_2}{28.02 \text{ g } N_2} \text{ x } \frac{3 \text{ mol } H_2}{1 \text{ mol } N_2} = 1.50 \text{ mol } H_2$$

e. $$1.50 \text{ mol } H_2 \text{ x } \frac{2 \text{ mol } NH_3}{3 \text{ mol } H_2} \text{ x } \frac{17.03 \text{ g } NH_3}{1 \text{ mol } NH_3} = 17.0 \text{ g } NH_3$$

4.93 a.

5 C atoms x 12.01 amu/atom C =	60.05 amu
11 H atoms x 1.008 amu/atom H =	11.09 amu
1 N atom x 14.01 amu/atom N =	14.01 amu
2 O atoms x 16.00 amu/atom O =	32.00 amu
1 S atom x 32.06 amu/atom S =	32.06 amu
	149.21 amu

The mass of a single unit of $C_5H_{11}NO_2S$ is 149.21 amu/formula unit. Therefore, the mass of a mole of $C_5H_{11}NO_2S$ formula units is 149.21 g/mol.

b. $1 \text{ mol } C_5H_{11}NO_2S \times \frac{2 \text{ mol O atoms}}{1 \text{ mol } C_5H_{11}NO_2S} \times \frac{6.02 \times 10^{23} \text{ O atoms}}{1 \text{ mol O atoms}}$

$= 1.20 \times 10^{24}$ O atoms

c. $1 \text{ mol } C_5H_{11}NO_2S \times \frac{2 \text{ mol O atoms}}{1 \text{ mol } C_5H_{11}NO_2S} \times \frac{16.00 \text{ g O}}{1 \text{ mol O atoms}} = 32.00 \text{ g O}$

d. $50.0 \text{ g } C_5H_{11}NO_2S \times \frac{1 \text{ mol } C_5H_{11}NO_2S}{149.21 \text{ g } C_5H_{11}NO_2S} \times \frac{2 \text{ mol O atoms}}{1 \text{ mol } C_5H_{11}NO_2S} \times$

$\frac{16.00 \text{ g O}}{1 \text{ mol O atoms}} = 10.7 \text{ g O}$

4.95 The formula mass of HgO is 216.59 g/mol.

$$1.00 \times 10^2 \text{ g HgO} \times \frac{1 \text{ mol HgO}}{216.59 \text{ g HgO}} \times \frac{1 \text{ mol } O_2}{2 \text{ mol HgO}} \times \frac{32.00 \text{ g } O_2}{1 \text{ mol } O_2} = 7.39 \text{ g } O_2$$

4.97 The balanced equation is:

$$2C_2H_2(g) + 5O_2(g) \xrightarrow{\Delta} 4CO_2(g) + 2H_2O(g)$$

The formula mass of C_2H_2 is 26.04 g/mol.

$$20.0 \text{ kg } C_2H_2 \times \frac{10^3 \text{ g } C_2H_2}{1 \text{ kg } C_2H_2} \times \frac{1 \text{ mol } C_2H_2}{26.04 \text{ g } C_2H_2} \times \frac{4 \text{ mol } CO_2}{2 \text{ mol } C_2H_2} \times \frac{44.01 \text{ g } CO_2}{1 \text{ mol } CO_2}$$

$= 6.77 \times 10^4 \text{ g } CO_2$

4.99 Step 1. Write down information about the reaction:

$C_{10}H_{20}(l) + H_2(g) \rightarrow C_{10}H_{22}(s)$

(excess) 1.00 g

Step 2. Convert the mass of hydrogen to moles of hydrogen:

$$1.00 \text{ g } H_2 \times \frac{1 \text{ mol } H_2}{2.016 \text{ g } H_2} = 0.496 \text{ mol } H_2$$

Step 3. The reaction states that decene and hydrogen react in a 1:1 mole ratio. Use this conversion factor to calculate the mass of product:

$$0.496 \text{ mol } H_2 \times \frac{1 \text{ mol } C_{10}H_{22}}{1 \text{ mol } H_2} \times \frac{143 \text{ g } C_{10}H_{22}}{1 \text{ mol } C_{10}H_{22}} = 70.6 \text{ g } C_{10}H_{22}$$

4.101 Step 1. Write down information about the reaction:

$$N_2O_4(l) + 2\,N_2H_4(l) \rightarrow 3\,N_2(g) + 4\,H_2O(g)$$

1.00 kg (excess)

Step 2. Convert the mass of N_2O_4 to moles of N_2O_4:

$$1.00 \text{ kg} \times \frac{10^3 \text{ g } N_2O_4}{1 \text{ kg } N_2O_4} \times \frac{1 \text{ mol } N_2O_4}{92.02 \text{ g } N_2O_4} = 10.9 \text{ mol } N_2O_4$$

Step 3. The reaction states that the ratio of moles of N_2O_4 to N_2 is 1:3. Use this conversion factor to calculate the mass of N_2:

$$10.9 \text{ mol } N_2O_4 \times \frac{2 \text{ mol } N_2}{1 \text{ mol } N_2O_4} \times \frac{28.02 \text{ g } N_2}{1 \text{ mol } N_2} = 9.13 \times 10^2 \text{ g } N_2$$

4.103 In Question 4.99, we found that the theoretical yield of decane was 70.6 g. Since only 65.4 g were obtained, the percent yield would be the fractional yield multiplied by 100%, or:

$$\frac{65.4 \text{ g}}{70.6 \text{ g}} \times 100\,\% = 92.6\,\%$$

4.105 In Question 4.101, we found that the theoretical yield of nitrogen was 9.13×10^2 g. Since only 75.0% was actually obtained, corresponding to a decimal fraction of 0.750,

$$9.13 \times 10^2 \text{ g } N_2 \times 0.750 = 6.85 \times 10^2 \text{ g } N_2$$

Chapter 5

States of Matter: Gases, Liquids, and Solids

Solutions to the Odd-Numbered Questions and Problems

In-Chapter Questions and Problems

5.1 a. $725 \text{ mm Hg} \times \frac{1 \text{ atm}}{760 \text{ mm Hg}} = 0.954 \text{ atm}$

b. $29.0 \text{ cm Hg} \times \frac{10 \text{ mm Hg}}{1 \text{ cm Hg}} \times \frac{1 \text{ atm}}{760 \text{ mm Hg}} = 0.382 \text{ atm}$

c. $555 \text{ torr} \times \frac{1 \text{ atm}}{760 \text{ torr}} = 0.730 \text{ atm}$

d. $95 \text{ psi} \times \frac{1 \text{ atm}}{14.7 \text{ psi}} = 6.5 \text{ atm}$ (two significant figures)

5.3 $PV = nRT$

Solving for pressure, P, $P = \frac{nRT}{V}$

$$P = \frac{nRT}{V}$$

$$n = 4.80 \text{ g } H_2 \times \frac{(1 \text{ mol } H_2)}{2.02 \text{ g } H_2} = 2.38 \text{ mol } H_2$$

$$R = \frac{0.0821 \text{ L}\cdot\text{atm}}{\text{K}\cdot\text{mol}}$$

$$T = 25°\text{C} + 273 = 298 \text{ K}$$

$$V = 20.0 \text{ L}$$

Substituting,

$$P = \frac{(2.38 \text{ mol } H_2)(0.0821 \text{ L}\cdot\text{atm/K}\cdot\text{mol})(298 \text{ K})}{20.0 \text{ L}} = 2.91 \text{ atm}$$

5.5 Radon (Rn) is a collection of atoms (recall that all atoms are inherently nonpolar), and nitrogen dioxide (NO_2) molecules are polar. Since nonpolar molecules are only weakly attracted to each other, they exhibit more ideal gas behavior.

5.7 Molecules with complex structures, which do not "slide" smoothly past each other, and polar molecules tend to have higher viscosities than less complex, less polar liquids.

5.9 Evaporation is the conversion of a liquid to a gas at a temperature lower than the boiling point of the liquid. Condensation is the conversion of a gas to a liquid at a temperature lower than the boiling point of the liquid.

5.11 $CO_2 < CH_3Cl < CH_3OH$

Only CH_3OH exhibits London forces, dipole-dipole interactions, and hydrogen bonding. Hence, CH_3OH has the strongest intermolecular forces; therefore the highest boiling point.

5.13 a. Ionic solids are held together by electrostatic forces between the positive and negative ions. They generally have high melting points and a tendency to be hard and brittle.

b. Examples include table salt (NaCl) and calcium chloride ($CaCl_2$). These ionic solids are used to salt the roads in the winter time.

End-of-Chapter Questions and Problems

5.15 A monometer can be used to measures O_2 gas pressure in terms of the height of a column of liquid (mercury, for example) that is supported by the force exerted on the surface of the liquid by the O_2 gas being measured.

5.17 a. $94.4 \text{ cm Hg} \times \frac{10 \text{ mm Hg}}{1 \text{ cm Hg}} \times \frac{1 \text{ atm}}{760 \text{ mm Hg}} = 1.24 \text{ atm}$

b. $72.5 \text{ torr} \times \frac{1 \text{ atm}}{760 \text{ torr}} = 0.0954 \text{ atm}$

c. $150 \text{ mm Hg} \times \frac{1 \text{ atm}}{760 \text{ mm Hg}} = 0.197 \text{ atm}$

d. $124 \text{ kPa} \times \frac{1 \text{ atm}}{101 \text{ kPa}} = 1.23 \text{ kPa}$

5.19 a. $54.0 \text{ cm Hg} \times \frac{10 \text{ mm Hg}}{1 \text{ cm Hg}} \times \frac{1 \text{ atm}}{760 \text{ mm Hg}} \times \frac{14.7 \text{ psi}}{1 \text{ atm}} = 10.4 \text{ psi}$

b. 155 torr x $\frac{1 \text{ atm}}{760 \text{ torr}}$ x $\frac{14.7 \text{ psi}}{1 \text{ atm}}$ = 3.00 psi

c. 800 mm Hg x $\frac{1 \text{ atm}}{760 \text{ mm Hg}}$ x $\frac{14.7 \text{ psi}}{1 \text{ atm}}$ = 15 psi

d. 1.50 atm x $\frac{14.7 \text{ psi}}{1 \text{ atm}}$ = 22.1 psi

5.21 In all cases, gas particles are much farther apart than similar particles in the liquid or solid state. In most cases, particles in the liquid state are, on average, farther apart than those in the solid state. Water is the exception; liquid water's molecules are closer together than they are in the solid state.

5.23 Gases are easily compressed simply because there is a great deal of space between particles; they can be pushed closer together (compressed) because the space is available.

5.25 Gas particles are in continuous, random motion. They are free (minimal attractive forces between particles) to roam, up to the boundary of their container.

5.27 Gases exhibit more ideal behavior at low pressures. At low pressures, gas particles are more widely separated and therefore the attractive forces between particles are less. The ideal gas model assumes negligible attractive forces between gas particles.

5.29 The kinetic molecular theory states that the average kinetic energy of the gas particles increases as the temperature increases. Kinetic energy is proportional to $(\text{velocity})^2$. Therefore, as the temperature increases, the gas particle velocity increases, and the rate of mixing increases.

5.31 Boyle's law states that the volume of a gas varies inversely with the gas pressure if the temperature and the number of moles of gas are held constant.

5.33 Volume will decrease according to Boyle's law. Volume is inversely proportional to the pressure exerted on the gas.

5.35 A volume of 5 L (ordinate) corresponds to a pressure of 1 atm (abscissa).

5.37 A volume of 2 L (ordinate) corresponds to a pressure of 2.5 atm (abscissa).

$PV = k_b$

(2.5 atm) (2 L) = 5 L·atm = k_b

5.39 P_i = 1.00 atm P_f = ? atm

V_i = 20.9 L V_f = 4.00 L

$P_iV_i = P_fV_f$

$P_f = \frac{P_iV_i}{V_f}$

$P_f = \frac{(1.00 \text{ atm})(20.9 \text{ L})}{4.00 \text{ L}}$

$P_f = 5.23$ atm

5.41 Charles's law states that the volume of a gas varies directly with the absolute temperature if pressure and number of moles of gas are constant.

5.43 The Kelvin scale is the only scale that is directly proportional to molecular motion, and it is the motion that determines the physical properties of gases.

5.45 No. The volume is proportional to the temperature in K, not Celsius.

5.47 $V_i = 2.00$ L $\qquad V_f = ?$ L

$T_i = 250°C \qquad T_f = 500°C$

$\frac{V_i}{T_i} = \frac{V_f}{T_f}$

$V_f = \frac{V_iT_f}{T_i}$

$V_f = \frac{(2.00 \text{ L})(500°\text{C} + 273 \text{ K})}{(250°\text{C} + 273 \text{ K})}$

$V_f = 2.96$ L

The change in volume, $\Delta V = V_f - V_i$

$\Delta V = 2.96 \text{ L} - 2.00 \text{ L}$

$\Delta V = 0.96 \text{ L}$

5.49 $V_i = 1.25$ L $\qquad V_f = ?$ L

$T_i = 20°C \qquad T_f = 80°C$

$$\frac{V_i}{T_i}=\frac{V_f}{T_f}$$

$$V_f=\frac{V_iT_f}{T_i}$$

$$V_f=\frac{(1.25\text{ L})(80°\text{C}+273\text{ K})}{(20°\text{C}+273\text{ K})}$$

$$V_f=\frac{(1.25\text{ L})(353\text{ K})}{(293\text{ K})}$$

$V_f = 1.51$ L

5.51 $V_i = 2.00$ L $\qquad V_f = 2.20$ L

$T_i = 68°\text{F}$ $\qquad T_f = ?$

First convert °F to °C

$$T_{°C} = \frac{T_{°F}-32}{1.8} = \frac{68°\text{F}-32}{1.8} = \frac{36}{1.8} = 20°\text{C}$$

Then convert °C to K

$T_K = T_{°C} + 273 = 20 + 273 = 293$ K

Charles's law can now be applied to solve for the outdoor temperature.

$$\frac{V_i}{T_i}=\frac{V_f}{T_f}$$

$$T_f=\frac{V_fT_i}{V_i}$$

$$T_f=\frac{(2.20\text{ L})(293\text{ K})}{(2.00\text{ L})} = 322\text{ K}$$

The outdoor temperature needs to be converted from K to °F.

First, K is converted to °C.

$T_{°C} = T_K - 273 = 322 - 273 = 49°\text{C}$

Finally, °C is converted to °F.

$T_{°F} = 1.8(T_{°C}) + 32 = [(1.8)(49)] + 32 = 120°F$

5.53 Examine each effect separately:

- Volume and temperature are directly proportional; increasing T increases V.
- Volume and pressure are inversely proportional; decreasing P increases V.

Therefore, both variables work together to increase the volume.

5.55 $$\frac{P_i V_i}{T_i} = \frac{P_f V_f}{T_f}$$

$$P_f V_f T_i = P_i V_i T_f$$

$$V_f = \frac{P_i V_i T_f}{P_f T_i}$$

5.57 $P_i = 1.00$ atm $\quad P_f = 125$ atm

$V_i = 2.25$ L $\quad V_f = ?$ L

$T_i = 16°C \quad T_f = 20°C$

Using the equation derived in Question 5.55,

$$V_f = \frac{P_i V_i T_f}{P_f T_i}$$

and substituting:

$$V_f = \frac{(1.00 \text{ atm})(2.25 \text{ L})(20°C + 273 \text{ K})}{(125 \text{ atm})(16°C + 273 \text{ K})}$$

$$V_f = \frac{(1.00 \text{ atm})(2.25 \text{ L})(293 \text{ K})}{(125 \text{ atm})(289 \text{ K})}$$

$V_f = 1.82 \times 10^{-2}$ L

5.59 Step 1. Summarize the data.

$P_i = 2.00$ atm $\quad P_f = ?$ atm

$V_i = 5.00$ L $\quad V_f = 7.0$ L

$T_i = 30$ °C $\quad T_f = 40$ °C

Step 2. The combined gas law expression is:

$$\frac{P_i V_i}{T_i} = \frac{P_f V_f}{T_f}$$

Step 3. Rearrange: $P_f = \frac{P_i V_i T_f}{V_f T_i}$

Step 4. Convert °C to K.

$T_i = 30°C = 273 + 30 = 303$ K

$T_f = 40°C = 273 + 40 = 313$ K

Step 5. Substituting gives

$$P_f = \frac{(2.00 \text{ atm})(5.00 \text{ L})(313 \text{ K})}{(7.0 \text{ L})(303 \text{ K})} = \frac{3130 \text{ atm}}{2121} = 1.48 \text{ atm}$$

which rounds to 1.5 atm (2 significant figures).

5.61 Avogadro's law states that equal volumes of any ideal gas contain the same number of moles if measured at constant temperature and pressure.

5.63 $n_i = 1.00 \text{ g He x } \frac{1 \text{ mol He}}{4.00 \text{ g He}} = 0.25 \text{ mol He}$

$V_i = 1.00$ L

$n_f = 6.00 \text{ g He x } \frac{1 \text{ mol He}}{4.00 \text{ g He}} = 1.50 \text{ mol He}$

$V_f = ?$ L

$$\frac{V_i}{n_i} = \frac{V_f}{n_f}$$

$$V_f = \frac{V_i n_f}{n_i}$$

$$V_f = \frac{(1.00 \text{ L})(1.50 \text{ mol He})}{(0.25 \text{ mol He})} = 6.00 \text{ L}$$

5.65 No. One mole of an ideal gas will occupy exactly 22.4 L; however, there is no completely ideal gas and careful measurement will show different volumes for gases exhibiting varying degrees of ideality.

5.67 Standard temperature is 273K.

5.69 $PV = nRT$

$P = 5.0$ atm

$V = 4.0$ L

$T = 32^{\circ}C + 273 = 305$ K

$R = 0.0821$ L x atm/K x mol

$$n = \frac{PV}{RT} = \frac{(5.0 \text{ atm})(4.0 \text{ L})}{(0.0821 \text{ L x atm/K x mol})(305K)} = 0.799 \text{ mol} \approx 0.80 \text{ mol}$$

5.71 PV = nRT

$$n = 44.0 \text{ g CO x} \frac{1 \text{ mol}}{28.0 \text{ g } CO_2} = 1.57 \text{ mol CO}$$

$T = 273$ K

$P = 1.00$ atm

$R = 0.0821$ L x atm/K x mol

$$V = \frac{nRT}{P} = \frac{(1.57 \text{ mol})(0.0821 \text{ L x atm/K x mol})(273K)}{(1.00 \text{ atm})} = 35.2 \text{ L}$$

5.73 At STP, one mol of CO occupies 22.4L. In addition, 1 mol of CO weighs 28.01 g. Therefore,

$$d = \frac{m}{V} = \frac{28.01 \text{ g}}{22.4\ L} = 1.25 \text{ g/L at STP}$$

5.75 $PV = nRT$

$$P = 725 \text{ mmHg x } \frac{1 \text{ atm}}{760 \text{ mmHg}} = 0.954 \text{ atm}$$

$V = 7.55$ L

$T = 45°C + 273 = 318$ K

$R = 0.0821$ L x atm/K x mol

$$n = \frac{PV}{RT} = \frac{(0/945 \text{ atm})(7.55 \text{ L})}{(0.0821 \text{ L x atm/K x mol})(318K)} = 0.276 \text{ mol}$$

5.77 $P_i = 750$ torr $\quad P_f = 1.00$ atm

$V_i = 65.0$ mL $\quad V_f = ?$ L

$T_i = 22°C \quad T_f = 273$ K

$$\frac{P_iV_i}{T_i} = \frac{P_fV_f}{T_f}$$

$$V_f = \frac{P_iV_iT_f}{P_fT_i}$$

$$V_f = \frac{\left(750 \text{ torr} \times \frac{1 \text{ atm}}{760 \text{ torr}}\right)\left(650 \text{ mL} \times \frac{1 \text{ L}}{10^3 \text{ mL}}\right)(273 \text{ K})}{(1.00 \text{ atm})(22°\text{C} + 273\text{K})}$$

$$V_f = \frac{(0.987 \text{ atm})(6.5 \text{ x } 10^{-2} \text{ L})(273 \text{ K})}{(1.00 \text{ atm})(295 \text{ K})}$$

$V_f = 5.94 \text{ x } 10^{-2}$ L

5.79 $PV = nRT$

$$T = \frac{\text{PV}}{\text{nR}}$$

$$n = 1.75 \text{ g } O_2 \text{ x } \frac{1 \text{ mol } O_2}{32.0 \text{ g } O_2} = 5.47 \text{ x } 10^{-2} \text{ mol } O_2$$

$$T = \frac{(1.00 \text{ atm})(2.00 \text{ L})}{(5.47 \text{ x } 10^{-2} \text{ mol})(0.0821 \text{ L}\cdot\text{atm/K}\cdot\text{mol})}$$

$T = 445$ K

$T = 445 \text{ K} - 273 = 172°C$

5.81 $PV = nRT$

$$V = \frac{nRT}{P}$$

$n = 4.00 \text{ mol}$ $R = 0.0821 \frac{\text{L}\cdot\text{atm}}{\text{K}\cdot\text{mol}}$ $T = 27°\text{C} + 273 = 300 \text{ K}$

$$P = 8.25 \text{ torr x } \frac{1 \text{ atm}}{760 \text{ torr}} = 1.09 \text{ x } 10^{-2} \text{ atm}$$

$$V = \frac{(4.00 \text{ mol})(0.0821 \text{ L}\cdot\text{atm/K}\cdot\text{mol})(300 \text{ K})}{1.09 \text{ x } 10^{-2} \text{ atm}}$$

$$V = 9.08 \text{ x } 10^3 \text{ L}$$

5.83 Dalton's law states that the total pressure of a mixture of gases is the sum of the partial pressures of the component gases.

5.85 $P_T = p_{N_2} + p_{F_2} + p_{He}$

$P_T = 0.40 \text{ atm} + 0.16 \text{ atm} + 0.18 \text{ atm}$

$P_T = 0.74 \text{ atm}$

5.87 $P_T = p_{He} + p_{Ne}$

$p_{Ne} = P_T - p_{He}$

$p_{Ne} = 0.56 \text{ atm} - 0.27 \text{ atm}$

$p_{Ne} = 0.29 \text{ atm}$

5.89 Limitations to the ideal gas model arise from interactive forces that are present between the individual atoms or molecules of a gas. These interactive forces are present in gases composed of polar molecules. The forces increase as the temperature of the gas decreases or the pressure of the gas increases.

5.91 When temperature increases, the attractive forces present in gases decrease. CO behaves more ideally at 50 K than at 5 K.

5.93 Intermolecular forces in liquids are considerably stronger than intermolecular forces in gases. Particles are, on average, much closer together in liquids and the strength of attraction is inversely proportional to the distance of separation.

5.95 The vapor pressure of a liquid increases as the temperature of the liquid increases.

5.97 Viscosity is the resistance to flow caused by intermolecular attractive forces. Complex molecules may become entangled and not slide smoothly across one another.

5.99 All molecules exhibit London dispersion forces. This is because electrons are in constant motion in all molecules.

5.101 Only methanol exhibits hydrogen bonding. Methanol has an oxygen atom bonded to a hydrogen atom, a necessary condition for hydrogen bonding.

5.103 Propylene glycol will have the greatest viscosity in the liquid state because this polar molecule has two hydrogen bonding sites.

5.105 Solids are essentially incompressible because the average distance of separation among particles in the solid state is small. There is literally no space for the particles to crowd closer together.

5.107 a. Ionic solids - high melting temperature, brittle
b. Covalent solids - high melting temperature, hard

5.109 Beryllium. Metallic solids are good electrical conductors. Carbon forms covalent solids that are poor electrical conductors.

5.111 Mercury. Mercury is a liquid at room temperature, whereas chromium is a solid at room temperature. Liquids have higher vapor pressures than solids.

Chapter 6

Solutions

Solutions to the Odd-Numbered Questions and Problems

In-Chapter Questions and Problems

6.1 A chemical analysis must be performed in order to determine the identity of all components, a qualitative analysis. If only one component is found, it is a pure substance; two or more components indicate a true solution.

6.3 After the container of soft drink is opened, CO_2 diffuses into the surrounding atmosphere; consequently, the partial pressure of CO_2 over the soft drink decreases and the equilibrium [CO_2 (g) $\leftrightarrow$ CO_2 (aq)] shifts to the left, lowering the concentration of CO_2 in the soft drink.

6.5 Henry's Law is used to solve this problem.
According to the problem, $k = 3.1 \times 10^{-2}$ mol/(L·atm) and $P = 6.0$ atm.
Substituting,

$$M = kP = \{3.1 \times 10^{-2} \text{ mol/(L·atm)}\}(6.0 \text{ atm}) = 0.186 \text{ mol/L} \approx 0.19\ M$$

6.7 $$M_{\text{HCl}} = \frac{\text{mol HCl}}{\text{L}_{\text{solution}}}$$

Solving for mol HCl,

$$\text{mol HCl} = (M_{\text{HCl}})(\text{L}_{\text{solution}})$$

$$\text{mol HCl} = (0.250 \text{ M})(5.00 \times 10^2 \text{ mL} \times \frac{1 \text{ L}}{10^3 \text{ mL}})$$

$$\text{mol HCl} = 0.125 \text{ mol HCl}$$

6.9 Pure water has the higher freezing point. The presence of a solute decreases the freezing point.

6.11 *For the potassium chloride in the intravenous solution:*

$$\%\ (\text{m/V}) = \frac{\text{grams of solute}}{\text{milliliters of solution}} \times 100\%$$

$$\text{Grams of solute} = \frac{\%(\text{m/V})\ (\text{mL of solution})}{100\%}$$

$$\text{Grams of KCl} = \frac{0.15\ \%(\text{m/V})(1000\ \text{mL of solution})}{100\%} = 1.5\ \text{g KCl}$$

The molar mass of KCl is 74.55g/mol. Substituting,

$$1.5\ \text{g KCl} \times \frac{1\ \text{mol KCl}}{74.55\ \text{g KCl}} = 0.020\ \text{mol KCl}$$

KCl is an ionic compound and produces an electrolytic solution containing K^+ and Cl^-. One mol of KCl yields two moles of product ions. Consequently,

$$\frac{0.020\ \text{mol KCl}}{\text{L}} \times \frac{2\ \text{mol particles}}{1\ \text{mol KCl}} = 4.0 \times 10^{-2}\ \frac{\text{mol particles}}{\text{L}}$$

For the glucose in the intravenous solution:

$$\text{Grams of glucose} = \frac{5\ \%(\text{m/V})(1000\ \text{mL of solution})}{100\%} = 50\ \text{g glucose}$$

The molar mass of glucose ($C_6H_{12}O_6$) is 180 g/mol. Substituting,

$$50\ \text{g glucose} \times \frac{1\ \text{mol glucose}}{180\ \text{g glucose}} = 0.28\ \text{mol glucose}$$

Glucose is a molecular solute and a nonelectrolyte. Consequently,

$$\frac{0.28\ \text{mol glucose}}{\text{L}} \times \frac{1\ \text{mol particles}}{1\ \text{mol glucose}} = 2.8 \times 10^{-1}\ \frac{\text{mol particles}}{\text{L}} \approx 3 \times 10^{-1}\ \text{mol particles/L}$$

6.13 Step 1. The chloride ion has a 1- charge (recall that chlorine is in Group VIIA of the periodic table; hence a 1- charge on the chloride ion).

Step 2. Two conversion factors are needed to solve this problem:

meq to eq and eq to mol of chloride ion

Step 3. Using the conversion factor for meq to eq,

$$\frac{110\ \text{meq}\ Cl^-}{\text{L}} \times \frac{1\ \text{eq}\ Cl^-}{10^3\ \text{meq}\ Cl^-} = 0.110\ \text{eq}\ Cl^-/\text{L}$$

Step 4. Rearranging the conversion factor for eq to mol of chloride ion,

$$\text{eq}\ Cl^-/\text{L} = \left(\frac{\text{eq}\ Cl^-}{\text{mol}\ Cl^-}\right)(M)$$

$$M = \left(\frac{\text{eq Cl}^-}{\text{L}}\right)\left(\frac{\text{mol Cl}^-}{\text{eq Cl}^-}\right)$$

Step 5. Substituting,

$$M = \left(\frac{0.110 \text{ eq Cl}^-}{\text{L}}\right)\left(\frac{1 \text{ mol Cl}^-}{1 \text{ eq Cl}^-}\right) = 0.110 \text{ mol/L}$$

End-of-Chapter Questions and Problems

6.15 A solution is described as clear if it efficiently transmits light, showing no evidence of suspended particles. The solution does not have to be colorless to meet these conditions. (Think of cherry-flavored soft drinks.)

6.17 a. $NaNO_3$ is an electrolyte. In solution, $NaNO_3$ dissociates to Na^+ and NO_3^-.
b. $C_6H_{12}O_6$ is a nonelectrolyte. It does not dissociate into ions.
c. $FeCl_3$ is an electrolyte. In solution, FeC13 dissociates to Fe^{+3} and three Cl^-.

6.19 A true solution contains more than one substance with the tiny particles (diameter less than 1×10^{-9} m) homogeneously intermingled. A colloidal dispersion consists of solute particles distributed throughout a solvent. But, the distribution is not homogenous. Particles with diameters of 1×10^{-9} m to 2×10^{-7} m are colloids. A suspension is a heterogeneous mixture that contains particles much larger than 2×10^{-7} m.

6.21 A saturated solution is one in which undissolved solute is in equilibrium with the solution. A supersaturated solution is a solution that is more concentrated than a saturated solution. In a supersaturated solution, excess solute may remain in solution for a time. A supersaturated solution is unstable and with time, the excess solute will precipitate, and the solution will revert to a saturated solution, which is stable.

6.23 A colloidal dispersion of albumin is not completely homogenous. The colloid particles scatter light (Tyndall effect). Saline solution is completely homogenous and the dissolved NaCl ions do not scatter light.

6.25 CCl_4 is more likely to form a solution in benzene (C_6H_6). The rule "like dissolves like" suggests that CCl_4 is soluble in benzene because both CCl_4 and benzene are nonpolar. Water is polar and would be a solvent for a polar solute.

6.27 Stream temperature is much lower in early spring than mid-August. Henry's Law predicts that the concentration of dissolved oxygen in the stream is greater at lower water temperatures. Trout require oxygen and thrive in early spring.

6.29 Henry's Law is used to solve this problem.
According to the problem, $k = 1.3 \times 10^{-3}$ mol/(L atm) and $P = 25$ atm.

Substituting,

$$M = kP = \{1.3 \times 10^{-3}\ \text{mol/(L atm)}\}(25\ \text{atm}) = 0.0325\ \text{mol/L} \approx 0.033\ \text{mol/L}$$

6.31 $\%\ (\text{m/V}) = \dfrac{\text{grams of solute}}{\text{milliliters of solution}} \times 100\%$

a. $\%\ (\text{m/V}) = \dfrac{33.0\ \text{g}\ C_6H_{12}O_6}{5.00 \times 10^2\ \text{mL soln}} \times 100\%$

$\%\ (\text{m/V}) = 6.60\ \%\ C_6H_{12}O_6$

b. $\%\ (\text{m/V}) = \dfrac{20.0\ \text{g NaCl}}{\left(1.00\ \text{L soln} \times \dfrac{10^3\ \text{mL soln}}{1\ \text{L soln}}\right)} \times 100\%$

$\%\ (\text{m/V}) = 2.00\ \%\ \text{NaCl}$

6.33 $\%\ (\text{m/V}) = \dfrac{\text{grams of solute}}{\text{milliliters of solution}} \times 100\%$

a. $\%\ (\text{m/V}) = \dfrac{50.0\ \text{g ethanol}}{5.00 \times 10^2\ \text{mL soln}} \times 100\%$

$\%\ (\text{m/V}) = 10.0\%\ \text{ethanol}$

b. $\%\ (\text{m/V}) = \dfrac{50.0\ \text{g ethanol}}{\left(1.00\ \text{L soln} \times \dfrac{10^3\ \text{mL soln}}{1\ \text{L soln}}\right)} \times 100\%$

$\%\ (\text{m/V}) = 5.00\ \%\ \text{ethanol}$

6.35 a. $\%\ (\text{m/m}) = \dfrac{21.0\ \text{g NaCl}}{1.00 \times 10^2\ \text{g soln}} \times 100\%$

$\%\ (\text{m/m}) = 21.0\%\ \text{NaCl}$

b. Use the density of the sodium chloride solution as a conversion factor to calculate the volume of the sodium chloride solution.

$$5.00 \times 10^2\ \text{mL soln} \times \frac{1.12\ \text{g soln}}{1\ \text{mL soln}} = 5.60 \times 10^2\ \text{g soln}$$

Then,

$$\%\text{ (m/m)} = \frac{21.0\text{ g NaCl}}{5.60 \times 10^2\text{ g soln}} \times 100\%$$

$$\%\text{ (m/m)} = 3.75\%\text{ NaCl}$$

6.37 $$\%\text{(m/V)} = \frac{\text{grams of solute}}{\text{mL of solution}} \times 100\%$$

$$\%\text{(m/V)} = \frac{14.6\text{ g KNO3}}{75.0\text{ mL of solution}} \times 100\%$$

$$\%\text{(m/V)} = 19.47\%\text{ KNO}_3 \approx 19.5\%\text{ KNO}_3$$

6.39 $$\%\text{(m/V)} = \frac{\text{grams of solute}}{\text{mL of solution}} \times 100\%$$

$$\text{Grams of solute} = \frac{\%\text{(m/V)(mL of solution)}}{100\%}$$

$$\text{Grams of sugar} = \frac{1.00\ \%\text{(m/V)(100 mL of solution)}}{100\%}$$

Grams of sugar = 1.00 g sugar

6.41 $$\%\text{ (m/m)} = \frac{\text{grams solute}}{\text{grams solution}} \times 100\%$$

Solve for g solute,

$$\text{g solute} = \frac{\%\text{ (m/m)(g soln)}}{100\%}$$

a. $$\text{g NaCl} = \frac{[0.900\%\text{ (m/m)}](2.50 \times 10^2\text{ g soln})}{100\%} = 2.25\text{ g NaCl}$$

b. Assume that the density of the solution is 1.00 g/mL; then

$$\text{g CH}_3\text{COONa} = \frac{[1.25\%\text{ (m/m)}](2.50 \times 10^2\text{ g soln})}{100\%} = 3.13\text{ g NaC}_2\text{H}_3\text{O}_2$$

6.43 $1.0 \text{ mg } Cu^{2+} \times \frac{1 \text{ g}}{10^3 \text{ mg}} = 1.0 \times 10^{-3} \text{ g } Cu^{2+}$

$0.50 \text{ kg} \times \frac{10^3 \text{ g}}{1 \text{ kg}} = 5.0 \times 10^2 \text{ g solution}$

$\text{ppt } Cu^{2+} = \frac{\text{g } Cu^{2+}}{\text{g solution}} \times 10^3 \text{ ppt} = \frac{1.0 \times 10^{-3} \text{ g } Cu^{2+}}{5.0 \times 10^2 \text{ g solution}} \times 10^3 \text{ ppt}$

$\text{ppt } Cu^{2+} = 2.0 \times 10^{-3} \text{ ppt}$

6.45 First, ensure that both concentration units are the same; convert %(m/m) to ppm:

$100\%(m/m) = 10^6 \text{ ppm}$

$0.04\%(m/m) \times \frac{10^6 \text{ ppm}}{100\%(m/m)} = 4 \times 10^2 \text{ ppm}$

$4 \times 10^2 > 50$ ppm; therefore, the 0.04 %(m/m) solution is more concentrated.

6.47 $M = \frac{\text{mol } HNO_3}{\text{L solution}}$

$M = \frac{2.5 \text{ mol } HNO_3}{5.0 \text{ L solution}}$

$M = 0.50\ M\ HNO_3$

6.49 $M = \frac{\text{mol}}{\text{L}} = \frac{2.25 \text{ mol}}{2.50 \text{ L}} = 0.900\ M$

6.51 Laboratory managers often purchase concentrated solutions for practical reasons such as economy and conservation of storage space.

6.53 $M = \frac{\text{mol solute}}{\text{L solution}}$

Solving for mol solute, mol solute = (M)(L solution)

a. $\text{mol NaCl} = (0.100\ M)\left(2.50 \times 10^2 \text{ mL soln} \times \frac{1 \text{ L soln}}{10^3 \text{ mL soln}}\right)$

$\text{mol NaCl} = 2.50 \times 10^{-2} \text{ mol NaCl}$

and, 2.50×10^{-2} mol NaCl x $\frac{58.44 \text{ g NaCl}}{1 \text{ mol NaCl}}$ = 1.46 g NaCl

b. mol $C_6H_{12}O_6$ = $(0.200\ M)\left(2.50 \times 10^2 \text{ mL soln} \times \frac{1 \text{ L soln}}{10^3 \text{ mL soln}}\right)$

mol $C_6H_{12}O_6$ = 5.00×10^{-2} mol $C_6H_{12}O_6$

and, 5.00×10^{-2} mol $C_6H_{12}O_6$ x $\frac{180.0 \text{ g } C_6H_{12}O_6}{1 \text{ mol } C_6H_{12}O_6}$ = 9.00 g $C_6H_{12}O_6$

6.55 $M = \frac{\text{mol}}{\text{L}}$; mol = (*M*)(L); mol = (0.500 *M*)(1.75 L) = 0.875 mol

Then (0.875 mol) $\frac{(180 \text{ g glucose})}{1 \text{ mol glucose}}$ = 157.5 g glucose ≈ 158 g glucose

6.57 $M = \frac{\text{mol solute}}{\text{L solution}}$

Rearranging,

$$\text{L solution} = \frac{\text{mol solute}}{M}$$

Substituting,

$$\text{L solution} = \frac{0.133 \text{ mol sucrose}}{0.500\ M \text{ sucrose solution}} = 0.266 \text{ L}$$

6.59 $(M_1)(V_1) = (M_2)(V_2)$

$M_1 = 1.00\ M$ $M_2 = 0.100\ M$

$V_1 = ?$ $V_2 = 0.500$ L

Solve for V_1,

$$V_1 = \frac{(M_2)(V_2)}{M_1}$$

Substitute,

$$V_1 = \frac{(0.100)(0.500)}{1.00} = 5.00 \times 10^{-2} \text{ L} = 50.0 \text{ mL}$$

6.61 $(M_1)(V_1) = (M_2)(V_2)$

Solve for M_1,

$$M_1 = \frac{(M_2)(V_2)}{V_1}$$

Substitute,

$$M_1 = \frac{(2.00)(5.000 \text{ x } 10^{-1} \text{ L})}{5.00 \text{ x } 10^{-2} \text{ L}} = 20.0\ M$$

6.63 $(M_1)(V_1) = (M_2)(V_2);\ M_1 = \frac{(M_1)(V_1)}{V_2} = \frac{(0.500\ M)(50.0 \text{ mL})}{(500.0 \text{ mL})}$

$M_1 = 0.0500\ M = 5.00 \text{ x } 10^{-2}\ M$

6.65 A colligative property is a solution property that depends on the concentration of solute particles rather than the identity of the particles.

6.67 Salt is an ionic substance that dissociates in water to produce positive and negative ions. These ions (or particles) lower the freezing point of water. If the concentration of salt particles is large, the freezing point may be depressed below the surrounding temperature, and the ice would melt.

6.69 Raoult's law states that when a solute is added to a solvent, the vapor pressure of the solvent decreases in proportion to the concentration of the solute.

6.71 One mole of $CaCl_2$ produces three moles of particle in solution, whereas one mole of NaCl produces two moles of particles in solution. Therefore, a one molar $CaCl_2$ solution contains a greater number of particles than a one molar NaCl solution and will produce a greater freezing-point depression.

6.73 a. Since urea, N_2H_4CO, is a covalent compound it is a nonelectrolyte. Therefore, one mole of urea produces one mol of particles, and the molality of particles in a 1.50 *m* solution is:

$$1 \text{ x } 1.50\ m = 1.50\ m \text{ particles}$$

Using the expression for freezing-point depression: $\Delta T_f = (1.86°\text{C}/\ m)\ m$ particles

And, substituting molality of particles, $\Delta T_f = (1.86°\text{C}/\ m)\ 1.50\ m = 2.79°\text{C}$

The solution freezing point is 2.79°C below 0.0°C, the freezing point of pure water. Therefore,

Freezing Point for 1.50 *m* urea solution = 0.00°C – 2.79°C = -2.79°C

b. Since LiBr is an ionic compound, one mole of LiBr produces two mol of ions. Therefore, the molality of ions in a 1.50 *m* LiBr solution is:

$2 \times 1.50\ m = 3.00\ m$ particles

Using the expression for freezing-point depression: $\Delta T_f = (1.86°C/\ m)\ m$ particles

And, substituting molality of particles, $\Delta\ T_f = (1.86°C/\ m)\ 3.00\ m = 5.58°C$
The solution freezing point is 5.58°C below 0.00°C, the freezing point of pure water. Therefore,

Freezing Point for 1.50 *m* LiBr solution = 0.00°C – 5.58°C = -5.58°C

Overview of Questions 6.75-6.77:

The *molar* concentration of both solutions is identical. Sodium chloride is an ionic compound that dissociates in water to produce two ions for each ion pair. Therefore, the molarity of particles is $2 \times 0.50\ M = 1.0\ M$. Sucrose, in contrast, is a covalent, nondissociating solute; the number of particles and molecules is identical. Consequently, the molarity of *particles* is 0.50 *M*. Armed with this information; we can now answer each question.

6.75 *Freezing Temperature of 0.50 m NaCl Solution*

Since one mole of NaCl produces two mol of ions, the molality of ions is:

$2 \times 0.50\ m = 1.00\ m$ particles

Using the expression for freezing-point depression: $\Delta T_f = (1.86°C/\ m)\ m$ particles

And, substituting molality of particles, $\Delta\ T_f = (1.86°C/\ m)\ 1.00\ m$ particles $= 1.86°C$

The solution freezing point is 1.86°C below 0.00°C, the freezing point of pure water. Therefore,

Freezing Point for NaCl Solution = 0.00°C – 1.86°C = -1.86°C

Freezing Temperature of 0.50 m Sucrose Solution

Sucrose is a nonelectrolyte. Therefore, the concentration of particles is equal to the molality of the sucrose solution. For sucrose, the molality is:

$1 \times 0.50\ m = 0.50\ m$ particles

Using the expression for freezing-point depression: $\Delta\ T_f = (1.86°C/\ m)\ m$ particles

And, substituting molality of particles, $\Delta\ T_f = (1.86°C/\ m)\ 0.50\ m = 0.93°C$

The solution freezing point is 0.93°C below 0.00°C, the freezing point of pure water. Therefore,

Freezing Point for Sucrose Solution = 0.00°C – 0.93°C = -0.93°C

6.77 The NaCl solution would have the lower vapor pressure because it has the larger number of particles per liter. The sucrose solution, with fewer particles per liter, would have the higher vapor pressure.

6.79 A $\rightarrow$ B (more dilute to more concentrated solution)

6.81 No net flow. (Both solutions have the same concentration of particles.)

6.83 $KNO_3 \xrightarrow{H_2O} K^+ + NO_3^-$

$$\frac{2 \text{ mol particles}}{1 \text{ mol } KNO_3} \times \frac{5.0 \times 10^{-4} \text{ mol } KNO_3}{\text{L solution}} = 1.0 \times 10^{-3} \frac{\text{mol particles}}{\text{L solution}}$$

6.85 Using our definition of osmotic pressure: $\pi = MRT$

The concentration, M, must be represented as osmolarity.

$$M = \frac{0.50 \text{ mol } KNO_3}{\text{L}} \times \frac{2 \text{ mol particles}}{1 \text{ mol } KNO_3} = \frac{1.0 \text{ mol particles}}{\text{L}}$$

Now, substituting in our osmotic pressure expression:

π = (1.0 mol particles/L) x [0.0821 (L•atm)/(K•mol particles)] x 298 K = 24 atm

(Note that there are two significant figures in the answer; the initial data, the KNO_3 concentration, has only two significant figures.)

6.87 $CaCl_2$ is 3 x 0.15 M = 0.45 M particles
NaCl is 2 x 0.15 M = 0.30 M particles; 0.45 M > 0.30 M; hypertonic

6.89 Glucose is 1 x 0.15 M = 0.15 M particles
NaCl is 2 x 0.15 M = 0.30 M particles; 0.15 M < 0.30 M; hypotonic

6.91 Water is often termed the "universal solvent" because it is a polar molecule and will dissolve, at least to some extent, most ionic and polar covalent compounds. The majority of our body mass is water and this water is an important part of the nutrient transport system due to its solvent properties. This is true in other animals and plants as well. Because of its ability to hydrogen bond, water has a high boiling point and a low vapor pressure. Also, water is abundant and easily purified.

6.93

6.95 The number of particles in solution is dependent on the degree of dissociation.

6.97 In dialysis, sodium ions move from a region of high concentration to a region of low concentration. If we wish to remove (transport) sodium ions from the blood, they can move to a region of lower concentration, the dialysis solution.

6.99 The shelf life is a function of the stability of the ammonia-water solution. The ammonia can react with the water to convert to the extremely soluble and stable ammonium ion. Also, ammonia and water are polar molecules. Polar interactions, particularly hydrogen bonding, are strong and contribute to the long-term solution stability.

6.101 In the Lewis structure of water, there are two lone pairs of electrons on the oxygen atom. Remember that this causes the geometry, or shape, of the water molecule to be bent.

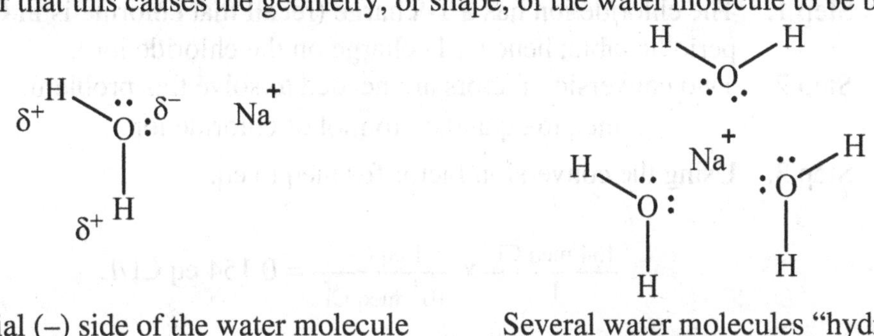

The partial (–) side of the water molecule attracts the positive sodium ion.

Several water molecules "hydrate" each sodium ion.

6.103 Polar; like dissolves like (H_2O is polar)

6.105 Elevated concentrations of sodium ions in the blood may cause confusion, stupor, or coma.

6.107 Elevated concentrations of sodium ions in the blood may occur whenever large amounts of water are lost. Diarrhea, diabetes, and certain high-protein diets are particularly problematic.

6.109

$$\frac{5.0\times10^{-2}\text{mol Ca}^{2+}}{\text{L}} \times \frac{2\text{ mol charge}}{1\text{ mol Ca}^{2+}} \times \frac{1\text{ eq Ca}^{2+}}{1\text{ mol charge}} = \frac{1.0\times10^{-1}\text{ eq Ca}^{2+}}{\text{L}}$$

6.111 a. Step 1. The sodium ion has a 1+ charge (recall that sodium is in Group IA of the periodic table; hence a 1+ charge on the sodium ion).

Step 2. Two conversion factors are needed to solve this problem:

meq to eq and eq to mol of sodium ion

Step 3. Using the conversion factor for meq to eq,

$$\frac{154 \text{ meq Na}^+}{\text{L}} \times \frac{1 \text{ eq Na}^+}{10^3 \text{ meq Na}^+} = 0.154 \text{ eq Na}^+/\text{L}$$

Step 4. Rearranging the conversion factor for eq to mol of sodium ion,

$$\text{eq Na}^+/\text{L} = \left(\frac{\text{eq Na}^+}{\text{mol Na}^+}\right)(M)$$

$$M = \left(\frac{\text{eq Na}^+}{\text{L}}\right)\left(\frac{\text{mol Na}^+}{\text{eq Na}^+}\right)$$

Step 5. Substituting,

$$M = \left(\frac{0.154 \text{ eq Na}^+}{\text{L}}\right)\left(\frac{1 \text{ mol Na}^+}{1 \text{ eq Na}^+}\right) = 0.154 \text{ mol/L}$$

b. Step 1. The chloride ion has a 1- charge (recall that chlorine is in Group VIIA of the periodic table; hence a 1- charge on the chloride ion).

Step 2. Two conversion factors are needed to solve this problem:

meq to eq and eq to mol of chloride ion

Step 3. Using the conversion factor for meq to eq,

$$\frac{154 \text{ meq Cl}^-}{\text{L}} \times \frac{1 \text{ eq Cl}^-}{10^3 \text{ meq Cl}^-} = 0.154 \text{ eq Cl}^-/\text{L}$$

Step 4. Rearranging the conversion factor for eq to mol of chloride ion,

$$\text{eq Cl}^-/\text{L} = \left(\frac{\text{eq Cl}^-}{\text{mol Cl}^-}\right)(M)$$

$$M = \left(\frac{\text{eq Cl}^-}{\text{L}}\right)\left(\frac{\text{mol Cl}^-}{\text{eq Cl}^-}\right)$$

Step 5. Substituting,

$$M = \left(\frac{0.154 \text{ eq Cl}^-}{\text{L}}\right)\left(\frac{1 \text{ mol Cl}^-}{1 \text{ eq Cl}^-}\right) = 0.154 \text{ mol/L}$$

6.113 $$\frac{40 \text{ meq K}^+}{1 \text{ L}} \times \frac{1 \text{ eq K}^+}{10^3 \text{ meq K}^+} \times \frac{1 \text{ mol charge}}{1 \text{ eq K}^+} \times \frac{1 \text{ mol K}^+}{1 \text{ mol charge}} = 4 \times 10^{-2} \frac{\text{mol K}^+}{\text{L}}$$

Chapter 7

Energy, Rate, and Equilibrium

Solutions to the Odd-Numbered Questions and Problems

In-Chapter Questions and Problems

7.1 a. Exothermic. ΔH° is *negative,* meaning that energy is *released* by the reaction.

b. Exothermic. 18.3 kcal of energy is shown as a *product* of the reaction.

7.3 He(g) has a greater entropy than Na(s). Gases have a greater degree of disorder than solids.

7.5 $\Delta G = \Delta H - T\Delta S$

Substituting signs:

$\Delta G = (+) - T(-)$

ΔG must always be positive. A positive value for ΔG indicates a nonspontaneous process.

7.7 $6.5 \times 10^2 \text{ cal} \times \frac{4.18 \text{ J}}{1 \text{ cal}} = 2.7 \times 10^3 \text{ J}$

7.9 Heat energy produced by the friction of striking the match provides the activation energy necessary for this combustion process.

7.11 If the enzyme catalyzed a process needed to sustain life, the substance interfering with that enzyme would be classified as a poison.

7.13 At a busy restaurant during lunchtime, approximately the same number of people will enter and exit the restaurant at any given moment. Throughout lunchtime, the number of people in the restaurant may be essentially unchanged, but the identity of the individuals in the restaurant is continually changing.

7.15 Measure the concentrations of products and reactants at a series of times until no further concentration change is observed.

7.17 A large value for the equilibrium constant favors product formation.

End-of-Chapter Questions and Problems

7.19 The first law of thermodynamics, the law of conservation of energy, states that the energy of the universe is constant.

7.21 An exothermic reaction is one in which energy is released during chemical change.

7.23 A fuel must release heat in the combustion (oxidation) process. The release of heat during the process characterizes an exothermic reaction.

7.25 Free energy is the combined contribution of entropy and enthalpy for a chemical reaction.

7.27 Enthalpy is a measure of heat energy.

7.29 a. Entropy increases. Conversion of a solid to a liquid results in an increase in disorder of the substance. Solids retain their shape while liquids will flow and their shape is determined by their container.
b. Entropy increases. Conversion of a liquid to a gas results in an increase in disorder of the substance. Gas particles move randomly with very weak interactions between particles, much weaker than those interactions in the liquid state.

7.31 Isopropyl alcohol quickly evaporates (liquid → gas) after being applied to the skin. Conversion of a liquid to a gas requires heat energy. The heat energy is supplied by the skin. When this heat is lost, the skin temperature drops.

7.33 $\Delta G = \Delta H - T\Delta S$

Substituting signs: $\Delta G = (-) - T(+)$

ΔG must always be negative. A negative value for ΔG indicates a spontaneous process.

7.35 Fuel value is the amount of energy per gram of food.

7.37 The temperature of the water (or solution) is measured in a calorimeter. If the reaction being studied is exothermic, released energy heats the water and the temperature increases. In an endothermic reaction, heat flows from the water to the reaction and the water temperature decreases.

7.39 Joule

7.41 Double-walled containers, used in calorimeters, provide a small airspace between the part of the calorimeter (inside wall) containing the sample solution and the outside wall, contacting the surroundings. This makes heat transfer more difficult.

7.43 The calorimetry expression is: $Q = m_s \times \Delta T_s \times SH_s$

Substituting:

$$Q = 2.00 \times 10^2 \text{ g } H_2O \times 6.00°C \times \frac{1.00 \text{ cal}}{\text{g } H_2O \text{ °C}} = 1.20 \times 10^3 \text{ cal}$$

7.45 $Q = m_s \times \Delta T_s \times SH_s$

The change in temperature is:

$\Delta T_s = T_{s\ final} - T_{s\ initial} = 25.0°C - 34.6°C = 9.6°C$

Substituting:

$Q = 2.50 \times 10^2 \text{ g } H_2O \times 9.6°C \times (1.00 \text{ cal/g}H_2O°C) = 2.4 \times 10^3 \text{ cal}$

2.4×10^3 calories are derived from 30 g of the nutrient substance (chips)

$$\frac{2.4 \times 10^3 \text{ cal}}{30 \text{ g substance}} = 80 \text{ cal/g substance}$$

fuel value = (80 cal/g substance) x (1 nutritional Cal/10^3 cal)

fuel value = 8×10^{-2} nutritional Calories/g substance

7.47 Decomposition of leaves and twigs to produce soil would be an extremely slow reaction.

7.49 The activated complex is the arrangement of reactants in an unstable transition state as a chemical reaction proceeds. The activated complex must form in order to convert reactants to products.

7.51 The rate of a reaction is the change in concentration of a reactant or product per unit time. The rate constant is the proportionality constant that relates rate and concentration. The order is the exponent of each concentration term in the rate equation.

7.53 An increase in concentration of reactants means that there are more molecules in a certain volume. The probability of collision is enhanced because they travel a shorter distance before meeting another molecule. The rate is proportional to the number of collisions per unit time.

7.55 A catalyst increases the rate of a reaction without itself undergoing change.

7.57 A catalyst speeds up a chemical reaction by facilitating the formation of the activated complex, thus lowering the activation energy, the energy barrier for the reaction.

7.59 See textbook Figure 7.11.

7.61 rate = $k[CH_4][O_2]$

If the rate constant doubles, the rate of the reaction will increase. This is because the rate is proportional to k.

7.63 Rate = $k[N_2O_4]^n$ Note: n must be experimentally determined.

7.65 The rate equation for the reaction is determined to be rate = $k[I]^2$.
It is known that the value of $k = 7.0 \times 10^9\ M^{-1}s^{-1}$. We are not given specific values for the [I], so we will begin by assuming [I] = 1 M and determine the rate. Substituting these values into the rate equation,

$$\text{Rate} = 7.0 \times 10^9\ M^{-1}s^{-1}\ [1\ M]^2 = 7.0 \times 10^9\ M/s$$

To determine the effect of doubling the concentration of I, we use [I] = 2 M and determine the rate.

$$\text{Rate} = 7.0 \times 10^9\ M^{-1}s^{-1}\ [2\ M]^2 = 28.0 \times 10^9\ M/s$$

Comparing the rates obtained, it can be concluded that doubling the concentration of I will increase the rate of the reaction four-fold.

7.67 LeChatelier's principle states that when a system at equilibrium is disturbed, the equilibrium shifts in the direction that minimizes the disturbance.

7.69 A physical equilibrium occurs between two phases of the same substance. A chemical equilibrium is a state of a chemical reaction in which the rates of the forward and reverse reactions are equal.

7.71 $K_{eq} = \frac{[\text{products}]}{[\text{reactants}]}$; A large K_{eq} indicates a large numerator; hence, products are favored.

7.73 a. *A slow reaction is an incomplete reaction* – False. A slow reaction may go to completion, but take a longer period of time.
b. *The rate of forward and reverse reactions is never the same* – False. The rate of forward and reverse reactions is equal in a dynamic equilibrium situation.

7.75 Increasing the pressure during the reaction, A(g) $\rightleftharpoons$ 2B(g), will result in an equilibrium shift to produce more of the reactant. Conversion of two moles of B to one mole of A decreases the volume. This will increase the concentration of A. The equilibrium position shifts to the left. Therefore, diagram (I) represents the system once equilibrium is reestablished.

7.77 A dynamic equilibrium has fixed concentrations of all reactants and products; these concentrations do not change with time. However, the process is dynamic because products and reactants are continuously being formed and consumed. The concentrations do not change because the *rates* of production and consumption are equal.

7.79 The position of the equilibrium addresses the question of whether products or reactants are favored. For the hypothetical equilibrium reaction: $A \rightleftharpoons B$

If the equilibrium position shifts to the left, this means that more A is formed, at the expense of B. If the equilibrium position shifts to the right, this means that more B is formed, at the expense of A.

7.81 Three factors that can shift the position of an equilibrium are concentration, heat, and pressure.

7.83 $K_{eq} = \frac{[NO_2]^2}{[N_2O_4]}$

7.85 The equilibrium constant expression for the reaction $N_2(g) + 3H_2(g) \rightleftharpoons 2NH_3(g)$ is:

$$K_{eq} = \frac{[NH_3]^2}{[N_2][H_2]^3}$$

7.87 The equilibrium constant expression for the reaction $S_2(g) + 2H_2(g) \rightleftharpoons 2H_2S(g)$ is:

$$K_{eq} = \frac{[H_2S]^2}{[S_2][H_2]^2}$$

7.89 $K_{eq} = \frac{[NH_3]^2}{[N_2][H_2]^3}$

Rearrange to solve for $[NH_3]^2$

$$[NH_3]^2 = K_{eq}[H_2]^3[N_2]$$

Then, solve for $[NH_3]$

$$[NH_3] = \{K_{eq}[H_2]^3[N_2]\}^{1/2}$$

Substitute,

$$[NH_3] = \{(0.59)(5.0 \times 10^{-3})^3 (8.0 \times 10^{-2})\}^{1/2} = 7.7 \times 10^{-5}\ M \text{ (two significant figures)}$$

7.91 a. Equilibrium shifts to the left. Increasing the temperature increases the energy, a product of the reaction.
b. No change; the number of moles of gaseous products and reactants are equal.
c. No change; a catalyst has no effect on the equilibrium position of the reaction.

7.93 a. PCl_3 increases. Addition of product shifts the equilibrium to the left, favoring reactants.

b. PCl_3 decreases. Added Cl_2 reacts with PCl_3 to produce products; the equilibrium shifts to the right.
c. PCl_3 decreases. Removal of product shifts the equilibrium to the right, favoring the formation of more products.
d. PCl_3 decreases. Decreasing the temperature removes heat from the system. Heat is a product; therefore the equilibrium shifts to the right.
e. PCl_3 remains the same. Addition of a catalyst has no effect on the equilibrium position.

7.95 To determine the effect of pressure on equilibrium concentrations, focus on the number of moles of substances in the gaseous state. An increase in pressure would shift the equilibrium to the side of the reaction that has the least number of moles of gas. For *each* mole of *reactant* (in the gaseous state), *two* moles of *product* (in the gaseous state) are formed. Therefore, this reaction equilibrium would shift to the *left* upon an increase in pressure, and the concentration of H_2, a product, would *decrease*.

7.97 $K_{eq} = \frac{[CO][H_2]}{[H_2O]}$

7.99 False. The position of equilibrium is not affected by a catalyst, only the rate at which equilibrium is attained.

7.101 Carbon dioxide is dissolved in cola. Heating shifts the equilibrium to the right.

$$CO_2\,(l) \rightleftharpoons CO_2\,(g)$$

Since gases are less soluble at elevated temperatures, the pressure build up from the carbon dioxide gas in the sealed bottle can lead to an explosion.

7.103 a. The equilibrium constant expression for the reaction

$$2SO_2(g) + O_2(g) \rightleftharpoons 2SO_3(g)$$

is: $K_{eq} = \frac{[SO_3]^2}{[SO_2]^2[O_2]}$

b. To solve this problem, the concentrations provided in the question are substituted into the equilibrium expression.

$$K_{eq} = \frac{[SO_3]^2}{[SO_2]^2[O_2]} = \frac{[0.60]^2}{[0.10]^2[0.12]} = \frac{0.36}{0.0012} = 300 \text{ or } 3.0 \times 10^2$$

(two significant figures)

7.105 According to the reaction equation, $2SO_2(g) + O_2(g) \rightleftharpoons 2SO_3(g)$, experimental conditions that would result in an increase of $SO_3(g)$ include adding $SO_2(g)$ or $O_2(g)$, or increasing the pressure.

Chapter 8

Acids and Bases and Oxidation-Reduction

Solutions to the Odd-Numbered Questions and Problems

In-Chapter Questions and Problems

8.1
a. $HClO_4$ is a Brønsted-Lowry acid.
b. HCOOH is a Brønsted-Lowry acid.
c. ClO_4^- is a Brønsted-Lowry base.
d. $C_6H_5COO^-$ is a Brønsted-Lowry base.

8.3
a. $HF(aq) + H_2O(l) \rightleftharpoons F^-(aq) + H_3O^+(aq)$

b. $C_6H_5COO^-(aq) + H_2O(l) \rightleftharpoons C_6H_5COOH\,(aq) + OH^-(aq)$

8.5
a. $HF(aq)$ (acid) + $H_2O\,(l)$ (base) $\rightleftharpoons$ $F^-(aq)$ (conjugate base) + $H_3O^+\,(aq)$ (conjugate acid)

b. $C_6H_5COO^-(aq)$ (base) + $H_2O(l)$ (acid) $\rightleftharpoons$ $C_6H_5COOH\,(aq)$ (conjugate acid) + $OH^-(aq)$ (conjugate base)

8.7 $[H_3O^+]\,[OH^-] = 1.0 \times 10^{-14}$

$$[OH^-]=\frac{1.0\times10^{-14}}{[H_3O^+]}=\frac{1.0\times10^{-14}}{2.5\times10^{-2}}=4.0\times10^{-13}\,M$$

8.9 HCl is a strong acid; consequently [HCl] = $[H_3O^+]$. For 1.0×10^{-3} *M* HCl, $[H_3O^+]$ is equal to 1.0×10^{-3} *M*.

$[H_3O^+]\,[OH^-] = 1.0 \times 10^{-14}$

$$[OH^-]=\frac{1.0\times10^{-14}}{[H_3O^+]}=\frac{1.0\times10^{-14}}{1.0\times10^{-3}}=1.0\times10^{-11}\,M$$

8.11 [acid] = 1.0×10^{-1} *M*
[conjugate base] = 1.0×10^{-1} *M*

$K_a = 1.8 \times 10^{-5}$
$pK_a = -\log K_a$
$pK_a = -\log(1.8 \times 10^{-5})$
$pK_a = 4.74$

Henderson-Hasselbalch equation: $pH = pK_a + \log\frac{[\text{conjugate base}]}{[\text{weak acid}]}$

Substituting,

$$pH = 4.74 + \log\frac{1.0 \times 10^{-1}\ M}{1.0 \times 10^{-1}\ M}$$

$pH = 4.74 + \log(1)$

$pH = 4.74$

Note that the log(1) = 0. So, the Henderson-Hasselbalch equation shows us that when the buffer concentrations of acid and conjugate base are equal, the pH will equal the pK_a.

8.13 [acid] = $2.00 \times 10^{-1}\ M$
[conjugate base] = $2.00 \times 10^{-1}\ M$
$K_a = 1.34 \times 10^{-5}$
$pK_a = -\log K_a$
$pK_a = -\log(1.34 \times 10^{-5})$
$pK_a = 4.87$

Henderson-Hasselbalch equation: $pH = pK_a + \log\frac{[\text{conjugate base}]}{[\text{weak acid}]}$

Substituting,

$$pH = 4.76 + \log\frac{2.00 \times 10^{-1}\ M}{2.00 \times 10^{-1}\ M}$$

$pH = 4.87$

8.15 The equilibrium reaction is:

$$CO_2 + H_2O \rightleftharpoons H_2CO_3 \rightleftharpoons H_3O^+ + HCO_3^-$$

An increase in the partial pressure of CO_2 is a stress on the left side of the equilibrium. The equilibrium will shift to the right in an effort to decrease the concentration of CO_2. This will cause the molar concentration of H_2CO_3 to increase.

8.17 The equilibrium reaction is:

$$CO_2 + H_2O \rightleftharpoons H_2CO_3 \rightleftharpoons H_3O^+ + HCO_3^-$$

In Question 8.15, the equilibrium shifts to the right. Therefore, the molar concentration of H_3O^+ should increase.

In Question 8.16, the equilibrium shifts to the left. Therefore, the molar concentration of H_3O^+ should decrease.

8.19 According to the equilibrium equation, $H_2CO_3(aq) + H_2O(l) \rightleftharpoons H_3O^+(aq) + HCO_3^-(aq)$, carbonic acid is the weak acid and the biocarbonate ion is its conjugate base.

Henderson-Hasselbalch expression: $pH = pK_a + \log\frac{[\text{conjugate base}]}{[\text{weak acid}]}$

Substituting,

$$pH = pK_a + \log\frac{[HCO_3^-]}{[H_2CO_3]}$$

8.21 $Ca \rightarrow Ca^{2+} + 2\,e^-$ (oxidation ½ reaction)
$S + 2\,e^- \rightarrow S^{2-}$ (reduction ½ reaction)
$Ca + S \rightarrow CaS$ (complete reaction)

8.23 Oxidizing agent: S (S oxidizes Ca.)
Reducing agent: Ca (Ca reduces S.)
Substance oxidized: Ca (Ca loses electrons.)
Substance reduced: S (S gains electrons.)

8.25 a. In order for $Cr^{3+}(aq)$ to form Cr(s) at the electrode to be electroplated, Cr^{3+} must gain 3 electrons and undergo reduction. Therefore, the electrode must have a negative charge.
b. The electrode at which reduction occurs is the cathode.
c. The reduction reaction that occurs is: $Cr^{3+}(aq) + 3\,e^- \rightarrow Cr(s)$

End-of-Chapter Questions and Problems

8.27 a. An Arrhenius acid is a substance that dissociates, producing hydrogen ions.
b. A Brønsted-Lowry acid is a substance that behaves as a proton donor.

8.29 The Brønsted-Lowry theory provides a broader view of acid-base theory than does the Arrhenius theory. Brønsted-Lowry emphasizes the role of the solvent in the dissociation process.

8.31 a. H_3O^+ is a Brønsted-Lowry acid.
b. OH^- is a Brønsted-Lowry base.
c. H_2O can behave as both acid and base (amphiprotic).

8.33 a. HOCOOH is a Brønsted-Lowry acid.
b. HCO_3^- can behave as both acid and base (amphiprotic).
c. CO_3^{2-} is a Brønsted-Lowry base.

8.35 a. $HNO_2(aq) + H_2O(l) \rightleftharpoons H_3O^+(aq) + NO_2^-(aq)$
b. $HCN(aq) + H_2O(l) \rightleftharpoons H_3O^+(aq) + CN^-(aq)$
c. $CH_3CH_2CH_2COO^-(aq) + H_2O(l) \rightleftharpoons CH_3CH_2CH_2COOH(aq) + OH^-(aq)$

8.37 $CN^- + H^+ \rightarrow HCN$
(base) (conjugate acid)

8.39 $HI \rightarrow H^+ + I^-$
(acid) (conjugate base)

8.41 HNO_3

8.43 CN^-

8.45 HF

8.47 a. HCN/CN^- and NH_4^+/NH_3
b. HCO_3^-/CO_3^{2-} and HCl/Cl^-

8.49 Concentration refers to the quantity of acid or base contained in a specified volume of solvent. Strength refers to the degree of dissociation of the acid or base.

8.51 a. Weak acid
b. Weak acid
c. Weak acid

8.53 $[H_3O^+][OH^-] = 1.0 \times 10^{-14}$

Solving for $[H_3O^+]$, $$[H_3O^+] = \frac{1.0 \times 10^{-14}}{[OH^-]}$$

a. Substituting $[OH^-] = 1.0 \times 10^{-7} M$

$$[H_3O^+] = \frac{1.0 \times 10^{-14}}{1.0 \times 10^{-7}} = 1.0 \times 10^{-7}\ M$$

b. Substituting $[OH^-] = 1.0 \times 10^{-3}\ M$

$$[H_3O^+] = \frac{1.0 \times 10^{-14}}{1.0 \times 10^{-3}} = 1.0 \times 10^{-11}\ M$$

8.55 $[H_3O^+][OH^-] = 1.0 \times 10^{-14}$

a. Solving for $[OH^-]$ and substituting $[H_3O^+] = 1.0 \times 10^{-4}\ M$,

$$[OH^-] = \frac{1.0\times10^{-14}}{[H_3O^+]} = \frac{1.0\times10^{-14}}{1.0\times10^{-4}} = 1.0\times10^{-10}\ M$$

b. Solving for [OH^-] and substituting [H_3O^+] = 1.0×10^{-2} *M*,

$$[OH^-] = \frac{1.0\times10^{-14}}{[H_3O^+]} = \frac{1.0\times10^{-14}}{1.0\times10^{-2}} = 1.0\times10^{-12}\ M$$

8.57 [H_3O^+] [OH^-] = 1.0×10^{-14}

Solving for [OH^-] and substituting [H_3O^+] = 6.0×10^{-4} *M*,

$$[OH^-] = \frac{1.0\times10^{-14}}{[H_3O^+]} = \frac{1.0\times10^{-14}}{6.0\times10^{-4}} = 1.7\times10^{-11}\ M$$

8.59 HCl is a strong acid; consequently [HCl] = [H_3O^+]. For HCl, [H_3O^+] is therefore equal to 1.0×10^{-1} *M*. CH_3COOH is a weak acid; consequently, [CH_3COOH] > [H_3O^+]. For CH_3COOH, [H_3O^+] is less than 1.0×10^{-1} *M*. Because pH = –log[H_3O^+], the solution of CH_3COOH has a greater pH than the solution of HCl. When comparing a strong acid with a weak acid, the weaker acid has a higher pH.

8.61 a. HCl is a strong acid; consequently [HCl] = [H_3O^+] and [H_3O^+] is therefore equal to 1.0×10^{-2} *M*.

pH = –log[H_3O^+]

pH = –log[1.0×10^{-2}]

pH = 2.00

b. HNO_3 is a strong acid; consequently [HNO_3] = [H_3O^+] and [H_3O^+] is therefore equal to 1.0×10^{-4} *M*.

pH = –log[H_3O^+]
pH = –log[1.0×10^{-4}]
pH = 4.00

8.63 a. pH = –log[H_3O^+] and, [H_3O^+] = 10^{-pH}

[H_3O^+] = 1.0×10^{-1} *M*

b. pH = –log [H_3O^+] and, [H_3O^+] = 10^{-pH}

[H_3O^+] = 1.0×10^{-5} *M*

8.65 KOH is a strong base; consequently [KOH] = [OH^-] and [OH^-] is therefore equal to 1.0×10^{-3} *M*.

Since [H_3O^+] [OH^-] = 1.0×10^{-14} and [H_3O^+] = $\frac{1.0 \times 10^{-14}}{[OH^-]}$

Substituting,

$$[H_3O^+] = \frac{1.0 \times 10^{-14}}{1.0 \times 10^{-3}} = 1.0 \times 10^{-11}\ M$$

and, $pH = -\log[H_3O^+]$

$pH = 11.00$

8.67 a. $pH = -\log [H_3O^+]$ and, $[H_3O^+] = 10^{-pH}$

Substituting the pH value given in the problem into the equation,

$$[H_3O^+] = 10^{-pH} = 10^{-1.30} = 5.0 \times 10^{-2}\ M$$

Then, using the expression, $[H_3O^+][OH^-] = 1.0 \times 10^{-14}$
Solve for $[OH^-]$

$$[OH^-] = \frac{1.0 \times 10^{-14}}{[H_3O^+]} = \frac{1.0 \times 10^{-14}}{5.0 \times 10^{-2}} = 2.0 \times 10^{-13}\ M$$

b. $pH = -\log [H_3O^+]$ and, $[H_3O^+] = 10^{-pH}$

Substituting the pH value given in the problem into the equation,

$$[H_3O^+] = 10^{-pH} = 10^{-9.70} = 2.0 \times 10^{-10}\ M$$

Then, using the expression, $[H_3O^+][OH^-] = 1.0 \times 10^{-14}$

Solve for $[OH^-]$

$$[OH^-] = \frac{1.0 \times 10^{-14}}{[H_3O^+]} = \frac{1.0 \times 10^{-14}}{2.0 \times 10^{-10}} = 5.0 \times 10^{-5}\ M$$

8.69 A neutralization reaction is one in which an acid and a base react to produce water and a salt (a "neutral" solution).

8.71 a. $pH = -\log [H_3O^+]$and, $[H_3O^+] = 10^{-pH}$

Substituting the pH value given in the problem into the equation,

$$[H_3O^+] = 10^{-pH} = 10^{-6.00} = 1.0 \times 10^{-6}\ M$$

Then, using the expression, $[H_3O^+][OH^-] = 1.0 \times 10^{-14}$

Solve for $[OH^-]$

$$[OH^-] = \frac{1.0 \times 10^{-14}}{[H_3O^+]} = \frac{1.0 \times 10^{-14}}{1.0 \times 10^{-6}} = 1.0 \times 10^{-8} \text{ M}$$

b. $pH = -\log [H_3O+]$ and, $[H_3O+] = 10^{-pH}$

Substituting the pH value given in the problem into the equation,

$[H_3O+] = 10^{-pH} = 10^{-5.20} = 6.3 \times 10^{-6} M$

Then, using the expression, $[H_3O+][OH^-] = 1.0 \times 10^{-14}$

Solve for $[OH^-]$

$$[OH^-] = \frac{1.0 \times 10^{-14}}{[H_3O^+]} = \frac{1.0 \times 10^{-14}}{6.3 \times 10^{-6}} = 1.6 \times 10^{-9} \text{ M}$$

c. $pH = -\log [H_3O+]$ and, $[H_3O+] = 10^{-pH}$

Substituting the pH value given in the problem into the equation,

$[H_3O+] = 10^{-pH} = 10^{-7.80} = 1.6 \times 10^{-8} M$

Then, using the expression, $[H_3O+][OH^-] = 1.0 \times 10^{-14}$

Solve for $[OH^-]$

$$[OH^-] = \frac{1.0 \times 10^{-14}}{[H_3O^+]} = \frac{1.0 \times 10^{-14}}{1.6 \times 10^{-8}} = 6.3 \times 10^{-7} \text{ M}$$

8.73 Remember, pH is a logarithmic function. We must compare $[H_3O^+]$.

$pH = -\log [H_3O+]$
$3.0 = -\log [H_3O+]$
$-3.0 = \log [H_3O+]$
$[H_3O+] = 1.0 \times 10^{-3} M$

and,

$6.0 = -\log [H_3O+]$
$-6.0 = \log [H_3O+]$
$[H_3O+] = 1.0 \times 10^{-6} M$

$$\frac{1.0 \times 10^{-3} M}{1.0 \times 10^{-6} M} = 1.0 \times 10^{3} \text{ or } 1{,}000$$

The statement is incorrect. The pH=3 solution is 1,000 times as acidic as the pH=6 solution!

8.75 $pH = -\log [H_3O+]$ and, $[H_3O+] = 10^{-pH}$

a. $pH = 5.0$
$[H_3O+] = 10^{-5.0}$
$[H_3O+] = 1 \times 10^{-5}\ M$

b. $pH = 12.0$
$[H_3O+] = 10^{-12.0}$
$[H_3O+] = 1 \times 10^{-12}\ M$

c. $pH = 5.5$
$[H_3O+] = 10^{-5.5}$
$[H_3O+] = 3.2 \times 10^{-6}\ M$

8.77 $pH = -\log [H_3O+]$

a. $[H_3O+] = 1.0 \times 10^{-6}\ M$
$pH = -\log[1.0 \times 10^{-6}]$
$pH = 6.00$

b. $[H_3O+] = 1.0 \times 10^{-8}\ M$
$pH = -\log[1.0 \times 10^{-8}]$
$pH = 8.00$

c. $[H_3O+] = 5.6 \times 10^{-4}\ M$
$pH = -\log[5.6 \times 10^{-4}]$
$pH = 3.25$

8.79 $pH = -\log[H_3O^+]$
$pH = -\log 7.5 \times 10^{-4}$
$pH = 3.12$

8.81 $[H_3O^+]\,[OH^-] = 1.0 \times 10^{-14}$

$$[H_3O^+] = \frac{1.0 \times 10^{-14}}{[OH^-]} = \frac{1.0 \times 10^{-14}}{[5.5 \times 10^{-4}]} = 1.8 \times 10^{-11}$$

$pH = -\log[H_3O^+]$
$pH = -\log 1.8 \times 10^{-11}$
$pH = 10.74$

8.83 Neutralization reactions require equal numbers of moles of H_3O^+ and OH^-. Therefore, 4 mol of HCl are needed to react with 4 mol of NaOH.

8.85 An indicator is a substance that is added to a solution and changes color as the solution reaches a certain pH. It is often used in the technique of titration to determine the equivalence point.

8.87 $HNO_3(aq) + NaOH(aq) \rightarrow H_2O(l) + NaNO_3(aq)$

8.89 $H^+(aq) + OH^-(aq) \rightarrow H_2O(l)$ or $H_3O^+(aq) + OH^-(aq) \rightarrow 2H_2O(l)$

8.91 H_2CO_3 can donate two protons.

8.93 Pertinent information for this titration includes:
Volume of the unknown acid solution, 15.00 mL
Volume of sodium hydroxide solution added, 22.50 mL
Concentration of the sodium hydroxide solution, 0.1200 *M*
From the balanced equation:

$HCl(aq) + NaO\,(aq) \rightarrow H_2O(l) + NaCl(aq)$

We know that HCl and NaOH react in a 1:1 ratio.

Then,

$$22.50 \text{ mL NaOH} \times \frac{1 \text{ L NaOH}}{10^3 \text{ mL NaOH}} = \frac{0.1200 \text{ mol NaOH}}{1 \text{ L NaOH}} = 2.700 \times 10^{-3}\ M \text{ NaOH}$$

Since HCl and NaOH undergo a 1:1 reaction

$$2.700 \times 10^{-3}\ M \text{ NaOH} \times \frac{1 \text{ mol HCl}}{1 \text{ mol NaOH}} = 2.700 \times 10^{-3}\ M \text{ HCl}$$

and, 2.700×10^{-3} *M* HCl are contained in 15.00 mL of HCl solution.

Thus,

$$\frac{2.700 \times 10^{-3} \text{ mol HCl}}{15.00 \text{ mL HCl}} \times \frac{10^3 \text{ mL HCl}}{1 \text{ L HCl}} = \frac{0.1800 \text{ mol HCl}}{\text{L HCl}} = 0.1800\ M \text{ HCl}$$

8.95 Volume of HCl is 20.00 mL
Concentration of HCl is 0.1000 *M*
Concentration of NaOH is 0.1500 *M*

$$20.00 \text{ mL HCl} \times \frac{1 \text{ L HCl}}{10^3 \text{ mL HCl}} \times \frac{0.1000 \text{ mol HCl}}{1 \text{ L HCl}} = 2.000 \times 10^{-3} \text{ mol HCl}$$

From the balanced equation (see Question 8.93), we know that HCl and NaOH react in a 1:1 ratio; therefore the number of moles HCl is equal to the number of moles NaOH, which is equal to 2.000 x 10^{-3} mol NaOH.

and,

$$2.000 \times 10^{-3} \text{ mol NaOH} \times \frac{1 \text{ L NaOH}}{0.1500 \text{ mol NaOH}} \times \frac{10^3 \text{ mL NaOH}}{1 \text{ L NaOH}} = 13.33 \text{ mL NaOH}$$

8.97 Step 1. $H_2CO_3(aq) + H_2O(l) \rightleftharpoons H_3O^+(aq) + HCO_3^-(aq)$

Step 2. $HCO_3^-(aq) + H_2O(l) \rightleftharpoons H_3O^+(aq) + CO_3^{2-}(aq)$

8.99 a. NH_3 is a weak base.
NH_4Cl is a salt formed from NH_3.
Therefore, NH_3 and NH_4Cl can form a buffer solution.

b. HNO_3 is a strong acid; strong acids are not suitable for buffer preparation; they are completely dissociated.
Therefore, HNO_3 and KNO_3 cannot form a buffer solution.

8.101 HCl/NaCl is not a buffer solution. It is a strong acid and a salt. The NaOH would produce a significant pH change. On the other hand, CH_3COOH/CH_3COONa is a buffer solution because it is a weak acid and its salt. Therefore, it would resist significant pH change upon addition of a strong acid.

8.103 The equilibrium reaction is:

$CO_2 + H_2O \rightleftharpoons H_2CO_3 \rightleftharpoons H_3O^+ + HCO_3^-$

A situation of high blood CO_2 levels and low pH is termed acidosis. A high concentration of CO_2 is a stress on the left side of the equilibrium. The equilibrium will shift to the right in an effort to decrease the concentration of CO_2. This will cause the molar concentration of carbonic acid (H_2CO_3) to increase.

8.105 a. Addition of strong acid is equivalent to adding H_3O^+. This is a stress on the right side of the equilibrium and the equilibrium will shift to the left. Consequently the $[CH_3COOH]$ increases.

b. Water, in this case, is a solvent and does not appear in the equilibrium expression. Hence, it does not alter the position of the equilibrium.

8.107 $K_a = \frac{[H_3O^+][\text{conjugate base}]}{[\text{acid}]}$

$$[H_3O^+] = \frac{[\text{acid}]K_a}{[\text{conjugate base}]}$$

$$[H_3O^+] = \frac{[0.200]5.80 \times 10^{-7}}{[0.500]}$$

$[H_3O^+] = 2.32 \times 10^{-7}\ M$

8.109 CH_3COO^-, a conjugate base, reacts with added H_3O^+ to maintain pH.

8.111 $pH = pK_a + \log \frac{[\text{conjugate base}]}{[\text{weak acid}]}$; $pH = -\log 1.8 \times 10^{-5} + \log \frac{[1.0]}{[1.0]} = 4.74$

8.113 $pH = pK_a + \log \frac{[\text{conjugate base}]}{[\text{weak acid}]}$; $7.40 = -\log 4.5 \times 10^{-7} + \log \frac{[HCO_3^-]}{[H_2CO_3]}$

$7.40 = 6.35 + \log \frac{[HCO_3^-]}{[H_2CO_3]}$

$1.05 = \log \frac{[HCO_3^-]}{[H_2CO_3]}$

Taking the antilog of both sides,

$11.2 = \frac{[HCO_3^-]}{[H_2CO_3]}$

8.115 During an oxidation process in an oxidation-reduction reaction, the species oxidized *loses* electrons.

8.117 During an oxidation-reduction reaction, the species *oxidized* is the reducing agent.

8.119

Cl_2 +	$2KI$	$\rightarrow 2KCl + I_2$
substance reduced	substance oxidized	
oxidizing agent	reducing agent	

8.121 $2I^- \rightarrow I_2 + 2e^-$ (oxidation ½ reaction)
$Cl_2 + 2e^- \rightarrow 2Cl^-$ (reduction ½ reaction)

8.123 An oxidation-reduction reaction must take place to produce electron flow in a voltaic cell.

8.125 Storage battery

Chapter 9

The Nucleus, Radioactivity, and Nuclear Medicine

Solutions to the Odd-Numbered Questions and Problems

In-Chapter Questions and Problems

9.1 Gamma radiation is very high-energy electromagnetic radiation. Other forms of electromagnetic radiation (in descending order of energy) are X-ray, ultraviolet, visible, infrared, microwave, and radiowave. [See Chapter 2, *GREEN CHEMISTRY: Practical Applications of Electromagnetic Radiation*, for more information.]

9.3 The mass number of neodymium is 144. Therefore, the sum of the mass numbers of the products must also be 144, and the unknown product must have a mass number of 140.

$^{144}_{60}\text{Nd} \rightarrow ? + ^{4}_{2}\text{He}$, $^{4}_{2}\text{He}$ is the α particle

Likewise, the atomic number of Nd is 60, and the sum of the unknown atom number plus the atomic number of the alpha particle must be 60. The unknown atomic number must be 58, because [60 = 58 + 2]. From the periodic table, it can be determined that the element with the atomic number 58 is cesium. The complete equation is:

$$^{144}_{60}\text{Nd} \rightarrow ^{140}_{58}\text{Ce} + ^{4}_{2}\text{He}$$

9.5 Iodine-131 decays by beta emission. Therefore, the balanced nuclear equation is:

$$^{131}_{53}\text{I} \rightarrow ^{131}_{54}\text{Xe} + ^{0}_{-1}\text{e}$$

9.7 A positron has a positive charge, and a beta particle has a negative charge.

9.9 $$200\ \mu\text{g} \xrightarrow[\text{half-life}]{\text{first}} 100\ \mu\text{g} \xrightarrow[\text{half-life}]{\text{second}} 50\ \mu\text{g} \xrightarrow[\text{half-life}]{\text{third}} 25\ \mu\text{g}$$

So, $3 \text{ half-lives} \times \dfrac{67 \text{ hours}}{1 \text{ half-life}} = 201 \text{ hours}$

9.11 The half-life of technetium-99m is 6 hours. The number (n) of half-lives elapsed is:

$$n = 12 \text{ hours} \times \frac{1 \text{ half-life}}{6 \text{ hours}} = 2 \text{ half-lives}$$

Then, assume that the original amount is x g.

$$x\ g \xrightarrow{\text{first half-life}} \frac{x}{2}\ g \xrightarrow{\text{second half-life}} \frac{x}{4}\ g$$

Therefore, 1/4 of the radioisotope remains after 2 half-lives, 12 hours.

9.13 Isotopes with short half-lives release their radiation rapidly. There is much more radiation per unit time observed with short half-life substances; hence, the signal is stronger and the sensitivity of the procedure is enhanced.

9.15 The rem takes into account the relative biological effect of the radiation in addition to the quantity of radiation. This provides a more meaningful estimate of potential radiation damage to human tissue.

End-of-Chapter Questions and Problems

9.17 Natural radioactivity is the spontaneous decay of a nucleus to produce high-energy particles or rays.

9.19 Alpha particles contain two protons and two neutrons; hence they have a charge of 2+.

9.21 A beta particle is an electron with a 1– charge.

9.23 Charge, $\alpha = 2+$, $\beta = 1-$

Mass, $\alpha = 4$ amu, $\beta = 0.000549$ amu

Velocity, $\alpha = 10\%$ of the speed of light, $\beta = 90\%$ of the speed of light

9.25 Chemical reactions involve joining, separating and rearranging atoms; valence electrons are critically involved. Nuclear reactions only involve changes in nuclear composition.

9.27 The nuclear symbol for an alpha particle is $^{4}_{2}He$.

9.29 A helium atom has two electrons; an α particle has no electrons.

9.31 Alpha particles, beta particles, and positrons are matter; gamma radiation is pure energy. Alpha particles are large and relatively slow moving. They are the least energetic and least penetrating. Beta particles and positrons are smaller, faster and more penetrating than alpha particles. Gamma radiation moves at the speed of light, is highly energetic, and is most penetrating.

9.33 The nuclear symbol for nitrogen-15 is $^{15}_{7}N$.

9.35 The nuclear symbol for uranium-235 is $^{235}_{92}U$.

9.37 Mass number – atomic number = number of neutrons

So $^{1}_{1}H$ has 1 – 1 = 0 neutrons, $^{2}_{1}H$ has 2 –1 = 1 neutron, $^{3}_{1}H$ has 3 – 1 = 2 neutrons. Each isotope contains one proton.

9.39 $^{60}_{27}Co \rightarrow ^{60}_{28}Ni + ^{0}_{-1}\beta + \gamma$

9.41 $^{23}_{11}Na + ^{2}_{1}H \rightarrow ^{24}_{11}Na + ^{1}_{1}H$

9.43 $^{218}_{92}U \rightarrow ^{214}_{90}Th + ^{4}_{2}He$, $^{4}_{2}He$ is the α particle

9.45 $^{140}_{55}Cs \rightarrow ^{140}_{56}Ba + ^{0}_{-1}e$

9.47 $^{209}_{83}Bi + ^{54}_{24}Cr \rightarrow ^{262}_{107}Bh + ^{1}_{0}n$

9.49 $^{27}_{12}Mg \rightarrow ^{27}_{13}Al + ^{0}_{-1}e$

9.51 $^{12}_{7}N \rightarrow ^{12}_{6}C + ^{0}_{+1}e$

9.53 Americium-241 decays by alpha decay. Therefore, the balanced nuclear equation is:

$$^{241}_{95}Am \rightarrow ^{237}_{93}Np + ^{4}_{2}He, \quad ^{4}_{2}He \text{ is the } \alpha \text{ particle}$$

9.55
- Nuclei for light atoms tend to be most stable if their neutron/proton ratio is close to 1.
- Nuclei with more than 84 protons tend to be unstable.
- Isotopes with a "magic number" of protons or neutrons (2, 8, 20, 50, 82, or 126 protons or neutrons) tend to be stable.
- Isotopes with even numbers of protons or neutrons tend to be more stable.

9.57 The half-life of sodium-24 is 15 h. The number (n) of half-lives elapsed is:

$$n = 225 \text{ h x } \frac{1 \text{ half-life}}{15 \text{ h}} = 15 \text{ half-lives (2 significant figures)}$$

9.59 $^{20}_{8}O$; Oxygen – 20 has 20 – 8 = 12 neutrons, an n/p of 12/8, or 1.5. The n/p is probably too high for stability even though it does have a "magic number" of protons and an even number of protons and neutrons.

9.61 $^{48}_{24}Cr$; Chromium – 48 has 48 – 24 = 24 neutrons, an n/p of 24/24, or 1.0. It also has an even number of protons and neutrons. It would probably be stable.

9.63 The half-life of iodine-131 is 8.1 days. The number (n) of half-lives elapsed is:

$$n = 24 \text{ days} \times \frac{1 \text{ half-life}}{8.1 \text{ days}} = 3.0 \text{ half-lives (2 significant figures)}$$

Then,

$$3.2 \text{ mg} \xrightarrow[\text{half-life}]{\text{first}} 1.6 \text{ mg} \xrightarrow[\text{half-life}]{\text{second}} 0.80 \text{ mg} \xrightarrow[\text{half-life}]{\text{third}} 0.40 \text{ mg}$$

Therefore, 0.40 mg of iodine-131 remain after 24 days.

9.65 The half-life of iron-59 is 45 days. The number (n) of half-lives elapsed is:

$$n = 135 \text{ days} \times \frac{1 \text{ half-life}}{45 \text{ days}} = 3.0 \text{ half-lives}$$

Then,

$$100 \text{ mg} \xrightarrow[\text{half-life}]{\text{first}} 50 \text{ mg} \xrightarrow[\text{half-life}]{\text{second}} 25 \text{ mg} \xrightarrow[\text{half-life}]{\text{third}} 13 \text{ mg}$$

Therefore, 13 mg of iron-59 remain after 135 days.

9.67 Radiocarbon dating is a process used to determine the age of objects. The ratio of the masses of the stable isotope, carbon-12, and unstable isotope, carbon-14, is measured. Using this value and the half-life of carbon-14, the age of the coffin may be calculated.

9.69 Fission splits nuclei to produce energy.

9.71 a. The fission process involves the breaking down of large, unstable nuclei into smaller, more stable nuclei. This process releases some of the binding energy in the form of heat and/or light.

b. The heat generated during the fission process could be used to generate steam, which is then used to drive a turbine to create electricity.

9.73 ${}^{3}_{1}H + {}^{1}_{1}H \rightarrow {}^{4}_{2}He + \text{energy}$

9.75 A "breeder" reactor creates the fuel that can be used by a conventional fission reactor during its fission process.

9.77 Chain reaction refers to the reaction in a fission reactor that involves neutron production and causes subsequent reactions accompanied by the production of more neutrons in a continuing process.

9.79 High operating temperatures is the greatest barrier to development of fusion reactors.

9.81 Radiation therapy provides sufficient energy to destroy molecules critical to the reproduction of cancer cells.

9.83 Natural radioactivity is a spontaneous process; artificial radioactivity is nonspontaneous and results from a nuclear reaction that produces an unstable nucleus.

9.85 a. Technetium-99m is used to study the heart (cardiac output, size, and shape), kidney (follow-up procedure for kidney transplant), and liver and spleen (size, shape, and presence of tumors).
b. Xenon-133 is used to locate regions of reduced ventilation and presence of tumors in the lung.

9.87 $^{108}_{47}Ag + ^{4}_{2}He \rightarrow ^{112}_{49}In$

$^{112}_{49}In$ is the intermediate isotope of indium.

9.89 Background radiation, radiation from natural sources, is emitted by the sun as cosmic radiation, and from naturally radioactive isotopes found throughout our environment.

9.91 The level decreases. Radiation level is inversely proportional to the square of the distance from the source.

9.93 Yes, it would lead to a positive effect. Potential damage is often directly proportional to the time of exposure.

9.95 Yes, it would lead to a positive effect. The operator of the robotic device could be located far from the source, no physical contact with the source is necessary, and barriers of lead or other shielding can isolate the control and the robot.

9.97 Yes. Concrete has a higher density than wood and thus serves as a better radiation shield.

9.99 Relative biological effect is a measure of the damage to biological tissue caused by different forms of radiation.

9.101 a. The curie is the amount of radioactive material needed to produce 3.7×10^{10} atomic disintegrations per second.
b. The roentgen is the amount of radioactive material needed to produce 2×10^{9} ion-pairs when passing through 1 cc of air at 0°C.
c. The becquerel is the amount of radioactive material needed to produce 1 atomic disintegration per second.

9.103 A film badge detects gamma radiation by darkening photographic film in proportion to the amount of radiation exposure over time. Badges are periodically collected and evaluated for their level of exposure. This mirrors the level of exposure of the personnel wearing the badges.

Chapter 10

An Introduction to Organic Chemistry: The Saturated Hydrocarbons

Solutions to the Odd-Numbered Questions and Problems

In-Chapter Questions and Problems

10.1 Passive studying includes reading over notes and passages from the text and perhaps memorizing material from lecture. It does not test your knowledge and will not prepare you for critical thinking questions that may arise on an exam. Active studying involves being engaged in the material both mentally and physically. Active study involves rewriting notes, making outlines of key concepts, creating study tools such as flash cards or concept maps. These are strategies that enable you to think deeply about the material and test your knowledge of the concepts.

10.3 The student could test the solubility of the substance in water and in an organic solvent, such as hexane. Solubility in hexane would suggest an organic substance; whereas solubility in water would indicate an inorganic compound. The student could also determine the melting and boiling points of the substance. If the melting and boiling points are very high, an inorganic substance would be suspected.

10.5 A primary carbon is directly bonded to one other carbon. A secondary carbon is bonded to two other carbons. A tertiary carbon is bonded to three other carbons. A quaternary carbon is bonded to four other carbons.

```
                                                                   1°H
a.                     b.     1°H                       c.         H-C-H
     2° 2°                   H-C-H                                  3° | 2°
   H H H H                     |                             H H  |  H H
 H-C-C-C-C-H              H    |  H                         H-C-C-C-C-C-H
 1°H H H H 1°             H-C-C-C-H                         1° H   H H H 1°
                          1° H 4° H 1°                             3°
                               |                                   |
                             H-C-H                               H-C-H
                              1°H                                1° H
```

10.7 a. The monobromination of propane will produce two products, as shown in the following two equations:

$$CH_3CH_2CH_3 + Br_2 \xrightarrow{\text{Light or heat}} CH_3CH_2CH_2Br + HBr$$

$$CH_3CH_2CH_3 + Br_2 \xrightarrow{\text{Light or heat}} CH_3CHBrCH_3 + HBr$$

b. The monochlorination of butane will produce two products, as shown in the following two equations:

$$CH_3CH_2CH_2CH_3 + Cl_2 \xrightarrow{\text{Light or heat}} CH_3CH_2CH_2CH_2Cl + HCl$$

$$CH_3CH_2CH_2CH_3 + Cl_2 \xrightarrow{\text{Light or heat}} CH_3CH_2CHClCH_3 + HCl$$

c. The monochlorination of cyclobutane:

$$\square + Cl_2 \xrightarrow{\text{Light}} \underset{\square}{\overset{Cl}{|}} + HCl$$

d. The monobromination of pentane will produce three products as shown in the following equations:

$$CH_3CH_2CH_2CH_2CH_3 + Br_2 \xrightarrow{\text{Light or heat}} CH_3CH_2CH_2CH_2CH_2Br + HBr$$

$$CH_3CH_2CH_2CH_2CH_3 + Br_2 \xrightarrow{\text{Light or heat}} CH_3CH_2CH_2CHBrCH_3 + HBr$$

$$CH_3CH_2CH_2CH_2CH_3 + Br_2 \xrightarrow{\text{Light or heat}} CH_3CH_2CHBrCH_2CH_3 + HBr$$

End-of-Chapter Questions and Problems

10.9 There are literally millions of organic compounds. It is impossible to memorize such a large number of molecules, including their names, structures, properties, and the chemical reactions they undergo. By learning the rules of nomenclature, the general features of each of the classes of organic molecules, and the types of reactions those classes of molecules undergo, it is possible to master the discipline of organic chemistry.

10.11 Flash cards are an active studying strategy. You must be engaged with the information to create the flash cards. Using them gives you an opportunity to learn the material and test your knowledge and understanding.

10.13 The number of organic compounds is nearly limitless because carbon forms stable covalent bonds with other carbon atoms in a variety of different patterns. In addition, carbon can form stable bonds with other elements and functional groups, producing many families of organic compounds, including alcohols, aldehydes, ketones, esters, ethers, amines, and amides. Finally, carbon can form double or triple bonds with other carbon atoms to produce organic molecules with different properties.

10.15 Because ionic substances often form three-dimensional crystals made up of many positive and negative ions, they generally have much higher melting and boiling points than covalent compounds.

10.17 a. LiCl > H_2O > CH_4

LiCl has the highest boiling point because it is an ionic compound. Water has the second highest boiling point because the water molecules can hydrogen bond with other water molecules. Methane has the lowest boiling point because it has the weakest intermolecular forces between molecules.

b. NaCl > C_3H_8 > C_2H_6

NaCl has the highest boiling point because it is an ionic compound. Both C_3H_8 and C_2H_6 are nonpolar compounds. Because there are more carbons in C_3H_8, it has a higher molecular mass and thus, a higher boiling point than C_2H_6.

10.19 a. LiCl would be a solid; H_2O would be a liquid; and CH_4 would be a gas. LiCl is a solid at room temperature because it is an ionic compound and there are strong attractive forces between the positive and negative ions in the three-dimensional crystal lattice structure. Water has the second highest boiling point because it is polar, and the water molecules can hydrogen bond with one another. Hydrogen bonds are not as strong as ionic attractions, thus water is a liquid at room temperature. Methane has the weakest intermolecular forces because it is a nonpolar compound. Because of its small size, it has very low London dispersion forces, and for this reason it is a gas at room temperature.

b. NaCl would be a solid; both C_3H_8 and C_2H_6 would be gases. NaCl has the strongest intermolecular forces because it is an ionic compound, and there are strong attractive forces between the positive and negative ions in the three-dimensional lattice structure. Both C_3H_8 and C_2H_6 are nonpolar compounds. In addition, both of these compounds have fewer than four carbons and as a result, have low London dispersion forces. Therefore, both C_3H_8 and C_2H_6 are gases at room temperature.

10.21 a. Water-soluble inorganic compounds are good electrolytes because they dissociate into ions in water. The ions conduct an electrical charge.
b. Inorganic compounds exhibit ionic bonding.
c. Organic compounds have lower melting points.
d. Inorganic compounds are more likely to be water soluble.
e. Organic compounds are flammable.

10.23 a. Pristane has the molecular formula $C_{19}H_{40}$. This can be determined from the name 2,6,10,14-tetramethylpentadecane. It can be concluded from the name that there are 19 carbons that make up pristine including a parent chain of 15 carbons and four methyl group carbons. The equation C_nH_{2n+2} is then used to determine the number of hydrogens that complete the molecular formula.

b. The chemical name 2,6,10,14-tetramethylpentadecane is used to draw both the line structure and the condensed formulas for pristine.

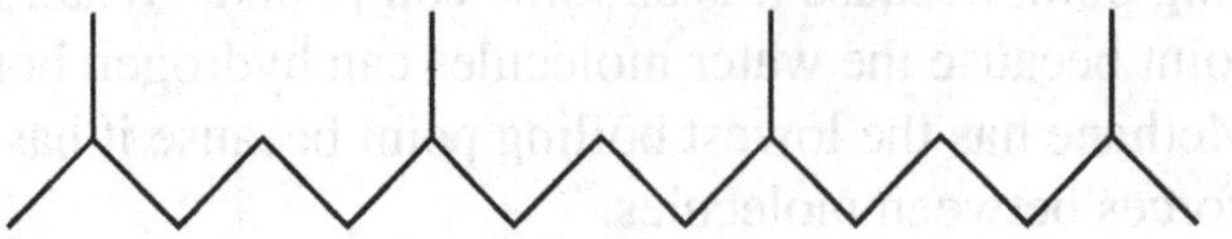

$CH(CH_3)_2(CH_2)_3CH(CH_3)(CH_2)_3CH(CH_3)(CH_2)_3CH(CH_3)_2$

c. The molecular mass of pristine can be calculated from the $C_{19}H_{40}$.

19 C	x	12.01	=	228.19
40 H	x	1.008	=	40.32
				268.51 g/mol

10.25 The condensed formula shows all the atoms in a molecule and places them in a sequential order that indicates which atoms are bonded to which.

a. $CH_3CH_2CH(CH_3)_2$ b. $CH_3CH_2C(CH_3)_2(CH_2)_2CH(CH_3)_2$

c. $CH_3CH_2C\ (CH_3)_2(CH_2)_3CH(CH_3)CH(CH_3)_2$

10.27 In line formulas, it is assumed that there is a carbon atom at any location where two or more lines intersect. It is also assume that there is a carbon at the end of any line and that each carbon in the structure is bonded to the correct number of hydrogen atoms.

a. 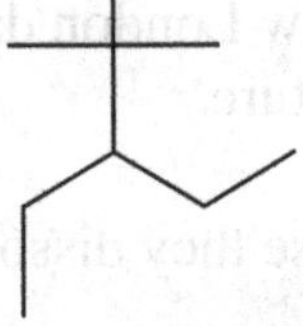 b. 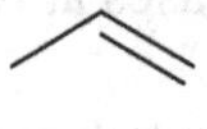 c.

10.29 a. $(CH_3)_3CCH(CH_2CH_3)_2$ b. CH_3CHCH_2 c. $CH_3CH_2CH_3$

10.31 In line formulas, it is assumed that there is a carbon atom at any location where two or more lines intersect. It is also assume that there is a carbon at the end of any line and that each carbon in the structure is bonded to the correct number of hydrogen atoms.

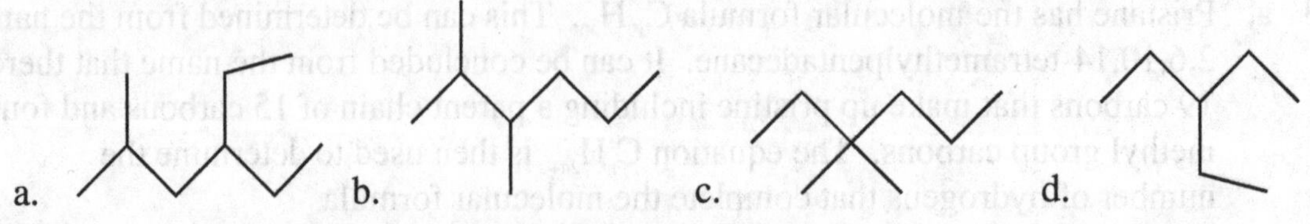

10.33 In line formulas, it is assumed that there is a carbon atom at any location where two or more lines intersect. It is also assumed that there is a carbon at the end of any line and that each carbon in the structure is bonded to the correct number of hydrogen atoms.

These line formulas can be determined from the molecular formulas that are provided.

a. Tricosane. This compound has twenty-three carbons.

Pentacosane. This compound has twenty-five carbons.

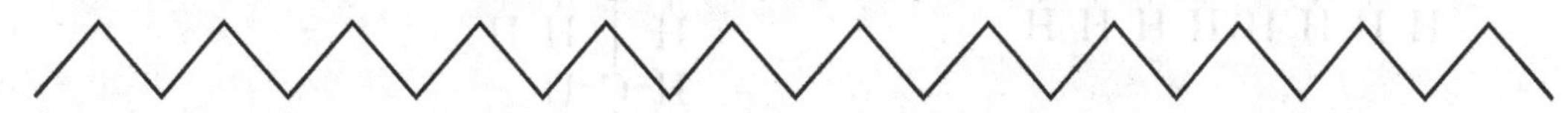

Heptacosane. This compound has twenty-seven carbons.

b. The molar mass of tricosane can be calculated from the molecular formula, $C_{23}H_{48}$.

23 C	x	12.01	=	276.23
48 H	x	1.008	=	48.384
				324.61 g/mol

The molar mass of pentacosane can be calculated from the molecular formula, $C_{25}H_{52}$.

25 C	x	12.01	=	300.25
52 H	x	1.008	=	52.416
				352.67 g/mol

The molar mass of heptacosane can be calculated from the molecular formula, $C_{27}H_{56}$.

27 C	x	12.01	=	324.27
56 H	x	1.008	=	56.448
				380.72 g/mol

10.35 The condensed formula shows all the atoms in a molecule and places them in a sequential order that indicates which atoms are bonded to which.

a. $CH_3CH(CH_3)CH(CH_3)CH_2CH_3$ b. $CH_3(CH_2)_3CH_3$

```
          H
          |
        H-C-H
    H     |    H    H    H
    |     |    |    |    |
H---C-----C----C----C----C---H
    |     |    |    |    |
    H     H    |    H    H
             H-C-H
               |
               H

    H    H    H    H    H
    |    |    |    |    |
H---C----C----C----C----C---H
    |    |    |    |    |
    H    H    H    H    H
```

10.37 The structural formula shows each atom and bond in a molecule.

a. 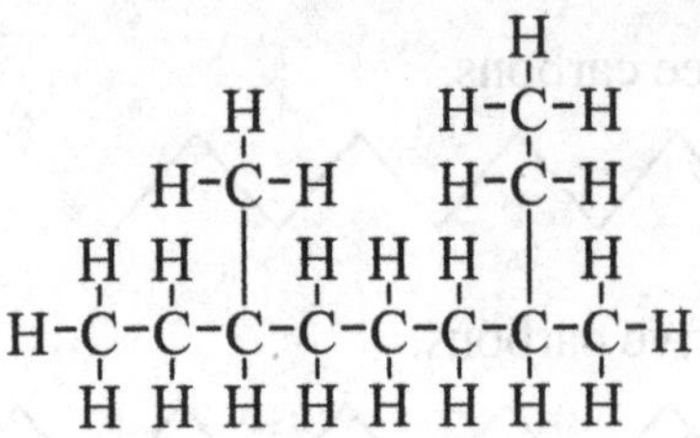
```
                    H
          H       H-C-H
        H-C-H     H-C-H
      H H |  H H H |  H
    H-C-C-C-C-C-C-C-C-H
      H H H H H H H H
```

b. 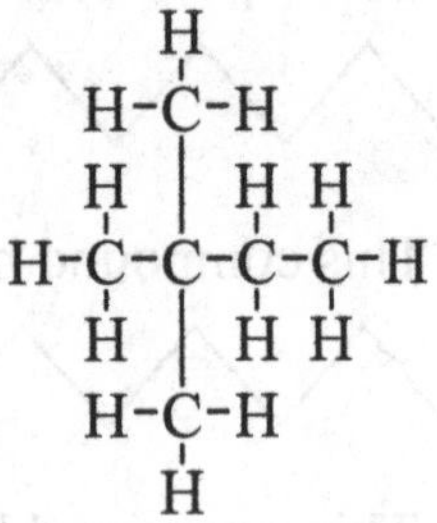
```
          H
        H-C-H
      H |  H H
    H-C-C-C-C-H
      H |  H H
        H-C-H
          H
```

10.39 a. ROH is a hydroxyl group (alcohol).

b. RNH_2 is an amino group (amine).

c. RCHO is a carbonyl group (aldehyde).

d. RCOOH is a carboxyl group (carboxylic acid).

e. RCOOR' is an ester group (ester).

f. ROR' is an ether group (ether).

g. RI is a halide (halogen atom).

10.41 a. An alkane: C_nH_{2n+2}

b. An alkyne: C_nH_{2n-2}

c. An alkene: C_nH_{2n}

d. A cycloalkane: C_nH_{2n}

e. A cycloalkene: C_nH_{2n-2}

10.43 Alkanes have only carbon-to-carbon and carbon-to-hydrogen single bonds, as in the molecule ethane:

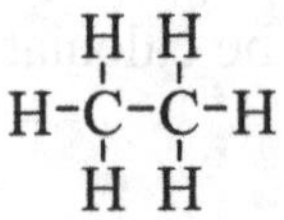

```
  H H
H-C-C-H
  H H
```

Alkenes have at least one carbon-to-carbon double bond, as in the molecule ethene:

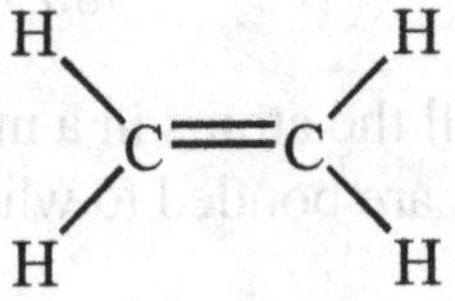

```
H       H
 \     /
  C = C
 /     \
H       H
```

Alkynes have at least one carbon-to-carbon triple bond, as in the molecule ethyne:

$$H-C\equiv C-H$$

10.45 Lisinopril

Amino group

H_2N

Carboxyl group

HO O

N

NH

Amino group

O

OH

O

Carboxyl group

Amide group

Aromatic ring

10.47 Aspartame:

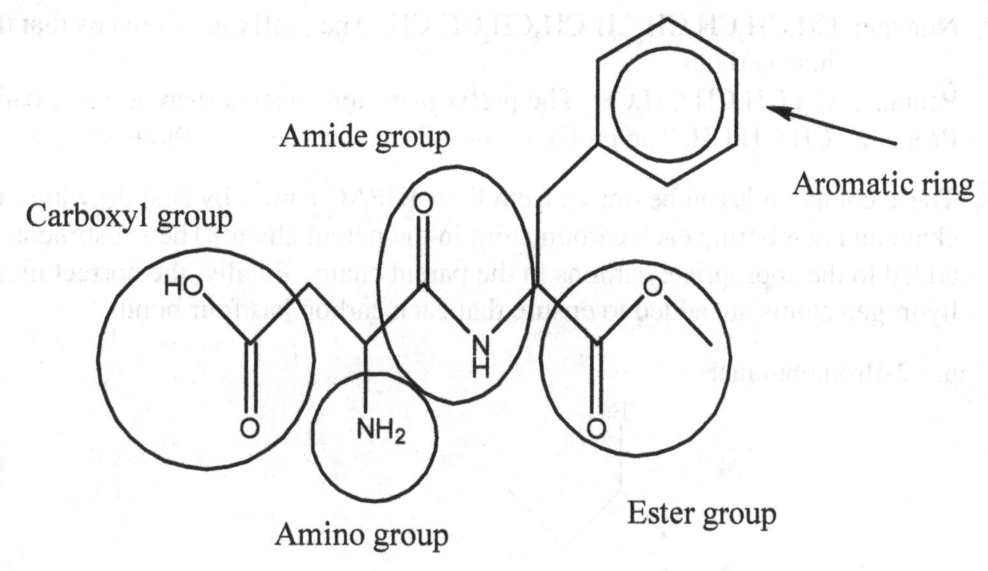

10.49 van der Waals forces are the attractive forces between neutral molecules. They include dipole-dipole attractions and London dispersion forces.

10.51 London dispersion forces result from the attraction of two molecules that experience short-lived dipoles as a result of transient shifts in the electron cloud. Larger molecules with more electrons exhibit a stronger the attraction. As a result, they will have higher melting and boiling points.

10.53 Hydrocarbons are nonpolar molecules, and hence are not soluble in water.

10.55 a. heptane > hexane > butane > ethane

The names of these hydrocarbons tells us how many carbons are present in the parent chain. Boiling points decrease as the length of the parent chain decreases because of the decrease in London dispersion forces between the hydrocarbons as they decrease in molar mass and length.

b. $CH_3CH_2CH_2CH_2CH_2CH_2CH_2CH_2CH_3$ > $CH_3CH_2CH_2CH_2CH_3$ > $CH_3CH_2CH_3$

Boiling points increase as the length of the parent chain increases because of the increase in London dispersion forces between the hydrocarbons as they increase in molar mass and length.

10.57 a. Heptane and hexane would be liquid at room temperature; butane and ethane would be gases. This can be predicted based on the name of these hydrocarbons because the name tells how many carbons are in the parent chain. Hydrocarbons with four or fewer carbons are gases at room temperature.

b. $CH_3CH_2CH_2CH_2CH_2CH_2CH_2CH_2CH_3$ and $CH_3CH_2CH_2CH_2CH_3$ would be liquids at room temperature; $CH_3CH_2CH_3$ would be a gas. Hydrocarbons with four or fewer carbons are gases at room temperature.

10.59 Nonane: $CH_3CH_2CH_2CH_2CH_2CH_2CH_2CH_2CH_3$ The prefix *non-* tells us that there are nine carbons.

Pentane: $CH_3CH_2CH_2CH_2CH_3$ The prefix *pent-* tells us that there are five carbons.

Propane: $CH_3CH_2CH_3$ The prefix *prop-* tells us that there are three carbons.

10.61 These compounds can be drawn from their IUPAC names by first drawing the parent chain and numbering each carbon atom in the parent chain. The substituents are then added to the appropriate carbons in the parent chain. Finally, the correct number of hydrogen atoms are added to ensure that each carbon has four bonds.

a. 2-Bromobutane:

Br

b. 2-Chloro-2-methylpropane:

Cl

c. 2,2-Dimethylhexane:

10.63 These compounds can be drawn from their IUPAC names by first drawing the parent chain and numbering each carbon atom in the parent chain. The substituents are then added to the appropriate carbons in the parent chain. Finally the correct number of hydrogen atoms are added to ensure that each carbon has four bonds.

a. 2,2-Dibromobutane:

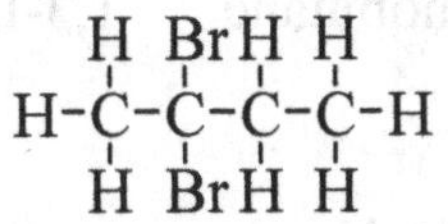

```
  H BrH H
  | |  | |
H-C-C-C-C-H
  | |  | |
  H BrH H
```

b. 2-Iododecane:

```
  H I H H H H H H H H
  | | | | | | | | | |
H-C-C-C-C-C-C-C-C-C-C-H
  | | | | | | | | | |
  H H H H H H H H H H
```

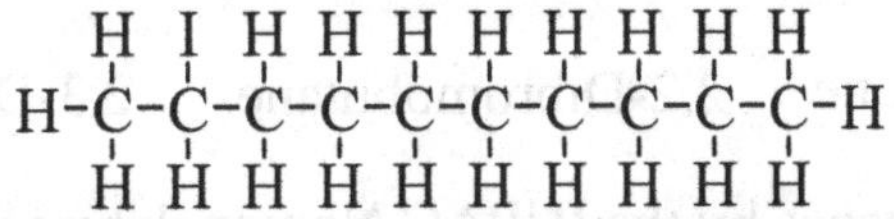

c. 1,2-Dichloropentane:

```
   H ClH H H
   | | | | |
Cl-C-C-C-C-C-H
   | | | | |
   H H H H H
```

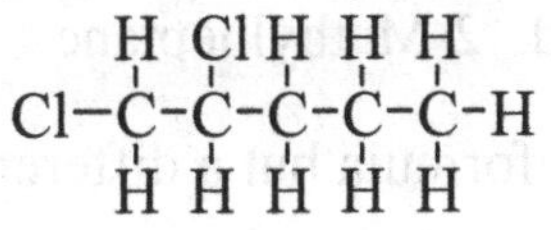

d. 1-Bromo-2-methylpentane:

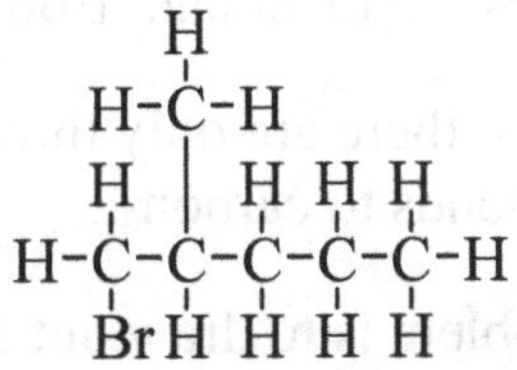

```
      H
      |
    H-C-H
  H |  H H H
  | |  | | |
H-C-C-C-C-C-H
  | |  | | |
  BrH  H H H
```

10.65 a. 3-Methylpentane
b. 2,5-Dimethylhexane
c. 1-Bromoheptane
d. 1-Chloro-3-methylbutane

10.67 a. 2-Chloropropane
b. 2-Iodobutane
c. 2,2-Dibromopropane
d. 1-Chloro-2-methylpropane
e. 2-Iodo-2-methylpropane

10.69 The structural formula shows each atom and bond in a molecule.

a. The straight-chain isomers of molecular formula C_4H_9Br:

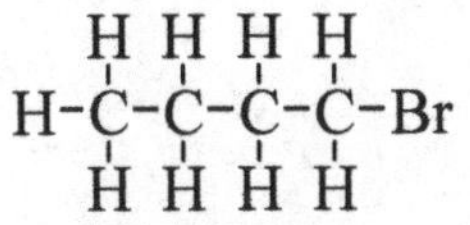

```
  H H H H              H H BrH
  | | | |              | | | |
H-C-C-C-C-Br         H-C-C-C-C-H
  | | | |              | | | |
  H H H H              H H H H

1-Bromobutane        2-Bromobutane
```

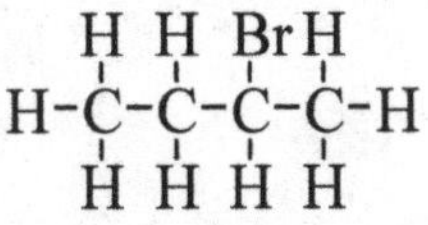

b. The straight-chain isomers of molecular formula $C_4H_8Br_2$:

```
  H H H Br        H H BrBr        H BrH H
  | | | |         | | | |         | | | |
H-C-C-C-C-Br    H-C-C-C-C-H     H-C-C-C-C-Br
  | | | |         | | | |         | | | |
  H H H H         H H H H         H H H H
```

1,1-Dibromobutane   1,2-Dibromobutane   1,3-Dibromobutane

```
  BrH H H         H H BrH         H Br BrH
  | | | |         | | | |         | | | |
H-C-C-C-C-Br    H-C-C-C-C-H     H-C-C-C-C-Br
  | | | |         | | | |         | | | |
  H H H H         H H BrH         H H H H
```

1,4-Dibromobutane   2,2-Dibromobutane   2,3-Dibromobutane

10.71 The rules for naming alkanes by the IUPAC Nomenclature system are used to derive these names.

a. 2-Chlorohexane
b. 1,4-Dibromobutane
c. 3-Chloropentane
d. 2-Methylheptane

10.73 Molecules having the same molecular formula but a different arrangement of atoms are called constitutional isomers.

a. The first pair of molecules are constitutional isomers: hexane and 2-methylpentane.
b. The second pair of molecules are identical. Both are heptane.

10.75 Structure "a" is incorrect because there are only three bonds to carbon-3. Structure "c" is incorrect because there are six bonds to carbon-2.

10.77 The first step in solving this problem is to draw out the structure as the name is provided. Then, using the IUPAC System of Nomenclature, rename the structure.

a. Incorrect: 3-Methylhexane
b. Incorrect: 2-Methylbutane
c. Incorrect: 3-Methylheptane
d. Correct

10.79 The structures drawn are used to determine if the name given is correct.

```
    CH3
    |
a. CH3CHCH2CHCH3
            |
            CH3
```

The name given in the problem is correct.

```
              CH3
              |
b. CH3CH2CH2CHCH2CH2CH3
```

The correct name is 4-methylheptane.

c. I——$CH_2CH_2CH_2CH_2CH_2$——I   The name given in the problem is correct.

d. $CH_3CH_2CH_2CH_2CH_2CHCH_2CH_2CH_3$ The correct name is 4-ethylnonane.

$|$

CH_2CH_3

Br Br

$|$ $|$

e. $CH_3CH_2CH_2CH_2CH_2CCH_2CH_3$ The name given in the problem is correct.

$|$

CH_3

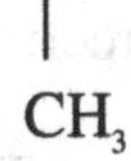

10.81 Cycloalkanes are a family of molecules having carbon-to-carbon bonds in a ring structure.

10.83 The general formula for a cycloalkane is C_nH_{2n}.

10.85 For cycloalkanes, the cycloalkane name is derived from the alkane with the same number of carbons and adding the prefix *cyclo-* to the parent name.

a. Chlorocyclopropane
b. *cis*-1,2-Dichlorocyclopropane (both chlorines are on the same side of the ring)
c. *trans*-1,2-Dichlorocyclopropane (both chlorines are on the opposite sides of the ring)
d. Bromocyclopropane

10.87 For each cycloalkane, begin by drawing the cycloalkane with the appropriate number of carbon atoms. Then number each carbon in the ring to give the lowest numbers to the substituents. Finally, add the names of the appropriate substituents.

a. 1-Bromo-2-methylcyclobutane:

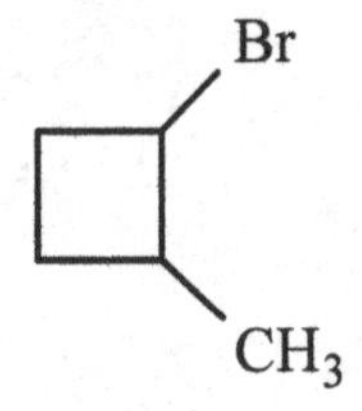

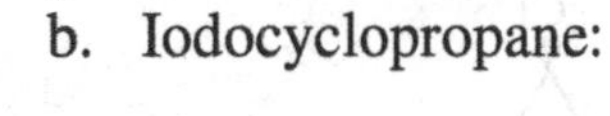

b. Iodocyclopropane:

c. 1-Bromo-3-chlorocyclopentane:

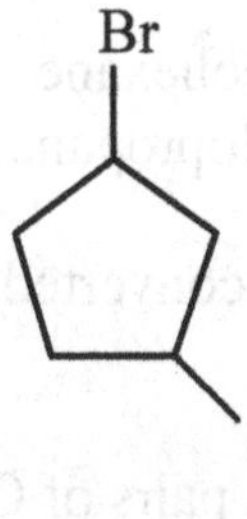

Br

Cl

d. 1,2-Dibromo-3-methylcyclohexane:

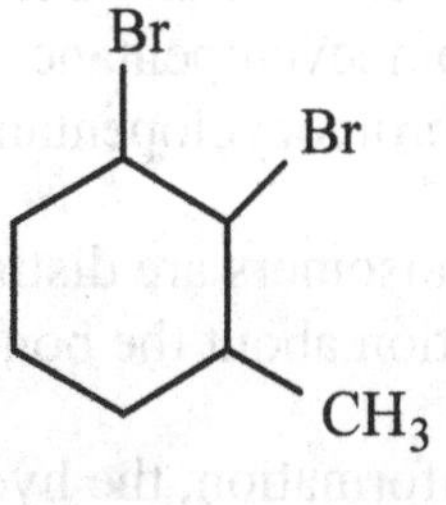

Br

Br

CH_3

10.89 The following are the three structural isomers of dichlorocyclopropane. These structures all have the same molecular formula, but different arrangement of atoms. You can confirm that they have different arrangements by naming each cycloalkane. Two of these compounds are geometric isomers.

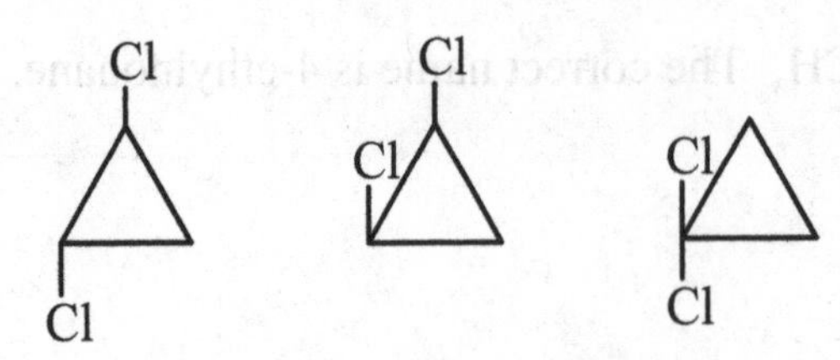

10.91 The first step in solving this problem is to draw out the structure as the name is provided. Then, using the IUPAC System of Nomenclature, rename the structure.

a. Incorrect: 1,2-Dibromocyclobutane
b. Incorrect: 1,2-Diethylcyclobutane
c. Correct
d. Incorrect: 1,2,3-Trichlorocyclohexane

10.93 For each cycloalkane, begin by drawing the cycloakane. Then number each carbon in the ring. Add the appropriate substituents. Make sure that *cis*-cycloalkanes have the same substituents on the same side of the ring and the *trans*-cycloalkanes have the same substituents on opposite sides of the ring.

a. *cis*-1,3-Dibromocyclopentane

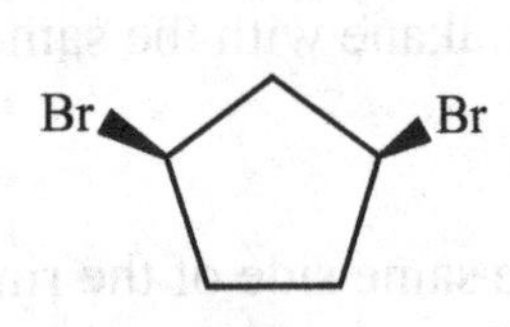

b. *trans*-1,2-Dimethylcyclobutane

CH_3

CH_3

c. *cis*-1,2-Dichlorocyclopropane

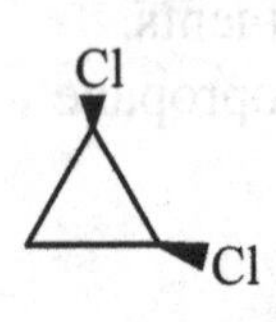

d. *trans*-1,4-Diethylcyclohexane

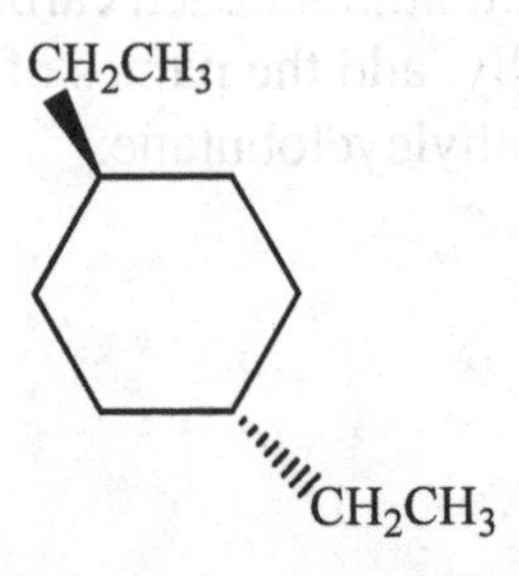

10.95 For cycloalkanes, the cycloalkane name is derived from the name of the alkane with the same number of carbons and adding the prefix *cyclo*-.

a. *cis*-1,2-Dibromocyclopentane
b. *trans*-1,3-Dibromocyclopentane
c. *cis*-1,2-Dimethylcyclohexane
d. *cis*-1,2-Dimethylcyclopropane

10.97 Conformational isomers are distinct isomeric structures that may be converted into one another by rotation about the bonds in the molecule.

10.99 In the chair conformation, the hydrogen atoms, and thus the electron pairs of C-H bonds, are farther from one another. As a result, there is less electron repulsion and the structure is more stable. In the boat conformation, the electron pairs are more crowded. This causes greater electron repulsion, producing a less stable conformation.

10.101 Combustion is the oxidation of hydrocarbons by burning in the presence of air to produce carbon dioxide and water.

10.103 To write a balanced equation for the combustion reaction with each of these hydrocarbons, first determine the molecular formula for each alkane. In a complete combustion reaction, the alkane reacts with oxygen to produce water and carbon dioxide. After you know the molecular formula for each reactant and product, you should be able to balance the equation to ensure that there are the same number of carbons, hydrogens, and oxygens on both sides of the equation.

a. $C_3H_8 + 5O_2 \rightarrow 4H_2O + 3CO_2$

b. $C_7H_{16} + 11O_2 \rightarrow 8H_2O + 7CO_2$

c. $C_9H_{20} + 14O_2 \rightarrow 10H_2O + 9CO_2$

d. $2C_{10}H_{22} + 31O_2 \rightarrow 22H_2O + 20CO_2$

10.105 Cetane is hexadecane. Therefore, its molecular formula is $C_{16}H_{34}$. In a complete combustion reaction, the alkane reacts with oxygen to produce water and carbon dioxide. The balanced equation for the complete combustion of cetane is:

$2C_{16}H_{34} + 49O_2 \rightarrow 32CO_2 + 34H_2O$

This can be checked by comparing the numbers of each atom on the reactant side and the product side of the equation. It can be seen that on both sides of this balanced equation, there are 32 carbons, 68 hydrogens, and 98 oxygens.

10.107 a. $8CO_2 + 10H_2O$ Complete combustion reactions produce carbon dioxide and water. This equation has been balanced.

b.

```
        CH3                       CH3
        |                         |
Br------C------CH3      +      CH3CHCH2Br + 2HBr
        |
        CH3
```

2-Bromo-2-methylpropane   1-Bromo-2-methylpropane

c. Cl_2 is the missing reactant, and light goes on the arrow. This is a monochlorination reaction.

10.109 The following molecules are all the constitutional isomers of C_6H_{14}:

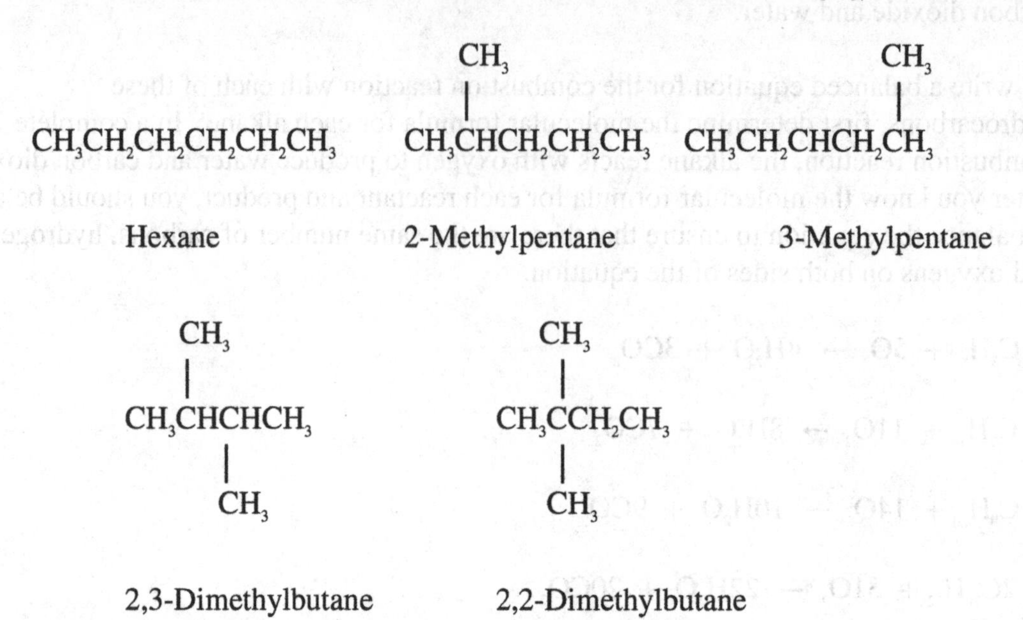

a. 2,3-Dimethylbutane produces only two monobrominated derivatives:
   1-bromo-2,3-dimethylbutane and 2-bromo-2,3-dimethylbutane.

b. Hexane produces three monobrominated products:
   1-bromohexane, 2-bromohexane, and 3-bromohexane.
   2,2-Dimethylbutane also produces three monobrominated products:
   1-bromo-2,2-dimethylbutane, 2-bromo-3,3-dimethylbutane, and 1-bromo-3,3-dimethylbutane.

c. 3-Methylpentane produces four monobrominated products:
   1-bromo-3-methylpentane, 2-bromo-3-methylpentane, 3-bromo-3-methylpentane, and 1-bromo-2-ethylbutane.

10.111 The hydrocarbon is cyclooctane having a molecular formula of C_8H_{16}. This can be deduced from the combustion products: $8CO_2 + 8H_2O$. The other reactant was oxygen. Because there are eight carbons in the products we can conclude that we started with eight carbons in our hydrocarbon.

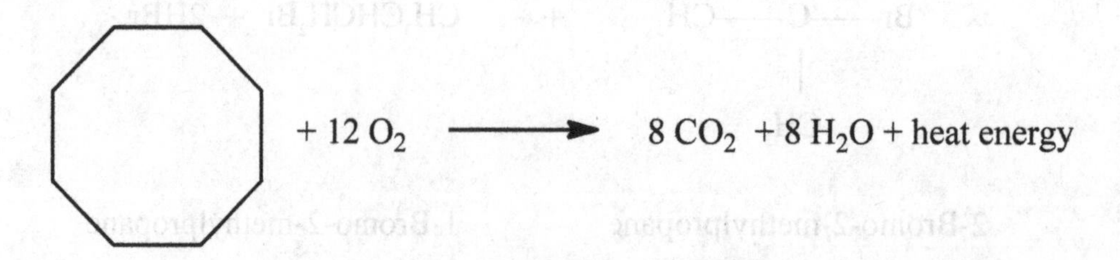

Chapter 11
The Unsaturated Hydrocarbons: Alkenes, Alkynes, and Aromatics
Solutions to the Odd-Numbered Questions and Problems

In-Chapter Questions and Problems

11.1 These condensed structural formulas are drawn from the IUPAC names. The first step in drawing these alkynes is to draw out the parent chain. Then, number each of the carbon atoms in the chain. Add the triple bond starting at the carbon indicated in the IUPAC name. Add the substituents at the carbons indicated in the name. Finally, add the appropriate number of hydrogens so that each carbon has four bonds.

a.

$CH_2BrCH_2C{\equiv}CCH_2CH_3$

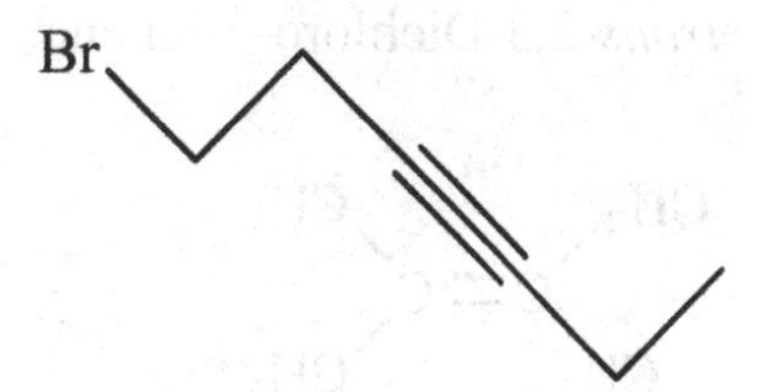

b.

$CH_3C{\equiv}CCH_3$

c.

$ClC{\equiv}CCl$ Cl—≡—Cl

d.

$HC{\equiv}C(CH_2)_7I$ I

11.3 If both groups are on the same side of the double bond, the molecule is a *cis* isomer. If the groups are on opposite sides of the double bond, the molecule is a *trans* isomer.

a.

H, CH_3CH_2 $\backslash$C=C/ H, CH_2CH_3	H, CH_3CH_2 $\backslash$C=C/ CH_2CH_3, H
cis-3-Hexene	*trans*-3-Hexene

b.

Br, CH_3 $\backslash$C=C/ Br, CH_3	Br, CH_3 $\backslash$C=C/ CH_3, Br
*cis-2,*3-Dibromo-2-butene	*trans-2,*3-Dibromo-2-butene

11.5 Molecule c can exist as *cis*- and *trans*-isomers because there are two different groups on each of the carbon atoms attached by the double bond.

11.7 a. *cis*-3-Octene. The hydrogens are on the same side of the double bond.

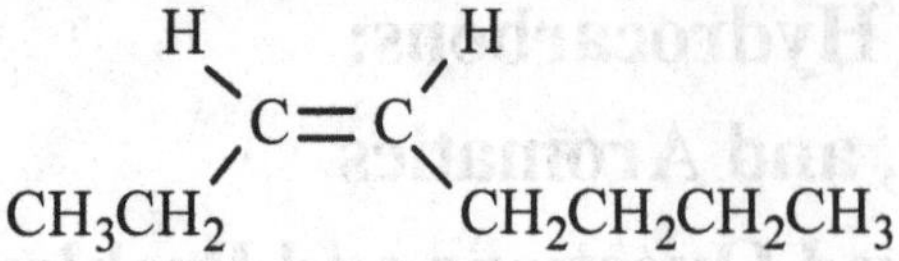

$$\begin{matrix} H & & H \\ & C{=}C & \\ CH_3CH_2 & & CH_2CH_2CH_2CH_3 \end{matrix}$$

b. *trans*-5-Chloro-2-hexene The hydrogens are on opposite sides of the double bond.

$$\begin{matrix} CH_3 & & H \\ & C{=}C & \\ H & & CH_2CHCH_3 \\ & & \quad\; Cl \end{matrix}$$

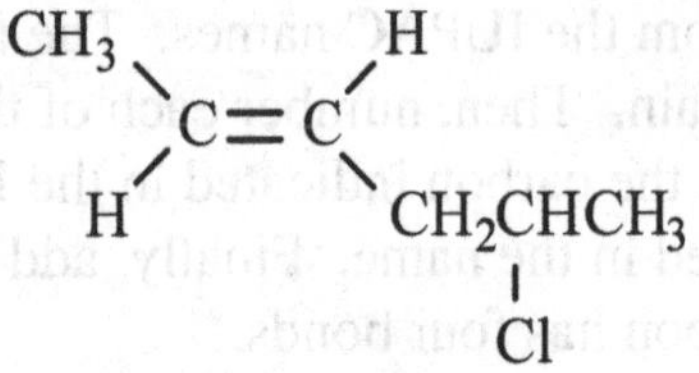

c. *trans*-2,3-Dichloro-2-butene. The chlorines are on opposite sides of the double bond.

$$\begin{matrix} CH_3 & & Cl \\ & C{=}C & \\ Cl & & CH_3 \end{matrix}$$

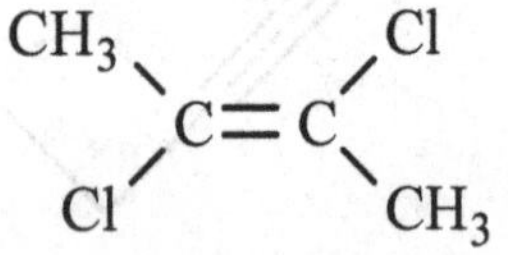

11.9 No, the hydrogenation of the *cis*- and *trans*- isomers of 2-pentene would produce the same product, pentane. Hydrogenation of an alkene produces an alkane.

11.11 To achieve complete hydrogenation of an alkyne, two moles of elemental hydrogen are required. In these answers, Ni is the catalyst listed. Note that Pt or Pd with the addition of heat or pressure would also yield the same result, and you would have an alkane as the product.

a. 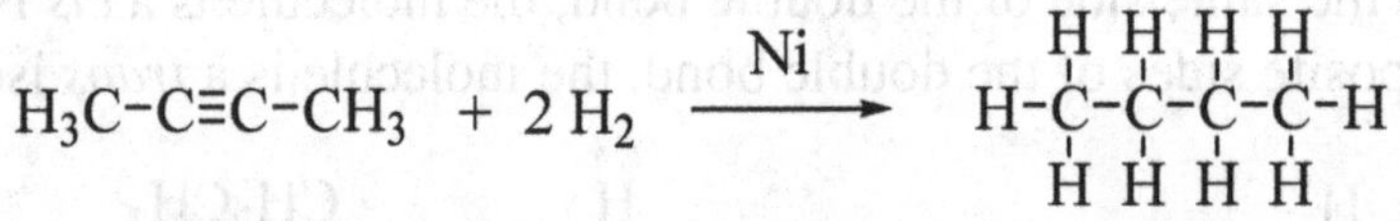 $H_3C{-}C{\equiv}C{-}CH_3 + 2\,H_2 \xrightarrow{Ni} H{-}\overset{H}{\underset{H}{C}}{-}\overset{H}{\underset{H}{C}}{-}\overset{H}{\underset{H}{C}}{-}\overset{H}{\underset{H}{C}}{-}H$

2-Butyne → Butane

b. 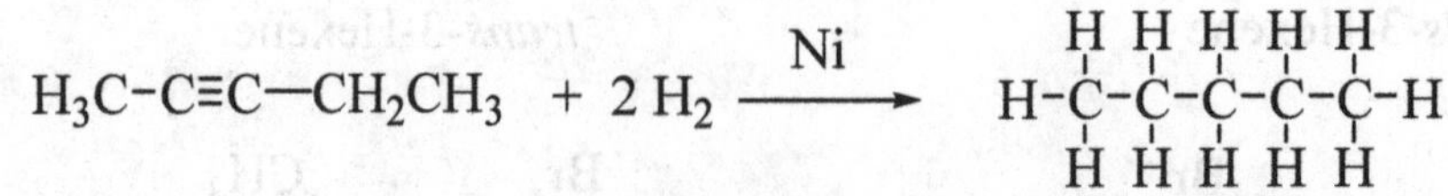 $H_3C{-}C{\equiv}C{-}CH_2CH_3 + 2\,H_2 \xrightarrow{Ni} H{-}\overset{H}{\underset{H}{C}}{-}\overset{H}{\underset{H}{C}}{-}\overset{H}{\underset{H}{C}}{-}\overset{H}{\underset{H}{C}}{-}\overset{H}{\underset{H}{C}}{-}H$

2-Pentyne → Pentane

11.13 Halogenation of an alkene does not need a catalyst. The halogen, Br_2, adds across the double bond to give you a disubstituted alkane.

a. $CH_3CH{=}CH_2 + Br_2 \longrightarrow$

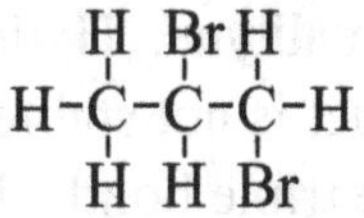

```
  H Br H
  |  |  |
H-C--C--C-H
  |  |  |
  H  H  Br
```

b. $CH_3CH{=}CHCH_3 + Br_2 \longrightarrow$

```
  H  H  Br H
  |  |  |  |
H-C--C--C--C-H
  |  |  |  |
  H  Br H  H
```

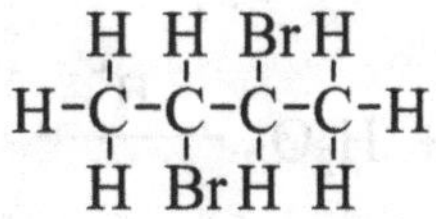

11.15 Halogenation of an alkyne does not need a catalyst. Because these are alkynes, two moles of halogen must be added to achieve complete halogenation. The halogen, Cl_2, adds across the triple bond to give you a tetrasubstituted alkane.

a. $CH_3C{\equiv}CCH_3 + 2\,Cl_2 \longrightarrow$

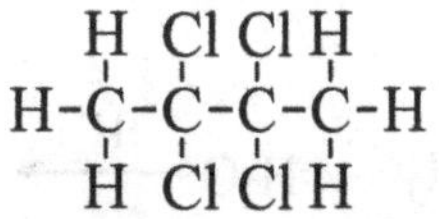

```
  H  Cl Cl H
  |  |  |  |
H-C--C--C--C-H
  |  |  |  |
  H  Cl Cl H
```

b. $CH_3C{\equiv}CCH_2CH_3 + 2\,Cl_2 \longrightarrow$

```
  H  Cl Cl H  H
  |  |  |  |  |
H-C--C--C--C--C-H
  |  |  |  |  |
  H  Cl Cl H  H
```

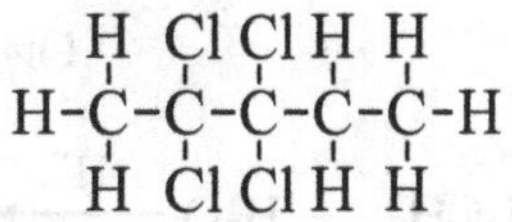

11.17 Hydration refers to the addition of a water molecule to an alkane. The reaction requires a trace amount of strong acid as a catalyst. The product is an alcohol. Markovnikov's rule tells us that the carbon of the carbon-carbon double bond that originally had the most hydrogens will receive the hydrogen atom added. The other carbon receives the –OH.

a.

$$CH_3CH{=}CHCH_3 + H_2O \xrightarrow{H+} CH_3CHOHCH_2CH_3$$

b.

$$H_2C{=}CHCH_2CH_2CH(CH_3)_2 + H_2O \xrightarrow{H+} CH_3CHOHCH_2CH_2CH(CH_3)_2$$

(Major product)

$$H_2C{=}CHCH_2CH_2CH(CH_3)_2 + H_2O \xrightarrow{H+} CH_2OHCH_2CH_2CH_2CH(CH_3)_2$$

(Minor product)

c. These products will be formed in approximately equal amounts:

$$CH_3CH_2CH_2CH{=}CHCH_2CH_3 + H_2O \xrightarrow{H+} CH_3CH_2CH_2CHOHCH_2CH_2CH_3$$

$$CH_3CH_2CH_2CH{=}CHCH_2CH_3 + H_2O \xrightarrow{H+} CH_3CH_2CH_2CH_2CHOHCH_2CH_3$$

d.

$$CH_3CHClCH{=}CHCHClCH_3 + H_2O \xrightarrow{H+} CH_3CHClCHOHCH_2CHClCH_3$$

Only product

11.19 Water can also add to an alkyne. Hydration of the alkyne is a more complex process because the initial product is not stable and is rapidly isomerized. Only one mole of water adds to the alkyne triple bond. This results in the formation of an enol. The enol then isomerizes to an aldehyde or a ketone.

```
                               H H                       H H
                       H+      | |                       | |
a. H3C-C≡CH + H2O  ------->  H-C-C=C-H   -------->     H-C-C-C-H
                               |   |                     | | ||
                               H   OH                    H H O
                                  Or
                               H OH                      H O H
                       H+      | |                       | || |
   H3C-C≡CH + H2O  ------->  H-C-C=C-H   -------->     H-C-C-C-H
                               |   |                     |    |
                               H   H                     H    H

                                 H H   H H                   H H   H H
                         H+      | |   | |                   | |   | |
b. H3C-C≡CCH2CH3 + H2O ----->  H-C-C=C-C-C-H   ------->    H-C-C-C-C-C-H
                                 |   | | |                   | | || | |
                                 H   OHH H                   H H O  H H
                                       Or
                                 H OH  H H                   H O H H
                         H+      | |   | |                   | || | |
   H3C-C≡CCH2CH3 + H2O ----->  H-C-C=C-C-C-H   ------->    H-C-C-C-C-C-H
                                 |   | | |                   |   | | |
                                 H   H H H                   H   H H H
```

11.21 a. The IUPAC name 1,3,5-trichlorobenzene tells us that there are chlorine atoms at positions 1, 3, and 5 on the benzene ring.

Cl

Cl Cl

b. The common name *ortho*-cresol tells us that there is a methyl group and a hydroxyl group and that they are on adjacent carbon atoms of the ring.

OH

CH_3

c. The IUPAC name 2,5-dibromophenol tells us that there are two Br atoms on a phenol molecule. Using the phenol –OH as carbon-1, place a Br at carbon-2 and carbon-5.

OH

Br

Br

d. The common name *p*-dinitrobenzene indicates that there are two $-NO_2$ groups and that they are separated from one another by two carbon atoms (*para*).

e. The name 2-nitroaniline tells us that there is a $-NO_2$ group on an aniline molecule (benzene with $-NH_3$ group). The $-NH_3$ group is located on carbon-1 by definition. Thus the $-NO_2$ group must be on an adjacent carbon atom.

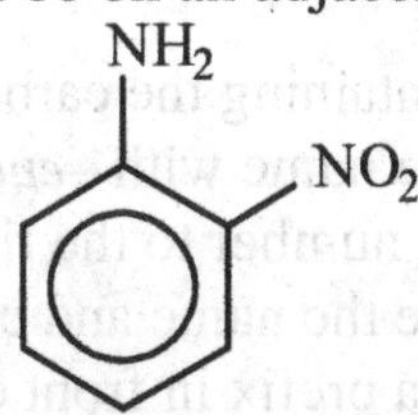

f. The name *meta*-nitrotoluene tells us that there is a $-NO_2$ group on a toluene molecule (benzene with $-CH_3$ group). The $-CH_3$ group is located on carbon-1 by definition. The prefix *meta* indicates that there is one carbon between the two substituents. Thus the $-NO_2$ group must be on carbon-3.

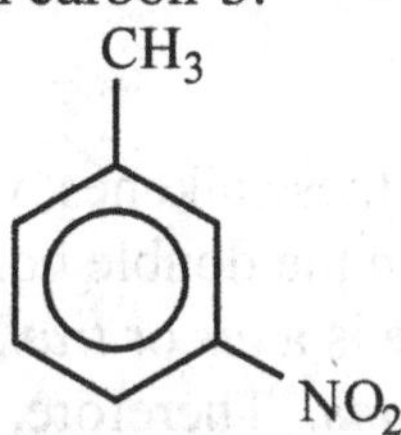

End-of-Chapter Questions and Problems

11.23 As the length of the hydrocarbon chain increases, the London dispersion forces between the molecules increase. The stronger these attractive forces between molecules are, the higher the boiling point will be.

11.25 The general formula for an alkane is C_nH_{2n+2}.

The general formula for an alkene is C_nH_{2n}.

The general formula for an alkyne is C_nH_{2n-2}.

11.27 Ethene is a planar molecule. All of the bond angles are 120°.

11.29 In alkanes, such as ethane, the four bonds around each carbon atom have tetrahedral geometry. The bond angles are 109.5°. In alkenes, such as ethene, each carbon is bonded by two single bonds and one double bond. The molecule is planar and each bond angle is approximately 120°.

11.31 Ethyne is a linear molecule. All of the bond angles are 180°.

11.33 In alkanes, such as ethane, the four bonds around each carbon atom have tetrahedral geometry. The bond angles are 109.5°. In alkenes, such as ethene, each carbon is bonded by two single bonds and one double bond. The molecule is planar and each bond angle is approximately 120°. In alkynes, such as ethyne, each carbon is bonded by one single bond and one triple bond. The molecular is linear and the bond angles are 180°.

11.35 The length of the hydrocarbon chain is directly proportional to the boiling point. Therefore, the shorter the hydrocarbon chain is, the lower the boiling point will be.
a. 2-Pentyne > Propyne > Ethyne
b. 3-Decene > 2-Butene > Ethene

11.37 Identify the longest carbon chain containing the carbon-to-carbon double or triple bond. Replace the *–ane* suffix of the alkane name with *–ene* for an alkene or *-yne* for an alkyne. Number the chain to give the lowest number to the first of the two carbons involved in the double or triple bond. Determine the name and carbon number of each substituent group and place that information as a prefix in front of the name of the parent compound.

11.39 Geometric isomers of alkenes differ from one another in the placement of substituents attached to each of the carbon atoms of the double bond. Of the pair of geometric isomers, the *cis*-isomer is the one in which identical groups are on the same side of the double bond.

11.41 To draw the condensed formulas for these alkenes, you should first draw the parent chain. After you number each carbon, add in the double bond as appropriate. Pay careful attention to whether or not the alkene is a *cis* or *trans* isomer. In each of these problems, the hydrogens will be either *cis* or *trans*. Therefore, you should add the hydrogens to the carbons that make up the double bond to indicate whether the hydrogens are *cis* or *trans*. The other substituents should be added. Finally, the remaining hydrogen atoms should be added to ensure that each carbon has four bonds.

a. 2-Methyl-2-hexene:

H_3C $CH_2CH_2CH_3$
\ /
C=C
/ \
H_3C H

b. *trans*-3-Heptene:

CH_3CH_2 H
\ /
C=C
/ \
H $CH_2CH_2CH_3$

c. *cis*-1-Chloro-2-pentene:

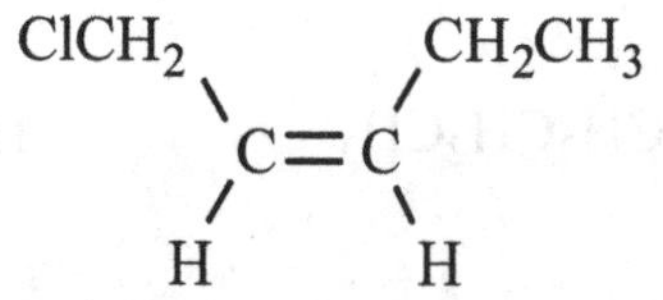

$$\begin{array}{ccc} ClCH_2 & & CH_2CH_3 \\ & C{=}C & \\ H & & H \end{array}$$

d. *cis*-2-Chloro-2-methyl-3-heptene:

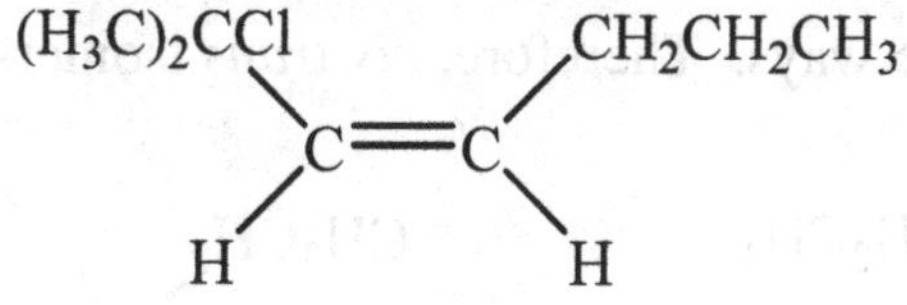

$$\begin{array}{ccc} (H_3C)_2CCl & & CH_2CH_2CH_3 \\ & C{=}C & \\ H & & H \end{array}$$

e. *trans*-5-Bromo-2,6-dimethyl-3-octene:

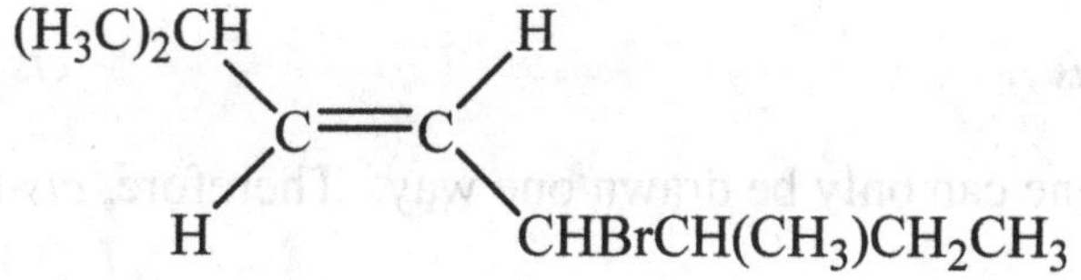

$$\begin{array}{ccc} (H_3C)_2CH & & H \\ & C{=}C & \\ H & & CHBrCH(CH_3)CH_2CH_3 \end{array}$$

11.43 The rules for naming according the IUPAC system of nomenclature for alkenes should be followed.

a. 3-Methyl-1-pentene
b. 7-Bromo-1-heptene
c. 5-Bromo-3-heptene
d. 1-*t*-Butyl-4-methylcyclohexene

11.45 a. 1,3,5-Trifluoropentane: $CH_2FCH_2CHFCH_2CH_2F$

b. *cis*-2-Octene:

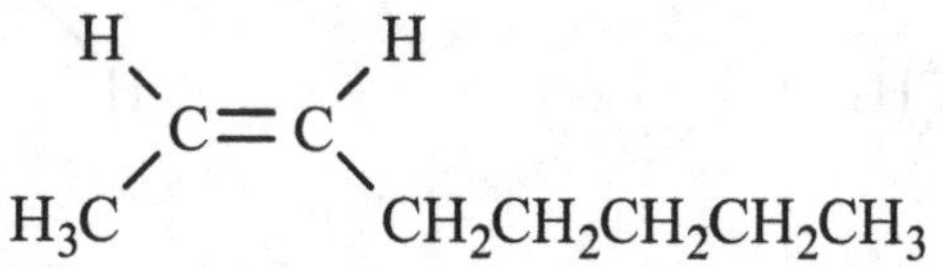

$$\begin{array}{ccc} H & & H \\ & C{=}C & \\ H_3C & & CH_2CH_2CH_2CH_2CH_3 \end{array}$$

c. Dipropylacetylene: $CH_3CH_2CH_2C{\equiv}CCH_2CH_2CH_3$

11.47 To determine which of the following alkenes can exist as *cis-trans* isomers, the structures for each alkene should be drawn.

a. 1-Heptene can only be drawn one way. Therefore, *cis-trans* isomers do not exist.

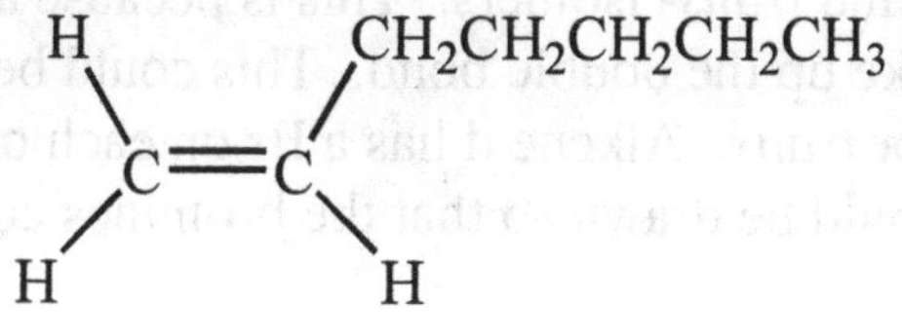

$$\begin{array}{ccc} H & & CH_2CH_2CH_2CH_2CH_3 \\ & C{=}C & \\ H & & H \end{array}$$

b. 2-Heptene can be drawn two ways. Therefore, *cis-trans* isomers do exist.

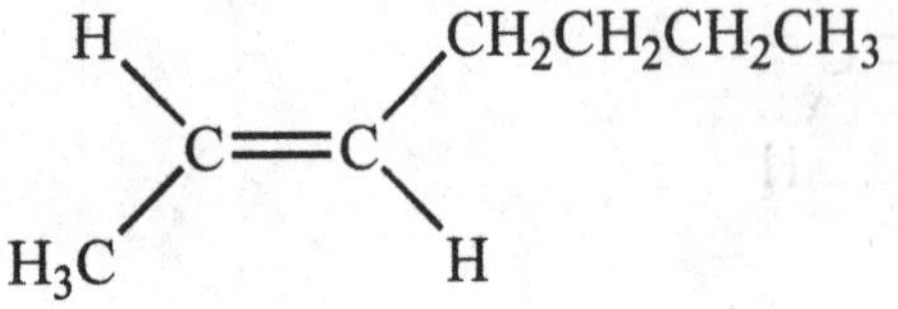

$H \quad CH_2CH_2CH_2CH_3$
$C{=}C$
$H_3C \quad H$

trans

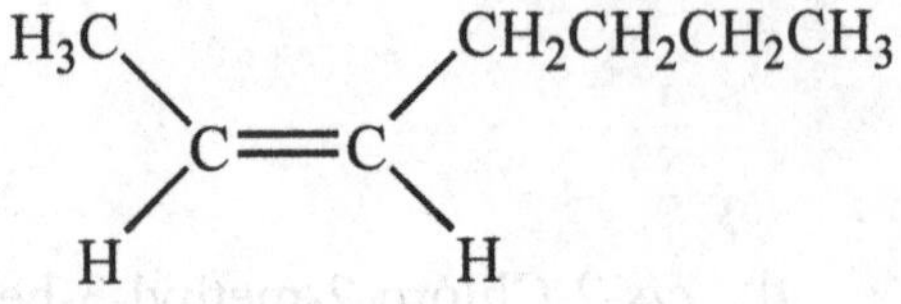

$H_3C \quad CH_2CH_2CH_2CH_3$
$C{=}C$
$H \quad H$

cis

c. 3-Heptene can be drawn two ways. Therefore, *cis-trans* isomers do exist.

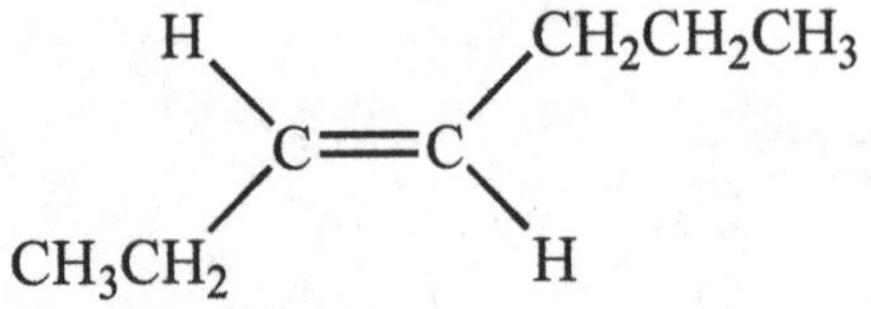

$H \quad CH_2CH_2CH_3$
$C{=}C$
$CH_3CH_2 \quad H$

trans

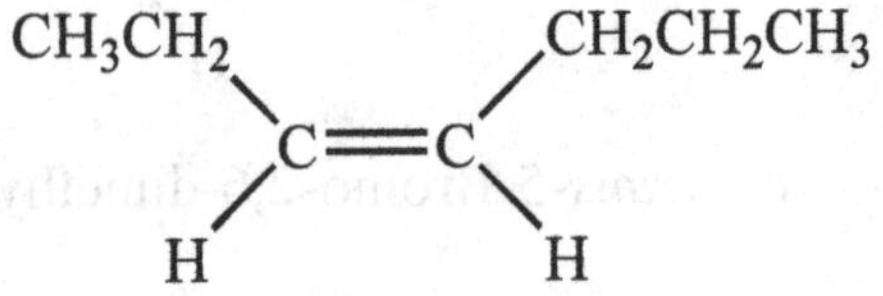

$CH_3CH_2 \quad CH_2CH_2CH_3$
$C{=}C$
$H \quad H$

cis

d. 2-Methyl-2-hexene can only be drawn one way. Therefore, *cis-trans* isomers do not exist.

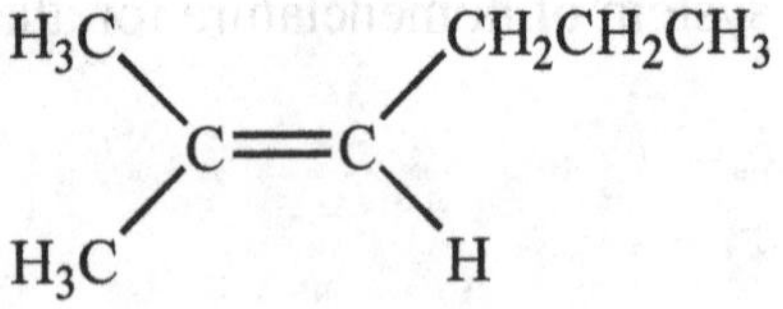

$H_3C \quad CH_2CH_2CH_3$
$C{=}C$
$H_3C \quad H$

e. 3-Methyl-2-hexene can be drawn two ways. Therefore, *cis-trans* isomers do exist.

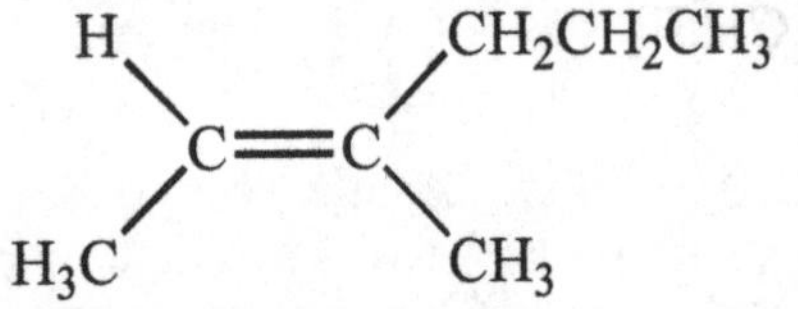

$H \quad CH_2CH_2CH_3$
$C{=}C$
$H_3C \quad CH_3$

trans

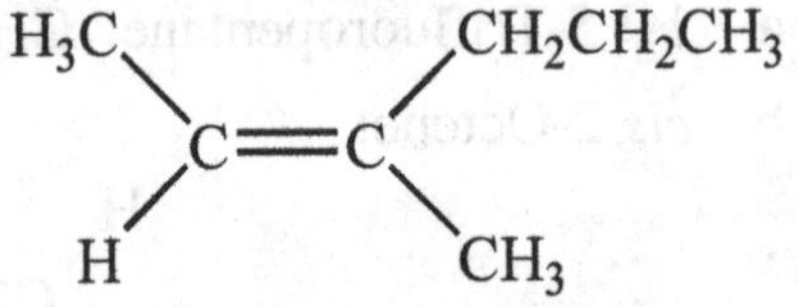

$H_3C \quad CH_2CH_2CH_3$
$C{=}C$
$H \quad CH_3$

cis

11.49 Alkenes b and c would not exhibit *cis-trans* isomerism. This is because alkene b has two methyl groups on one of the carbons of the double bond. Alkene c has four different substituents on the two carbons that make up the double bond. It would need to have a substituent that is the same on each of the carbons of the double bond to have *cis-* and *trans-* isomers.

11.51 Alkenes b and d can exist as both *cis-* and *trans-* isomers. This is because alkene b has a methyl on each of the carbons that make up the double bond. This could be drawn so that the methyl groups could be either *cis* or *trans*. Alkene d has a Br on each of the carbons that make up the double bond. This could be drawn so that the bromines could be either *cis* or *trans*.

11.53 These compounds can be named using the rules for the IUPAC system of nomenclature of alkenes. The prefix in front of the *–ene* indicates the number of double bonds in the compound.

a. 1,5-Nonadiene
b. 1,4,7-Nonatriene
c. 2,5-Octadiene
d. 4-Methyl-2,5-heptadiene

11.55 In a hydrogenation reaction of an alkene, a hydrogen molecule is added across the double bond. Pt, Pd or Ni with heat or pressure is needed for this reaction to occur. An alkane will be the product.

$$R_2C{=}CR_2 + H_2 \xrightarrow[\text{heat or pressure}]{\text{Pt, Pd, or Ni}} R{-}\underset{R}{\overset{H}{C}}{-}\underset{H}{\overset{R}{C}}{-}R$$

11.57 In a halogenation reaction of an alkene, either Br_2 or Cl_2 is added across the double bond. The halogen is represented by X in this general equation.

$$R_2C{=}CR_2 + X_2 \longrightarrow R{-}\underset{R}{\overset{X}{C}}{-}\underset{X}{\overset{R}{C}}{-}R$$

11.59 In a hydration reaction of an alkene, a water molecule is added across the double bond. Hydrogen adds to one carbon of the double bond, and –OH adds to the other carbon. An acid catalyst is required. The product is an alcohol.

$$R_2C{=}CR_2 + H_2O \xrightarrow{H^+} R{-}\underset{R}{\overset{H}{C}}{-}\underset{OH}{\overset{R}{C}}{-}R$$

11.61 The primary difference between complete hydrogenation of an alkene and an alkyne is that 2 moles of H_2 are required for the complete hydrogenation of an alkyne.

11.63 a. 1-Heptene reacts with water in the presence of acid to produce an alcohol.

$$CH_2{=}CH(CH_2)_4CH_3 \xrightarrow{H^+} CH_3CHOH(CH_2)_4CH_3 + CH_2OH(CH_2)_5CH_3$$

1-Heptene → 2-Heptanol (Major Product) + 1-Heptanol (Minor Product)

b. Both the *cis*-2-heptene and *trans*-2-heptene undergo hydrohalogenation to produce the same products, monobrominated heptenes.

$$CH_3CH{=}CH(CH_2)_3CH_3 + HBr \rightarrow CH_3CH_2CHBr(CH_2)_3CH_3 + CH_3CHBr(CH_2)_4CH_3$$

2-Heptene → 3-Bromoheptane + 2-Bromoheptane

c. 3-Heptene reacts with hydrogen to form the alkane, heptane.

$$CH_3CH_2CH{=}CH(CH_2)_2CH_3 + H_2 \xrightarrow[\text{heat or pressure}]{\text{Pt, Pd, or Ni}} CH_3(CH_2)_5CH_3$$

3-Heptene → Heptane

d. 2-Methyl-2-hexene reacts with HCl according to Markovnikov's Rule.

$$(CH_3)_2C{=}CH(CH_2)_2CH_3 + HCl \rightarrow (CH_3)_2CCl(CH_2)_3CH_3$$

2-Methyl-2-hexene → 2-Chloro-2-methylhexane (Major Product)

+

$$(CH_3)_2CHCHCl(CH_2)_2CH_3$$

3-Chloro-2-methylhexane (Minor Product)

11.65 a. H_2
This reaction is a hydrogenation.

b. H_2O
This reaction is a hydration.

c. HBr
This reaction is a hydrohalogenation.

d. $19\ O_2 \rightarrow 12\ CO_2 + 14\ H_2O$
This reaction is a combustion.

e. Cl_2
This reaction is a monochlorination.

f. This reaction is a hydration.

11.67 Complete hydrogenation and halogenation reactions with alkynes require two moles of hydrogen or the halogen.

a.

$$H_3C{-}C{\equiv}C{-}CH_3 + 2H_2 \xrightarrow[\text{heat or pressure}]{\text{Pt, Pd, or Ni}} H_3C{-}CH_2{-}CH_2{-}CH_3$$

2-Butyne

b.

$$CH_3CH_2{-}C{\equiv}C{-}CH_3 + 2\ X_2 \longrightarrow CH_3CH_2{-}CX_2{-}CX_2{-}CH_3$$

2-Pentyne

11.69 a. This reaction produces an alkane. Reactant is *cis*-2-butene; only product is butane.

b. This reaction produces an alcohol. Markovinkov's Rule is followed.
Reactant is 1-butene; major product is 2-butanol.

c. This reaction adds chlorine to each carbon that was in the alkene's double bond.
Reactant is 2-butene; only product is 2,3-dichlorobutane.

d. This reaction adds a bromine to one carbon of the double bond. It adds a hydrogen to the other carbon. Markovinkov's Rule is followed.
Reactant is 1-pentene; major product is 2-bromopentane.

11.71 a. Hydrogenation of 4-chlorocyclooctene

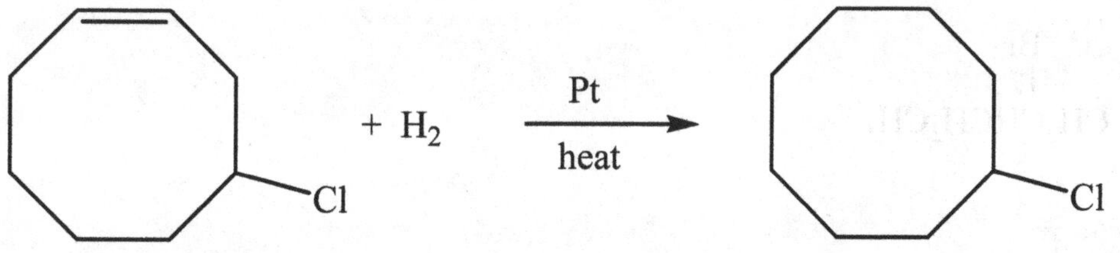

Cl + H_2 —(Pt, heat)→ Cl

b. Halogenation of 1,3-cyclooctadiene

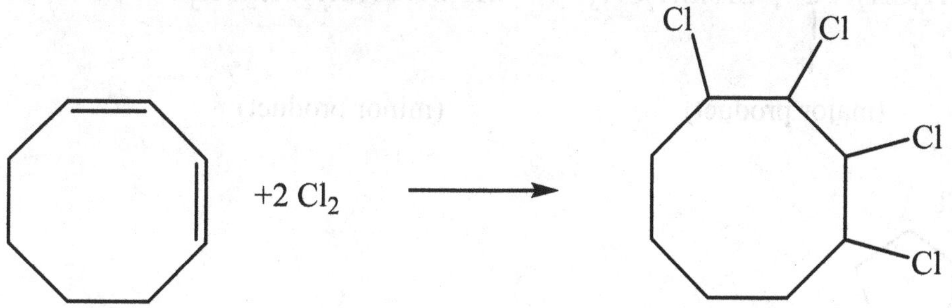

+2 Cl_2 → Cl Cl Cl Cl

c. Hydration of 3-methylcyclobutene

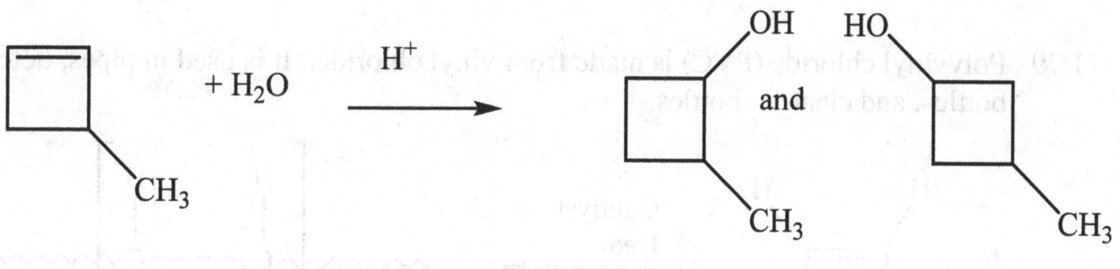

+ H_2O —(H^+)→ OH CH_3 and HO CH_3

CH_3

d. Hydrohalogenation of cyclopentene

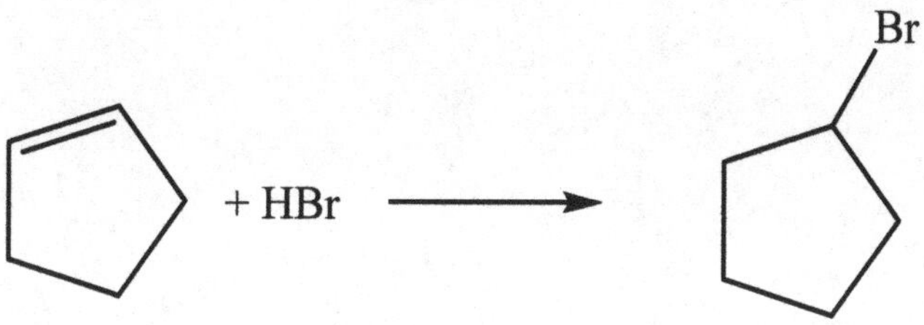

+ HBr → Br

11.73

$CH_2=CHCH_2CH_2CH_3$ $\quad$ $CH_3CH=CHCH_2CH_3$

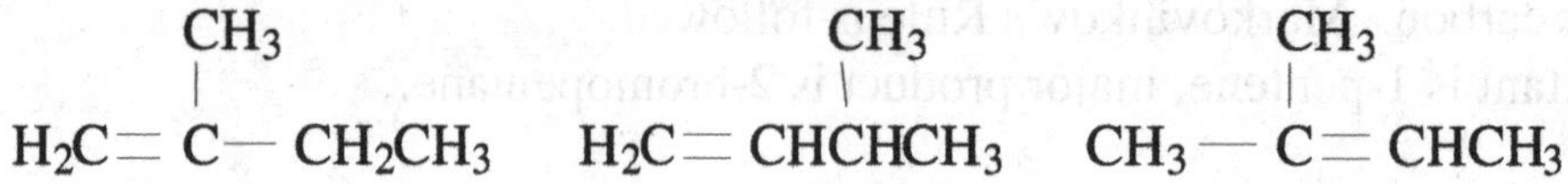

$H_2C=C(CH_3)-CH_2CH_3$ $\quad$ $H_2C=CHCH(CH_3)CH_3$ $\quad$ $CH_3-C(CH_3)=CHCH_3$

11.75 a.

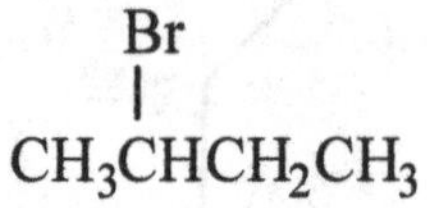

$CH_3CH(Br)CH_2CH_3$

b.

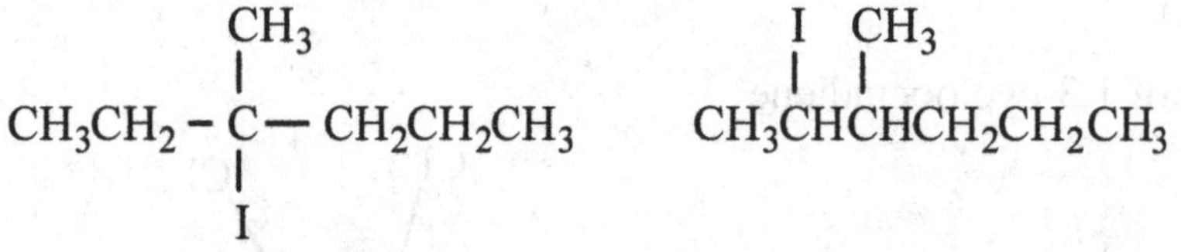

$CH_3CH_2-C(CH_3)(I)-CH_2CH_2CH_3$ (major product) $\quad$ $CH_3CH(I)CH(CH_3)CH_2CH_2CH_3$ (minor product)

c.

Cl

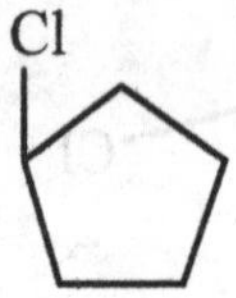

11.77 A polymer is a macromolecule composed of repeating structural units called *monomers*.

11.79 Polyvinyl chloride (PVC) is made from vinyl chloride. It is used in pipes, detergent bottles, and cleanser bottles.

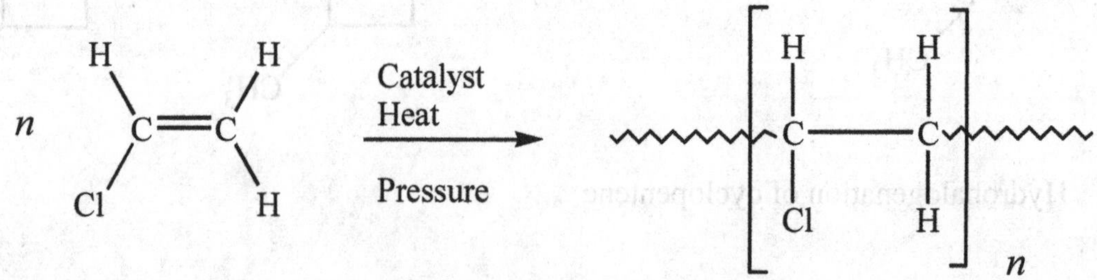

$$n\ \mathrm{CH(Cl)=CH_2} \xrightarrow[\text{Pressure}]{\text{Catalyst, Heat}} \sim\!\left[\mathrm{CH(Cl)-CH_2}\right]_n\!\sim$$

11.81 The IUPAC name for (a) is 2-pentene, for (b) is 3-bromo-1-propene, and for (c) is 3,4-dimethylcyclohexene.

a. These products will be formed in approximately equal amounts.

$$CH_3CH{=}CHCH_2CH_3 + H_2O \xrightarrow{H+} CH_3CHOHCH_2CH_2CH_3$$

$$CH_3CH{=}CHCH_2CH_3 + H_2O \xrightarrow{H+} CH_3CH_2CHOHCH_2CH_3$$

b.

$$CH_2BrCH{=}CH_2 + H_2O \xrightarrow{H^+} CH_2BrCH_2CH_2OH + CH_2BrCHOHCH_3$$

Minor product (CH₂BrCH₂CH₂OH); Major product (CH₂BrCHOHCH₃)

c. These products will be formed in approximately equal amounts.

3,4-dimethylcyclohexene + $H_2O \xrightarrow{H^+}$ 2,3-dimethylcyclohexanol (OH, CH_3, CH_3)

3,4-dimethylcyclohexene + $H_2O \xrightarrow{H^+}$ 3,4-dimethylcyclohexanol (HO, CH_3, CH_3)

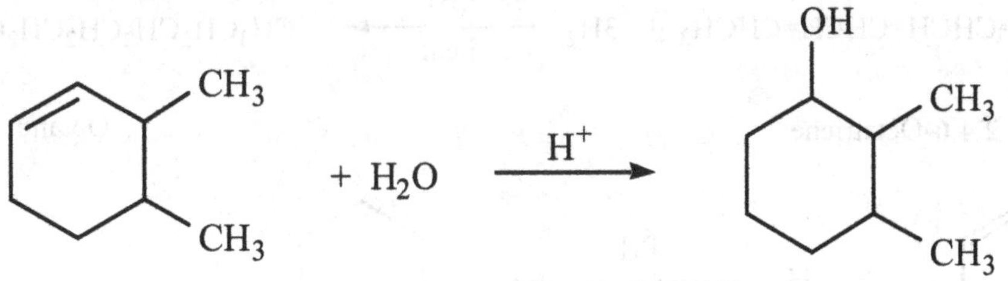

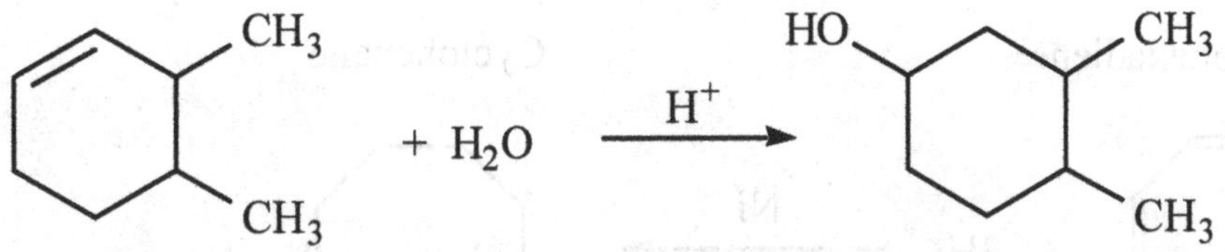

11.83 a. This is the minor product of this reaction.

$$H_2C{=}CHCH_2CH(CH_3)_2 + H_2O \xrightarrow{H^+} CH_2OHCH_2CH_2CH(CH_3)_2$$

b.

$$CH_3CH{=}CHCH_2CH_2CH_3 + HBr \longrightarrow CH_3CH_2CHBrCH_2CH_2CH_3$$

OR

$$CH_3CH_2CH{=}CHCH_2CH_3 + HBr \longrightarrow CH_3CH_2CHBrCH_2CH_2CH_3$$

c.

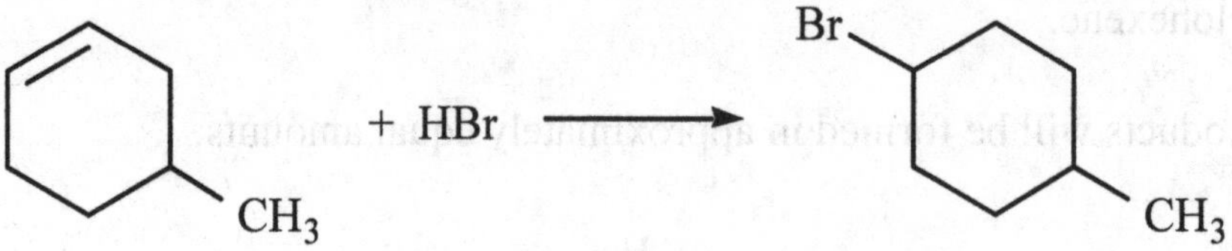

$$\text{4-methylcyclohexene} + HBr \longrightarrow \text{1-bromo-4-methylcyclohexane}$$

d.

$$\text{3-ethylcyclopentene} + H_2O \xrightarrow{H^+} \text{2-ethylcyclopentanol}$$

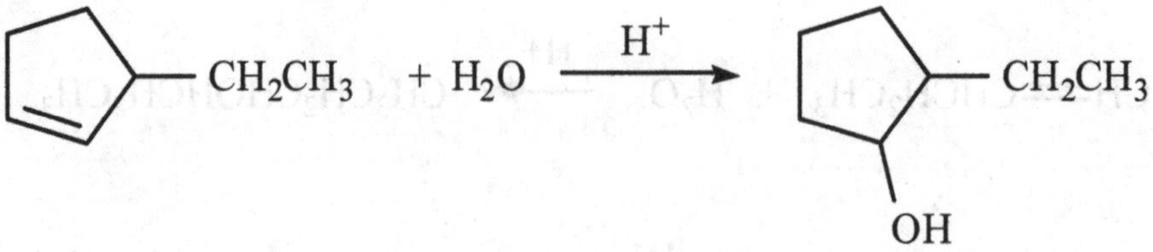

11.85

a. $CH_2{=}CHCH_2CH{=}CHCH_3 + 2H_2 \xrightarrow[\text{heat}]{\text{Pt}} CH_3CH_2CH_2CH_2CH_2CH_3$

1,4-Hexadiene Hexane

b. $CH_3CH{=}CHCH{=}CHCH{=}CHCH_3 + 3H_2 \xrightarrow[\text{heat}]{\text{Ni}} CH_3CH_2CH_2CH_2CH_2CH_2CH_2CH_3$

2,4,6-Octatriene Octane

c. $\text{1,3-Cyclohexadiene} + 2H_2 \xrightarrow[\text{pressure}]{\text{Pd}} \text{Cyclohexane}$

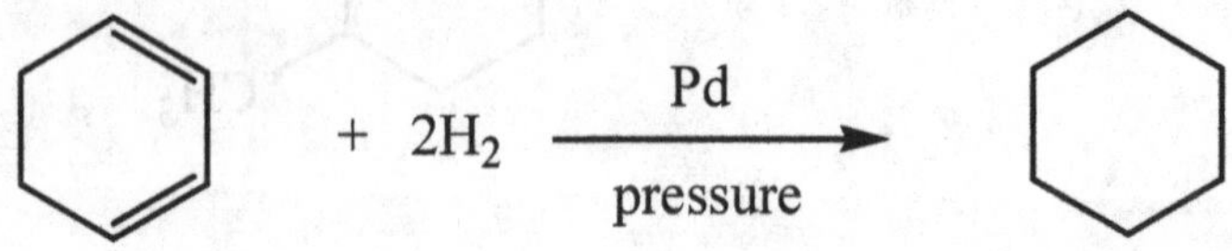

1,3-Cyclohexadiene Cyclohexane

d. $\text{1,3,5-Cyclooctatriene} + 3H_2 \xrightarrow[\text{heat}]{\text{Ni}} \text{Cyclooctane}$

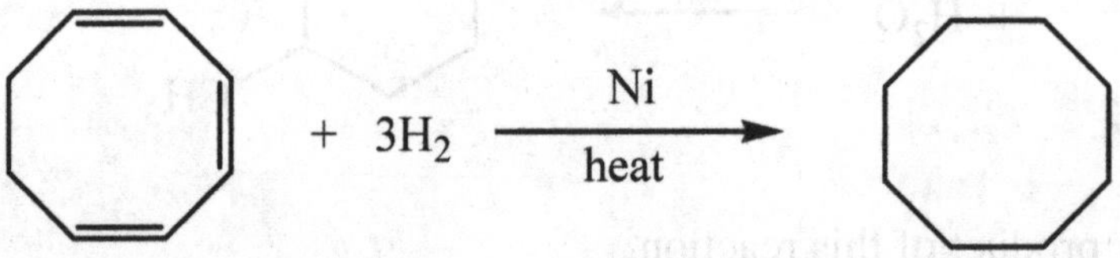

1,3,5-Cyclooctatriene Cyclooctane

11.87 The term *aromatic hydrocarbon* was first used as a term to describe the pleasant-smelling resins of tropical trees.

11.89 Resonance hybrids are molecules for which more than one valid Lewis structure can be written.

11.91 a. 2,4-Dibromotoluene:

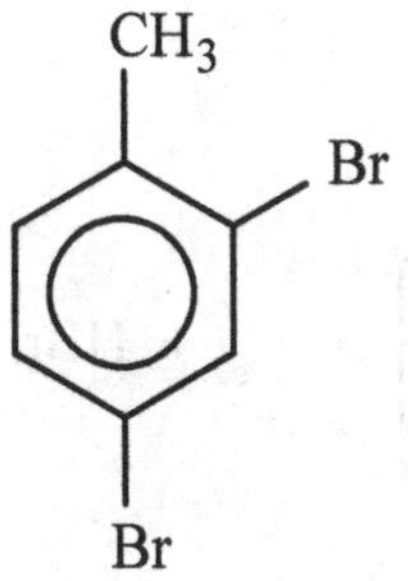

CH_3

Br

Br

b. 1,2,4-Triethylbenzene:

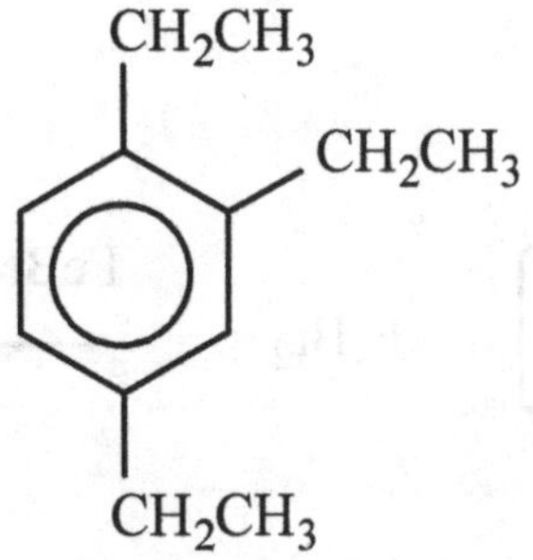

CH_2CH_3

CH_2CH_3

CH_2CH_3

c. Isopropylbenzene:

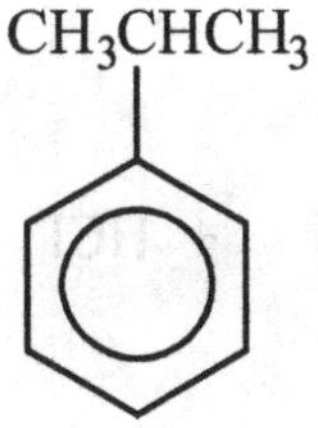

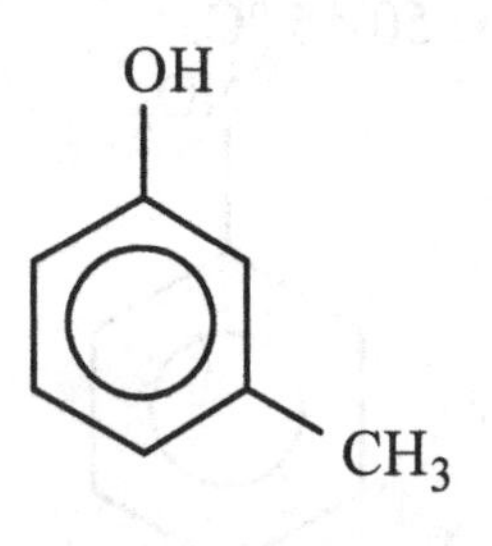

d. 2-Bromo-5-chlorotoluene:

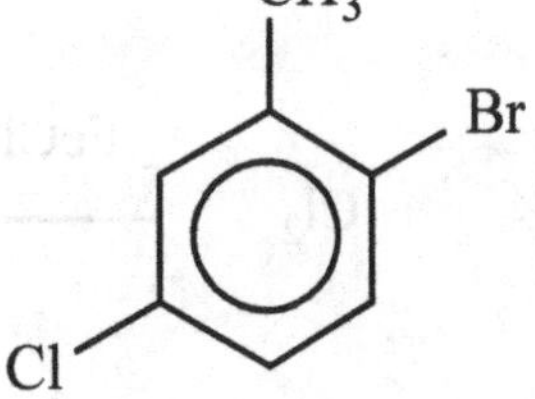

CH_3

Br

Cl

11.93 a.

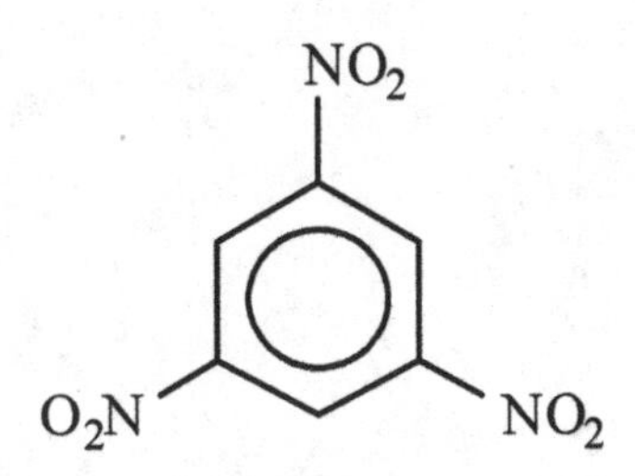

b.

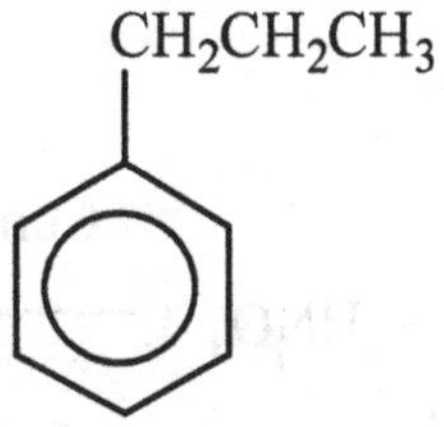

$CH_2CH_2CH_3$

c.

NO_2

O_2N

NO_2

d.

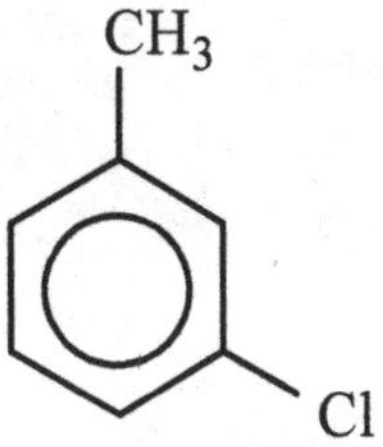

CH_3

Cl

11.95 Kekulé proposed that single and double carbon-carbon bonds alternate around the benzene ring. To explain why benzene does not react like other unsaturated compounds, he proposed that the double and single bonds shift positions rapidly.

11.97 An addition reaction involves addition of a molecule to a double or triple bond in an unsaturated molecule. In a substitution reaction, one chemical group replaces another.

11.99 a. Benzene can react by substitution with Br_2. The reaction requires an iron halide as a catalyst.

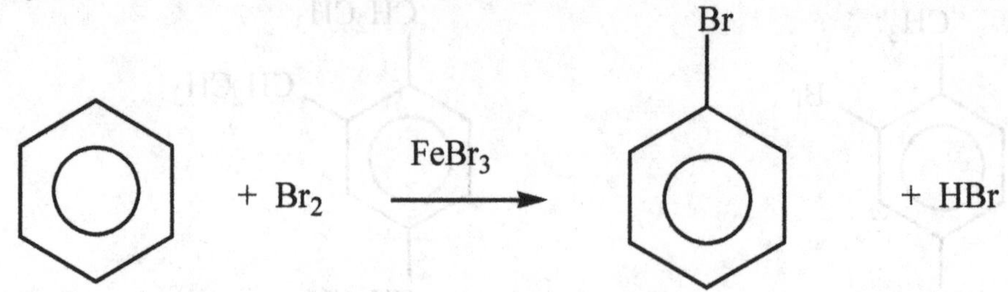

b. Benzene can react by substitution with Cl_2. The reaction requires an iron halide as a catalyst.

Cl

+ Cl_2 $\xrightarrow{FeCl_3}$ + HCl

c. Benzene can undergo nitration with concentrated nitric acid dissolved in concentrated sulfuric acid. The reaction requires the temperature to be 50-55 °C.

NO_2

+ HNO_3 $\xrightarrow[50\text{-}55\ ^{o}C]{\text{Concentrated } H_2SO_4}$ + H_2O

11.101

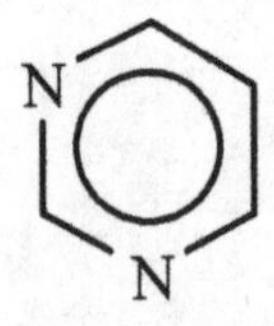

Pyrimidine

11.103

N N N N H

Purine

Chapter 12
Alcohols, Phenols, Thiols, and Ethers
Solutions to the Odd-Numbered Questions and Problems

In-Chapter Questions and Problems

12.1 These alcohols have been named according to the IUPAC system of nomenclature rules. To derive the correct structures, begin by drawing the parent chain. After each carbon has been numbered, the hydroxyl group should be added to the appropriate carbon. The substituents and hydrogens should be added as appropriate so that each carbon atom has a total of four bonds.

a. 2-Methyl-1-propanol

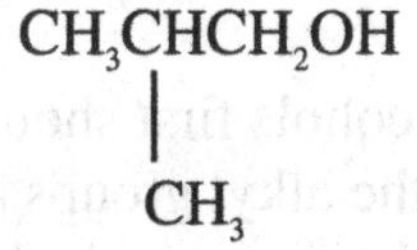

CH_3CHCH_2OH
|
CH_3

b. 2-Chlorocyclopentanol

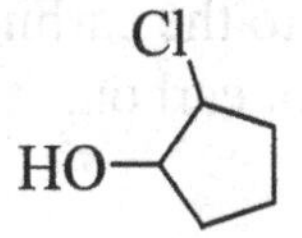

Cl

HO—

c. 2,4-Dimethylcyclohexanol

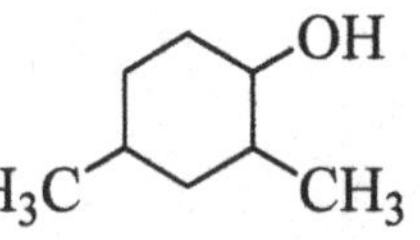

OH

H_3C CH_3

d. 2,3-Dichloro-3-hexanol

OH
|
$CH_3CHCCH_2CH_2CH_3$
| |
Cl Cl

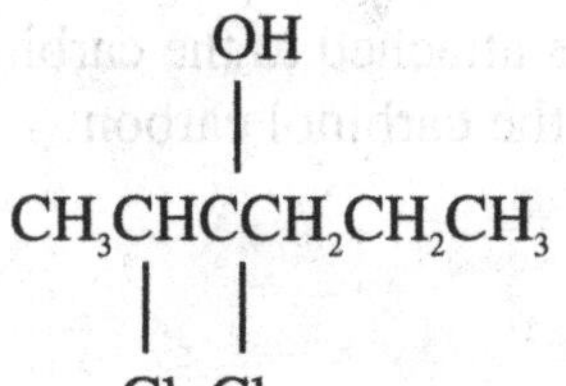

12.3 The alcohol formed by the reduction of butanal is 1-butanol. The common name for 1-butanol is butyl alcohol. It is a primary alcohol.

12.5 a. Ethanol is a primary alcohol. Primary alcohols have only one alkyl group attached to the carbinol carbon. The product is ethene.

b. 2-Propanol is a secondary alcohol. Secondary alcohols have two alkyl groups attached to the carbinol carbon. The product is propene.

c. 4-Methyl-3-hexanol is a secondary alcohol. Secondary alcohols have two alkyls groups attached to the carbinol carbon. The products are 3-methyl-3-hexene and 4-methyl-2-hexene.

d. 2-Methyl-2-propanol is a tertiary alcohol. Tertiary alcohols have three alkyls groups attached to the carbinol carbon. The product is 2-methyl-1-propene.

12.7 a. The reactant is 2-butanol, and the product is butanone.
b. The reactant is 2-pentanol, and the product is 2-pentanone.

12.9 Simple phenols are somewhat soluble in water because they have a polar hydroxyl group.

12.11 Ethers have much lower boiling points than alcohols because ether molecules cannot hydrogen bond to one another.

End-of-Chapter Questions and Problems

12.13 The longer the hydrocarbon tail of an alcohol becomes, the less water soluble it will be.

12.15 The carbinol carbon is the one to which the hydroxyl group is bonded.

12.17 Drawing out the structure of these alcohols first should help you classify the alcohol because it should be easier to count the alkyl groups attached to the carbinol carbon. Primary alcohols have only one alkyl group attached to the carbinol carbon. Secondary alcohols have two alkyl groups attached to the carbinol carbon. Tertiary alcohols have three alkyl groups attached to the carbinol carbon.
a. Primary alcohol
b. Secondary alcohol
c. Tertiary alcohol
d. Tertiary alcohol
e. Tertiary alcohol

12.19 Primary alcohols have only one alkyl group attached to the carbinol carbon. Secondary alcohols have two alkyl groups attached to the carbinol carbon. Tertiary alcohols have three alkyl groups attached to the carbinol carbon.
a. Primary alcohol
b. Secondary alcohol
c. Primary alcohol
d. Primary alcohol
e. Secondary alcohol

12.21 Primary alcohols have only one alkyl group attached to the carbinol carbon. Secondary alcohols have two alkyl groups attached to the carbinol carbon. Tertiary alcohols have three alkyl groups attached to the carbinol carbon.
a. 2-Nonanol: $CH_3CHOHCH_2CH_2CH_2CH_2CH_2CH_2CH_3$
2-Nonanol is a secondary alcohol.
b. 2-Heptanol: $CH_3CHOHCH_2CH_2CH_2CH_2CH_3$
2-Heptanol is a secondary alcohol.
c. 2-Undecanol: $CH_3CHOHCH_2CH_2CH_2CH_2CH_2CH_2CH_2CH_2CH_3$
2-Undecanol is a secondary alcohol.

12.23 $a < d < c < b$

Compound b has the highest boiling point because it is a diol and can form two hydrogen bonds. Compound c has the next highest boiling point because it has one hydroxyl group. That –OH allows it to form one hydrogen bond. Compound d has a higher boiling point than compound a because compound d contains an ether, which is polar. This allows it to have stronger intermolecular forces than compound a.

12.25 If a molecule has an –OH, it will be more soluble than a hydrocarbon. When comparing two alcohols, the alcohol that has the fewest carbon atoms will be more soluble.

a. CH_3CH_2OH

b. $CH_3CH_2CH_2CH_2OH$

c. CH_3CHCH_3
   |
   OH

12.27 The IUPAC rules for the nomenclature of alcohols require you to name the parent compound, that is the longest continuous carbon chain bonded to the –OH group. Replace the *–e* ending of the parent alkane with *–ol* of the alcohol. Number the parent chain so that the carbon bearing the hydroxyl group has the lowest possible number. Name and number all other substituents. If there are more than one hydroxyl groups, the *–ol* ending will be modified to reflect the number. If there are two –OH groups, the suffix *–diol* is used; if it has three –OH groups, the suffix *–triol* is used, etc.

12.29 You may find it easier to name a compound in the IUPAC system if you draw out its structural formula from the condensed formula provided.

a. 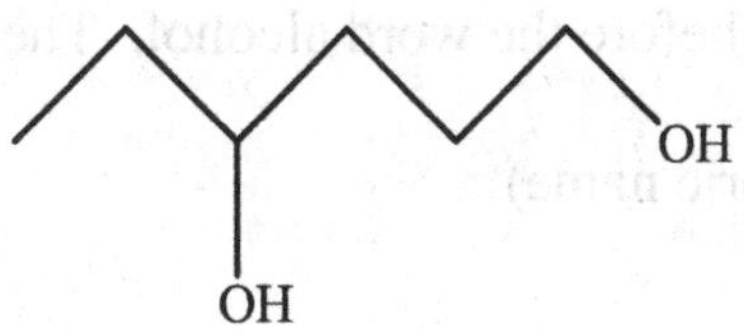 OH OH

1,4-hexanediol

b. 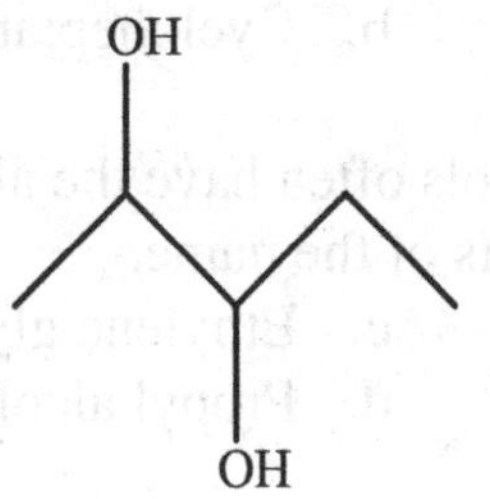 OH OH

2,3-pentanediol

c. 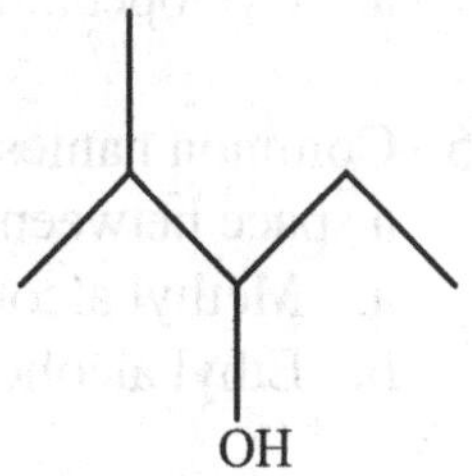 OH

2-methyl-3-pentanol

12.31 Complete structural formulas show all of the bonds between each atom. With a line structure the carbons are represented by intersecting points between lines. The end of a line also represents a carbon. It is assumed that each carbon has the appropriate number of hydrogens to give it a total of four bonds. For the attached hydroxyl groups, both the oxygen atom and its hydrogen must be written in.

a.

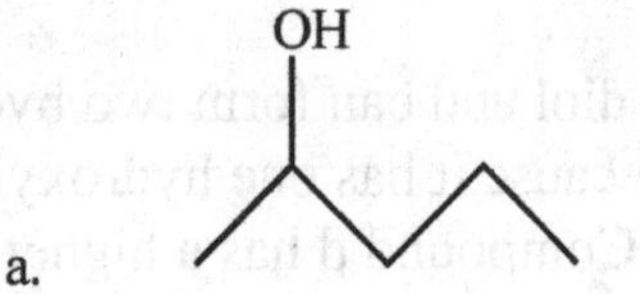

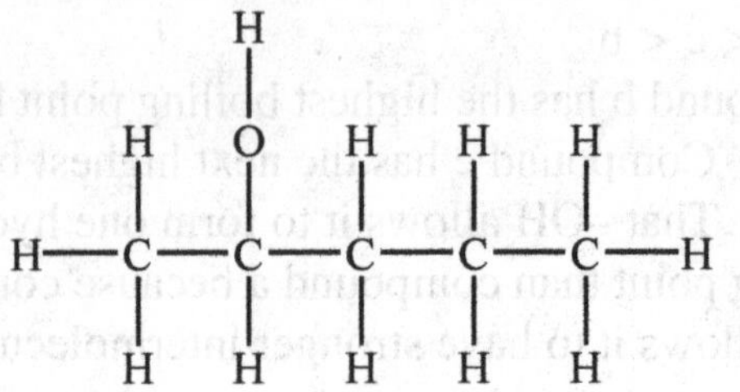

b.

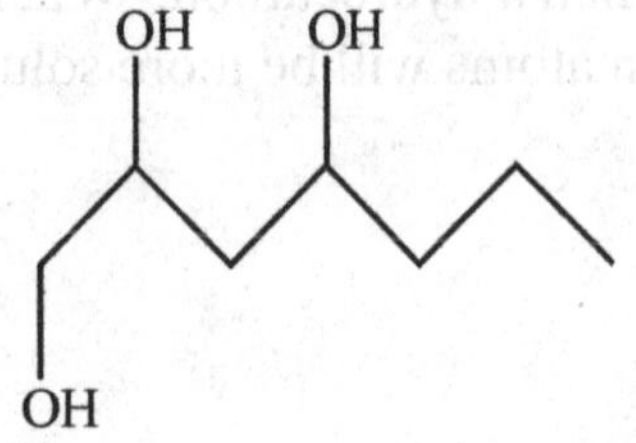

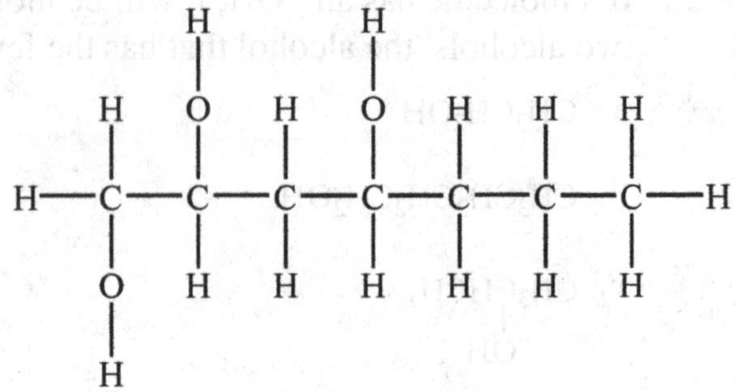

c.

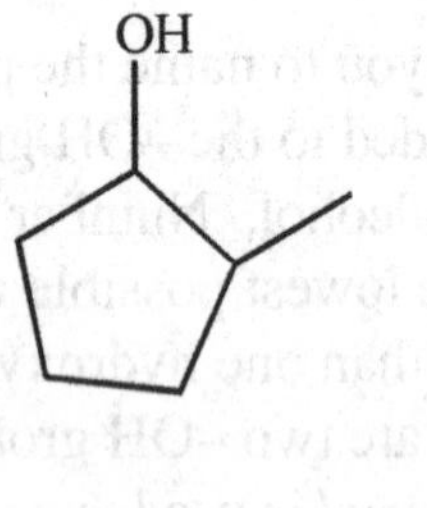

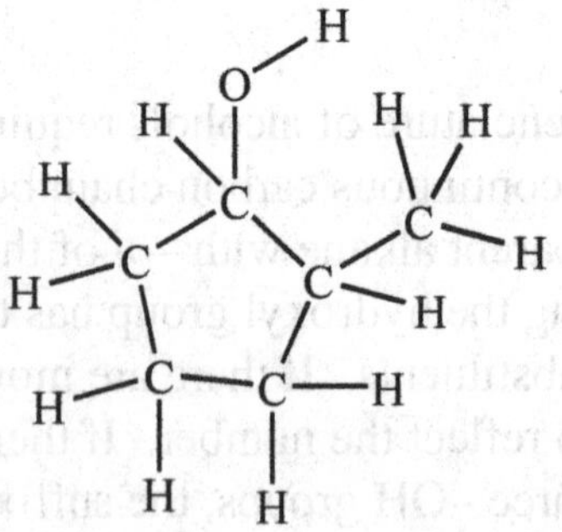

12.33 When there is only one substituent on a cycloalkane, a number is not needed to indicate the position of that substituent. These cyclic compounds all have a hydroxyl. Therefore, they all have the cyclic alcohol as the parent when deriving their IUPAC names.

a. Cyclopentanol b. Cycloheptanol c. 3-Methylcyclohexanol

12.35 Common names for alcohols often have the alkyl name before the word alcohol. There is a space between these parts of the name.

a. Methyl alcohol
b. Ethyl alcohol
c. Ethylene glycol (historic name)
d. Propyl alcohol

12.37 a. 4-Methyl-2-hexanol: $CH_3CHOHCH_2CH(CH_3)CH_2CH_3$
b. Isobutyl alcohol: $CH(CH_3)_2CH_2OH$
c. 1,5-Pentanediol: $CH_2OH(CH_2)_3CH_2OH$
d. 2-Nonanol: $CH_3CHOH(CH_2)_6CH_3$
e. 1,3,5-Cyclohexanetriol:

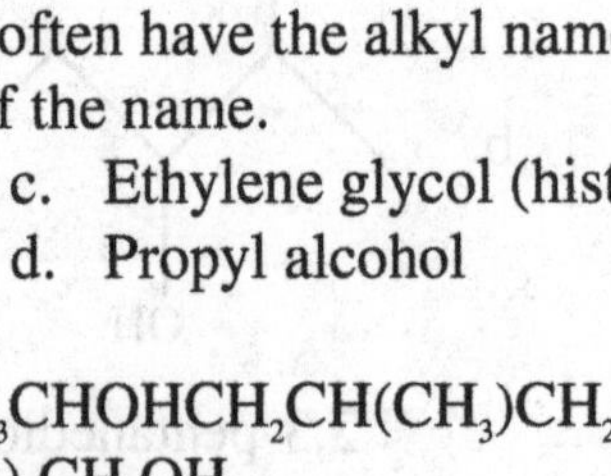

12.39 Denatured alcohol is 100% ethanol to which benzene or methanol is added. The additive makes the ethanol unfit to drink and prevents illegal use of pure ethanol.

12.41 Fermentation is the anaerobic degradation of sugar that involves no net oxidation. The alcohol fermentation, carried out by yeast, produces ethanol and carbon dioxide.

12.43 When the ethanol concentration in a fermentation reaches 12-13%, the yeast producing the ethanol are killed by it. To produce a liquor of higher alcohol concentration, the product of the original fermentation must be distilled.

12.45 Hydration of an alkene results in the production of an alcohol because the hydrogen adds to one carbon of the double bond, and the –OH adds to the other carbon of the double bond. Markovnikov's rule should be followed if the alkene is unsymmetrical.

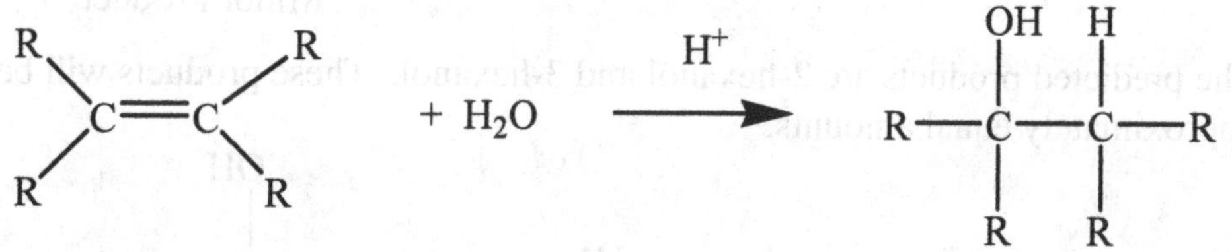

$$R_2C{=}CR_2 + H_2O \xrightarrow{H^+} R{-}C(OH)(R){-}C(H)(R){-}R$$

Alkene

Alcohol

12.47 Dehydration of an alcohol results in an alkene. Water is lost from the alcohol as the –OH and the hydrogen leave and the double bond forms.

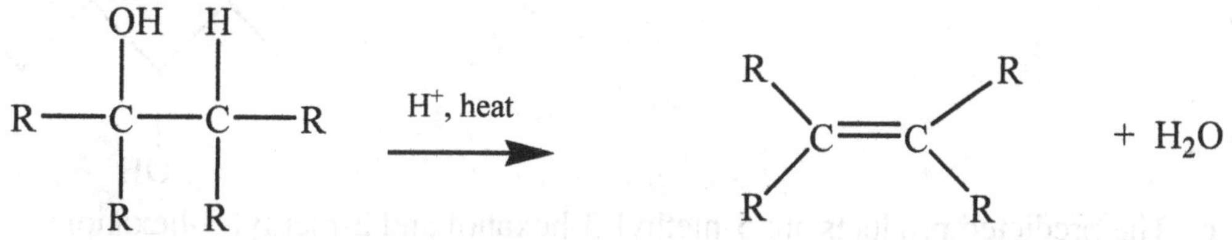

$$R{-}C(OH)(R){-}C(H)(R){-}R \xrightarrow{H^+,\ heat} R_2C{=}CR_2 + H_2O$$

Alcohol

Alkene

12.49 Oxidation of a secondary alcohol results in the formation of a ketone.

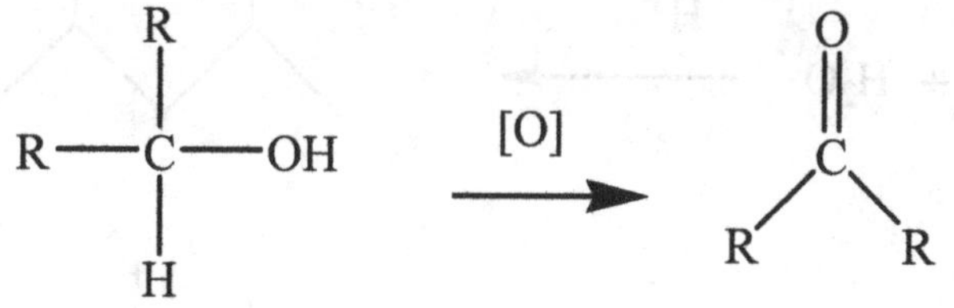

$$R{-}C(R)(H){-}OH \xrightarrow{[O]} R{-}C(=O){-}R$$

Secondary alcohol

Ketone

12.51

a. The predicted products are 1-hexanol (minor) and 2-hexanol (major).

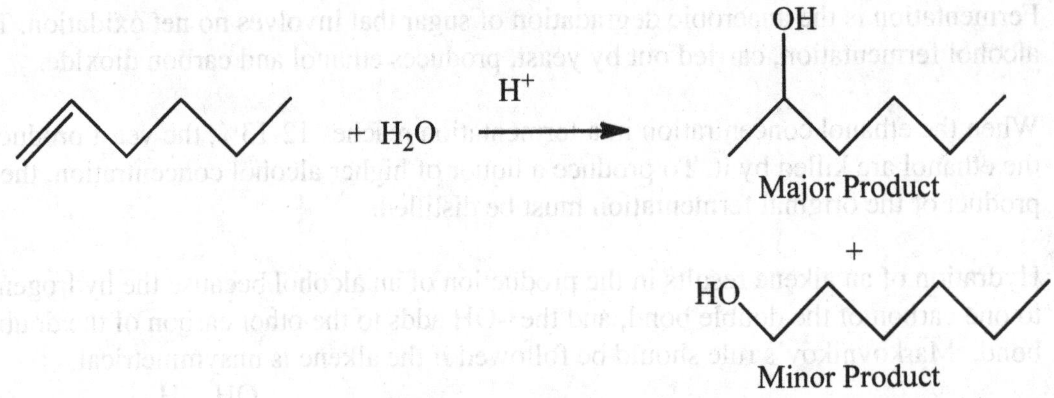

$$+ H_2O \xrightarrow{H^+}$$

OH

Major Product

+

HO

Minor Product

b. The predicted products are 2-hexanol and 3-hexanol. These products will be formed in approximately equal amounts.

$$+ H_2O \xrightarrow{H^+}$$

OH

+

OH

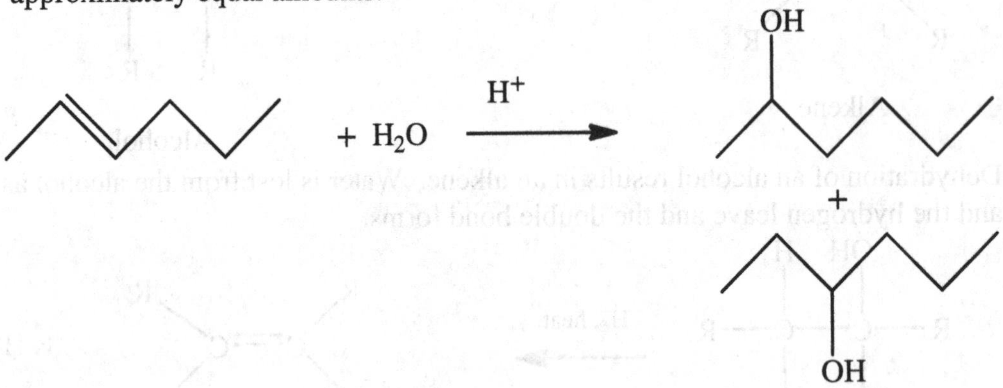

c. The predicted products are 5-methyl-3-hexanol and 2-methyl-3-hexanol. These products will be formed in approximately equal amounts.

$$+ H_2O \xrightarrow{H^+}$$

OH

+

OH

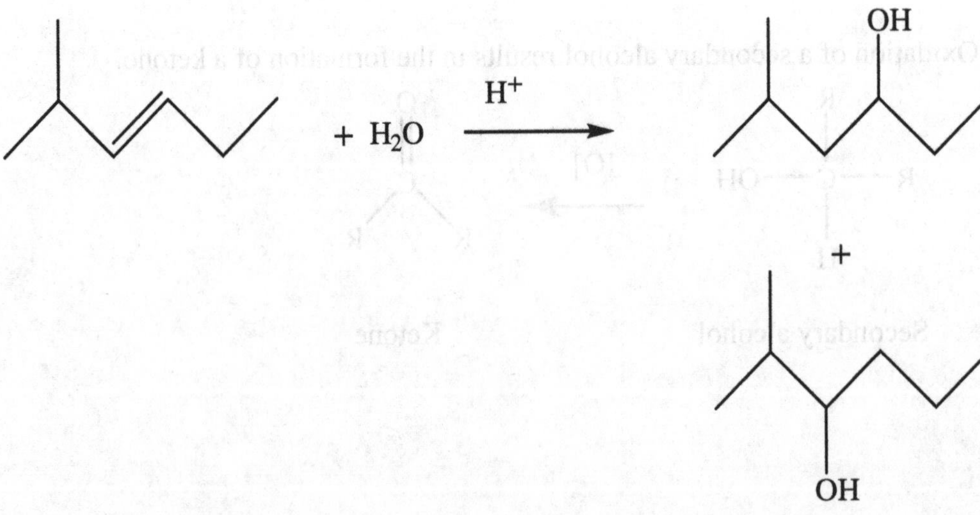

d. The predicted products are 2,2-dimethyl-4-heptanol and 2,2-dimethyl-3-heptanol. These products will be formed in approximately equal amounts.

$$+ H_2O \xrightarrow{H^+}$$ OH

+

OH

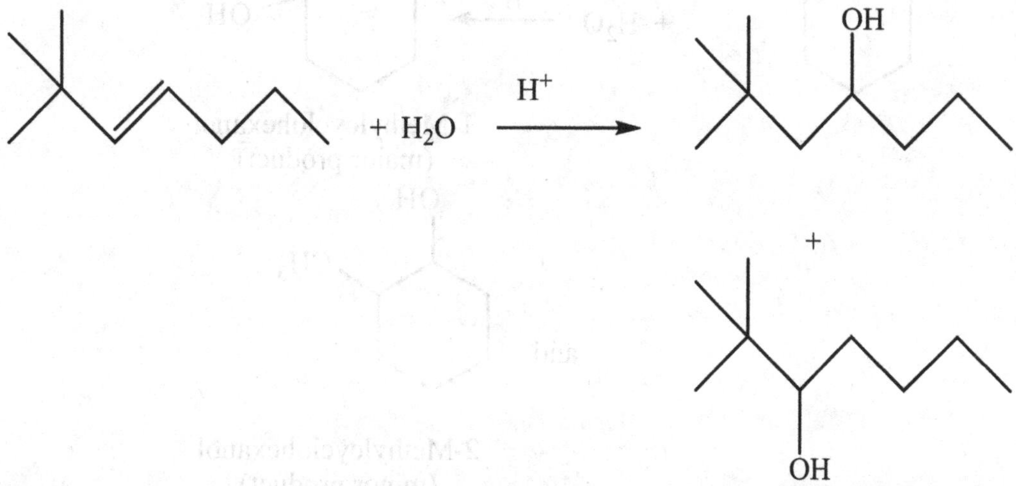

12.53 a. These products will be formed in approximately equal amounts.

$$CH_3CH{=}CHCH_2CH_2CH_3 + H_2O \xrightarrow{H^+} CH_3\overset{OH}{\overset{|}{C}}HCH_2CH_2CH_2CH_3$$

2-Hexene 2-Hexanol

and 3-Hexanol $CH_3CH_2\overset{OH}{\overset{|}{C}}HCH_2CH_2CH_3$

b. There is only one product that forms.

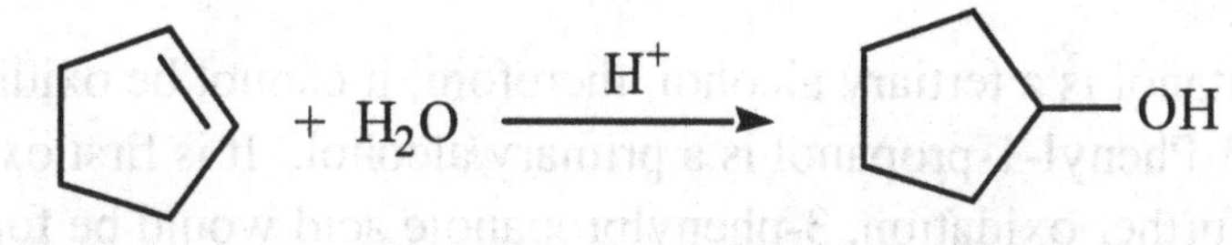

$$+ H_2O \xrightarrow{H^+}$$ OH

Cyclopentene Cyclopentanol

c. The products form according to Markovnikov's rule.

$$CH_2{=}CHCH_2CH_2CH_2CH_2CH_2CH_3 + H_2O \xrightarrow{H^+} CH_3\overset{OH}{\overset{|}{C}}HCH_2CH_2CH_2CH_2CH_2CH_3$$

2-Octanol (major product)

and 1-Octanol (minor product) $\overset{OH}{\overset{|}{C}}H_2CH_2CH_2CH_2CH_2CH_2CH_2CH_3$

d. The products form according to Markovnikov's rule.

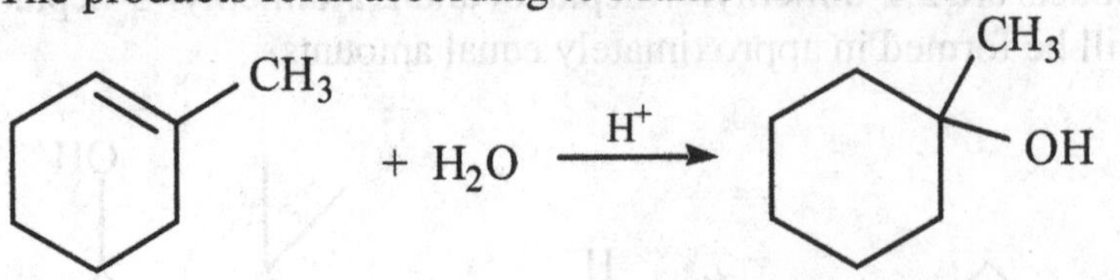

$$\text{1-methylcyclohexene} + H_2O \xrightarrow{H^+} \text{1-Methylcyclohexanol (major product)} \text{ and } \text{2-Methylcyclohexanol (minor product)}$$

1-Methylcyclohexanol
(major product)

and

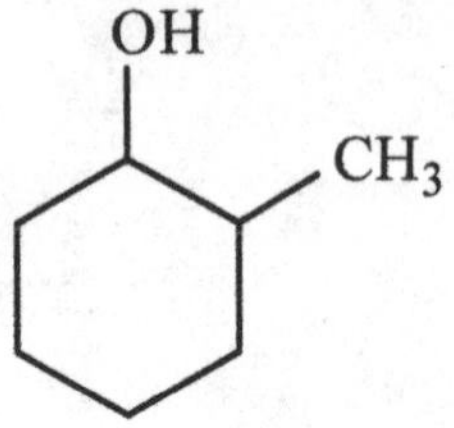

2-Methylcyclohexanol
(minor product)

12.55 a. Butanone. 2-Butanol is a secondary alcohol, therefore; it oxidizes to a ketone.
b. N.R. 2-Methyl-2-hexanol is a tertiary alcohol, therefore; it cannot be oxidized.
c. Cyclohexanone Cyclohexanol is a secondary alcohol, therefore it oxidizes to a ketone.
d. N.R. 1-Methyl-1-cyclopentanol is a tertiary alcohol, therefore; it cannot be oxidized.

12.57 a. 3-Pentanone. 3-Pentanol is a secondary alcohol, therefore; it oxidizes to a ketone.
b. Propanal. Propanol is a primary alcohol. It is first oxidized to an aldehyde, but upon further oxidation, propanoic acid would be formed.
c. 4-Methyl-2-pentanone. 4-Methyl-2-pentanol is a secondary alcohol, therefore; it oxidizes to a ketone.
d. N.R. 2-Methyl-2-butanol is a tertiary alcohol, therefore; it cannot be oxidized.
e. 3-Phenylpropanal. 3-Phenyl-1-propanol is a primary alcohol. It is first oxidized to an aldehyde, but upon further oxidation, 3-phenylpropanoic acid would be formed.

12.59

$$CH_3CH_2OH \xrightarrow{\text{liver enzymes}} H_3C{-}C({=}O){-}H$$

Ethanol — Ethanal

The product, ethanal, is responsible for the symptoms of a hangover.

12.61 The reaction in which a water molecule is added to 1-butene is a hydration reaction.

$$CH_2{=}CHCH_2CH_3 + H_2O \xrightarrow{H^+} CH_3CH(OH)CH_2CH_3$$

2-Butanol is the major product of this reaction. 1-Butanol would also be formed as a minor product.

12.63

$$CH_3CH{=}CH_2 + H_2O \xrightarrow{H^+} CH_3\overset{\overset{\displaystyle OH}{|}}{C}HCH_3 \xrightarrow{[O]} CH_3\overset{\overset{\displaystyle O}{\|}}{C}CH_3$$

Propene (propylene) — 2-Propanol (isopropanol) — Propanone (acetone)

12.65

a. Hexanal + H_2 $\xrightarrow{\text{catalyst}}$ 1-Hexanol

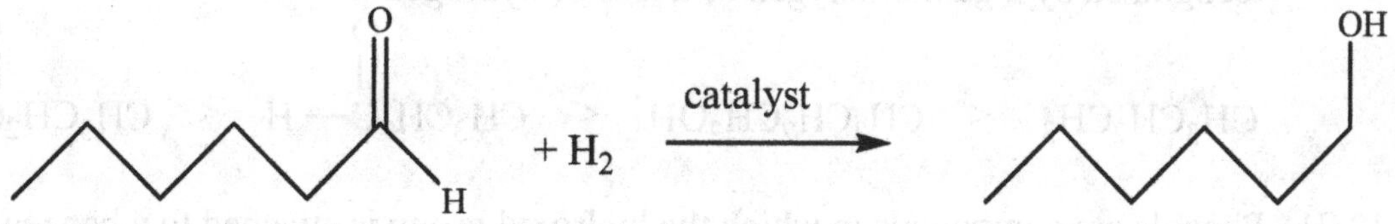

b. 2-Hexanone + H_2 $\xrightarrow{\text{catalyst}}$ 2-Hexanol

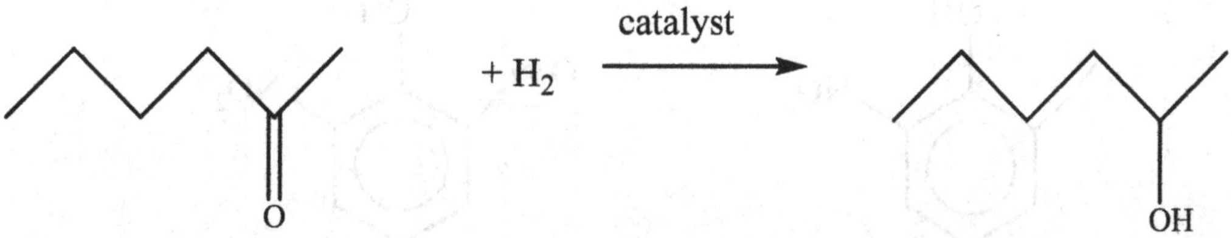

c. 2-Methylbutanal + H_2 $\xrightarrow{\text{catalyst}}$ 2-Methyl-1-butanol

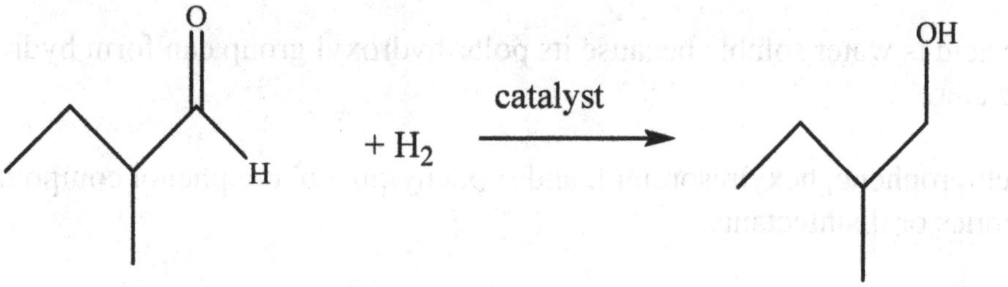

d. 6-Ethyl-2-octanone $+ H_2$ $\xrightarrow{\text{catalyst}}$ 6-Ethyl-2-octanol

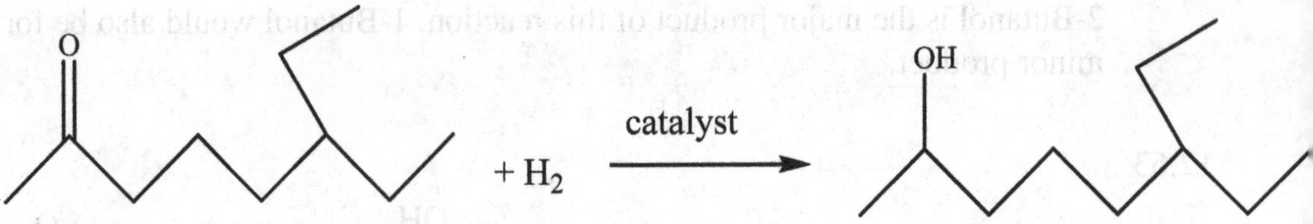

12.67 When discussing inorganic compounds, oxidation is a loss of electrons, whereas reduction is a gain of electrons.

12.69 The most reduced compound is the alkane. A primary alcohol can be oxidized to an aldeyhyde. An aldehyde can be oxidized further to a carboxylic acid. Oxidation may be recognized by a gain of oxygen or a loss of hydrogen.

$$CH_3CH_2CH_3 < CH_3CH_2CH_2OH < CH_3CH_2C(=O)-H < CH_3CH_2C(=O)-OH$$

12.71 Phenols are compounds in which the hydroxyl group is attached to a benzene ring.

12.73 Picric acid: 2,4,6,-Trinitrotoluene:

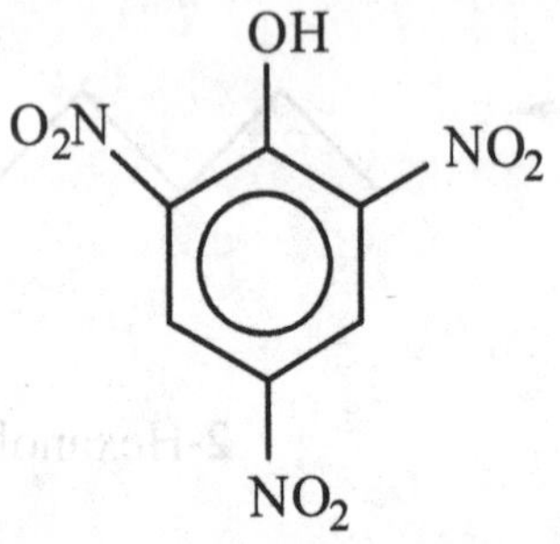

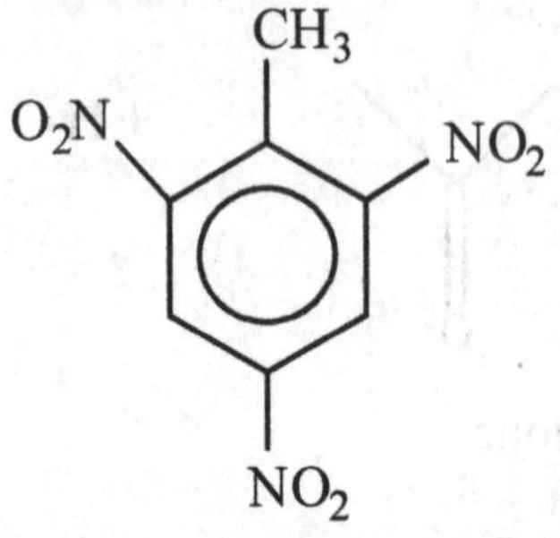

OH, O_2N, NO_2, NO_2 — CH_3, O_2N, NO_2, NO_2

Picric acid is water soluble because its polar hydroxyl group can form hydrogen bonds with water.

12.75 Hexachlorophene, hexylresorcinol, and *o*-phenylphenol are phenol compounds used as antiseptics or disinfectants.

12.77 Ethers have much lower boiling points than alcohols of similar molar mass, but higher boiling points than alkanes of similar molar mass. The boiling points are higher than alkanes because the R-O-R bond is polar. However, there is no –OH group, so ether molecules cannot hydrogen bond to one another. For this reason they have lower boiling points than alcohols of similar molar mass.

12.79 Alcohols of molecular formula $C_4H_{10}O$

$$CH_3CH_2CH_2CH_2OH \quad CH_3\overset{\overset{\displaystyle OH}{|}}{C}HCH_2CH_3 \quad CH_3\underset{\underset{\displaystyle CH_3}{|}}{C}HCH_2OH \quad CH_3-\overset{\overset{\displaystyle OH}{|}}{\underset{\underset{\displaystyle CH_3}{|}}{C}}-CH_3$$

Ethers of molecular formula $C_4H_{10}O$

$$CH_3-O-CH_2CH_2CH_3 \quad CH_3CH_2-O-CH_2CH_3 \quad CH_3-O-\underset{\underset{\displaystyle CH_3}{|}}{C}HCH_3$$

12.81 a. The reactant alcohol is ethanol (ethyl alcohol). There is only one possible ether formed from this reaction, ethoxyethane (diethyl ether).

$$CH_3CH_2-O-CH_2CH_3 \; + H_2O$$

b. The reactant alcohols are methanol (methyl alcohol) and ethanol (ethyl alcohol). The three products are ethoxyethane (diethyl ether), methoxymethane (dimethyl ether), and methoxyethane (ethyl methyl ether).

$$CH_3CH_2-O-CH_2CH_3 \; + \; CH_3-O-CH_3$$

$$+ \; CH_3-O-CH_2CH_3 \; + H_2O$$

c. The reactant alcohols are 2-propanol (isopropyl alcohol) and ethanol (ethyl alcohol). The products are methoxymethane (dimethyl ether), 2-methoxypropane (methyl isopropyl ether), and 2-isopropoxy propane (diisopropyl ether).

$$CH_3-O-CH_3 \; + \; CH_3-O-\overset{\overset{\displaystyle CH_3}{|}}{C}HCH_3$$

$$+ \; CH_3\overset{\overset{\displaystyle CH_3}{|}}{C}H-O-\overset{\overset{\displaystyle CH_3}{|}}{C}HCH_3 \; + H_2O$$

d. The reactant is cyclopentanol (cyclopentyl alcohol). There is only one possible ether formed from this reaction, cyclopentoxy cyclopentane (dicyclopentyl ether).

(cyclopentyl)–O–(cyclopentyl) $+ H_2O$

12.83 In the IUPAC system of naming ethers, the longest carbon chain is named as an alkane. Then the –OR substituent is named as an alkoxy group. The location of the alkoxy substituent on the alkane is indicated by the carbon number.

a. 2-Ethoxypentane
b. 2-Methoxybutane
c. 1-Ethoxybutane
d. Methoxycyclopentane

12.85 a. Dibutyl ether has two butyl groups joined an oxygen atom.

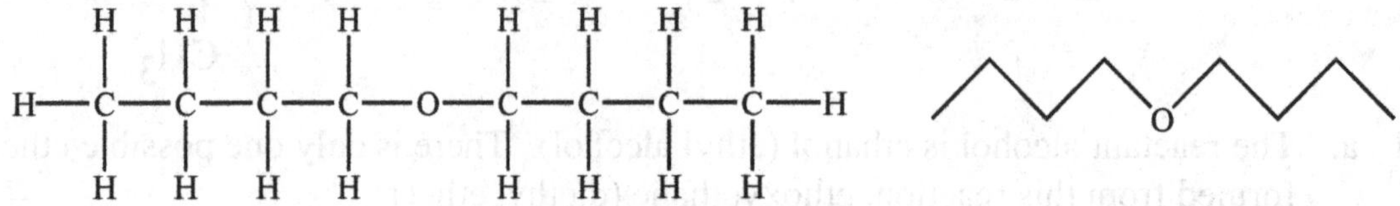

```
    H   H   H   H       H   H   H   H
    |   |   |   |       |   |   |   |
H — C — C — C — C — O — C — C — C — C — H
    |   |   |   |       |   |   |   |
    H   H   H   H       H   H   H   H
```

b. Ethyl heptyl ether has an ethyl group and a heptyl group joined by an oxygen atom.

```
    H   H   H   H   H   H   H       H   H
    |   |   |   |   |   |   |       |   |
H — C — C — C — C — C — C — C — O — C — C — H
    |   |   |   |   |   |   |       |   |
    H   H   H   H   H   H   H       H   H
```

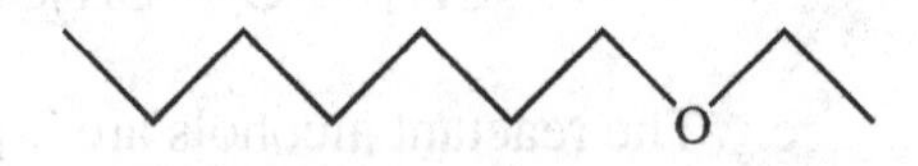

c. Propyl pentyl ether has a propyl group and a pentyl group joined by an oxygen atom.

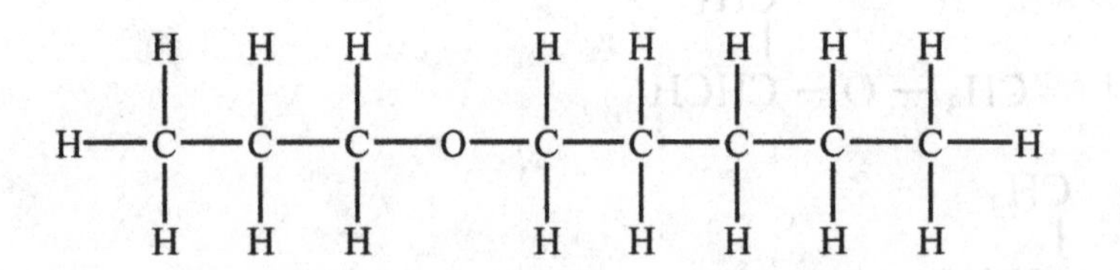

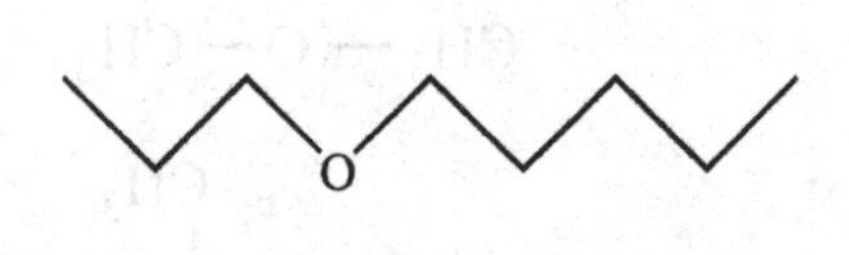

d. *t*-Butyl hexyl ether has a *tert*-butyl group and a hexyl group joined by an oxygen atom.

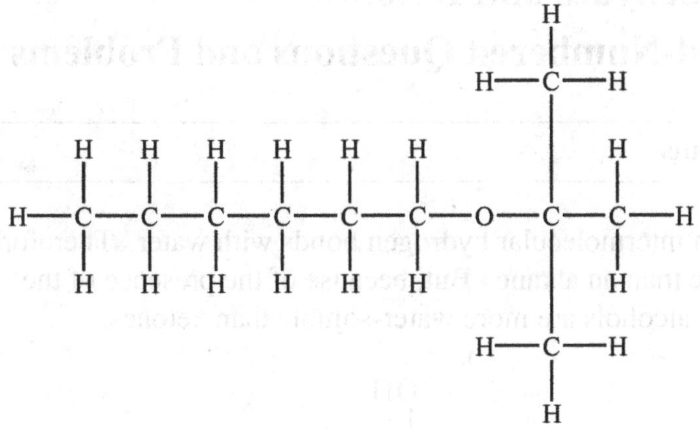

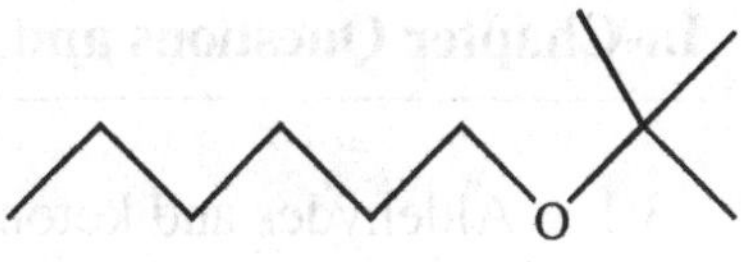

```
                                          H
                                          |
                                       H—C—H
                                          |
   H   H   H   H   H   H                  H
   |   |   |   |   |   |                  |
H—C—C—C—C—C—C—O—C—C—H
   |   |   |   |   |   |                  |
   H   H   H   H   H   H                  H
                                          |
                                       H—C—H
                                          |
                                          H
```

O

12.87 Thiols contain the sulfhydryl group (-SH). The sulfhydryl group is similar to the hydroxyl group (-OH) of alcohols, except that a sulfur atom replaces the oxygen atom.

12.89 Cystine:

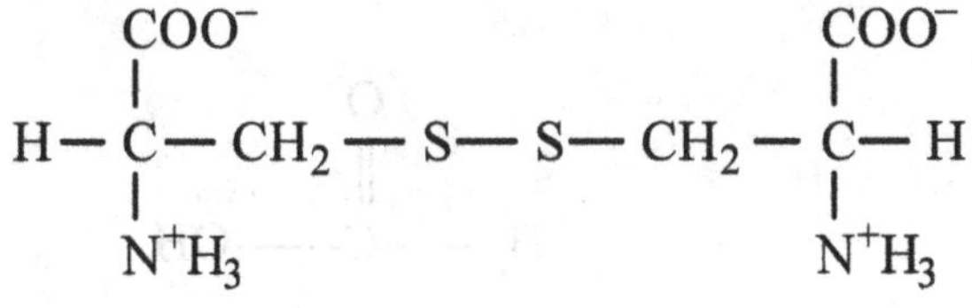

```
        COO⁻                        COO⁻
         |                           |
    H — C — CH₂ — S — S — CH₂ — C — H
         |                           |
        N⁺H₃                        N⁺H₃
```

12.91 The IUPAC rules for naming thiols are similar to those for naming alcohols, except that the full name of the alkane is retained. The suffix *–thiol* follows the name of the parent compound.

a. 1-Propanethiol
b. 2-Butanethiol
c. 2-Methyl-2-butanethiol
d. 1,4-Cyclohexanedithiol

Chapter 13

Aldehydes and Ketones

Solutions to the Odd-Numbered Questions and Problems

In-Chapter Questions and Problems

13.1 Aldehydes and ketones form intermolecular hydrogen bonds with water. Therefore a ketone is more water-soluble than an alkane. But, because of the presence of the hydroxyl group on alcohols, alcohols are more water-soluble than ketones.

a. $CH_3-C(=O)-CH_3$

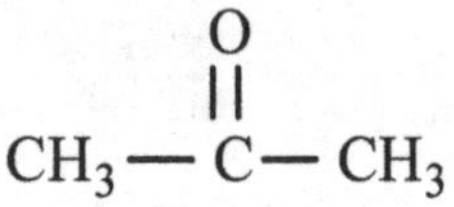

b. $CH_3CH(OH)CH_2CHCH_3$

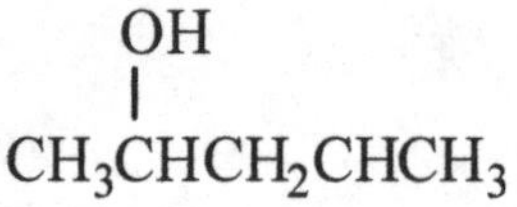

13.3 Carboxylic acids form intermolecular hydrogen bonds, whereas, aldehydes and ketones do not. Therefore, carboxylic acids have higher boiling points.

a. $CH_3CH_2-C(=O)-OH$

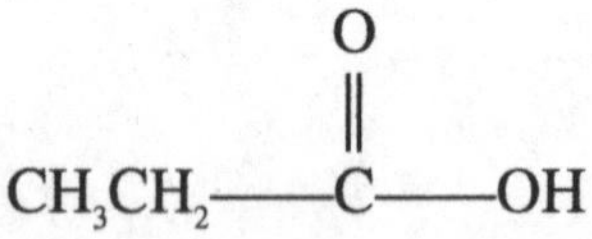

b. $CH_3-C(=O)-OH$

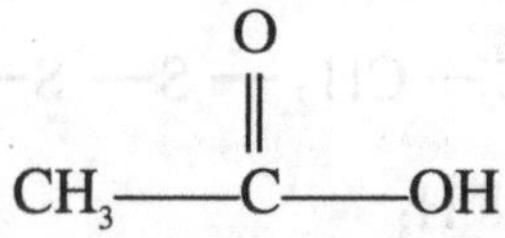

13.5 When drawing the condensed formula of an aldehyde from the IUPAC name, it is important that you number the carbons in the parent chain so that the carbonyl carbon is carbon number one.

a. 2,3-Dichloropentanal

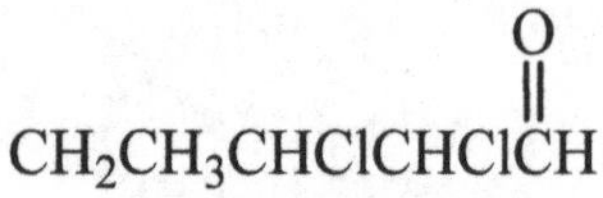

$CH_2CH_3CHClCHClCH(=O)$

b. 2-Bromobutanal

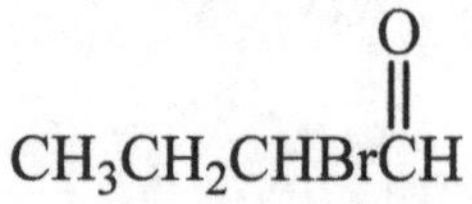

$CH_3CH_2CHBrCH(=O)$

c. 4-Methylhexanal

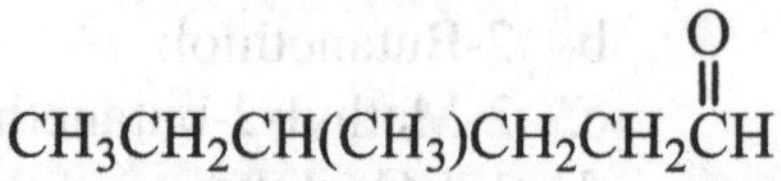

$CH_3CH_2CH(CH_3)CH_2CH_2CH(=O)$

d. Butanal

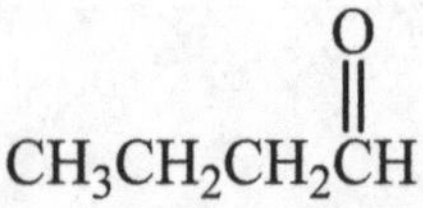

$CH_3CH_2CH_2CH(=O)$

e. 2,4-Dimethylpentanal

$CH_3CH(CH_3)CH_2CH(CH_3)CH(=O)$

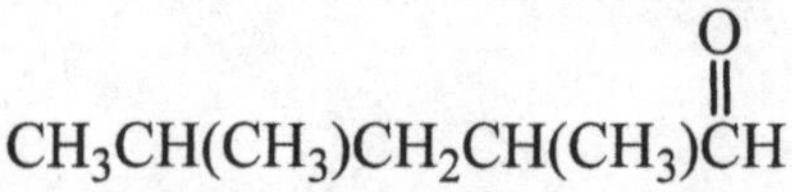

13.7 When naming ketones using the IUPAC Nomenclature System, make sure that the longest carbon chain is numbered to give the carbonyl carbon the lowest possible number.

a. 3-Iodobutanone
b. 4-Methyl-2-octanone
c. 3-Methylbutanone
d. 2-Methyl-3-pentanone
e. 2-Fluoro-3-pentanone

13.9 Ethanal (acetaldehyde) is the aldehyde synthesized from ethanol in the liver.

$$CH_3-\overset{\overset{\displaystyle O}{\|}}{C}-H$$

13.11
a. Reduction. An aldehyde is reduced to a primary alcohol.
b. Reduction. A carboxylic acid is reduced to an aldehyde.
c. Reduction. A ketone is reduced to a secondary alcohol.
d. Oxidation. An secondary alcohol is oxidized to a ketone.
e. Reduction. Both carbonyl groups are reduced to hydroxyl groups.

13.13
a. Hemiacetal There is a hydroxyl, a hydrogen, and an alkoxy group.
b. Aetal There are two alkoxy groups.
c. Acetal There is a hydrogen and there are two alkoxy groups.
d. Hemiacetal There is a hydroxyl and an alkoxy group.

End-of-Chapter Questions and Problems

13.15 As the carbon chain length increases, the compounds become less polar and more hydrocarbon-like. As a result, their solubility in water decreases.

13.17 A good solvent should dissolve a wide range of compounds. Simple ketones are considered to be universal solvents because they have both a polar carbonyl group and nonpolar side chains. As a result, they dissolve organic compounds and are also miscible in water.

13.19 Hydrogen bonds form between the carbonyl group of aldehydes and ketones and water.

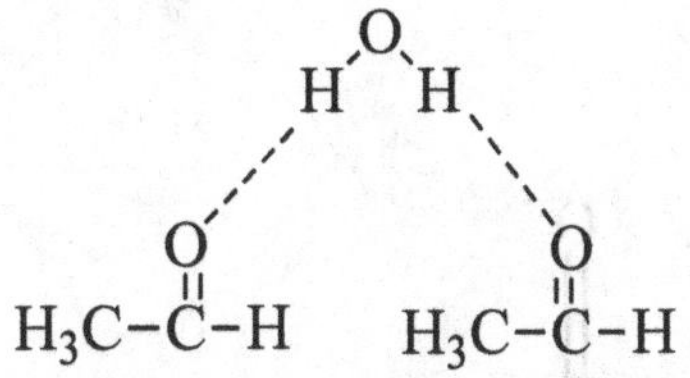

O
H H

O O
H₃C–C–H H₃C–C–H

13.21 Alcohols have higher boiling points than aldehydes or ketones of comparable molecular mass because alcohol molecules can form intermolecular hydrogen bonds with one another. Aldehydes and ketones cannot form intermolecular hydrogen bonds with one another.

13.23 a.

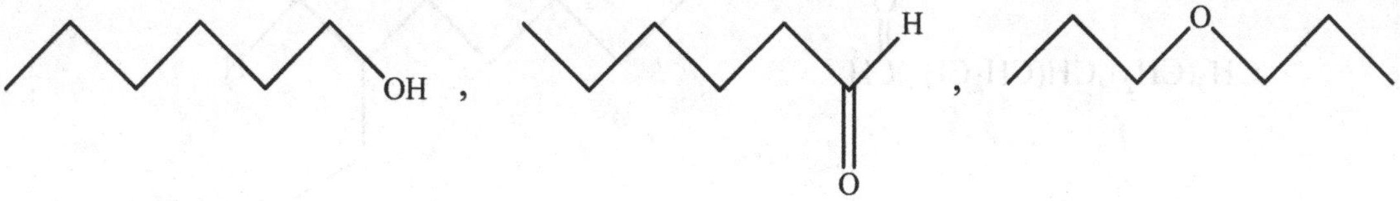

OH , H O , O

Highest Lowest

b.

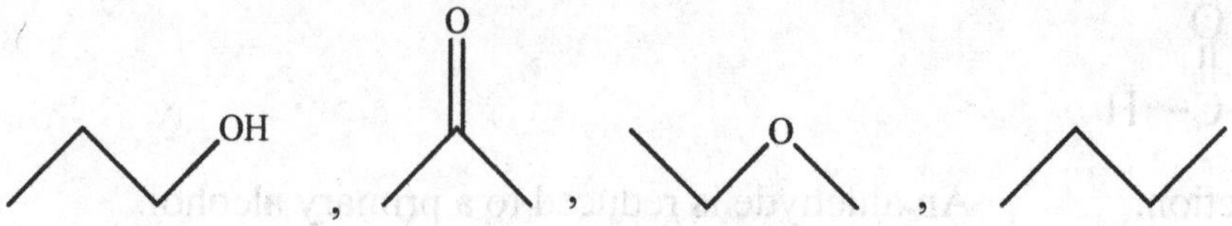

OH, O (ketone), O (ether), (alkane)

Highest ... Lowest

13.25 To name an aldehyde using the IUPAC nomenclature system, identify and name the longest carbon chain containing the carbonyl group. Replace the final *–e* of the alkane name with *– al*. Number and name all substituents as usual. Remember that the carbonyl carbon is always carbon-1 and does not need to be numbered in the name of the compound.

13.27 The common names of aldehydes are derived from the same Latin roots as the corresponding carboxylic acids. For instance, methanal is formaldehyde; ethanal is acetaldehyde; propanal is propionaldehyde, etc.

Substituted aldehydes are named as derivatives of the straight-chain parent compound. Greek letters are used to indicate the position of substituents. The carbon nearest the carbonyl group is the α-carbon, the next is the β-carbon, and so on.

13.29 In drawing these structures from the IUPAC name, begin with the carbon chain that represents the parent name. The carbonyl of the aldehyde will be carbon number one.

a. Ethanal

$CH_3\overset{O}{\overset{\|}{C}}H$

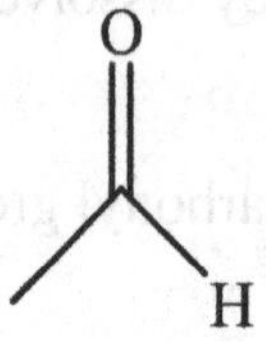

b. 3,4-Dimethylpentanal

$(CH_3)_2CHCH(CH_3)CH_2\overset{O}{\overset{\|}{C}}H$

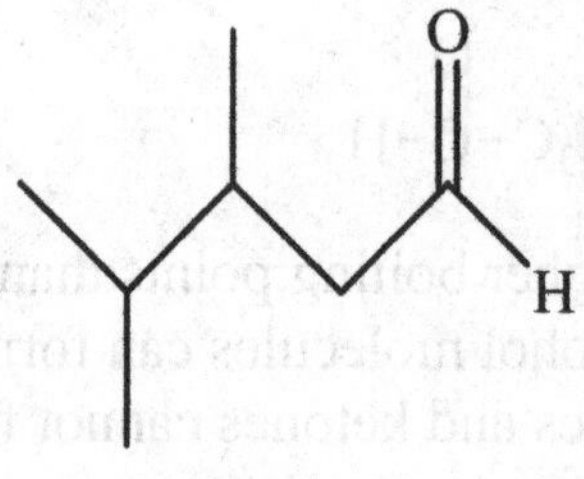

c. 2-Ethylheptanal

$CH_3(CH_2)_4CH(CH_2CH_3)\overset{O}{\overset{\|}{C}}H$

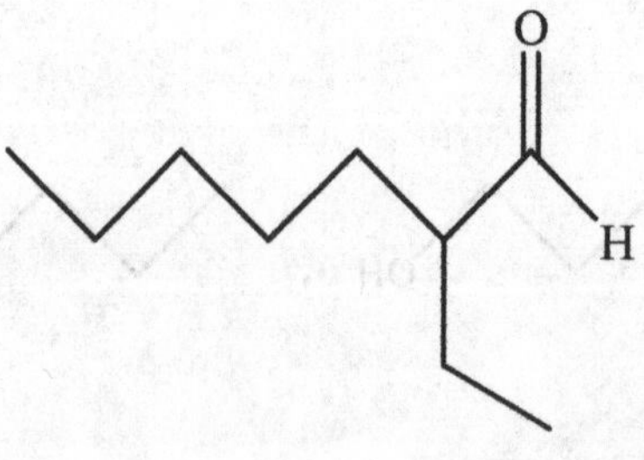

d. 5,7-Dichloroheptanal

$CH_2ClCH_2CHCl(CH_2)_3\overset{\overset{O}{\|}}{C}H$

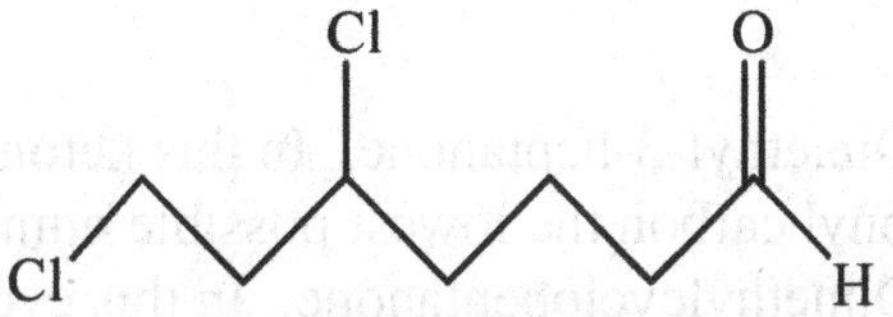

13.31 The common names of ketones are derived by naming the alkyl groups that are bonded to the carbonyl carbons. These are used as prefixes followed by the word ketone.

a. Ethyl isopropyl ketone

$CH_3CH(CH_3)\overset{\overset{O}{\|}}{C}CH_2CH_3$

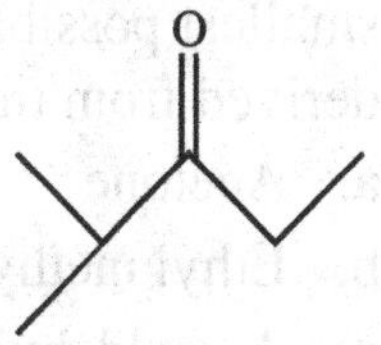

b. Ethyl propyl ketone

$CH_3(CH_2)_2\overset{\overset{O}{\|}}{C}CH_2CH_3$

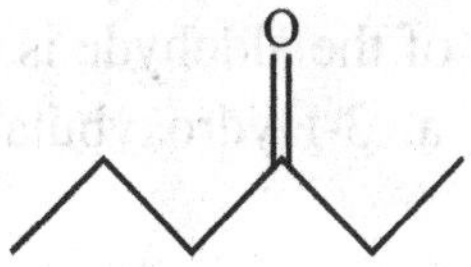

c. Dibutyl ketone

$CH_3(CH_2)_3\overset{\overset{O}{\|}}{C}(CH_2)_3CH_3$

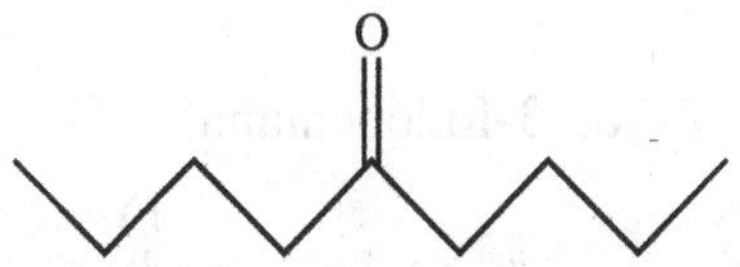

e. Heptyl hexyl ketone

$CH_3(CH_2)_5\overset{\overset{O}{\|}}{C}(CH_2)_6CH_3$

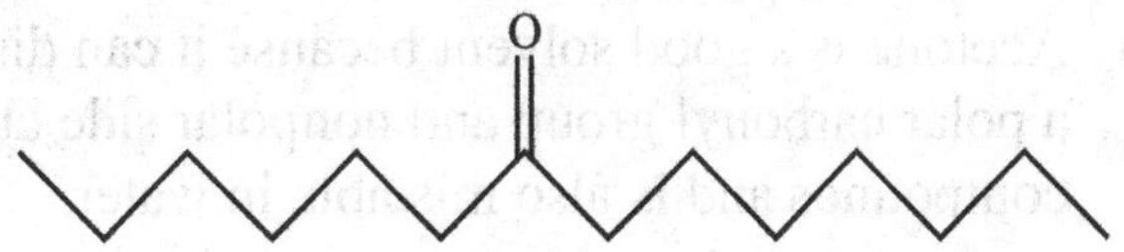

13.33 The longest chain is numbered to give the carbonyl carbon the lowest possible number.

a. Butanone

b. 2-Ethylhexanal

13.35 a. 3-Nitrobenzaldehyde. The carbon on the benzene ring that is bonded to the carbonyl of the aldehyde is carbon number one.

b. 3,4-Dihydroxycyclopentanone. In this cyclic compound, the carbonyl carbon is carbon number one because the rules state that the carbonyl carbon should have the lowest possible number.

13.37 7-Hydroxy-3,7-dimethyloctanal. The carbonyl carbon of an aldehyde is always carbon-one.

13.39 a. 4,6-Dimethyl-3-heptanone. In this ketone, the longest chain is numbered to give the carbonyl carbon the lowest possible number.
b. 3,3-Dimethylcyclopentanone. In this cyclic compound, the carbonyl carbon is carbon number one because the rules state that the carbonyl carbon should have the lowest possible number.

13.41 The common names of ketones are derived by naming the alkyl groups that are bonded to the carbonyl carbons. These are used as prefixes followed by the word ketone. The smallest possible ketone, acetone is an exception. The common names of aldehydes are derived from the same Latin roots as the corresponding carboxylic acids.
a. Acetone
b. Ethyl methyl ketone
c. Acetaldehyde
d. Propionaldehyde
e. Methyl isopropyl ketone

13.43 When drawing the structure of an aldehyde from the IUPAC name, the carbonyl carbon of the aldehyde is carbon number one.

a. 3-Hydroxybutanal

$$CH_3CHOHCH_2\overset{O}{\overset{\|}{C}}H$$

b. 2-Methylpentanal

$$CH_3(CH_2)_2CH(CH_3)\overset{O}{\overset{\|}{C}}H$$

c. 4-Bromohexanal

$$CH_3CH_2CHBr(CH_2)_2\overset{O}{\overset{\|}{C}}H$$

d. 3-Iodopentanal

$$CH_3CH_2CHICH_2\overset{O}{\overset{\|}{C}}H$$

e. 2-Hydroxy-3-methylheptanal

$$CH_3(CH_2)_3CH(CH_3)CHOH\overset{O}{\overset{\|}{C}}H$$

13.45 Acetone is a good solvent because it can dissolve a wide range of compounds. It has both a polar carbonyl group and nonpolar side chains. As a result, it dissolves organic compounds and is also miscible in water.

13.47 The oxidation of ethanol to ethanal occurs in the liver and is catalyzed by the enzyme alcohol dehydrogenase.

13.49 In organic molecules, oxidation may be recognized as a gain of oxygen or a loss of hydrogen. An aldehyde may be oxidized to form a carboxylic acid as in the following example in which ethanal is oxidized to produce ethanoic acid.

$$\underset{\text{Ethanal}}{H_3C-\overset{O}{\overset{\|}{C}}-H} \xrightarrow{[O]} \underset{\text{Ethanoic acid}}{H_3C-\overset{O}{\overset{\|}{C}}-OH}$$

13.51 Addition reactions of aldehydes or ketones are those in which a second molecule is added to the double bond of the carbonyl group.

13.53 The following equation represents the oxidation of an aldehyde. The product is a carboxylic acid.

$$\underset{\text{Aldehyde}}{R-\overset{\overset{\large O}{\|}}{C}-H} \xrightarrow{[O]} \underset{\text{Carboxylic acid}}{R-\overset{\overset{\large O}{\|}}{C}-OH}$$

13.55 The following general equation represents the addition of alcohol molecule to an aldehyde:

$$\underset{\text{Aldehyde}}{R-\overset{\overset{\large O}{\|}}{C}-H} + R'OH \xrightleftharpoons{H^+} \underset{\text{Hemiacetal}}{R-\overset{\overset{\large OH}{|}}{\underset{\underset{\large H}{|}}{C}}-OR'}$$

The following general equation represents the addition of one alcohol molecule to a ketone:

$$\underset{\text{Ketone}}{R-\overset{\overset{\large O}{\|}}{C}-R} + R'OH \xrightleftharpoons{H^+} \underset{\text{Hemiacetal}}{R-\overset{\overset{\large OH}{|}}{\underset{\underset{\large R}{|}}{C}}-OR'}$$

13.57

a.

4-Methyl-2-heptanol $\xrightarrow{[O]}$ 4-Methyl-2-heptanone

b.

3,4-Dimethyl-1-pentanol $\xrightarrow{[O]}$ 3,4-Dimethylpentanal

c.

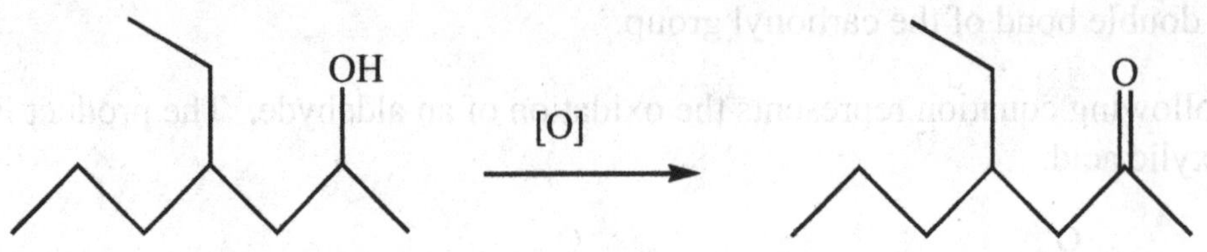

4-Ethyl-2-heptanol $\xrightarrow{[O]}$ 4-Ethyl-2-heptanone

d.

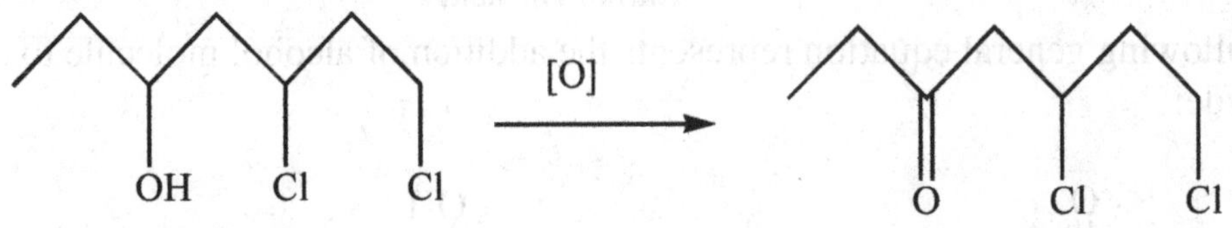

5,7-Dichloro-3-heptanol $\xrightarrow{[O]}$ 5,7-Dichloro-3-heptanone

13.59 The generalized equation for the oxidation of a primary alcohol shows that a primary alcohol is first oxidized to an aldehyde. It can then be further oxidized to a carboxylic acid.

$$R-CH_2OH \xrightarrow{[O]} R-\overset{\overset{\displaystyle O}{\|}}{C}-H \xrightarrow{[O]} R-\overset{\overset{\displaystyle O}{\|}}{C}-OH$$

Primary alcohol — Aldehyde — Carboxylic Acid

13.61 a. Reduction reaction. An aldehyde is reduced to an alcohol.

$$H_3C-\overset{\overset{\displaystyle O}{\|}}{C}-H + H_2 \xrightarrow{Pt} CH_3CH_2OH$$

Ethanal — Ethanol

b. Reduction reaction. A ketone is reduced to an alcohol.

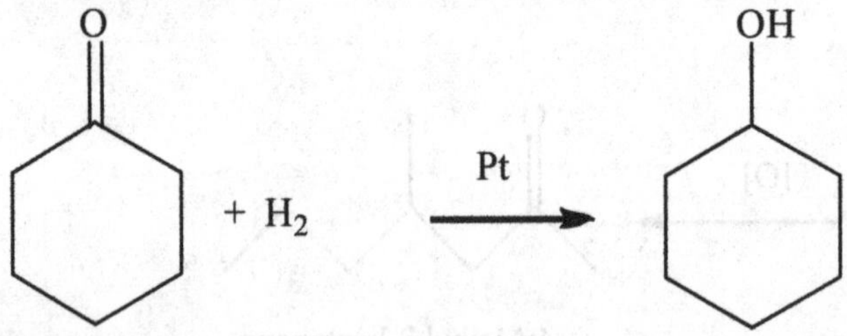

Cyclohexanone $+ H_2 \xrightarrow{Pt}$ Cyclohexanol

c. Oxidation reaction. A secondary alcohol is oxidized to a ketone.

$$H_3C-\underset{\underset{\displaystyle H}{|}}{\overset{\overset{\displaystyle OH}{|}}{C}}-CH_3 \xrightarrow{[O]} H_3C-\overset{\overset{\displaystyle O}{\|}}{C}-CH_3$$

2-Propanol — Propanone

13.63 Hydrogenation of an aldehyde produces an alcohol.

a. 3-Methylpentanal is reduced.

$$CH_3CH_2CH(CH_3)CH_2CHO + H_2 \xrightarrow{Pt} CH_3CH_2CH(CH_3)CH_2CH_2OH$$

b. 2-Hydroxypropanal is reduced.

$$CH_3CH(OH)CHO + H_2 \xrightarrow{Pt} CH_3CH(OH)CH_2OH$$

c. 2,3-Dimethylpentanal is reduced.

$$CH_3CH_2CH(CH_3)CH(CH_3)CHO + H_2 \xrightarrow{Pt} CH_3CH_2CH(CH_3)CH(CH_3)CH_2OH$$

d. 4-Chloropentanal is reduced.

$$CH_3CHClCH_2CH_2CHO + H_2 \xrightarrow{Pt} CH_3CHClCH_2CH_2CH_2OH$$

13.65 a. The following equation represents the hydrogenation of butanal. The product is 1-butanol.

$$\underset{\text{Butanal}}{CH_3CH_2CH_2\overset{O}{\overset{\|}{C}}H} + H_2 \xrightarrow{Pt} \underset{\text{1-Butanol}}{CH_3CH_2CH_2CH_2OH}$$

b. The following equation represents the hydrogenation of 3-methylpentanal. The product is 3-methyl-1-pentanol.

$$\underset{\text{3-Methylpentanal}}{CH_3CH_2\underset{CH_3}{\underset{|}{C}}HCH_2\overset{O}{\overset{\|}{C}}H} + H_2 \xrightarrow{Pt} \underset{\text{3-Methyl-1-pentanol}}{CH_3CH_2\underset{CH_3}{\underset{|}{C}}H\,CH_2CH_2OH}$$

c. The following equation represents the hydrogenation of 2-methylpropanal. The product is 2-methyl-1-propanol.

$$\underset{\substack{|\\ CH_3}}{CH_3CH}\overset{\overset{O}{\|}}{C}H + H_2 \xrightarrow{Pt} \underset{\substack{|\\ CH_3}}{CH_3CH}CH_2OH$$

13.67 Only (c) 3-methylbutanal and (f) acetaldehyde would give a positive Tollens' test. The Tollen's test is specific for aldehydes.

13.69 A hemiacetal is formed when an aldehyde reacts with one molecule of alcohol.

a.

$$CH_3CH_2-\overset{\overset{O}{\|}}{C}-H + CH_3CH_2OH \longrightarrow CH_3CH_2-\underset{\underset{OCH_2CH_3}{|}}{\overset{\overset{OH}{|}}{C}}-H$$

b.

$$CH_3-\overset{\overset{O}{\|}}{C}-H + CH_3CH_2OH \longrightarrow CH_3-\underset{\underset{OCH_2CH_3}{|}}{\overset{\overset{OH}{|}}{C}}-H$$

13.71 An acetal is formed when two molecules of alcohol react with an aldehyde.

a.

$$CH_3CH_2\overset{\overset{O}{\|}}{C}-H + 2CH_3OH \longrightarrow CH_3CH_2\underset{\underset{OCH_3}{|}}{\overset{\overset{OCH_3}{|}}{C}}-H + H_2O$$

b.

$$CH_3\overset{\overset{O}{\|}}{C}-H + 2CH_3OH \longrightarrow CH_3\underset{\underset{OCH_3}{|}}{\overset{\overset{OCH_3}{|}}{C}}-H + H_2O$$

13.73 a. Pentanal is oxidized to pentanoic acid.

$$CH_3CH_2CH_2CH_2\overset{\overset{O}{\|}}{C}H \xrightarrow{[O]} CH_3CH_2CH_2CH_2\overset{\overset{O}{\|}}{C}OH$$

b. Hexanal is oxidized to hexanoic acid.

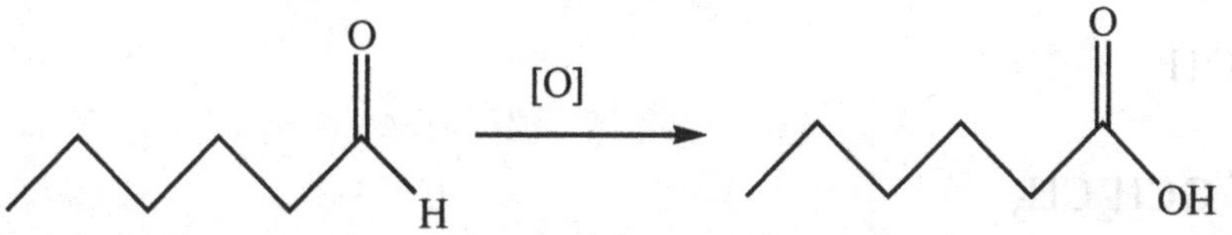

$$CH_3CH_2CH_2CH_2CH_2CHO \xrightarrow{[O]} CH_3CH_2CH_2CH_2CH_2COOH$$

c. Heptanal is oxidized to heptanoic acid.

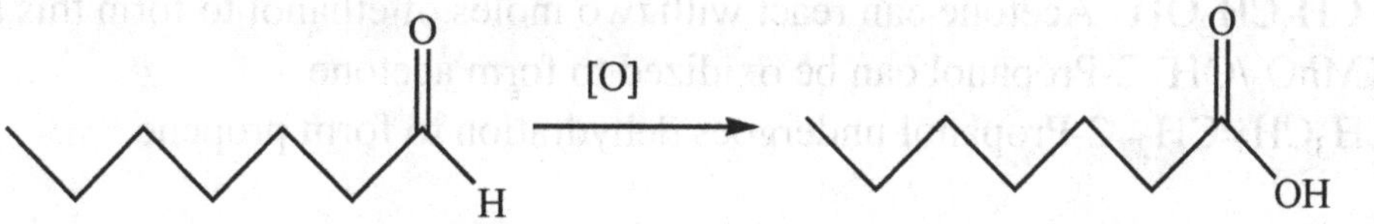

$$CH_3(CH_2)_5CHO \xrightarrow{[O]} CH_3(CH_2)_5COOH$$

d. Octanol is oxidized to octanoic acid.

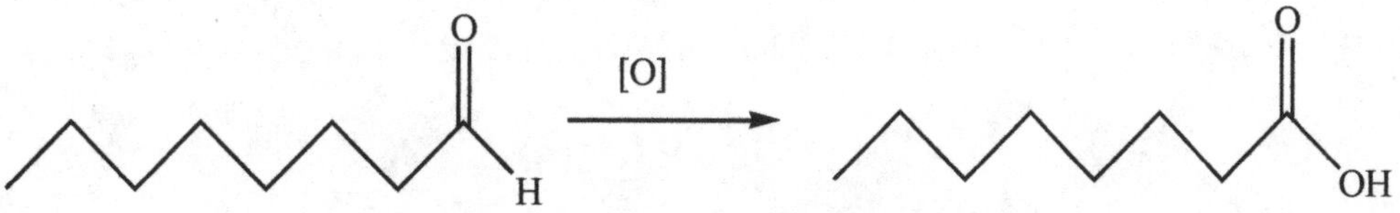

$$CH_3(CH_2)_6CHO \xrightarrow{[O]} CH_3(CH_2)_6COOH$$

13.75 a. Methanal is produced by the oxidation of methanol.
b. Propanal is produced by the oxidation of 1-propanol.

13.77 a. False. Aldehydes, but not ketones can be oxidized to produce carboxylic acids.
b. True.
c. False. Tertiary alcohols cannot be oxidized.
d. False. Alcohols can be produced by the reduction of an aldehyde or a ketone.

13.79 The keto form is a ketone. The enol form is an alkene and an alcohol.

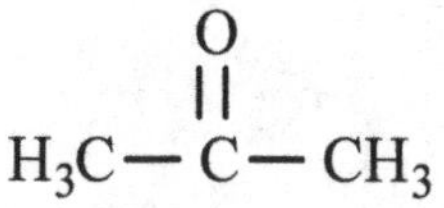

$$H_3C-\overset{O}{\overset{||}{C}}-CH_3$$

Keto form of Propanone

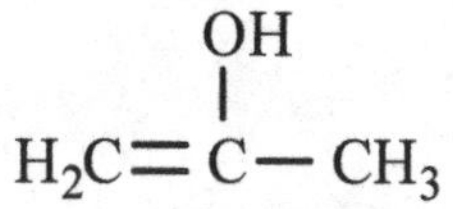

$$H_2C=\overset{OH}{\overset{|}{C}}-CH_3$$

Enol form of Propanone

13.81 a. The compound that forms when ethanol reacts with 2-penanone is a hemiacetal.

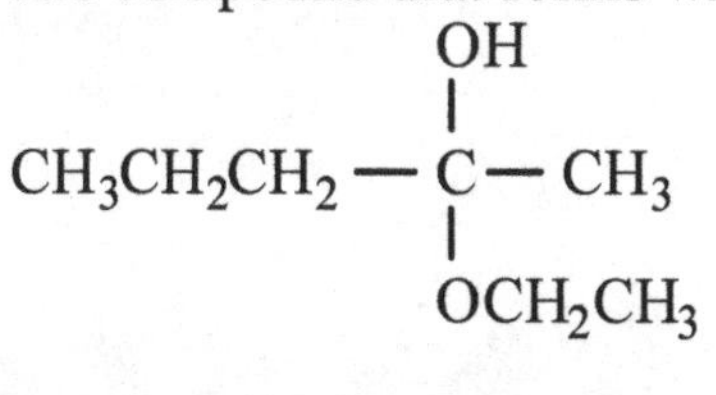

$$CH_3CH_2CH_2-\underset{OCH_2CH_3}{\underset{|}{\overset{OH}{\overset{|}{C}}}}-CH_3$$

b. The compound that forms when ethanol reacts with this ketone is a hemiacetal.

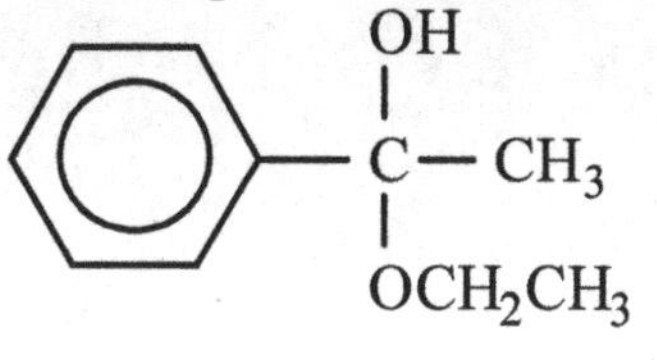

$$C_6H_5-\underset{OCH_2CH_3}{\underset{|}{\overset{OH}{\overset{|}{C}}}}-CH_3$$

c. The compound that forms when ethanol reacts with cyclopentanone is a hemiacetal.

$$\text{cyclopentane ring with C bearing OH and } OCH_2CH_3$$

13.83 (1) $2\ CH_3CH_2OH$ Acetone can react with two moles of ethanol to form this acetal.
(2) $KMnO_4/OH^-$ 2-Propanol can be oxidized to form acetone.
(3) $CH_3CH{=}CH_2$ 2-Propanol undergoes dehydration to form propene.

Chapter 14
Carboxylic Acids and Carboxylic Acid Derivatives
Solutions to the Odd-Numbered Questions and Problems

In-Chapter Questions and Problems

14.1 a. Ketone. Carboxylic acids can form intermolecular hydrogen bonds, therefore, they have higher boiling points.
b. Ketone. Alcohols can form intermolecular hydrogen bonds, therefore, they have higher boiling points.
c. Alkane. Carboxylic acids can form intermolecular hydrogen bonds, therefore, they have higher boiling points.

14.3 The carboxyl group consists of two very polar groups, the carbonyl group and the hydroxyl group. Thus, carboxylic acids are very polar, in addition to which, they can hydrogen bond to one another. Aldehydes are polar, as a result of the carbonyl group, but cannot hydrogen bond to one another. As a result, carboxylic acids have higher boiling points than aldehydes of the same carbon chain length.

14.5 The carboxylic acid derivatives of cycloalkanes are named by adding the suffix carboxylic acid to the name of the substituted cycloalkane.
a. 3-Methylcyclohexanecarboxylic acid
b. 2-Ethylcyclopentanecarboxylic acid

14.7 a. This ester is undergoing acidic hydrolysis.

CH_3COOH + $CH_3CH_2CH_2OH$
Ethanoic acid, 1-Propanol

b. This ester is undergoing basic hydrolysis.

$CH_3CH_2CH_2CH_2CH_2COO^-K^+$ + $CH_3CH_2CH_2OH$
Potassium hexanoate, 1-Propanol

c. This ester is undergoing basic hydrolysis.

$CH_3CH_2CH_2CH_2COO^-Na^+$ + CH_3OH
Sodium pentanoate, Methanol

d. This ester is undergoing acidic hydrolysis.

$CH_3CH_2CH_2CH_2CH_2COOH$ + $CH_3CH(OH)CH_2CH_2CH_3$
Hexanoic acid, 2-Pentanol

14.9 Acid anhydrides are typically formed by reacting acid chloride with a carboxylate ion.

a.

$$\underset{\text{3-Methylbutanoyl chloride}}{CH_3CH(CH_3)CH_2C(=O){-}Cl} \xrightarrow[\text{3-Methylbutanoate ion}]{CH_3CH(CH_3)CH_2C(=O){-}O^-} \underset{\text{3-Methylbutanoic anhydride}}{CH_3CH(CH_3)CH_2C(=O){-}O{-}C(=O){-}CH_2CH(CH_3)CH_3} + Cl^-$$

b.

$$\underset{\text{Methanoyl chloride}}{H{-}C(=O){-}Cl} \xrightarrow[\text{Ethanoate ion}]{CH_3C(=O){-}O^-} \underset{\text{Ethanoic methanoic anhydride}}{H{-}C(=O){-}O{-}C(=O){-}CH_3} + Cl^-$$

End-of-Chapter Questions and Problems

14.11 Aldehydes are polar as a result of the carbonyl group, but cannot hydrogen bond to one another. Alcohols are polar and can hydrogen bond as a result of the polar hydroxyl group. The carboxyl group of the carboxylic acids consists of both of these polar groups: the carbonyl group and the hydroxyl group. Thus, carboxylic acids are more polar than either aldehydes or alcohols, in addition to which, they can hydrogen bond to one another. As a result, carboxylic acids have higher boiling points than aldehydes or alcohols of comparable molar mass.

14.13 a. Pentanoic acid. Carboxylic acids can form intermolecular hydrogen bonds. Therefore, they have higher boiling points than aldehydes.

b. 2-Pentanol. Alcohols can form intermolecular hydrogen bonds. Therefore, they have higher boiling points than ketones.

c. 2-Pentanol. Alcohols can form intermolecular hydrogen bonds. Therefore, they have higher boiling points than alkanes.

14.15 Carboxylic acids have two polar functional groups, and as a result they form strong intermolecular hydrogen bonds. Alcohols have one polar functional group and can also form intermolecular hydrogen bonds. Butanal forms dipole-dipole interactions, but it cannot form hydrogen bonds. 2-Methylbutane does not have a polar functional group. Therefore, it can only form London dispersion force interactions. Stronger intermolecular interactions lead to higher boiling points.

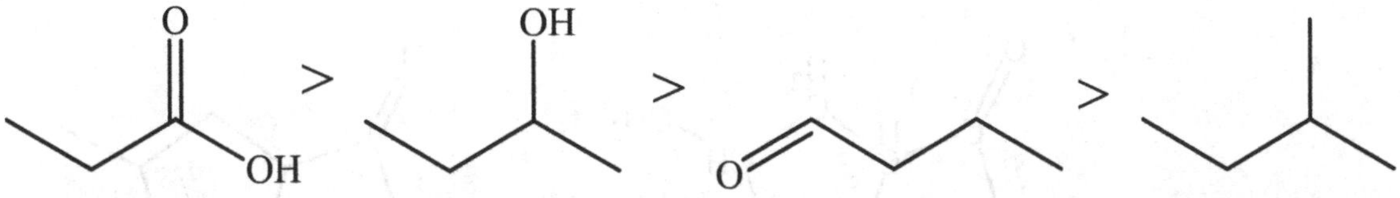

O OH > OH > O >

Propanoic acid 2-Butanol Butanal 2-Methylbutane

14.17 a. Heptanoic acid. Carboxylic acids have two polar groups and can form hydrogen bonds.
b. 1-Propanol. Alcohols can form hydrogen bonds, whereas, aldehydes cannot.
c. Pentanoic acid. Carboxylic acids can form hydrogen bonds. Esters cannot.
d. Butanoic acid. Carboxylic acids have two polar groups and can form hydrogen bonds.

14.19 The smaller carboxylic acids are water soluble. They have sharp, sour tastes and unpleasant aromas.

14.21 Citric acid is found naturally in citrus fruits. It is added to foods to give them a tart flavor or to act as a food preservative and antioxidant.

14.23 Glutaric acid is useful in the synthesis of condensation polymers because it has an odd number of carbons in the chain, which reduces the elasticity of the polymer.

14.25 Determine the name of the parent compound, that is, the longest carbon chain containing the carboxyl group. Change the *–e* ending of the alkane name to *–oic* acid. Number the chain so that the carboxyl carbon is carbon-1. Then, name and number the substituents.

14.27 The IUPAC name for adipic acid is hexanedioic acid. Adipic acid is a natural food additive that reduces spoilage by lowering the pH and thereby inhibiting the growth of bacteria and fungi.

14.29 The carboxyl carbon is carbon number one in the parent chain.
a. 3-Methylhexanoic acid: $CH_3(CH_2)_2CH(CH_3)CH_2COOH$

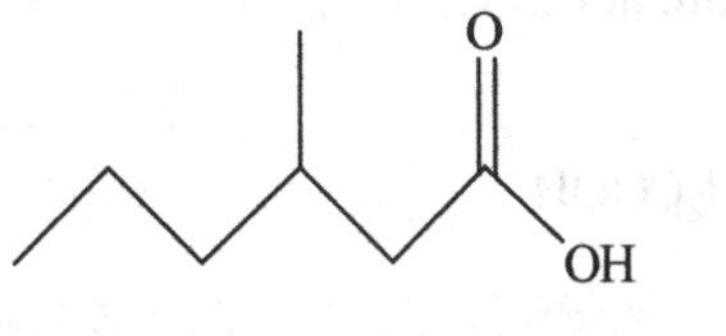

O OH

b. 2-Ethyl-2-methylpentanoic acid: $CH_3(CH_2)_2C(CH_3)(CH_2CH_3)COOH$

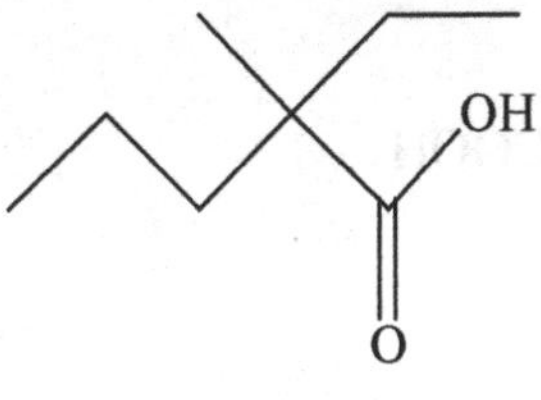

OH O

c. 3-Methylcyclopentanecarboxylic acid

```
  O        H2
   \\      C
    C   H /  \    CH3
     \ C        CH
HO /   |         |
      H2C —— CH2
```

14.31 Names derived using the IUPAC Nomenclature System use the parent name for the longest carbon chain containing the carboxyl carbon. The carboxyl carbon is carbon number one. The *–e* of the parent alkane is then replaced with *–oic acid.* Names derived using the common system of nomenclature are named based on the common names provided in the textbook. Greek letters are used to indicate the position of the substituent.

a. IUPAC name: Methanoic acid
   Common name: Formic acid
b. IUPAC name: 3-Methylbutanoic acid
   Common name: β-Methylbutyric acid
c. IUPAC name: Cyclopentanecarboxylic acid
   Common name: Cyclovalericcarboxylic acid

14.33 There are two carboxylic acids of the molecular formula, $C_4H_8O_2$.

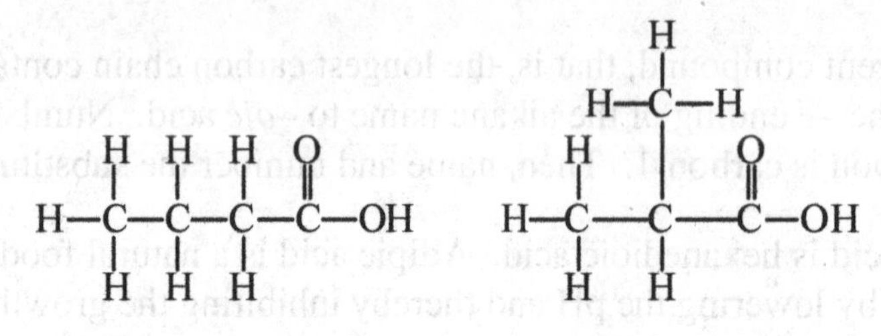

Butanoic acid — Methylpropanoic acid

14.35 When drawing the structure of a carboxylic acid from the IUPAC name, the carboxyl carbon is carbon number one.

a. 4,4-Dimethylhexanoic acid:

$$CH_3CH_2C(CH_3)_2CH_2CH_2COOH$$

b. 3-Bromo-4-methylpentanoic acid:

$$CH_3CH(CH_3)CH(Br)CH_2COOH$$

c. 2,3-Dinitrobenzoic acid:

$$\text{2,3-dinitrobenzoic acid: benzene ring with COOH, } NO_2 \text{ and } O_2N$$

d. 3-Methylcyclohexanecarboxylic acid:

$$\text{cyclohexane ring with COOH and } H_3C$$

14.37 Names derived using the IUPAC Nomenclature System use the parent name for the longest carbon chain containing the carboxyl carbon. The carboxyl carbon is carbon number one. The *–e* of the parent alkane is then replaced with *–oic acid.* Names derived using the common system of nomenclature are named based on the common names provided in the textbook. Greek letters are used to indicate the position of the substituent.

a. IUPAC name: 2-Hydroxypropanoic acid
   Common name: α-Hydroxypropionic acid
b. IUPAC name: 3-Hydroxybutanoic acid
   Common name: β-Hydroxybutyric acid
c. IUPAC name: 4,4-Dimethylpentanoic acid
   Common name: γ,γ-Dimethylvaleric acid
d. IUPAC name: 3,3-Dichloropentanoic acid
   Common name: β,β-Dichlorovaleric acid

14.39 Aromatic carboxylic acids have the parent name benzoic acid. The carbon that is bonded to the carboxylic acid is carbon number one.

a. 3-Bromobenzoic acid (or *meta*-bromobenzoic acid or *m*-bromobenzoic acid)
b. 2-Ethylbenzoic acid (or *ortho*-ethylbenzoic acid or *o*-bromobenzoic acid)
c. 4-Hydroxybenzoic acid (or *para*-hydroxybenzoic acid or *p*-hydroxybenzoic acid)

14.41 In organic molecules, oxidation may be recognized as a gain of oxygen or a loss of hydrogen. An aldehyde may be oxidized to form a carboxylic acid as in the following example in which ethanal is oxidized to produce ethanoic acid.

$$\underset{\text{Ethanal}}{H_3C{-}\overset{\overset{\displaystyle O}{\|}}{C}{-}H} \xrightarrow{[O]} \underset{\text{Ethanoic acid}}{H_3C{-}\overset{\overset{\displaystyle O}{\|}}{C}{-}OH}$$

14.43 The following general equation represents the dissociation of a carboxylic acid.

$$R{-}\overset{\overset{\displaystyle O}{\|}}{C}{-}OH \rightleftharpoons R{-}\overset{\overset{\displaystyle O}{\|}}{C}{-}O^- + H^+$$

14.45 When a strong base is added to a carboxylic acid, neutralization occurs.

14.47 Soaps are made from water, a strong base, and natural fats or oils. The fats and oils are triesters of glycerol. In the presence of the strong base, the ester bonds are hydrolyzed and the salts of the long chain fatty acids are formed. The salts of fatty acids are soaps.

14.49 a. 1-Pentanol is a primary alcohol. Therefore it is first oxidized to an aldehyde. It can then be further oxidized to a carboxylic acid.

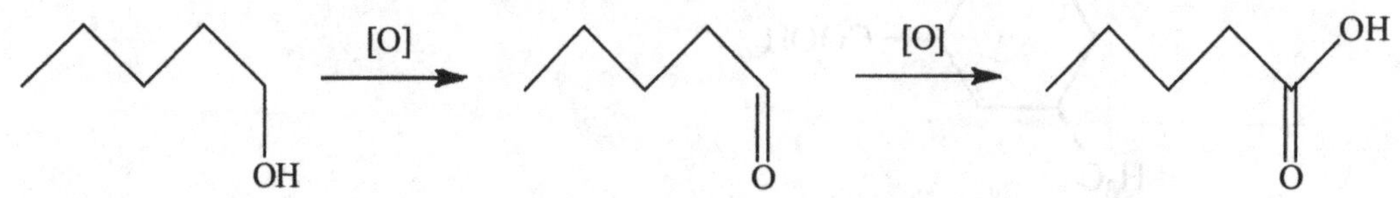

b. Butanal is an aldehyde. It can be oxidized to the carboxylic acid, butanoic acid.

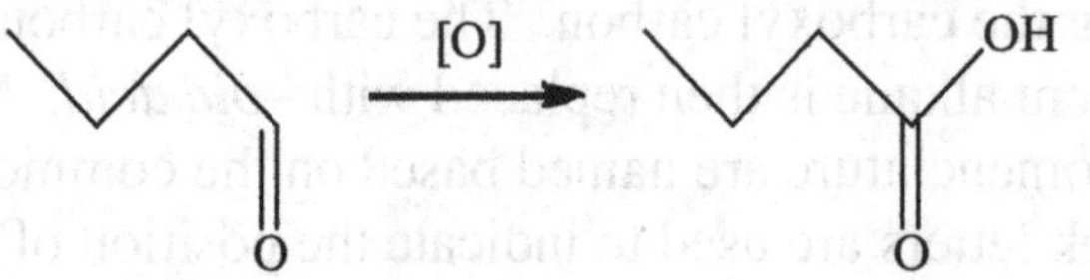

[O]
OH
O
O

c. Butanone is a ketone. It cannot be oxidized.

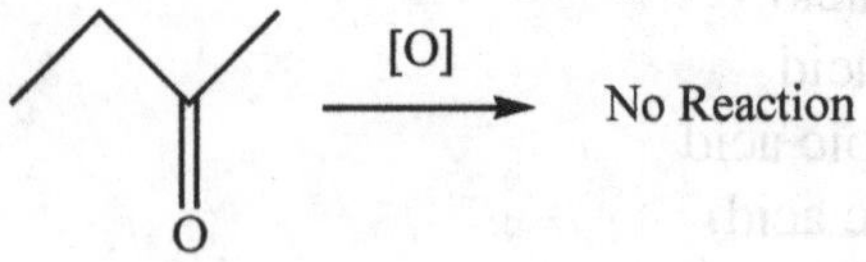

[O]
No Reaction
O

14.51 a. CH_3COOH. An aldehyde is oxidized to a carboxylic acid.

b. Under acid condition, an alcohol and a carboxylic acid can form an ester and water.

$$CH_3CH_2CH_2-\overset{O}{\overset{\|}{C}}-O-CH_3 \quad + \quad H_2O$$

c. CH_3OH. An alcohol and a carboxylic acid can react to form an ester

14.53 a. The oxidation of 1-pentanol in the presence of an oxidizing agent yields pentanal.

b. Continued oxidation of pentanal yields pentanoic acid.

14.55 a. The acidic proton is removed by the hydroxide ion to form a carboxylate ion. The carboxylate anion and the sodium cation form the carboxylic acid salt.

OH
+ NaOH ⟶
O^-Na^+
+ H_2O
O
O

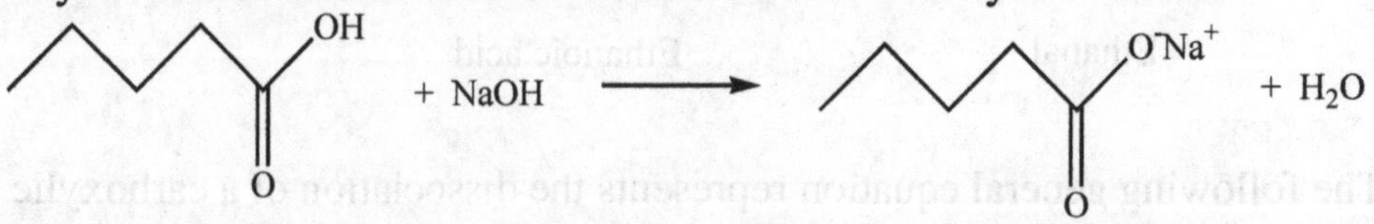

b. The acidic proton is removed by the hydroxide ion to form a carboxylate ion. The carboxylate anion and the potassium cation form the carboxylic acid salt.

$$CH_3CH_2CH_2CH_2COOH + KOH \longrightarrow CH_3CH_2CH_2CH_2COO^-K^+ + H_2O$$

c. The acidic protons are removed by the two hydroxide ions to form carboxylate ions. The carboxylate anions and the calcium cation form the carboxylic acid salt.

$$2\ CH_3CH_2CH_2CH_2COOH + Ca(OH)_2 \longrightarrow (CH_3CH_2CH_2CH_2COO^-)_2Ca^{2+} + 2H_2O$$

14.57 The structure of the calcium salt of propionic acid is $[CH_3CH_2COO^-]_2Ca^{+2}$. The common name of this salt is calcium propionate and the IUPAC name is calcium propanoate. Notice that two propionate anions are combined with one calcium cation. This is because calcium has a +2 charge.

14.59 Esters are slightly polar as a result of the polar carbonyl group within the structure.

14.61 Esters are formed from the reaction of a carboxylic acid with an alcohol. The name is derived by using the alkyl or aryl portion of the alcohol IUPAC name as the first name. The *–ic acid* ending of the IUPAC name of the carboxylic acid is replaced with *–ate* and follows the name of the alkyl or aryl group.

14.63 The IUPAC names of these compounds tell you that they are all esters.

a. Methyl benzoate:

$$C_6H_5-\overset{O}{\overset{\|}{C}}-O-CH_3$$

b. Butyl decanoate:

$$CH_3CH_2CH_2CH_2CH_2CH_2CH_2CH_2CH_2-\overset{O}{\overset{\|}{C}}-O-CH_2CH_2CH_2CH_3$$

c. Methyl propionate:

$$CH_3CH_2-\overset{O}{\overset{\|}{C}}-O-CH_3$$

d. Ethyl propionate:

$$CH_3CH_2-\overset{O}{\overset{\|}{C}}-O-CH_2CH_3$$

14.65 When naming esters according to the IUPAC System of Nomenclature, the alkyl portion of the alcohol name is listed first. The *–ic* acid ending of the name of the carboxylic acid is replaced with the *–ate* and follows the first name.

a. Ethyl ethanoate
b. Methyl propanoate
c. Methyl 3-methylbutanoate
d. Cyclopentyl benzoate

14.67 The following equation shows the general reaction for the preparation of an ester. Esters are commonly formed by the reaction between a carboxylic acid and an alcohol in the presence of acid and heat.

$$\underset{\text{Carboxylic acid}}{R{-}\overset{\overset{O}{\|}}{C}{-}OH} + \underset{\text{Alcohol}}{R{-}OH} \xrightleftharpoons{H^+,\ \text{heat}} \underset{\text{Ester}}{R{-}\overset{\overset{O}{\|}}{C}{-}OR} + \underset{\text{Water}}{H_2O}$$

14.69 The following equation shows the general reaction for the acid-catalyzed hydrolysis of an ester. Acidic hydrolysis of an ester produces a carboxylic acid and an alcohol.

$$\underset{\text{Ester}}{R{-}\overset{\overset{O}{\|}}{C}{-}OR} + \underset{\text{Water}}{H_2O} \xrightleftharpoons{H^+,\ \text{heat}} \underset{\text{Carboxylic acid}}{R{-}\overset{\overset{O}{\|}}{C}{-}OH} + \underset{\text{Alcohol}}{R{-}OH}$$

14.71 A hydrolysis reaction is the cleavage of any bond by the addition of a water molecule.

14.73 In reaction a, an ester is formed by the reaction of a carboxylic acid with an alcohol in the presence of acid and heat. In reaction b, an ester undergoes acidic hydrolysis to produce a carboxylic acid and an alcohol. In reaction c, a carboxylic acid reacts with an alcohol to produce an ester. In reaction d, an ester undergoes basic hydrolysis to produce a carboxylate anion and an alcohol.

a. $CH_3CH_2CH_2{-}\overset{\overset{O}{\|}}{C}{-}O{-}CH_2CH_3$

b. $CH_3CH_2{-}\overset{\overset{O}{\|}}{C}{-}OH \ + \ CH_3CH_2OH$

c. $CH_3CH_2CH_2OH$

d. $CH_3CH_2\underset{\underset{Br}{|}}{C}HCH_2\overset{\overset{O}{\|}}{C}{-}O^- \ + \ CH_3CH_2OH$

14.75 a. Isobutyl methanoate is made from isobutyl alcohol (IUPAC name 2-methyl-1-propanol) and methanoic acid.

$$\underset{\text{Isobutyl alcohol}}{CH_3\underset{|}{\overset{CH_3}{C}}HCH_2OH} + \underset{\text{Methanoic acid}}{HCOOH} \rightarrow \underset{\text{Isobutyl methanoate}}{H\overset{O}{\overset{\|}{C}}OCH_2\overset{CH_3}{\overset{|}{C}}HCH_3}$$

Isobutyl alcohol is an allowed starting material, but methanoic acid is not. However, it can easily be produced by the oxidation of its corresponding alcohol, methanol:

$$\underset{\text{Methanol}}{CH_3OH} \xrightarrow{[O]} \underset{\text{Methanal}}{HCHO} \xrightarrow{[O]} \underset{\text{Methanoic acid}}{HCOOH}$$

b. Pentyl butanoate is made from pentanol and butanoic acid.

$$\underset{\text{Pentanol}}{CH_3(CH_2)_3CH_2OH} + \underset{\text{Butanoic acid}}{CH_3CH_2CH_2COOH} \rightarrow \underset{\text{Pentyl butanoate}}{CH_3CH_2CH_2\overset{O}{\overset{\|}{C}}OCH_2(CH_2)_3CH_3}$$

Pentanol is an allowed starting material, but butanoic acid is not. However, it can easily be produced by the oxidation of its corresponding alcohol, 1-butanol:

$$\underset{\text{Butanol}}{CH_3CH_2CH_2CH_2OH} \xrightarrow{[O]} \underset{\text{Butanal}}{CH_3CH_2CH_2CHO} \xrightarrow{[O]} \underset{\text{Butanoic acid}}{CH_3CH_2CH_2COOH}$$

14.77 Saponification is a reaction in which soap is produced. More generally, it is the hydrolysis of an ester in the presence of a base. The following reaction shows the base-catalyzed hydrolysis of an ester:

$$CH_3(CH_2)_{14}-\overset{O}{\overset{\|}{C}}-O-CH_3 + NaOH \longrightarrow CH_3(CH_2)_{14}-\overset{O}{\overset{\|}{C}}-O^- Na^+ + CH_3OH$$

14.79

$$C_6H_4(OH)-\overset{O}{\overset{\|}{C}}-OH + CH_3OH \xrightarrow{H^+} C_6H_4(OH)-\overset{O}{\overset{\|}{C}}-OCH_3 + H_2O$$

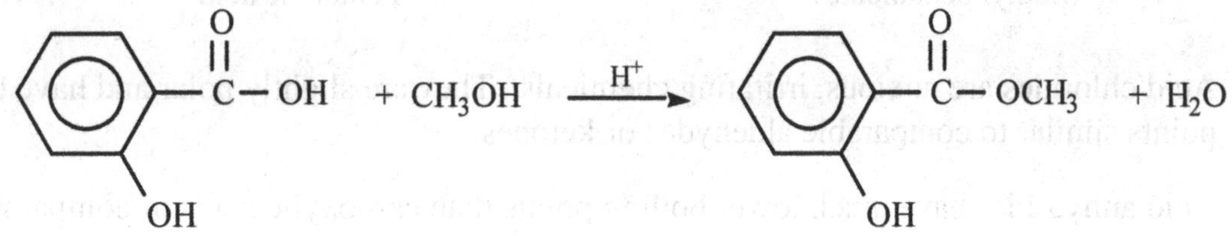

14.81 Compound A is

$$CH_3CH_2CH_2CH_2-\overset{\overset{\large O}{\|}}{C}-O-CH_3$$

An ester reacts with water, acid and heat to produce an alcohol and a carboxylic acid. Compound B is

$$CH_3CH_2CH_2CH_2-\overset{\overset{\large O}{\|}}{C}-OH$$

Carboxylic acids are acidic.
Compound C is CH_3OH

14.83 Acid catalyzed hydrolysis of an ester produces a carboxylic acid and an alcohol.

a.

$$\underset{\text{Propyl propanoate}}{CH_3CH_2-\overset{\overset{\large O}{\|}}{C}-OCH_2CH_2CH_3} \xrightleftharpoons{H^+,\ heat} \underset{\text{Propanoic acid}}{CH_3CH_2-\overset{\overset{\large O}{\|}}{C}-OH} + \underset{\text{1-Propanol}}{CH_3CH_2CH_2OH}$$

b.

$$\underset{\text{Butyl methanoate}}{H-\overset{\overset{\large O}{\|}}{C}-OCH_2CH_2CH_2CH_3} \xrightleftharpoons{H^+,\ heat} \underset{\text{Methanoic acid}}{H-\overset{\overset{\large O}{\|}}{C}-OH} + \underset{\text{1-Butanol}}{CH_3CH_2CH_2CH_2OH}$$

c.

$$\underset{\text{Ethyl methanoate}}{H-\overset{\overset{\large O}{\|}}{C}-OCH_2CH_3} \xrightleftharpoons{H^+,\ heat} \underset{\text{Methanoic acid}}{H-\overset{\overset{\large O}{\|}}{C}-OH} + \underset{\text{Ethanol}}{CH_3CH_2OH}$$

d.

$$\underset{\text{Methyl pentanoate}}{CH_3CH_2CH_2CH_2-\overset{\overset{\large O}{\|}}{C}-OCH_3} \xrightleftharpoons{H^+,\ heat} \underset{\text{Pentanoic acid}}{CH_3CH_2CH_2CH_2-\overset{\overset{\large O}{\|}}{C}-OH} + \underset{\text{Methanol}}{CH_3OH}$$

14.85 Acid chlorides are noxious, irritating chemicals. They are slightly polar and have boiling points similar to comparable aldehydes or ketones.

14.87 Acid anhydrides have much lower boiling points than carboxylic acids of comparable molar mass. They are also less soluble in water, and often react with it.

14.89 a. Decanoic anhydride:

$$CH_3(CH_2)_8-\overset{\overset{\large O}{||}}{C}-O-\overset{\overset{\large O}{||}}{C}-(CH_2)_8CH_3$$

b. Acetic anhydride:

$$CH_3-\overset{\overset{\large O}{||}}{C}-O-\overset{\overset{\large O}{||}}{C}-CH_3$$

14.91

a. $CH_3(CH_2)_6COCl$

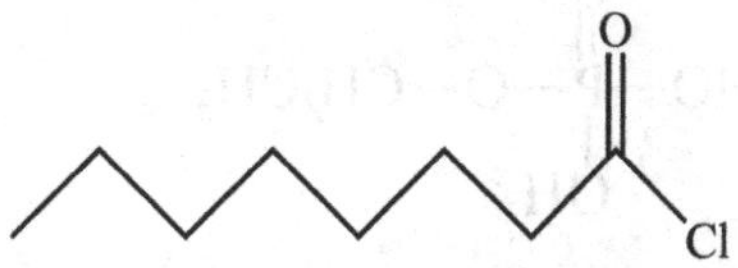

b. $CH_3(CH_2)_2COCl$

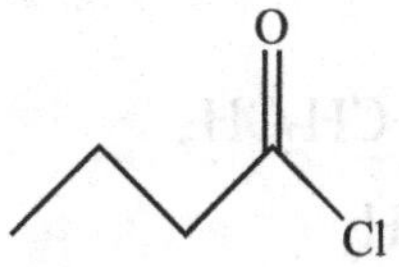

c. $CH_3(CH_2)_7COCl$

14.93 The following equation represents the synthesis of methanoic anhydride:

$$\underset{\text{Methanoate anion}}{H\overset{\overset{\large O}{||}}{C}O^-} + \underset{\text{Methanoic chloride}}{H\overset{\overset{\large O}{||}}{C}-Cl} \rightarrow \underset{\text{Methanoic anhydride}}{H\overset{\overset{\large O}{||}}{C}-O-\overset{\overset{\large O}{||}}{C}H}$$

14.95 An acid anhydride reacts with an alcohol to form an ester and a carboxylic acid.

a.

$$CH_3CH_2OH + CH_3CH_2-\overset{\overset{\large O}{||}}{C}-O-\overset{\overset{\large O}{||}}{C}-CH_2CH_3 \longrightarrow CH_3CH_2-\overset{\overset{\large O}{||}}{C}-OCH_2CH_3 + CH_3CH_2-\overset{\overset{\large O}{||}}{C}-OH$$

b.

$$CH_3CH_2OH + CH_3-\overset{\overset{\large O}{||}}{C}-O-\overset{\overset{\large O}{||}}{C}-CH_3 \longrightarrow CH_3-\overset{\overset{\large O}{||}}{C}-O-CH_2CH_3 + CH_3-\overset{\overset{\large O}{||}}{C}-OH$$

c.

$$CH_3CH_2OH + H-\overset{O}{\overset{\|}{C}}-O-\overset{O}{\overset{\|}{C}}-H \longrightarrow H-\overset{O}{\overset{\|}{C}}-OCH_2CH_3 + H-\overset{O}{\overset{\|}{C}}-OH$$

14.97 The chemical formula for phosphoric acid is H_3PO_4. In all of these phosphate esters, phosphorus has five bonds.

a. Monoester $$HO-\overset{O}{\overset{\|}{P}}(OH)-O-CH_2CH_3$$

b. Diester $$HO-\overset{O}{\overset{\|}{P}}(OCH_2CH_3)-O-CH_2CH_3$$

c. Triester $$CH_3CH_2O-\overset{O}{\overset{\|}{P}}(OCH_2CH_3)-O-CH_2CH_3$$

14.99 ATP is the molecule used to store the energy released in metabolic reactions. The energy is stored in the phosphoanhydride bonds between two phosphoryl groups. The energy is released when the bond is broken. A portion of the energy can be transferred to another molecule if the phosphoryl group is transferred from ATP to the other molecule.

14.101

$$CH_3-\overset{O}{\overset{\|}{C}}-S-\text{COENZYME A}$$

14.103 The structure of nitroglycerine:

$$\begin{array}{c} H \\ | \\ H-C-O-NO_2 \\ | \\ H-C-O-NO_2 \\ | \\ H-C-O-NO_2 \\ | \\ H \end{array}$$

Chapter 15
Amines and Amides
Solutions to the Odd-Numbered Questions and Problems

In-Chapter Questions and Problems

15.1 A secondary amine has two alkyl groups attached to the nitrogen. The nitrogen can hydrogen bond with a hydrogen of water. The nitrogen also has a hydrogen attached that can hydrogen bond with the oxygen of water.

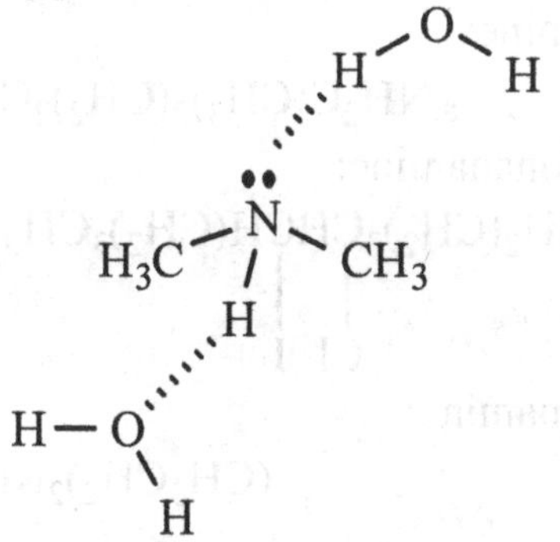

15.3 Aniline is a primary amine in which the nitrogen is attached to a benzene ring.

a. *N*-Methylaniline:

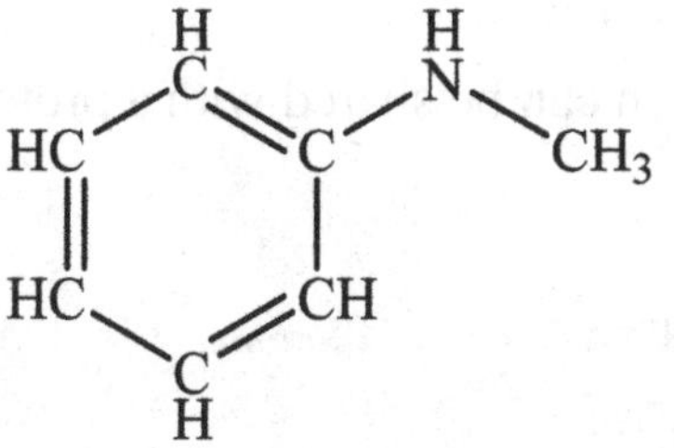

b. *N,N*-Dimethylaniline:

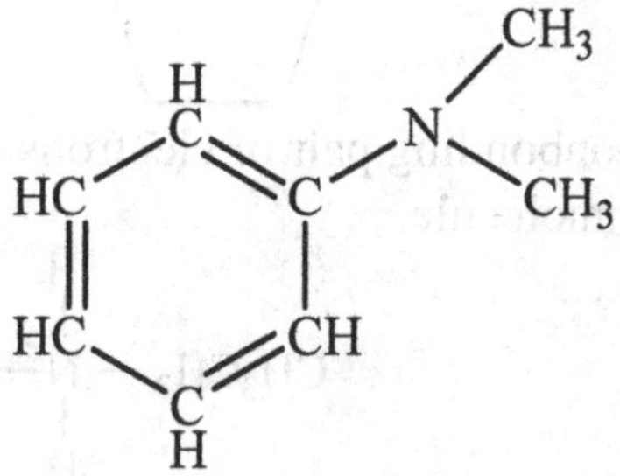

c. *N*-Ethylaniline:

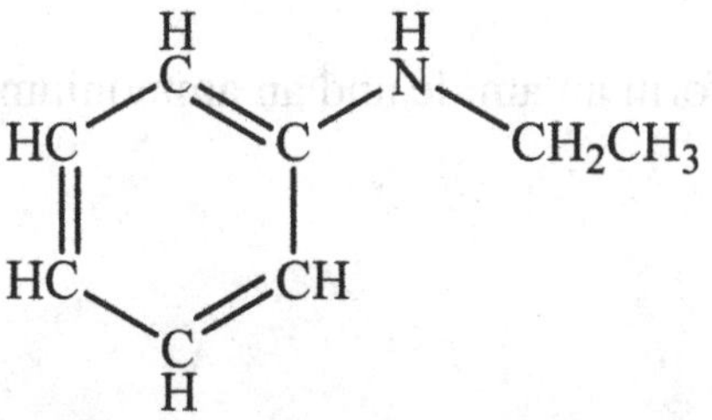

d. *N*-Isopropylaniline:

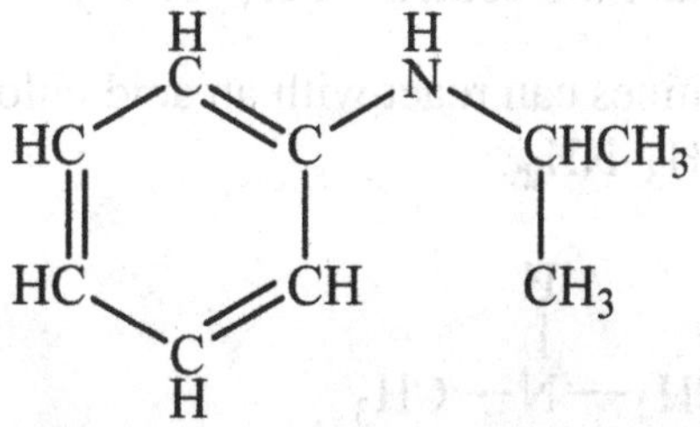

15.5 The names of these amines were derived by using the rules for systematic nomenclature for amines. The prefix *N-alkyl* indicates that the alkyl group is directly bonded to the amine nitrogen. The parent name is determined from the longest continuous carbon chain.

a. 2-Propanamine:

$$\begin{array}{c} NH_2 \\ | \\ CH_3CHCH_3 \end{array}$$

b. 3-Octanamine:

$$\begin{array}{c} \quad NH_2 \\ \quad | \\ CH_3CH_2CH(CH_2)_4CH_3 \end{array}$$

c. *N*-Ethyl-2-heptanamine:

$$\begin{array}{c} NHCH_2CH_3 \\ | \\ CH_3CH(CH_2)_4CH_3 \end{array}$$

d. 2-Methyl-2-pentanamine:

$$NH_2C(CH_3)_2(CH_2)_2CH_3$$

e. 4-Chloro-5-iodo-1-nonanamine:

$$\begin{array}{c} NH_2(CH_2)_3CHCH(CH_2)_3CH_3 \\ \quad | \quad | \\ \quad Cl \quad I \end{array}$$

f. *N,N*-Diethyl-1-pentanamine:

$$(CH_3CH_2)_2N(CH_2)_4CH_3$$

15.7 a. Alkylammonium salts are produced by reacting an amine with an acid.

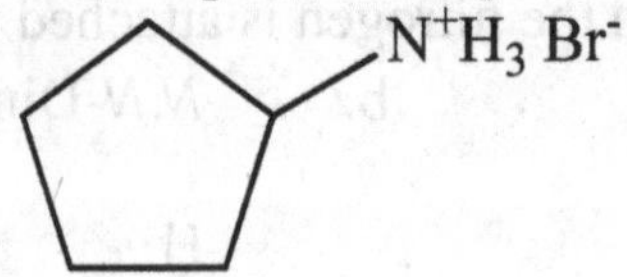

b. The nonbonding pair of electrons of the nitrogen can be shared with a proton from a water molecule.

$$\begin{array}{c} H \\ |\,+ \\ CH_3CH_2-N-CH_3 \quad + OH^- \\ | \\ H \end{array}$$

c. The nonbonding pair of electrons of the nitrogen can be shared with a proton from a water molecule. $CH_3-N^+H_3 + OH^-$

15.9 An amines can react with an acid chloride to form an amide and an ammonium salt.

a. CH_3-NH_2

b.

$$\begin{array}{c} H \\ | \\ CH_3-N-CH_3 \end{array}$$

End-of-Chapter Questions and Problems

15.11 The nitrogen atom is more electronegative than the hydrogen atom in amines; thus, the N-H bond is polar and hydrogen bonding can occur between primary or secondary amine molecules. Thus, amines have a higher boiling point than alkanes, which are nonpolar. Because nitrogen is not as electronegative as oxygen, the N-H bond is not as polar as the O-H. As a result, intermolecular hydrogen bonds between primary and secondary amine molecules are not as strong as the hydrogen bonds between alcohol molecules. Thus, alcohols have a higher boiling point.

15.13 In systematic nomenclature, primary amines are named by determining the name of the parent compound, the longest continuous carbon chain containing the amine group. The *–e* ending of the alkane chain is replaced with *–amine*. Thus, an alkane becomes an alkanamine. The parent chain is then numbered to give the carbon bearing the amine group the lowest possible number. Finally, all substituents are named and numbered and added as prefixes to the "alkanamine" name.

15.15 Amphetamines elevate blood pressure and pulse rate. They also decrease the appetite.

15.17 a. 1-Pentanamine would be more soluble in water because it has a polar amine group that can form hydrogen bonds with water molecules. Hexane is an alkane and therefore is nonpolar and water insoluble.

b. 2-Butanamine would be more soluble in water because it has a polar amine group that can form hydrogen bonds with water molecules. Cyclopentane is an alkane and therefore is nonpolar and water insoluble.

15.19 Triethylamine molecules cannot form hydrogen bonds with one another, but 1-hexanamine molecules are able to do so. As a result of the greater intermolecular attraction between 1-hexanamine molecules, it has a higher boiling point.

15.21 According to the systematic nomenclature rules for amines, the parent name is determined from the longest continuous carbon chain containing the amine group. The parent chain is numbered to give the carbon bearing the amine group the lowest possible number.

a. 2-Butanamine
b. 3-Hexanamine
c. Cyclopentanamine
d. 2-Methyl-2-propanamine

15.23 a. Diethylamine

$CH_3CH_2NHCH_2CH_3$

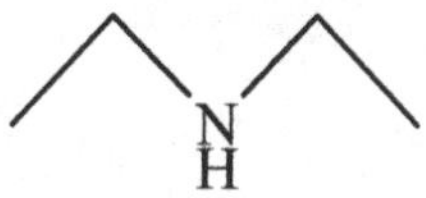

N
H

b. Butylamine

$CH_3(CH_2)_3NH_2$

NH_2

c. 3-Decanamine

NH_2
|
$CH_3(CH_2)_6CHCH_2CH_3$

NH_2

d. 3-Bromo-2-pentanamine

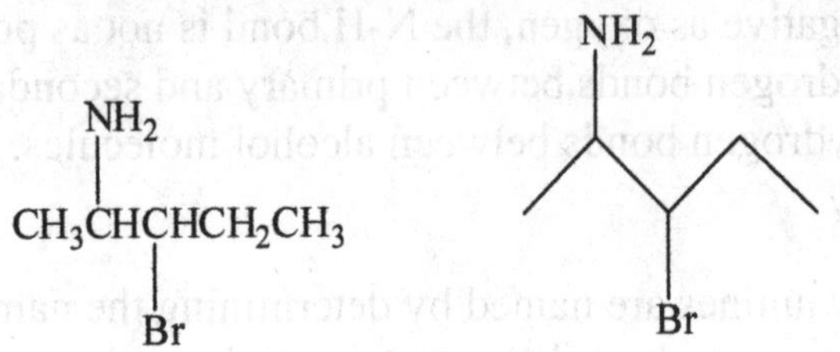

$$\begin{array}{c} NH_2 \\ | \\ CH_3CHCHCH_2CH_3 \\ | \\ Br \end{array}$$

NH_2

Br

e. Triphenylamine

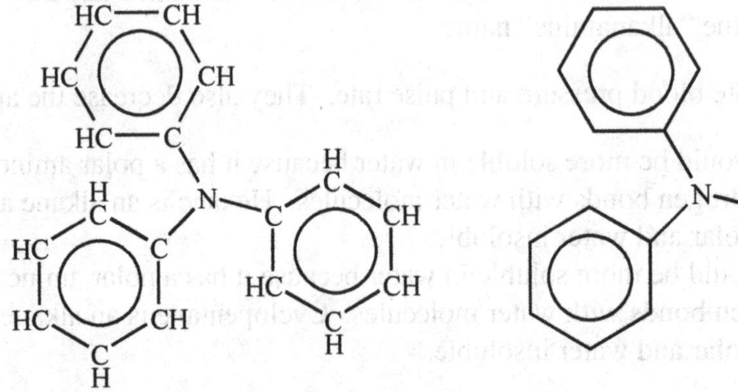

HC—CH

HC CH

HC—C

H C N C H C CH

HC C

HC CH

HC CH

C H

C H

N

15.25 a. 3-Hexanamine: $CH_3CH_2CH(NH_2)(CH_2)_2CH_3$

b. Hexylpentylamine: $CH_3(CH_2)_5NH(CH_2)_4CH_3$

c. Cyclobutanamine:

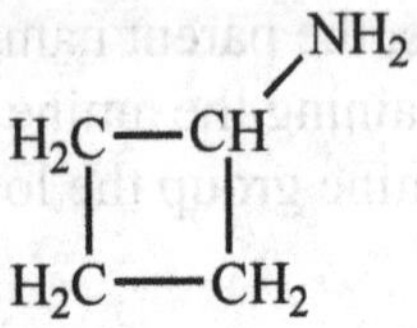

NH_2

H_2C—CH

H_2C—CH_2

d. 2-Methylcyclopentanamine:

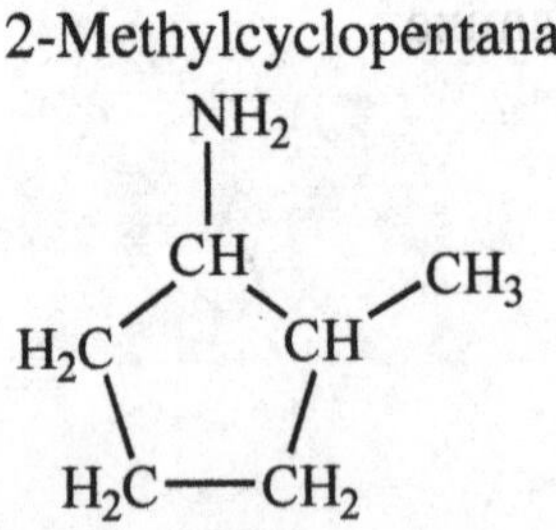

NH_2

CH CH_3

H_2C CH

H_2C—CH_2

e. Triethylammonium chloride.

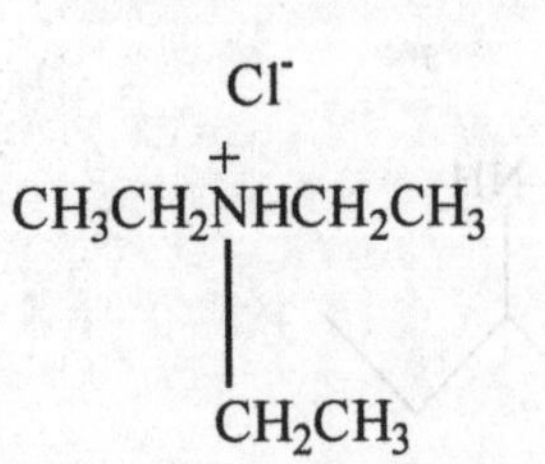

$$\begin{array}{c} Cl^- \\ + \\ CH_3CH_2NHCH_2CH_3 \\ | \\ CH_2CH_3 \end{array}$$

15.27 Condensed formulas for the eight isomeric amines with the formula $C_4H_{11}N$:

$CH_3CH_2CH_2CH_2NH_2$
1-Butanamine
(Primary amine)

$CH_3CH(NH_2)CH_2CH_3$
2-Butanamine
(Primary amine)

$CH_3CH(CH_3)CH_2NH_2$
2-Methyl-1-propanamine
(Primary amine)

$CH_3-C(CH_3)(NH_2)-CH_3$
2-Methyl-2-Propanamine
(Primary amine)

$CH_3CH_2-N(CH_3)-CH_3$
N,N-Dimethylethanamine
(Tertiary amine)

$CH_3CH_2-NH-CH_2CH_3$
N-Ethylethanamine
(Secondary amine)

$CH_3CH(NH-CH_3)CH_3$
N-Methyl-2-propanamine
(Secondary amine)

$CH_3CH_2CH_2-NH-CH_3$
N-Methyl-1-propanamine
(Secondary amine)

15.29 a. Cyclohexanamine is a primary amine. The nitrogen is attached to one alkyl group.
b. Dibutylamine is a secondary amine. The nitrogen is attached to two alkyl groups.
c. 2-Methyl-2-heptanamine is a primary amine. The nitrogen is attached to one alkyl group.
d. Tripentylamine is a tertiary amine. The nitrogen is attached to three alkyl groups.

15.31 a. Reduction of a nitro compound can produce an aromatic primary amine.

$H_3C-C_6H_4-NO_2 \xrightarrow{[H]} H_3C-C_6H_4-NH_2$

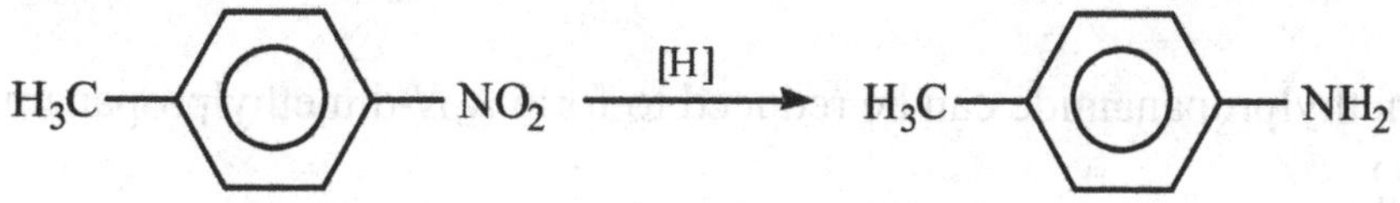

b. Reduction of a nitro compound can produce an aromatic primary amine.

$HO-C_6H_4-NO_2 \xrightarrow{[H]} HO-C_6H_4-NH_2$ (OH ortho to the nitro/amino group)

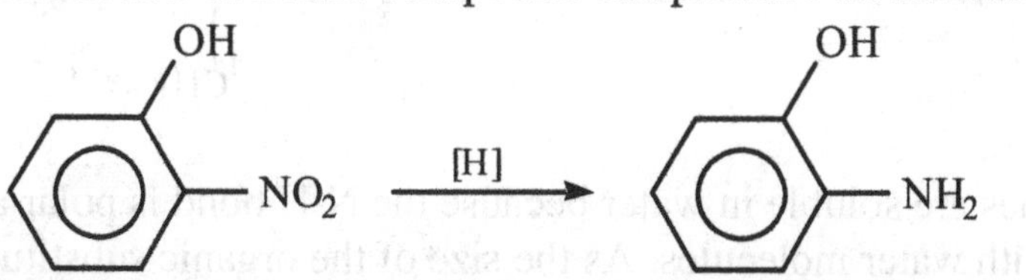

c. Reduction of a nitro compound can produce an aromatic primary amine.

$C_6H_5-NO_2 \xrightarrow{[H]} C_6H_5-NH_2$

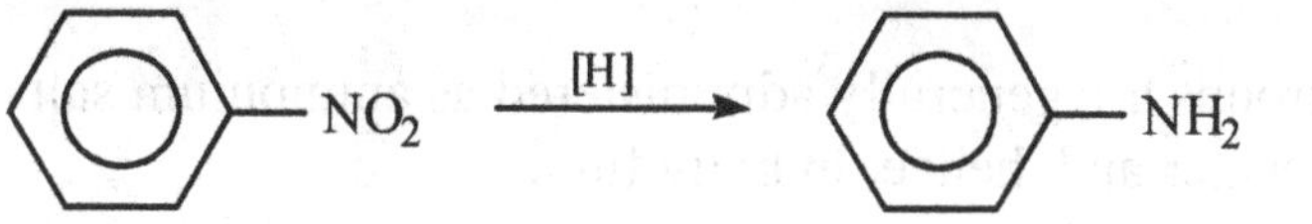

d. Reduction of an amide can produce an amine.

$$\text{C}_6\text{H}_{11}\text{-C(=O)NH}_2 \xrightarrow{[H]} \text{C}_6\text{H}_{11}\text{-CH}_2\text{NH}_2$$

15.33 a. H_2O Water can donate a proton to the amine nitrogen to form an ammonium ion.

b. HBr An acid can donate a proton to the amine nitrogen to form an ammonium salt.

c. $CH_3CH_2CH_2\text{–}N^+H_3$ Water can donate a proton to the amine nitrogen to form an ammonium ion.

d.

$$CH_3CH_2-\overset{+}{N}H_2-CH_2CH_3 \quad Cl^-$$

An acid can donate a proton to the amine nitrogen to form an ammonium salt.

15.35 a. Hexanamide can be reduced to form hexanamine, a primary amine.

$$CH_3CH_2CH_2CH_2CH_2\overset{O}{\overset{\|}{C}}-NH_2 \xrightarrow{[H]} CH_3CH_2CH_2CH_2CH_2CH_2NH_2$$

b. *N*-Methylbutanamide can be reduced to form *N*-methylbutanamine.

$$CH_3CH_2CH_2\overset{O}{\overset{\|}{C}}-\overset{H}{N}-CH_3 \xrightarrow{[H]} CH_3CH_2CH_2CH_2-\overset{H}{N}-CH_3$$

c. *N,N*-Dimethylpropanamide can be reduced to form *N,N*-dimethylpropanamine.

$$CH_3CH_2\overset{O}{\overset{\|}{C}}-N(CH_3)-CH_3 \xrightarrow{[H]} CH_3CH_2CH_2-N(CH_3)-CH_3$$

15.37 Lower molar mass amines are soluble in water because the N-H bond is polar and can form hydrogen bonds with water molecules. As the size of the organic substituents becomes larger, the entire molecule becomes more hydrocarbonlike and, thus, more hydrophobic overall.

15.39 Drugs containing amine groups are generally administered as ammonium salts because the salt is more soluble in water and, hence, in body fluids.

15.41 Putrescine (1,4-Butanediamine): $H_2N–CH_2–CH_2–CH_2–CH_2–NH_2$

Cadaverine (1,5-Pentanediamine): $H_2N–CH_2–CH_2–CH_2–CH_2–CH_2–NH_2$

15.43 The structures of pyridine and indole are provided in your textbook.

a.

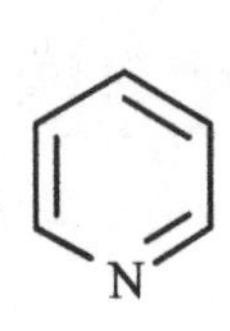

Pyridine

Indole

b. The pyridine ring is found in vitamin B_6, which is a water-soluble vitamin required for the synthesis and degradation of amino acids.

The indole ring is found in lysergic acid diethylamide, which is a hallucinogenic drug. It is also found in strychnine, which has been used as a rat poison.

15.45 Morphine has been used as a pain reliever or analgesic. Codeine is used as an analgesic and cough suppressant. Cocaine has been used as an anesthetic for the sinuses and eyes. Quinine is used to treat malaria. Vitamin B_6 is a water-soluble vitamin required by the body.

15.47 Amides have very high boiling points because the amide group consists of two very polar functional groups, the carbonyl group and the amino group. Strong intermolecular hydrogen bonding between the N-H bond of one amide and the C=O group of a second amide results in very high boiling points.

15.49 The IUPAC names of amides are derived from the IUPAC names of the carboxylic acids from which they are derived. The *–oic acid* ending of the carboxylic acid is replaced with the *–amide* ending.

15.51 Barbiturates are often called "downers" because they act as sedatives. They are sometimes used as anticonvulsants for epileptics and people suffering from other disorders that manifest as neurosis, anxiety, or tension.

15.53 Amide names are derived from the names of carboxylic acids from which they were made. Substituents on the nitrogen are placed as prefixes and are indicated by *N-* followed by the name of the substituent.

a. IUPAC name: Propanamide
Common name: Propionamide

b. IUPAC name: Pentanamide
Common name: Valeramide

c. IUPAC name: *N,N*-Dimethylethanamide
Common name: *N,N*-Dimethylacetamide

15.55 Knowing that amide names are derived from the names of carboxylic acids from which they were made and that substituents on the nitrogen are placed as prefixes and are indicated by *N*- followed by the name of the substituent, it is best to begin drawing the structure of an amide by drawing out the carbon skeleton of the alkylamide.

a. Propanamide:

$$CH_3CH_2\overset{O}{\overset{\|}{C}}NH_2$$

b. *N,N*-Diethylbutanamide:

$$CH_3(CH_2)_2\overset{O}{\overset{\|}{C}}N(CH_2CH_3)_2$$

c. 2,3-Diethylpentanamide:

$$(CH_3CH_2)_2CHCH(CH_2CH_3)\overset{O}{\overset{\|}{C}}NH_2$$

d. *N*-Methylhexanamide:

$$CH_3(CH_2)_4\overset{O}{\overset{\|}{C}}NHCH_3$$

15.57 a. Ethanamide:

$$H_3C-\overset{O}{\overset{\|}{C}}-NH_2$$

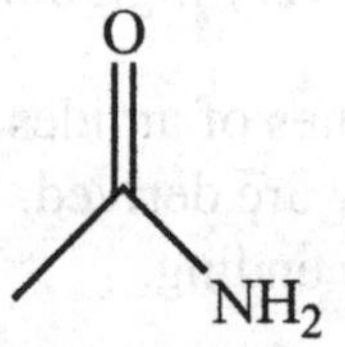

O

NH_2

b. *N*-Methylpropanamide:

$$CH_3CH_2-\overset{O}{\overset{\|}{C}}-NHCH_3$$

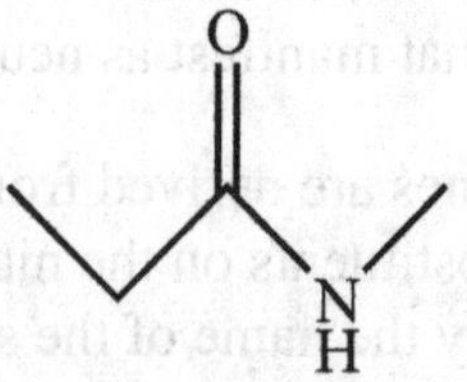

O

N
H

c. *N,N*-Diethylbenzenamide:

$$C_6H_5-\overset{O}{\overset{\|}{C}}-N(CH_2CH_3)-CH_2CH_3$$

d. 3-Bromo-4-methylhexanamide:

$$CH_3CH_2\overset{CH_3}{\overset{|}{C}H}\underset{Br}{\underset{|}{C}H}CH_2-\overset{O}{\overset{\|}{C}}-NH_2$$

e. *N,N*-Dimethylacetamide:

$$CH_3-\overset{O}{\overset{\|}{C}}-\underset{CH_3}{\underset{|}{N}}-CH_3$$

15.59 *N,N*-Diethyl-*m*-toluamide:

$$m\text{-}CH_3C_6H_4-\overset{O}{\overset{\|}{C}}-\overset{CH_2CH_3}{\overset{|}{N}}-CH_2CH_3$$

Hydrolysis of this compound would release the carboxylic acid *m*-toluic acid and the amine *N*-ethylethanamine (diethylamine).

15.61 Amides are not proton acceptors (bases) because the highly electronegative carbonyl oxygen has a strong attraction for the nitrogen lone pair of electrons. As a result, they cannot "hold" a proton. They are not proton donors because there are no hydrogen atoms bonded to the oxygen.

15.63 Lidocaine hydrochloride:

Amide group →

$$(2,6\text{-}(CH_3)_2C_6H_3)-NH-\overset{O}{\overset{\|}{C}}-CH_2-\underset{CH_2CH_3}{\underset{|}{\overset{+}{N}H}}-CH_2CH_3 \quad Cl^-$$

15.65 Penicillin BT:

$$CH_3(CH_2)_3SCH_2-\overset{\overset{\displaystyle O}{||}}{C}-\underset{\underset{\displaystyle H}{|}}{N}-\text{(β-lactam ring with C=O and N, fused to S-containing ring bearing } 2\,CH_3 \text{ and } COOH)$$

Carboxyl group (COOH); Amide group (ring C=O–N); Amide group (C(=O)–NH)

15.67 a. An amide can undergo acid hydrolysis to produce a carboxylic acid and an ammonium ion.

$$CH_3-\overset{\overset{\displaystyle O}{||}}{C}-NH-CH_3 + H_3O^+ \longrightarrow CH_3COOH + CH_3\overset{+}{N}H_3$$

N-Methylethanamide → Ethanoic acid + Methylammonium ion

b. An amide can undergo acid hydrolysis to produce a carboxylic acid and an ammonium ion.

$$CH_3CH_2CH_2-\overset{\overset{\displaystyle O}{||}}{C}-NH-CH_3 + H_3O^+ \longrightarrow CH_3CH_2CH_2COOH + CH_3\overset{+}{N}H_3$$

N-Methylbutanamide → Butanoic acid + Methylammonium ion

c. An amide can undergo acid hydrolysis to produce a carboxylic acid and an ammonium ion.

$$CH_3\overset{\overset{\displaystyle CH_3}{|}}{C}HCH_2-\overset{\overset{\displaystyle O}{||}}{C}-NH-CH_2CH_3 + H_3O^+ \longrightarrow CH_3\overset{\overset{\displaystyle CH_3}{|}}{C}HCH_2COOH + CH_3CH_2\overset{+}{N}H_3$$

N-Ethyl-3-methylbutanamide → 3-Methylbutanoic acid + Ethylammonium ion

15.69 a. An acid anhydride can react with two amines to produce an amide and a carboxylic acid salt.

$$CH_3CH_2-\overset{\overset{\displaystyle O}{||}}{C}-O-\overset{\overset{\displaystyle O}{||}}{C}-CH_2CH_3$$

b. An acid chloride can react with two amines to form an amide and an alkylammonium salt.

$$CH_3CH_2-\overset{\overset{\displaystyle O}{||}}{C}-NH_2 + NH_4^+Cl^-$$

c. An acid chloride can react with two amines to form an amide and an alkylammonium salt.

$$CH_3CH_2CH_2-\overset{\overset{\displaystyle O}{||}}{C}-Cl + 2\,CH_3CH_2NH_2$$

15.71 A primary (1°) amide is the product of the reaction between ammonia and an acid chloride. A primary amide has only one carbon, the carbonyl carbon, bonded to the nitrogen and has the following general structure:

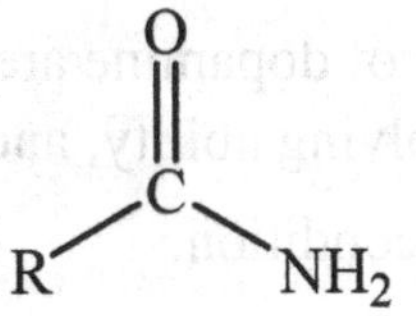

$$R-C(=O)-NH_2$$

15.73 A tertiary (3°) amide is the product of a reaction between a secondary amine and an acid chloride. A tertiary amide has three carbon atoms bonded to the nitrogen. One is the carbonyl carbon from the acid chloride and the other two are from the secondary amine reactant. The following is the general structure:

$$R^1-C(=O)-N(R^2)-R^2$$

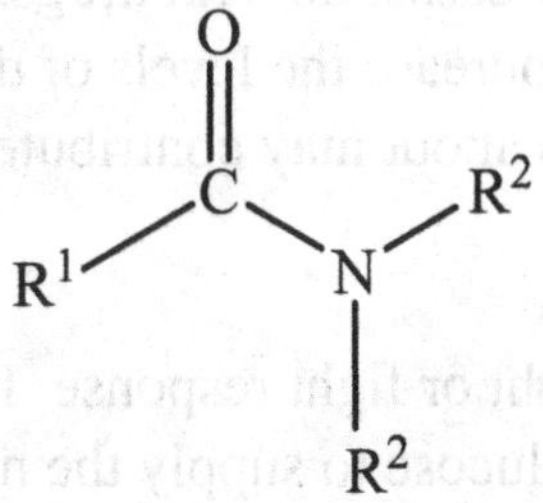

15.75 Note that under physiological conditions, the structure would have a protonated amino group and a carboxylate group. The following is the general structure of an amino acid:

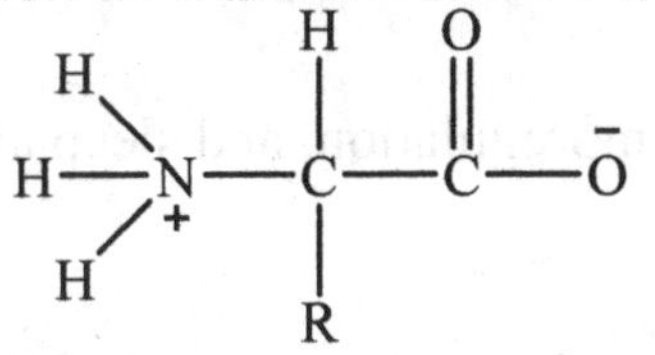

$$H_3N^+-CH(R)-C(=O)-O^-$$

15.77 Within the cell, the actual structures would have a protonated amino group and a carboxylate group.

Glycine:

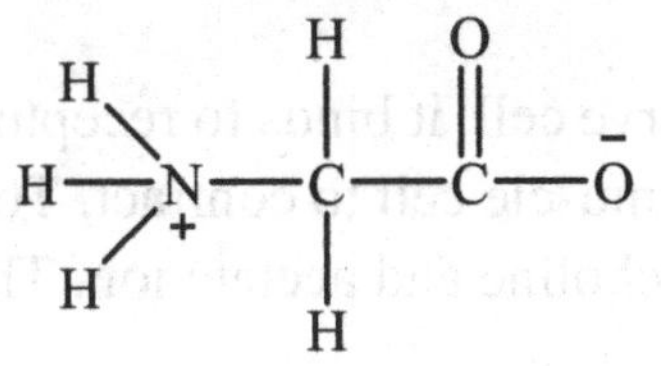

$$H_3N^+-CH_2-C(=O)-O^-$$

Alanine:

$$H_3N^+-CH(CH_3)-C(=O)-O^-$$

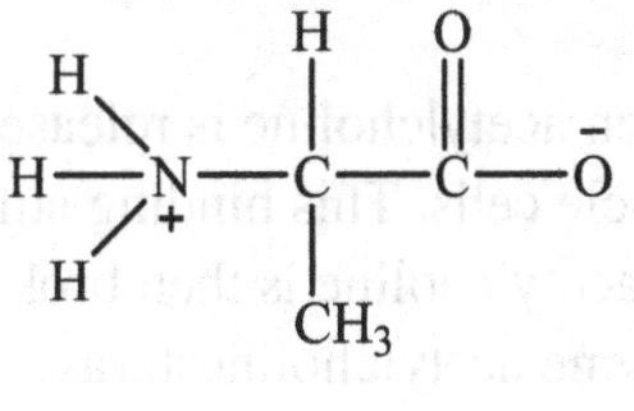

15.79 The structure of alanine is shown below. An asterisk denotes the chiral carbon.

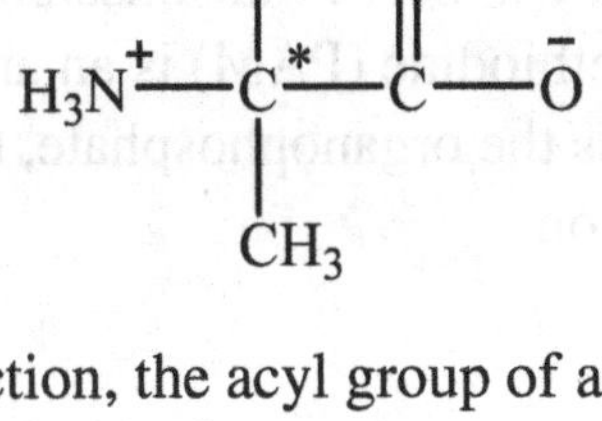

$$H_3\overset{+}{N}-\overset{*}{C}H(CH_3)-C(=O)-O^-$$

15.81 In an acyl group transfer reaction, the acyl group of an acid chloride is transferred from the Cl of the acid chloride to the N of an amine or ammonia. The product is an amide.

15.83 A neurotransmitter is a chemical that carries messages or signals from a nerve to a target cell.

15.85 a. Symptoms that result from a deficiency of dopamine are: tremors, monotonous speech, loss of memory and problem-solving ability, and loss of motor function.

b. Parkinson's disease is the name of this condition.

c. Symptoms that result from an excess of dopamine are schizophrenia and intense satiety sensations.

15.87 In proper amounts, dopamine causes a pleasant, satisfied feeling. This feeling becomes intense as the amount of dopamine increases. Several drugs, including cocaine, heroin, amphetamines, alcohol, and nicotine increase the levels of dopamine. It is thought that the intense satiety response this brings about may contribute to addiction to these substances.

15.89 Epinephrine is a component of the flight or fight response. It stimulates glycogen breakdown to provide the body with glucose to supply the needed energy for this stress response.

15.91 The amino acid tryptophan is the starting material from which serotonin is made.

15.93 The perception of pain, thermoregulation, and sleep are physiological responses that are affected by serotonin.

15.95 Histamine promotes the itchy skin rash associated with poison ivy and insect bites, the respiratory symptoms characteristic of hay fever, and the secretion of stomach acid.

15.97 Inhibitory neurotransmitters

15.99 When acetylcholine is released from a nerve cell, it binds to receptors on the surface of muscle cells. This binding stimulates the muscle cell to contract. To stop the contraction, the acetylcholine is then broken down to choline and acetate ion. This is catalyzed by the enzyme acetylcholinesterase.

15.101 Organophosphates inactivate acetylcholinesterase by binding covalently to it. Since acetylcholine is not broken down, nerve transmission continues, resulting in muscle spasm. Pyridine aldoxime methiodide (PAM) is an antidote to organophosphate poisoning because it displaces the organophosphate, thereby allowing acetylcholinesterase to function.

Chapter 16
Carbohydrates
Solutions to the Odd-Numbered Questions and Problems

In-Chapter Questions and Problems

16.1 If you know the functional groups and the kinds of reactions that they undergo, you will be able to carry that information into your understanding of biochemistry. Remembering that biological molecules are really just large organic molecules, you can recognize the functional groups and predict the ways they will react in biochemical reactions.

16.3 It is currently recommended that 45–55% of the calories in the diet should be carbohydrates. Of that amount, the World Health Organization recommends that no more than 5% should be simple sugars.

16.5 An aldose is a sugar with an aldehyde functional group. A ketose is a sugar with a ketone functional group.

16.7 A chiral carbon is a carbon that has four different groups bonded to it.

a.

CH_3
=O
H–*–OH
CH_2OH

CH_3
=O
HO–*–H
CH_2OH

b.

CHO
H–*–OH
H–*–OH
HO–*–H
CH_2OH

CHO
HO–*–H
HO–*–H
H–*–OH
CH_2OH

c.

CH_2OH
=O
HO–*–H
H–*–OH
H–*–OH
CH_2OH

CH_2OH
=O
H–*–OH
HO–*–H
HO–*–H
CH_2OH

16.9 If the –OH group is on the right, the molecule is in the D-configuration. If the –OH group is on the left, the molecule is in the L-configuration. The configuration for each of the molecules that were listed in Practice Problem 16.2 are:

a. D- b. L- c. D-

16.11 In D-galactose, the –OH group on the next to last carbon (on the bottom) is on the right.

CHO
H–OH
HO–H
HO–H
H–OH
CH_2OH

D-Galactose

16.13 α-Amylase and β-amylase are digestive enzymes that break down the starch amylose. α-Amylase cleaves glycosidic bonds of the amylose chain at random, producing shorter polysaccharide chains. β-Amylase sequentially cleaves maltose (a disaccharide of glucose) from the reducing end of the polysaccharide chain.

End-of-Chapter Questions and Problems

16.15 The first step of the study cycle is to **preview** the chapter before class. This allows you an opportunity to identify questions you may need to address in class. The second step is to **attend** class as an active participant. Ask questions and participate in class discussions. The third step is to **review** your notes as soon as possible after class. This will allow you to fill in any gaps that exist in your understanding and to make notes on additional questions that may arise. The fourth step is to study. Repetition is the key to success. Establish your goal for each study session. Use active study skills by making flash cards and developing concept maps. And finally, **assess** your learning. Test yourself by solving problems and explaining concepts to others. This will help you know what you have learned thoroughly and what you need to spend more time on.

16.17 You might make a set of flash cards with the structures of the monosaccharides on one side and the names on the other. You will learn more if you include the linear structures, the Haworth structures, and the Fischer Projections for each of the monosaccharides you are studying.

16.19 By inspecting the structure of glucose, you see that a carbonyl group at C-1 is going to react with a hydroxyl group at C-5. This reaction between a carbonyl group and a hydroxyl group is one that you learned in the organic chemistry portion of the course. In Section 13.4 you learned that the reaction between an aldehyde and an alcohol yields a hemiacetal.

16.21 A monosaccharide is the simplest sugar and consists of a single saccharide unit. A monosaccharide cannot be broken down into a simpler molecule by hydrolysis. A disaccharide is made up of two monosaccharides joined covalently by a glycosidic bond. Hydrolysis of a disaccharide yields two monosaccharides.

16.23 Mashed potato flakes, rice, and corn starch would contain amylose and amylopectin, both of which are polysaccharides. A candy bar contains sucrose, a disaccharide. Orange juice contains fructose, a monosaccharide. It may also contain sucrose if the label indicates that sugar has been added.

16.25 Four kilocalories of energy are released for each gram of carbohydrate "burned" or oxidized.

16.27 D-Galactose, an aldohexose D-Fructose (a ketohexose)

```
        O
        ||
        C—H                      CH2OH
        |                          |
    H—C—OH                        C=O
        |                          |
   HO—C—H                     HO—C—H
        |                          |
   HO—C—H                      H—C—OH
        |                          |
    H—C—OH                     H—C—OH
        |                          |
      CH2OH                      CH2OH
```

16.29 An aldose is a sugar that contains an aldehyde (carbonyl) group.

16.31 A tetrose is a sugar with a four-carbon backbone.

16.33 A ketopentose is a sugar with a five-carbon backbone and containing a ketone.

16.35 a. β-D-Glucose b. β-D-Fructose c. α-D-Galactose

16.37 Aldotrioses have an aldehyde group (aldose) and a total of three carbons (triose). There are two aldotrioses of molecular formula C_3H_6O.

```
        O                  O
        ||                 ||
        C—H                C—H
        |                  |
    H—C—OH            HO—C—H
        |                  |
      CH2OH              CH2OH

 D-Glyceraldehyde   L-Glyceraldehyde
```

16.39 Stereoisomers are pairs of molecules having the same structural formula and bonding pattern but differing in the arrangement of the atoms in space.

16.41 A chiral carbon is a carbon atom that is bonded to four different groups.

16.43 A polarimeter is an instrument that is used to measure the optical activity of molecules. It has a monochromatic light source. Light waves from the source are directed through a polarizer. The light that emerges is plane polarized. This passes through the sample and into the analyzer, which will measure the degree to which the plane of light has been rotated by the sample. If the plane of light is not altered by the sample, the compound is optically inactive. However, if the plane of light is rotated either clockwise or counterclockwise, the sample is optically active.

16.45 A Fischer Projection is a two-dimensional drawing of a molecule that shows a chiral carbon at the intersection of two lines. Horizontal lines at the intersection represent bonds projecting out of the page and vertical lines represent bonds that project into the page.

16.47 Diastereomers are a pair of stereoisomers that are not enantiomers.

16.49

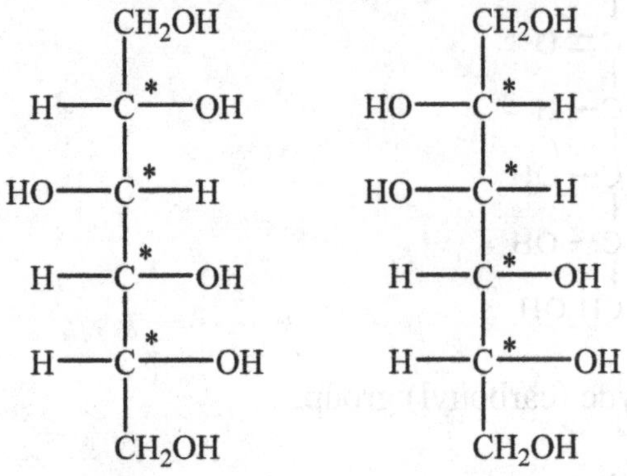

Sorbitol

CH_2OH
H—C*—OH
HO—C*—H
H—C*—OH
H—C*—OH
CH_2OH

Mannitol

CH_2OH
HO—C*—H
HO—C*—H
H—C*—OH
H—C*—OH
CH_2OH

16.51 Dextrose is a common name used for D-glucose.

16.53 D- and L-Glyceraldehyde are a pair of enantiomers; that is, they are nonsuperimposable mirror images of one another. In D-glyceraldehyde, the hydroxyl group on the chiral carbon farthest from the aldehyde group (C-2) is on the right of the structure. In L-glyceraldehyde, the hydroxyl group on C-2 is on the left of the structure.

16.55 a.

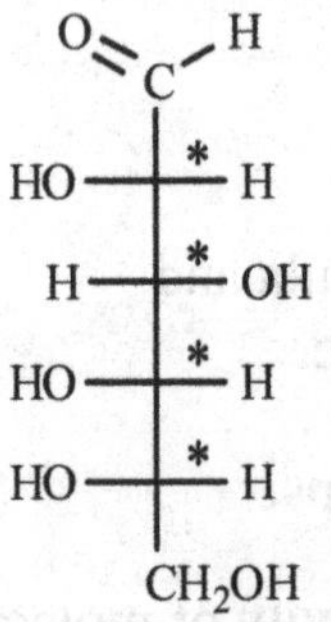

O=C—H
HO—*—H
H—*—OH
HO—*—H
HO—*—H
CH_2OH

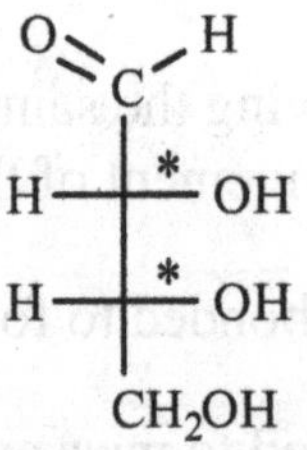

b.

O=C—H
H—*—OH
H—*—OH
CH_2OH

c.

O=C—H
HO—*—H
H—*—OH
HO—*—H
H—*—OH
HO—*—H
CH_2OH

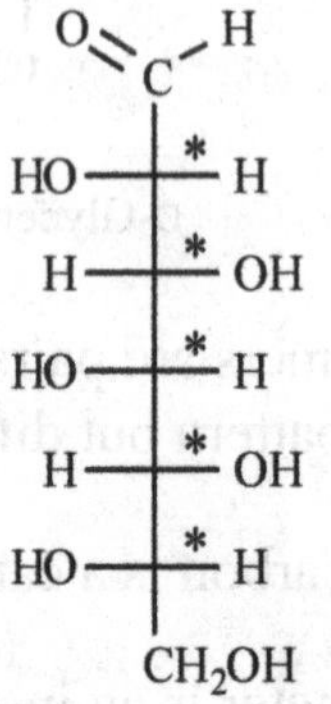

16.57 a. There are two chiral carbons in this molecule. But, each of the two chiral carbons is bonded to the same four nonidentical groups. Therefore, there are only three possible stereoisomers. The stereoisomers are A, C, and D. Compounds A and B are identical and they are meso compounds because they have an internal plane of symmetry. Compounds C and D are enantiomers. Compound A is a diastereomer to both compounds C and D.

A	B	C	D
HC=O	HC=O	HC=O	HC=O
H—OH	HO—H	HO—H	H—OH
H—OH	HO—H	H—OH	HO—H
HC=O	HC=O	CH=O	HC=O

b. There are two chiral carbons in this molecule. Each of the two chiral carbons is bonded to different nonidentical groups. Therefore, there are four possible stereoisomers. The two pairs of enationmers are A with B and C with D. Compounds A and B are both diastereomers to compounds C and D.

A	B	C	D
CH_2OH	CH_2OH	CH_2OH	CH_2OH
H—Br	Br—H	Br—H	H—Br
H—CH_3	H_3C—H	H—CH_3	H_3C—H
CH_2OH	CH_2OH	CH_2OH	CH_2OH

16.59

a. There are two chiral carbons in this molecule. But, each of the two chiral carbons is bonded to the same four nonidentical groups. Therefore, there are only three possible stereoisomers. The stereoisomers are A, C, and D. Compounds A and B are identical and they are meso compounds because they have an internal plane of symmetry. Compounds C and D are enantiomers. Compound A is a diastereomer to both compounds C and D.

A	B	C	D
CH_2CH_3	CH_2CH_3	CH_2CH_3	CH_2CH_3
H—OH	HO—H	HO—H	H—OH
H—OH	HO—H	H—OH	HO—H
CH_2CH_3	CH_2CH_3	CH_2CH_3	CH_2CH_3

b. There are two chiral carbons in this molecule. Because each of the two chiral carbons is bonded to four different nonidentical groups, there are four possible stereoisomers. Compounds A and B are enantiomers, and compounds C and D are enantiomers. Compounds A and B are diastereomers to both compounds C and D.

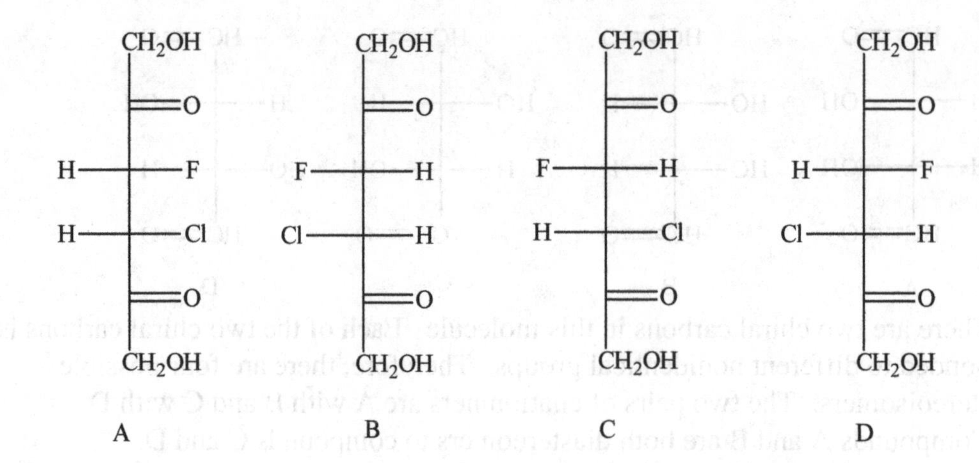

16.61 Anomers are isomers that differ in the arrangement of bonds around the hemiacetal carbon.

16.63 A hemiacetal is a member of the family of organic compounds formed in the reaction of one molecule of alcohol with an aldehyde. They have the following general structure:

OH
R—C—OR
H

16.65 When the carbonyl group at C-1 of D-glucose reacts with the C-5 hydroxyl group, a new chiral carbon is created (C-1). In the α-isomer of the cyclic sugar, the C-1 hydroxyl group is below the ring and in the β-isomer the C-1 hydroxyl group is above the ring.

16.67 β-Maltose and α-lactose would give positive Benedict's tests. Glycogen would give only a weak reaction because it is a long polymer and thus there are fewer reducing ends for a given mass of the carbohydrate.

16.69 Enantiomers are stereoisomers that are nonsuperimposable mirror images of one another. For instance, D-glyceraldyhyde and L-glyceraldehyde are enationmers.

CHO
H—C—OH
CH_2OH

D-Glyceraldehyde

CHO
HO—C—H
CH_2OH

L-Glyceraldehyde

16.71 The linear structure of an aldehyde sugar forms a cyclic structure by formation of an intramolecular hemiacetal. The carbonyl group of the monosaccharide, in this case, an aldose, reacts with a hydroxyl group on one of the other carbon atoms. The product is a cyclic intramolecular hemiacetal.

16.73 A disaccharide is a simple carbohydrate composed of two monosaccharides.

16.75 A glycosidic bond is the bond formed between the hydroxyl group of the C-1 carbon of one sugar and a hydroxyl group of another sugar.

16.77 The structure of β-maltose is:

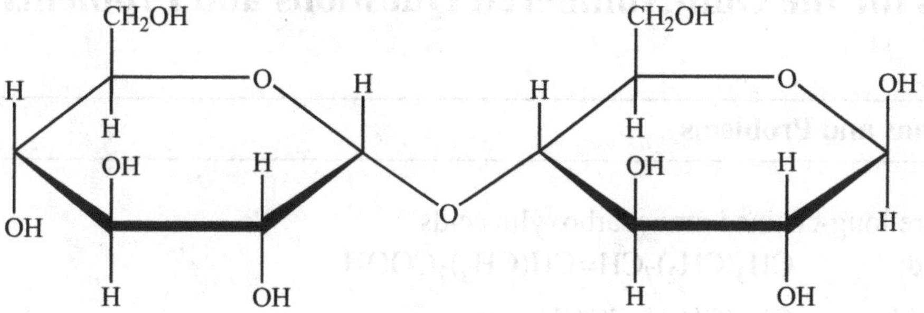

16.79 Milk is the major source of lactose.

16.81 Since galactose is one of the two monosaccharides making up lactose (milk sugar), eliminating milk and milk products from the diet allows the patient to avoid most of the ill effects of galactosemia.

16.83 Lactose intolerance is the inability to produce the enzyme lactase that hydrolyzes the milk sugar lactose into its component monosaccharides, glucose and galactose. As a result of the undigested lactose in the intestine, uncomfortable symptoms, including abdominal cramps and diarrhea, occur.

16.85 A polymer is a very large molecule formed by the combination of many small molecules called monomers.

16.87 The major storage form of sugar in plants is starch.

16.89 Homopolysaccharides are a class of polysaccharides that are composed of a single monosaccharide.

16.91 Starch, glycogen and cellulose are examples of homopolysaccharides. These homopolysaccharides are all made up of glucose.

16.93 Both amylose and cellulose are linear polymers of glucose units. However, the glucose units of amylose are joined by $\alpha(1 \rightarrow 4)$ glycosidic bonds, and those of cellulose are bonded together by $\beta(1 \rightarrow 4)$ glycosidic bonds.

16.95 The major physiological purpose of glycogen is to serve as a storage molecule for glucose. This represents an energy reservoir for the body. Glycogen synthesis and degradation in the liver are involved in regulation of blood glucose levels.

16.97 α-Amylase and β-amylase are produced in the salivary glands and in the pancreas.

Chapter 17
Lipids and Their Functions in Biochemical Systems
Solutions for the Odd-Numbered Questions and Problems

In-Chapter Questions and Problems

17.1 Fatty acids are long-chained monocarboxylic acids.

a. Oleic acid: $CH_3(CH_2)_7CH{=}CH(CH_2)_7COOH$

b. Lauric acid: $CH_3(CH_2)_{10}COOH$

c. Linoleic acid: $CH_3(CH_2)_4CH{=}CH{-}CH_2{-}CH{=}CH(CH_2)_7COOH$

d. Stearic acid: $CH_3(CH_2)_{16}COOH$

17.3 Myristic acid is the common name for tetradecanoic acid. Carboxylic acids can react with alchols to form esters.

$$CH_3(CH_2)_{12}\overset{O}{\overset{\|}{C}}{-}OH + CH_3CH_2OH \xrightarrow[\text{heat}]{H^+} CH_3(CH_2)_{12}\overset{O}{\overset{\|}{C}}{-}O{-}CH_2CH_3$$

Tetradecanoic acid Ethanol Ethyl tetradecanoate

17.5 The IUPAC name for pentyl butanoate identifies it as an ester. Acid hydrolysis of an ester results in the formation of an alcohol and a carboxylic acid.

$$CH_3CH_2CH_2\overset{O}{\overset{\|}{C}}{-}O{-}CH_2(CH_2)_3CH_3 \xrightarrow[\text{heat}]{H^+} CH_3CH_2CH_2\overset{O}{\overset{\|}{C}}{-}OH + CH_3(CH_2)_3CH_2OH$$

Pentyl butanoate Butanoic acid Pentanol

17.7 The IUPAC name for butyl acetate identifies it as an ester. Base hydrolysis of an ester results in the formation of an alcohol and a carboxylate salt.

$$CH_3\overset{O}{\overset{\|}{C}}{-}O{-}CH_2(CH_2)_2CH_3 + KOH \rightarrow CH_3\overset{O}{\overset{\|}{C}}{-}O^-K^+ + CH_3(CH_2)_2CH_2OH$$

Butyl acetate Potassium acetate Butanol

17.9 Hydrogenation of an unsaturated fatty acid produces a saturated fatty acid. The IUPAC name for linolenic acid is all *cis*-9,12,15-octadecatrienoic acid, which allows you to draw the structure.

$$CH_3CH_2CH{=}CHCH_2CH{=}CHCH_2CH{=}CH(CH_2)_7COOH \quad + \ 3\,H_2$$

All *cis*-9,12,15-Octadecatrienoic acid

$$\downarrow \ Ni$$

$$CH_3(CH_2)_{16}COOH$$

Octadecanoic acid

17.11 Esterification of glycerol may occur at one, two, or three of the hydroxyl positions and produce a mono, di, or triglyceride.

a. Mono, di, and triglycerides of oleic acid:

$$CH_3(CH_2)_7CH{=}CH(CH_2)_7-\overset{\overset{\displaystyle O}{\|}}{C}-O-CH_2$$
$$|$$
$$CH-OH$$
$$|$$
$$CH_2-OH$$

$$CH_3(CH_2)_7CH{=}CH(CH_2)_7-\overset{\overset{\displaystyle O}{\|}}{C}-O-CH_2$$
$$|$$
$$CH_3(CH_2)_7CH{=}CH(CH_2)_7-\overset{\overset{\displaystyle O}{\|}}{C}-O-CH$$
$$|$$
$$CH_2-OH$$

$$CH_3(CH_2)_7CH{=}CH(CH_2)_7-\overset{\overset{\displaystyle O}{\|}}{C}-O-CH_2$$
$$|$$
$$CH_3(CH_2)_7CH{=}CH(CH_2)_7-\overset{\overset{\displaystyle O}{\|}}{C}-O-CH$$
$$|$$
$$CH_3(CH_2)_7CH{=}CH(CH_2)_7-\overset{\overset{\displaystyle O}{\|}}{C}-O-CH_2$$

b. Mono, di, and triglycerides of capric acid:

$$CH_3(CH_2)_8-\overset{\overset{\displaystyle O}{\|}}{C}-O-CH_2$$
$$|$$
$$CH-OH$$
$$|$$
$$CH_2-OH$$

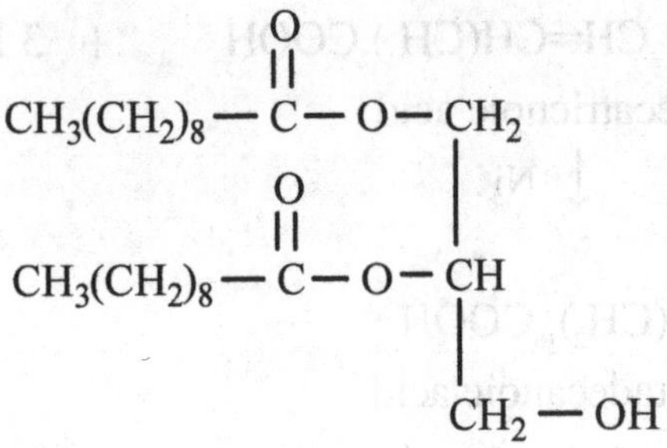

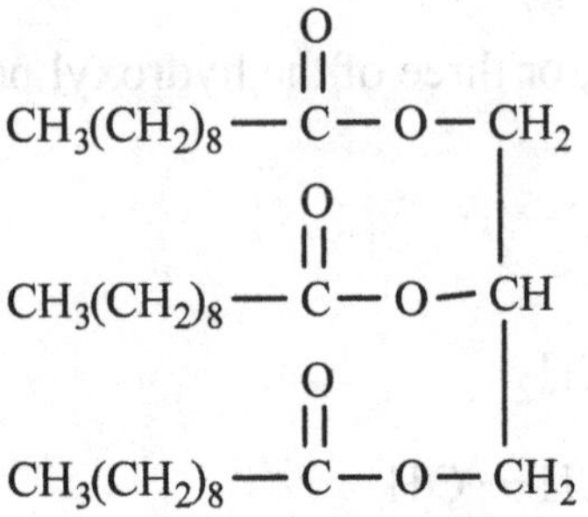

$$
\begin{array}{l}
CH_3(CH_2)_8-\overset{O}{\overset{\|}{C}}-O-CH_2 \\
\qquad\qquad\qquad\qquad\quad | \\
CH_3(CH_2)_8-\overset{O}{\overset{\|}{C}}-O-CH \\
\qquad\qquad\qquad\qquad\quad | \\
\qquad\qquad\qquad\qquad\quad CH_2-OH
\end{array}
$$

$$
\begin{array}{l}
CH_3(CH_2)_8-\overset{O}{\overset{\|}{C}}-O-CH_2 \\
\qquad\qquad\qquad\qquad\quad | \\
CH_3(CH_2)_8-\overset{O}{\overset{\|}{C}}-O-CH \\
\qquad\qquad\qquad\qquad\quad | \\
CH_3(CH_2)_8-\overset{O}{\overset{\|}{C}}-O-CH_2
\end{array}
$$

17.13 Structure of the steroid nucleus:

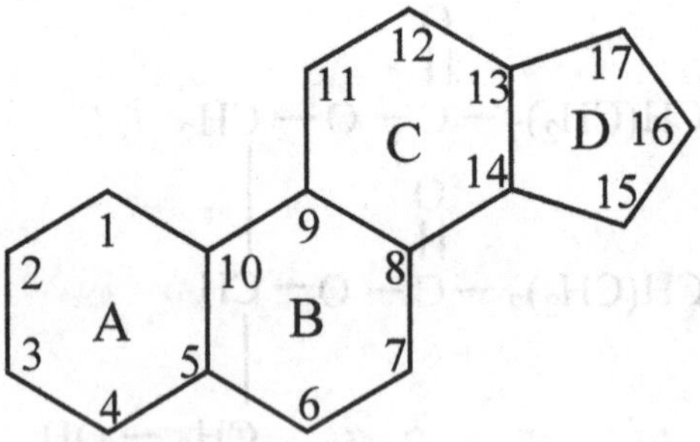

17.15 Cholesterol is carried in low density lipoprotein (LDL) particles in the plasma. These bind to specific LDL receptors within cell membranes. This binding stimulates receptor-mediated endocytosis, the invagination of the cell membrane that draws the LDL particles into the cytoplasm of the cell. The process encases the LDL within a vesicle or endosome in the cytoplasm.

End-of-Chapter Questions and Problems

17.17 The four main groups of lipids are fatty acids, glycerides, nonglyceride lipids, and complex lipids.

17.19 Lipid-soluble vitamins are transported into cells of the small intestine in association with dietary fat molecules. Thus, a diet low in fat reduces the amount of vitamins A, D, E, and K that enters the body.

17.21 A saturated fatty acid is one in which the hydrocarbon tail has only carbon-to-carbon single bonds. Thus, each carbon atom is bonded to the maximum number of hydrogen atoms. An unsaturated fatty acid has at least one carbon-to-carbon double bond.

17.23 As the length of the hydrocarbon chains of fatty acids increases, the melting points increase.

17.25 The melting points of fatty acids increase as the length of the hydrocarbon chains increase. This is because the intermolecular attractive forces, including the London dispersion forces, increase as the length of the hydrocarbon chain increases.

17.27 Fatty acids are long-chained monocarboxylic acids.

a. Decanoic acid:

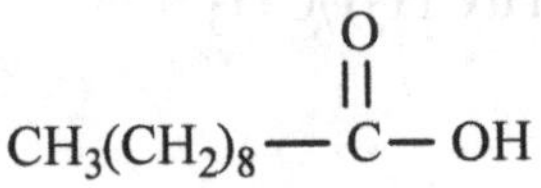

$$CH_3(CH_2)_8-\overset{\overset{\displaystyle O}{\|}}{C}-OH$$

b. Stearic acid:

$$CH_3(CH_2)_{16}-\overset{\overset{\displaystyle O}{\|}}{C}-OH$$

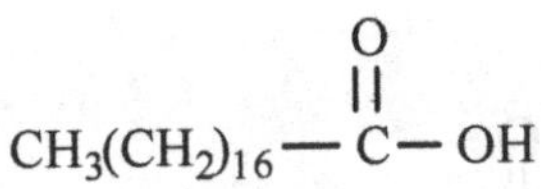

17.29 The common and IUPAC names are based on the number of carbons in the fatty acid and the location and number of the double bonds.

a. IUPAC name: Hexadecanoic acid
Common name: Palmitic acid

b. IUPAC name: Dodecanoic acid
Common name: Lauric acid

17.31 Esterification of glycerol with three molecules of myristic acid will produce a triglyceride.

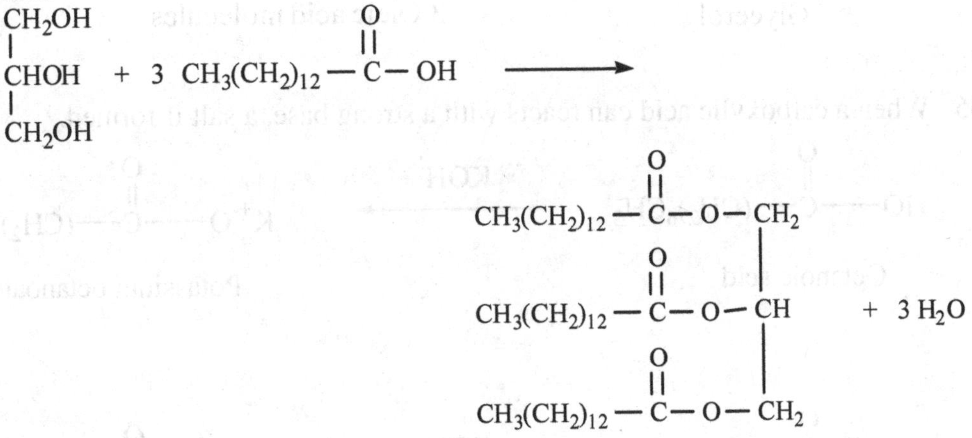

$$\begin{array}{l} CH_2OH \\ | \\ CHOH \\ | \\ CH_2OH \end{array} + 3\ CH_3(CH_2)_{12}-\overset{\overset{\displaystyle O}{\|}}{C}-OH \longrightarrow$$

$$\begin{array}{l} CH_3(CH_2)_{12}-\overset{\overset{\displaystyle O}{\|}}{C}-O-CH_2 \\ \qquad\qquad\qquad\qquad\quad | \\ CH_3(CH_2)_{12}-\overset{\overset{\displaystyle O}{\|}}{C}-O-CH \\ \qquad\qquad\qquad\qquad\quad | \\ CH_3(CH_2)_{12}-\overset{\overset{\displaystyle O}{\|}}{C}-O-CH_2 \end{array} + 3\ H_2O$$

17.33 This triglyceride was formed by the esterification of glycerol with three oleic acid molecules. When it is hydrolyzed, glycerol and three oleic acid molecules are produced.

$$\begin{array}{l} H-\overset{H}{\underset{|}{\overset{|}{C}}}-O-\overset{O}{\overset{\|}{C}}-(CH_2)_7CH{=}CH(CH_2)_7CH_3 \\ H-\underset{|}{\overset{|}{C}}-O-\overset{O}{\overset{\|}{C}}-(CH_2)_7CH{=}CH(CH_2)_7CH_3 \quad + \; 3H_2O \\ H-\underset{H}{\overset{|}{C}}-O-\overset{O}{\overset{\|}{C}}-(CH_2)_7CH{=}CH(CH_2)_7CH_3 \end{array}$$

$$\downarrow \text{H+, heat}$$

$$\begin{array}{lll} H-\overset{H}{\underset{|}{\overset{|}{C}}}-OH & + & HO-\overset{O}{\overset{\|}{C}}-(CH_2)_7CH{=}CH(CH_2)_7CH_3 \\ H-\underset{|}{\overset{|}{C}}-OH & + & HO-\overset{O}{\overset{\|}{C}}-(CH_2)_7CH{=}CH(CH_2)_7CH_3 \\ H-\underset{H}{\overset{|}{C}}-OH & + & HO-\overset{O}{\overset{\|}{C}}-(CH_2)_7CH{=}CH(CH_2)_7CH_3 \end{array}$$

Glycerol　　　　　　　　　3 Oleic acid molecules

17.35 When a carboxylic acid can reacts with a strong base, a salt if formed.

$$HO-\overset{O}{\overset{\|}{C}}-(CH_2)_6CH_3 \xrightarrow{KOH} K^+{}^-O-\overset{O}{\overset{\|}{C}}-(CH_2)_6CH_3$$

Octanoic acid　　　　　　　　　Potassium octanoate

$$HO-\overset{O}{\overset{\|}{C}}-(CH_2)_{16}CH_3 \xrightarrow{KOH} K^+{}^-O-\overset{O}{\overset{\|}{C}}-(CH_2)_{16}CH_3$$

Stearic acid　　　　　　　　　Potassium stearate

17.37

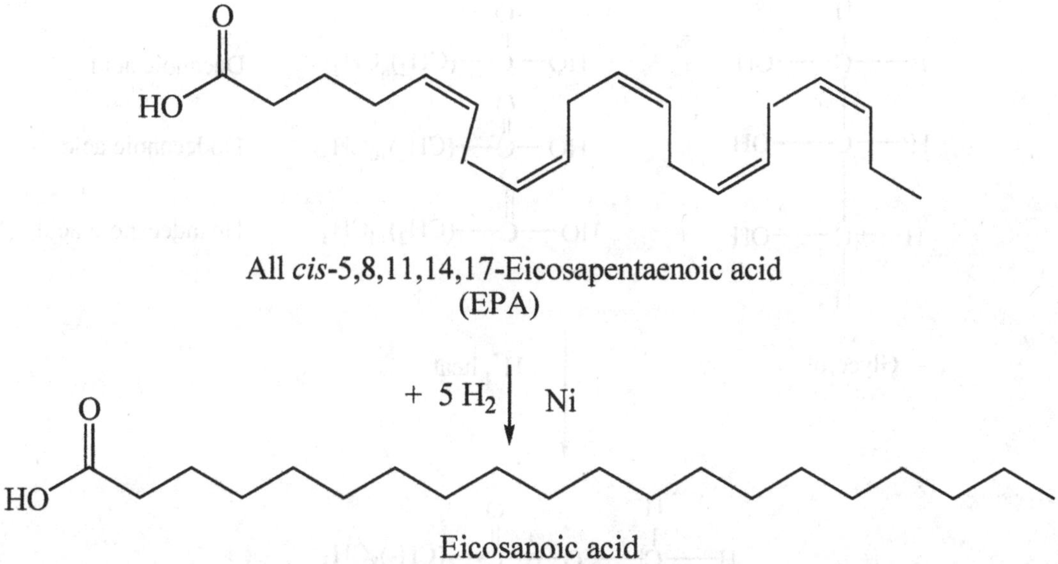

All *cis*-5,8,11,14,17-Eicosapentaenoic acid
(EPA)

$+ 5 H_2$ Ni

Eicosanoic acid

17.39 Regardless of where the acids are on the initial triglyceride, basic hydrolysis will yield the same products.

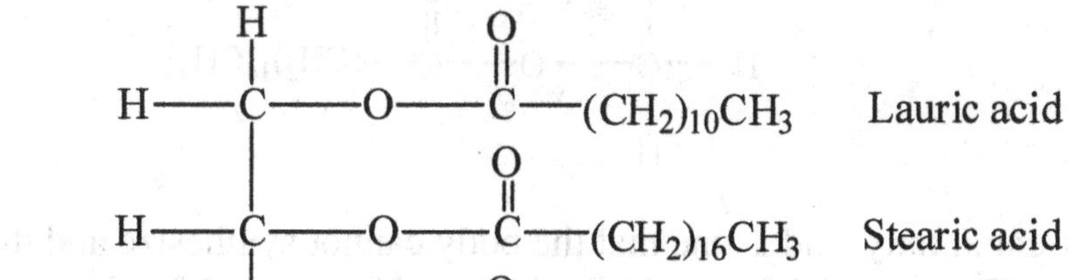

$$H_2C-O-C(=O)-(CH_2)_{10}CH_3 \quad \text{Lauric acid}$$
$$HC-O-C(=O)-(CH_2)_{16}CH_3 \quad \text{Stearic acid}$$
$$H_2C-O-C(=O)-(CH_2)_8CH_3 \quad \text{Capric acid}$$

NaOH

$$H_2C-OH \quad + \quad Na^+ {}^-O-C(=O)-(CH_2)_{10}CH_3$$
$$HC-OH \quad + \quad Na^+ {}^-O-C(=O)-(CH_2)_{16}CH_3$$
$$H_2C-OH \quad + \quad Na^+ {}^-O-C(=O)-(CH_2)_8CH_3$$

17.41

$$\begin{array}{l} \text{H—C(H)—OH} + \text{HO—C(=O)—(CH}_2\text{)}_8\text{CH}_3 \quad \text{Decanoic acid} \\ \text{H—C—OH} + \text{HO—C(=O)—(CH}_2\text{)}_{10}\text{CH}_3 \quad \text{Dodecanoic acid} \\ \text{H—C(H)—OH} + \text{HO—C(=O)—(CH}_2\text{)}_{14}\text{CH}_3 \quad \text{Hexadecanoic acid} \end{array}$$

Glycerol

$\xrightarrow{H^+,\ \text{heat}}$

$$\begin{array}{l} \text{H—C(H)—O—C(=O)—(CH}_2\text{)}_8\text{CH}_3 \\ \text{H—C—O—C(=O)—(CH}_2\text{)}_{10}\text{CH}_3 \quad + 3\,H_2O \\ \text{H—C(H)—O—C(=O)—(CH}_2\text{)}_{14}\text{CH}_3 \end{array}$$

17.43 An essential fatty acid is one that the body cannot synthesize and thus must be supplied in the diet. The essential fatty acid linoleic acid is required for the synthesis of arachidonic acid, a precursor for the synthesis of the prostaglandins, a group of hormonelike molecules.

17.45 Aspirin effectively decreases the inflammatory response by inhibiting the synthesis of all prostaglandins. Aspirin works by inhibiting cyclooxygenase, the first enzyme in prostaglandin biosynthesis. This inhibition results from the transfer of an acetyl group from aspirin to the enzyme. Since cyclooxygenase is found in all cells, synthesis of all prostaglandins is inhibited.

17.47 Prostaglandins stimulate smooth muscle contraction, especially uterine contractions during labor. They enhance fever and swelling associated with the inflammatory response. Some prostaglandins cause bronchial dilation. Others inhibit secretion of acid into the stomach and stimulate the secretion of a mucous layer that protects the stomach lining.

17.49 The name of these fatty acids arises from the position of the double bond nearest the terminal *methyl group* of the molecule. The terminal methyl group is designated omega (ω). In ω-3 fatty acids the double bond nearest the ω methyl group is three carbons along the chain. In ω-6 fatty acids, the nearest double bone is six carbons from the end.

17.51 Omega-3 fatty acids reduce the risk of cardiovascular disease by decreasing blood clot formation, blood triglyceride levels, and growth of atherosclerotic plaque.

17.53 The decrease in blood clot formation, along with the reduced blood triglyceride levels and decreased atherosclerotic plaque result in improved arterial health. This, in turn, results in lower blood pressure and a decreased risk of sudden death and heart arrhythmias.

17.55 Omega-3 fatty acids are precursors of prostaglandins that exhibit anti-inflammatory effects. On the other hand, omega-6 fatty acids are precursors to prostaglandins that have inflammatory effects. To reduce the inflammatory response contribution to cardiovascular disease, it is logical to increase the amount of omega-3 fatty acids in the diet and to decrease the amount of omega-6 fatty acids.

17.57 A glyceride is a lipid ester that contains the glycerol molecule and from 1 to 3 fatty acids.

17.59 An emulsifying agent is a molecule that aids in the suspension of triglycerides in water. They are amphipathic molecules, such as lecithin, that serve as bridges holding together the highly polar water molecules and the nonpolar triglycerides.

17.61 A triglyceride with three saturated fatty acid tails would be a solid at room temperature. The long, straight fatty acid tails would stack with one another because of strong intermolecular and intramolecular London dispersion force attractions.

17.63 This triglyceride was formed from the esterification of glycerol. Hexadecanoic acid reacted at C-1. *Trans*-9-hexadecenoic acid reacted at C-2. *Cis*-9-hexadecenoic acid reacted at C-3. Three molecules of water were also produced when this triglyceride was formed.

$CH_3(CH_2)_{14}-C(=O)-O-CH_2$

$H(CH_3(CH_2)_4CH_2)C=C(H)CH_2(CH_2)_6-C(=O)-O-CH$ (trans)

$H(CH_3(CH_2)_4CH_2)C=C(H)CH_2(CH_2)_6-C(=O)-O-CH_2$ (cis)

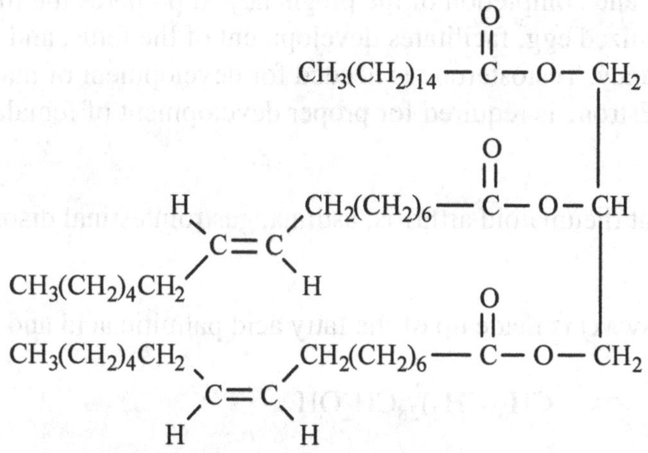

17.65 Glycerol-3-phosphate reacted with capric acid at C-1 and with lauric acid at C-2 to form this phosphatidate.

$$
\begin{array}{l}
\qquad\qquad\qquad\quad\; \overset{\displaystyle O}{\|} \\
CH_3(CH_2)_8 - C - O - CH_2 \\
\qquad\qquad\qquad\quad\; \overset{\displaystyle O}{\|} \qquad\;\; | \\
CH_3(CH_2)_{10} - C - O - CH \\
\qquad\qquad\qquad\qquad\qquad\;\; | \qquad\quad \overset{\displaystyle O}{\|} \\
\qquad\qquad\qquad\qquad\qquad CH_2 - O - P - O^- \\
\qquad\qquad\qquad\qquad\qquad\qquad\qquad\;\; | \\
\qquad\qquad\qquad\qquad\qquad\qquad\qquad O^-
\end{array}
$$

17.67 Triglycerides consist of three fatty acids esterified to the three hydroxyl groups of glycerol. In phospholipids, there are only two fatty acids esterified to glycerol. A phosphoryl group is esterified (phosphoester linkage) to the third hydroxyl group.

17.69 A sphingolipid is one that is not derived from glycerol, but rather from sphingosine, a long-chain, nitrogen-containing (amino) alcohol. Like phospholipids, sphingolipids are amphipathic.

17.71 A glycosphingolipid or glycolipid is one that is built on a ceramide backbone structure. Ceramide is a fatty acid derivative of sphingosine.

17.73 Sphingomyelins are important structural lipid components of nerve cell membranes. They are found in the myelin sheath that surrounds and insulates cells of the central nervous system.

17.75 Cholesterol is readily soluble in the hydrophobic region of biological membranes. It is involved in regulating the fluidity of the membrane.

17.77 Progesterone is the most important hormone associated with pregnancy. It is needed for the successful initiation and completion of the pregnancy. It prepares the lining of the uterus to accept the fertilized egg, facilitates development of the fetus, and suppresses ovulation during pregnancy. Testosterone is needed for development of male secondary sexual characteristics. Estrone is required for proper development of female secondary sexual characteristics.

17.79 Cortisone is used to treat rheumatoid arthritis, asthma, gastrointestinal disorders, and many skin conditions.

17.81 Myricyl palmitate (beeswax) is made up of the fatty acid palmitic acid and the alcohol myricyl alcohol:

$$CH_3(CH_2)_{28}CH_2OH$$

17.83 Isoprenoids are a large, diverse collection of lipids that are synthesized from the isoprene unit:

$$
\begin{array}{l}
\qquad\quad CH_3 \\
\qquad\quad\;\; | \\
CH_2{=}C\text{-}CH{=}CH_2
\end{array}
$$

17.85 Some biologically important terpenes include the steroids and bile salts, lipid-soluble vitamins, certain plant hormones, and chlorophyll.

17.87 The four major types of plasma lipoproteins are chylomicrons, high density lipoproteins, low density lipoproteins, and very low density lipoproteins.

17.89 The terms *good* and *bad* cholesterol refer to two classes of lipoprotein complexes. The high density lipoproteins, or HDL, are considered to be good cholesterol because a correlation has been made between elevated levels of HDL and a reduced incidence of atherosclerosis. Low density lipoproteins, or LDL, are considered to be bad cholesterol because evidence suggests that a high level of LDL is associated with increased risk of atherosclerosis.

17.91 Atherosclerosis results when cholesterol and other substances coat the arteries causing a narrowing of the passageways. As the passageways become narrower, greater pressure is required to provide adequate blood flow. This results in higher blood pressure (hypertension).

17.93 If the LDL receptor is defective, it cannot function to remove cholesterol-bearing LDL particles from the blood. The excess cholesterol, along with other substances, will accumulate along the walls of the arteries causing atherosclerosis.

17.95 The basic structure of a biological membrane is a bilayer of phospholipid molecules arranged so that the hydrophobic hydrocarbon tails are packed in the center and the hydrophilic head groups are exposed on the inner and outer surfaces.

17.97 A peripheral membrane protein is bound to only one surface of the membrane, either inside or outside the cell.

17.99 Cholesterol is freely soluble in the hydrophobic layer of a biological membrane. It moderates the fluidity of the membrane by disrupting the stacking of the fatty acid tails of membrane phospholipids.

17.101 Specific membrane proteins on human and mouse cells were labeled with red and green fluorescent dyes, respectively. The cells were fused into single-celled hybrids and were observed using a microscope with an ultraviolet light source. The ultraviolet light caused the dyes to fluoresce. Initially, the dyes were localized in regions of the membrane representing the original human or mouse cell. Within an hour, the proteins were evenly distributed throughout the membrane of the fused cell.

17.103 If the fatty acyl tails of membrane phospholipids are converted from saturated to unsaturated, the fluidity of the membrane will increase. Each carbon-to-carbon double bond that is added will introduce a "kink" into the fatty acyl tail. As a result, the tails cannot pack together as they would if they were saturated. The result is weaker London dispersion forces and increased fluidity.

Chapter 18
Protein Structure and Function
Solutions to the Odd-Numbered Questions and Problems

In-Chapter Questions and Problems

18.1 The abbreviations and the structures for the common amino acids are found in the textbook.

a. Glycine (Gly or G): H_3N^+—C(H)(H)—COO^-

b. Proline (Pro or P): COO^- on C(H) bonded to $^+NH_2$, ring H_2C—CH_2—$C(H_2)$

c. Threonine (Thr or T): H_3N^+—C(H)(COO^-)—C(H)(OH)—CH_3

d. Aspartate (Asp or D): H_3N^+—C(H)(COO^-)—CH_2—C(=O)O^-

e. Lysine (Lys or K): H_3N^+—C(H)(COO^-)—CH_2—CH_2—CH_2—CH_2—N^+H_3

18.3 In forming these peptides, an amide bond forms between the $-COO^-$ group of one amino acid and the α-N^+H_3 of another amino acid. The amino acid sequence is very specific to the method of naming peptides.

a. Alanyl-phenylalanine:

H_3N^+—CH(CH_3)—C(=O)—NH—CH($CH_2C_6H_5$)—C(=O)—O^-

b. Lysyl-alanine:

```
        H   O       H   O
        |   ||      |   ||
  H3+N— C — C — N — C — C — O-
        |       |   |
       CH2      H   CH3
        |
       CH2
        |
       CH2
        |
       CH2
        |
      +NH3
```

c. Phenylalanyl-tyrosyl-leucine:

```
        H   O       H   O       H   O
        |   ||      |   ||      |   ||
  H3+N— C — C — N — C — C — N — C — C — O-
        |       |   |       |   |
       CH2      H  CH2      H  CH2
        |           |           |
    (phenyl)    (phenyl)      H3CHC
                    |           |
                   OH          CH3
```

18.5 The primary structure of a protein is the amino acid sequence of the protein chain. Regular, repeating folding of the peptide chain caused by hydrogen bonding between the amide nitrogens and carbonyl oxygens of the peptide bond is the secondary structure of a protein. The two most common types of secondary structure are the α-helix and the β-pleated sheet. Tertiary structure is the further folding of the regions of α-helix and β-pleated sheet into a compact, globular structure. Formation and maintenance of the tertiary structure result from weak attractions between amino acid R groups. The binding of two or more peptides to produce a functional protein defines the quaternary structure.

18.7 Oxygen is efficiently transferred from hemoglobin to myoglobin in the muscle because myoglobin has a greater affinity for oxygen.

18.9 High temperature disrupts the hydrogen bonds and other weak interactions that maintain protein structure.

18.11 Vegetables vary in amino acid composition. Most vegetables do not provide all of the amino acid requirements of the body. By eating a variety of different vegetables, all the amino acid requirements of the human body can be met.

End-of-Chapter Questions and Problems

18.13 An enzyme is a protein that serves as a biological catalyst, speeding up biological reactions.

18.15 A transport protein is a protein that transports materials across the cell membrane or throughout the body.

18.17 Enzymes speed up reactions that might take days or weeks to occur on their own. They also catalyze reactions that might require very high temperatures or harsh conditions if carried out in the laboratory. In the body, these reactions occur quickly under physiological conditions.

18.19 Transferrin is a transport protein that carries iron from the liver to the bone marrow, where it is used to produce the heme group for hemoglobin and myoglobin. Hemoglobin transports oxygen in the blood.

18.21 Egg albumin is a nutrient protein that serves as a source of protein for the developing chick. Casein is the nutrient storage protein in milk, providing protein, a source of amino acids, for mammals.

18.23 The general structure of an L-α-amino acid has a carbon in the center that is referred to as the alpha carbon. Bonded to the alpha carbon, are an amino group, a carboxyl group, a hydrogen, and a side chain, R. The R group varies for each amino acid. The general structure of an L-α-amino acid is shown here.

$$\begin{array}{c} COO^- \\ | \\ H_3{}^+N - C - H \\ | \\ R \end{array}$$

18.25 A zwitterion is a neutral molecule with equal numbers of positive and negative charges. Under physiological conditions, amino acids are zwitterions.

18.27 A chiral carbon is one that has four different atoms or groups of atoms attached to it.

18.29 Interactions between the R groups of the amino acids in a polypeptide chain are important for the formation and maintenance of the tertiary and quaternary structures of proteins.

18.31 The following are the structures of the amino acids that have polar, neutral side chains.

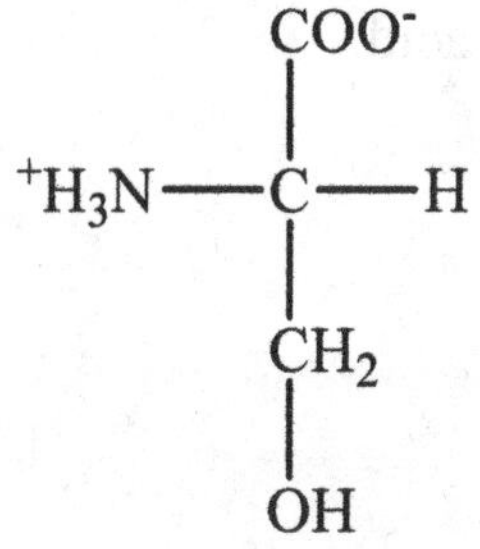

L-Serine

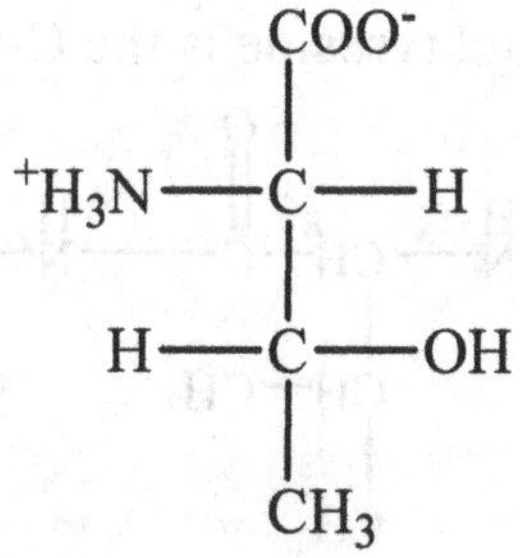

L-Threonine

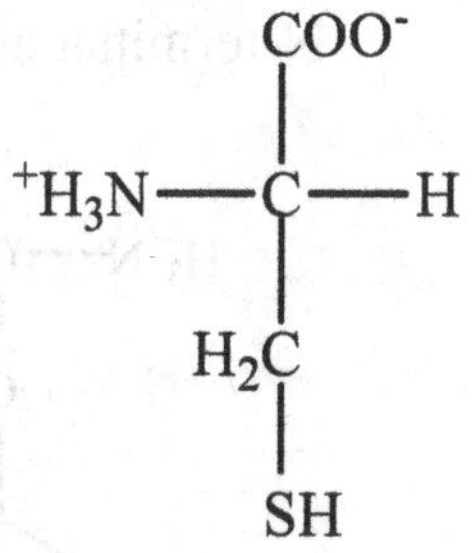

L-Cysteine

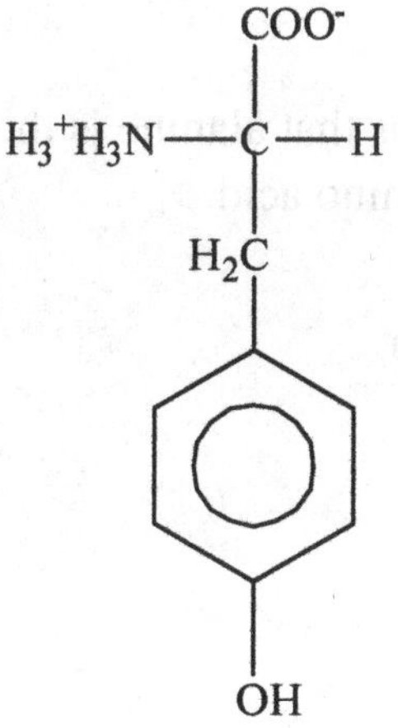

L-Tyrosine

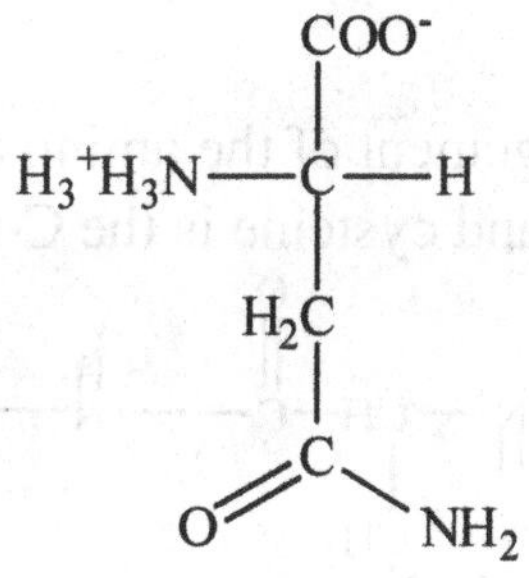

L-Asparagine

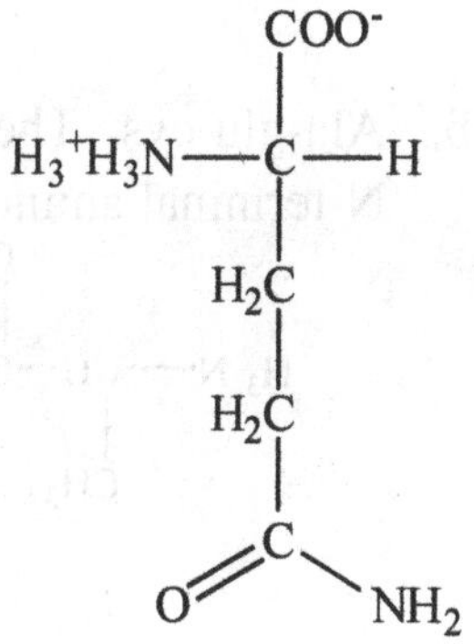

L-Glutamine

18.33 A peptide bond is an amide bond between two amino acids in a peptide chain.

18.35 Linus Pauling and his colleagues carried out X-ray diffraction studies of protein. Interpretation of the pattern formed when X-rays were diffracted by a crystal of pure protein led Pauling to conclude that peptide bonds are both planar (flat) and rigid and that the N-C bonds are shorter that expected. In other words, they deduced that the peptide bond has a partially double bond character because it exhibits resonance. There is no free rotation about the amide bond because the carbonyl group of the amide bond has a strong attraction for the amide nitrogen lone pair of electrons. This can best be described using a resonance model:

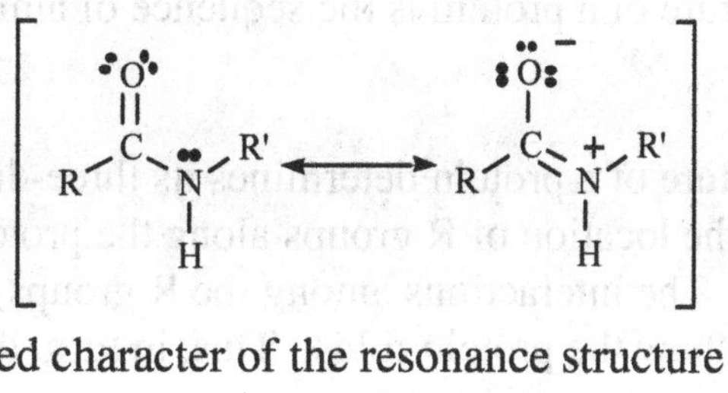

The partially double bonded character of the resonance structure restricts free rotation.

18.37 a. Phe-val-tyr. The arrangement of the amino acids tells us that phenylalanine is the N-terminal amino acid and tyrosine is the C-terminal amino acid.

```
                O                   O                  O
                ||      H           ||     H           ||
H3+N—CH—C———N—CH—C———N—CH—C—O-
       |               |                 |
      CH2             CH—CH3            CH2
       |               |                 |
  (benzene ring)      CH3          (benzene ring)
                                         |
                                         OH
```

b. Ala-glu-cys. The arrangement of the amino acids tells us that alanine is the N-terminal amino acid and cysteine is the C-terminal amino acid.

```
            O               O                O
            ||              ||       H       ||
H3+N—CH—C———N—CH—C———N—CH—C—O-
       |        H   |                 |
      CH3          CH2               CH2
                    |                 |
                   CH2                SH
                    |
                   C=O
                    |
                    O-
```

c. Asn-leu-gly. The arrangement of the amino acids tells us that asparagine is the N-terminal amino acid and glycine is the C-terminal amino acid.

```
            O             O            O
            ||            ||     H     ||
H3+N—CH—C———N—CH—C———N—CH—C—O-
       |        H   |             |
      CH2          CH2            H
       |            |
      C=O          CH—CH3
       |            |
      NH2          CH3
```

18.39 The primary structure of a protein is the sequence of amino acids bonded to one another by peptide bonds.

18.41 The primary structure of a protein determines its three-dimensional shape and biological function because the location of R groups along the protein chain is determined by the primary structure. The interactions among the R groups, based on their location in the chain, will govern how the protein folds. This, in turn, dictates its three-dimensional structure and biological function.

18.43 When leucine reacts with arginine to form leu-arg, a water molecule is also formed.

```
            O                      O                        O         O
            ||                     ||                       ||        ||
H3+N—CH—C—O-   H3+N—CH—C—O-        H3+N—CH—C—N—CH—C—O-
      |                 |                    |    H   |
     CH2       +       CH2      ——→         CH2      CH2      + H2O
      |                 |                    |        |
     CH—CH3            CH2                  CH—CH3   CH2
      |                 |                    |        |
     CH3               CH2                  CH3      CH2
                        |                             |
                        NH                            NH
                        |                             |
                        C═NH2+                        C═NH2+
                        |                             |
                        NH2                           NH2
```

18.45 The secondary structure of a protein is the folding of the primary structure into an α-helix or β-pleated sheet.

18.47 a. α-Helix
b. β-Pleated sheet

18.49 A fibrous protein is one that is composed of peptides arranged in long sheets or fibers.

18.51 A parallel β-pleated sheet is one in which the hydrogen-bonded peptide chains have their amino termini aligned head to head.

18.53 The tertiary structure of a protein is the globular, three-dimensional structure of a protein that results from folding the regions of secondary structure.

18.55 The oxidation of cysteine forms a disulfide bond. This dimeric amino acid is called cystine.

```
          COO-
           |
   H3+N—C—H
           |
          CH2
           |
           S
           |
           S
           |
          CH2
           |
   H3+N—C—H
           |
          COO-
```

18.57 The tertiary structure is a level of folding of a protein chain that has already undergone secondary folding. The regions of α-helix and β-pleated sheet are folded into a globular structure.

18.59 Quaternary protein structure is the aggregation of two or more folded peptide chains to produce a functional protein.

18.61 A glycoprotein is a protein with sugars as prosthetic groups. Glycoproteins are often receptors on the cell surface.

18.63 Hydrogen bonding maintains the secondary structure of a protein and contributes to the stability of the tertiary and quaternary levels of structure.

18.65 The partially double bonded character of the resonance structure restricts free rotation.

$$\left[R-\overset{\overset{\textstyle :\ddot{O}}{\|}}{C}-\underset{\underset{\textstyle H}{|}}{\ddot{N}}-R' \longleftrightarrow R-\overset{\overset{\textstyle :\ddot{O}:^{-}}{|}}{C}=\underset{\underset{\textstyle H}{|}}{\overset{+}{N}}-R' \right]$$

18.67 The code for the primary structure of a protein is carried in the genetic information (DNA). The sequence of nucleotides in the DNA dictates the sequence of amino acids in a protein.

18.69 The function of hemoglobin is to carry oxygen from the lungs to oxygen-demanding tissues throughout the body.

18.71 Hemoglobin is a protein composed of four subunits—two α-globin and two β-globin subunits. Each subunit holds a heme group, which in turn carries a Fe^{3+} ion.

18.73 The function of the heme group in hemoglobin and myoglobin is to bind to molecular oxygen.

18.75 Because carbon monoxide binds tightly to the heme groups of hemoglobin, it is not easily removed or replaced by oxygen. As a result, the effects of oxygen deprivation (suffocation) occur.

18.77 When sickle cell hemoglobin (HbS) is deoxygenated, the amino acid valine fits into a hydrophobic pocket on the surface of another HbS molecule. Many such sickle cell hemoglobin molecules polymerize into long rods that cause the red blood cell to sickle. In normal hemoglobin, glutamic acid is found in place of the valine. This negatively charged amino acid will not "fit" into the hydrophobic pocket.

18.79 When individuals have one copy of the sickle cell gene and one copy of the normal gene, they are said to carry the sickle cell trait. These individuals will not suffer serious side effects, but may pass the trait to their offspring. Individuals with two copies of the sickle cell globin gene exhibit all the symptoms of the disease and are said to have sickle cell anemia.

18.81 Albumin is the most abundant protein in the blood. It makes up approximately 55% of the blood protein.

18.83 Albumin in the blood can serve as a carrier for Ca^{2+} because it contains acidic amino acids. The negative charges in the acidic amino acids can form salt bridges (or ionic bonds) with Ca^{2+}. Albumin can also serve as a carrier for fatty acids because it contains basic amino acids. The positive charges in the basic amino acids can form salt bridges with the anionic fatty acids.

18.85 Denaturation is the process by which the organized structure of a protein is disrupted, resulting in a completely disorganized, nonfunctional form of the protein.

18.87 Heat is an effective means of sterilization because it destroys the proteins of microbial life-forms, including fungi, bacteria, and viruses.

18.89 The low pH of the yogurt denatures the proteins of microbial contaminants, inhibiting their growth.

18.91 An essential amino acid is one that must be provided in the diet because it cannot be synthesized in the body.

18.93 A complete protein is one that contains all of the essential and nonessential amino acids.

18.95 Chymotrypsin catalyzes the hydrolysis of peptide bonds on the carbonyl side of aromatic amino acids.

Site of chymotrypsin-catalyzed hydrolysis

$H_3^+N-CH(CH_2C_6H_5)-C(=O)-NH-CH(CH_3)-C(=O)-O^- + H_2O \longrightarrow H_3^+N-CH(CH_2C_6H_5)-C(=O)-O^- + H_3^+N-CH(CH_3)-C(=O)-O^-$

Phenylalanyl alanine

Phenylalanine

Alanine

18.97 In a vegetarian diet, vegetables are the only source of dietary protein. Since most individual vegetable sources do not provide all the needed amino acids, vegetables must be mixed to provide all essential and nonessential amino acids required for biosynthesis

18.99 Synthesis of digestive enzymes must be carefully controlled because the active enzyme would digest, and thus destroy, the cell that produces it.

Chapter 19
Enzymes
Solutions to the Odd-Numbered Questions and Problems

In-Chapter Questions and Problems

19.1 a. Pyruvate kinase catalyzes the transfer of a phosphoryl group from phosphoenolpyruvate to adenosine diphosphate.

$$H_2C{=}C(O\text{-}P){-}C(=O){-}O^- \xrightarrow[\text{Pyruvate kinase}]{ADP \rightarrow ATP} H_3C{-}C(=O){-}C(=O){-}O^-$$

Phosphoenolpyruvate Pyruvate

b. Alanine transaminase catalyzes the transfer of an amino group from alanine to α- ketoglutarate, producing pyruvate and glutamate.

Alanine + α-Ketoglutarate $\xrightarrow{\text{Alanine transaminase}}$ Pyruvate + Glutamate

Alanine: $H_3^+N{-}CH(CH_3){-}C(=O){-}O^-$

α-Ketoglutarate: $^-O{-}C(=O){-}C(=O){-}CH_2{-}CH_2{-}C(=O){-}O^-$

Pyruvate: $H_3C{-}C(=O){-}C(=O){-}O^-$

Glutamate: $H_3^+N{-}CH(COO^-){-}CH_2{-}CH_2{-}C(=O){-}O^-$

c. Triose phosphate isomerase catalyzes the isomerization of the ketone dihydroxyacetone phosphate to the aldehyde glyceraldehyde-3-phosphate.

$$H_2C(O\text{-}P){-}C(=O){-}CH_2{-}OH \xrightarrow{\text{Triose phosphate isomerase}} H_2C(O\text{-}P){-}CH(OH){-}C(=O){-}H$$

Dihydroxyacetone phosphate Glyceraldehyde-3-phosphate

d. Pyruvate dehydrogenase catalyzes the oxidation and decarboxylation of pyruvate, producing acetyl coenzyme A and CO_2.

$$\underset{\text{Pyruvate}}{H_3C-\overset{\overset{\large O}{\|}}{C}-\overset{\overset{\large O}{\|}}{C}-O^-} + \underset{\text{Coenzyme A}}{H\text{-}S\text{-}CoA} \xrightarrow{\text{Pyruvate dehydrogenase}} \underset{\text{Acetyl coenzyme A}}{H_3C-\overset{\overset{\large O}{\|}}{C}\sim S\text{-}CoA} + CO_2$$

19.3 The substrates for these enzymes can be deduced by the names of the enzyme.
a. Sucrose. This is the substrate for sucrase.
b. Pyruvate This is the substrate for pyruvate decarboxylase.
c. Succinate This is the substrate for succinate dehydrogenase.

19.5 The induced fit model assumes that the enzyme is flexible. Both the enzyme and the substrate are able to change shape to form the enzyme-substrate complex. The lock-and-key model assumes that the enzyme is inflexible (the lock) and the substrate (the key) fits into a specific rigid site (the active site) on the enzyme to form the enzyme-substrate complex.

19.7 An enzyme might distort a bond, thereby catalyzing bond breakage. An enzyme could bring two reactants into close proximity and in the proper orientation for the reaction to occur. Finally, an enzyme could alter the pH of the microenvironment of the active site, thereby serving as a transient donor or acceptor of H^+.

19.9 Water-soluble vitamins are required by the body for the synthesis of coenzymes that are required for the function of a variety of enzymes.

19.11 A decrease in pH will change the degree of ionization of the R groups within a peptide chain. This disturbs the weak interactions that maintain the structure of an enzyme, which may denature the enzyme. Less drastic alterations in the charge of R groups in the active site of the enzyme can inhibit enzyme-substrate binding or destroy the catalytic ability of the active site.

19.13 Irreversible inhibitors bind very tightly, sometimes even covalently, to an R group in enzyme active sites. They generally inhibit many different enzymes. The loss of enzyme activity impairs normal cellular metabolism, resulting in death of the cell or of the individual.

19.15 A structural analog is a molecule that has a structure and charge distribution very similar to that of the natural substrate of an enzyme. Generally they are able to bind to the enzyme active site. This inhibits enzyme activity because the normal substrate must compete with the structural analog to form an enzyme-substrate complex.

19.17 Chymotrypsin cleaves peptide bonds on the carbonyl side of aromatic amino acids and hydrophobic amino acids such as methionine.

a. ala-phe-ala

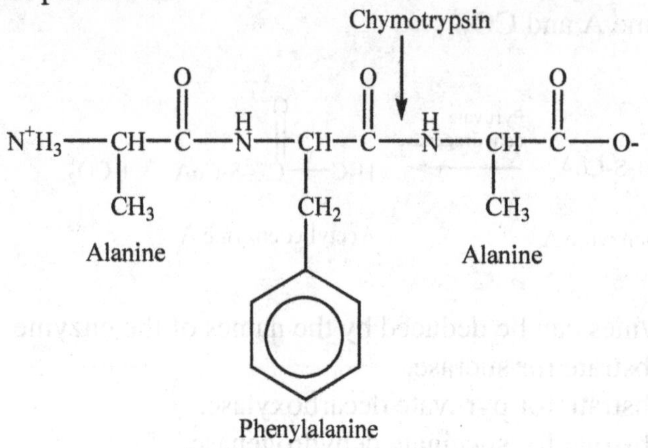

Chymotrypsin

$N^{+}H_3-CH(CH_3)-C(=O)-NH-CH(CH_2C_6H_5)-C(=O)\downarrow NH-CH(CH_3)-C(=O)-O^{-}$

Alanine — Phenylalanine — Alanine

b. tyr-ala-tyr

Chymotrypsin

$N^{+}H_3-CH(CH_2C_6H_4OH)-C(=O)\downarrow NH-CH(CH_3)-C(=O)-NH-CH(CH_2C_6H_4OH)-C(=O)-O^{-}$

Tyrosine — Alanine — Tyrosine

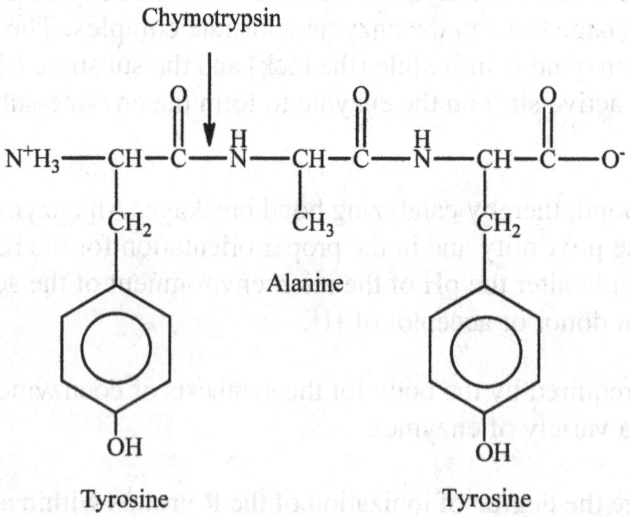

19.19 Chymotrypsin acts specifically at peptide bonds on the carbonyl side of the peptide bond. The C-terminal amino acids of the peptides released by bond cleavage are met, tyr, trp, and phe. Elastase cleaves peptide bonds on the carbonyl side of glycine and alanine. The structural formula for val-phe-ala-gly-leu is shown below. In addition, the bonds cleaved by chymotrypsin and elastase are also indicated.

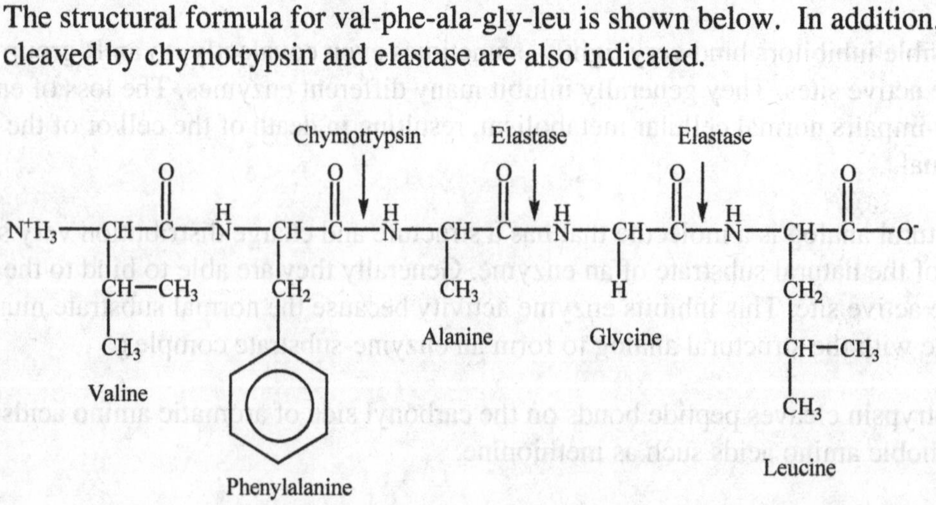

Chymotrypsin — Elastase — Elastase

$N^{+}H_3-CH(CH(CH_3)_2)-C(=O)-NH-CH(CH_2C_6H_5)-C(=O)\downarrow NH-CH(CH_3)-C(=O)\downarrow NH-CH(H)-C(=O)\downarrow NH-CH(CH_2CH(CH_3)_2)-C(=O)-O^{-}$

Valine — Phenylalanine — Alanine — Glycine — Leucine

End-of-Chapter Questions and Problems

19.21 The common name of an enzyme is often derived from the name of the substrate and/or the type of reaction that it catalyzes.

19.23 The names of enzymes are often derived from the name of the substrate.

1. Urease. The substrate is urea.
2. Peroxidase. The substrate is hydrogen peroxide.
3. Lipase. The substrate is a lipid.
4. Aspartase. The substrate is aspartic acid.
5. Glucose-6-phosphatase. The substrate is glucose-6-phosphate.
6. Sucrase. The substrate is sucrose.

19.25 The names of enzymes are often derived from the reaction of the substrate that is catalyzed by the enzyme.

a. Citrate decarboxylase catalyzes the cleavage of a carboxyl group from citrate.
b. Adenosine diphosphate phosphorylase catalyzes the addition of a phosphate group to adenosinde diphosphate.
c. Oxalate reductase catalyzes the reduction of oxalate.
d. Nitrite oxidase catalyzes the oxidation of nitrite.
e. *cis-trans* Isomerase catalyzes the interconversion of *cis* and *trans* isomers.

19.27 A substrate is the reactant in an enzyme-catalyzed reaction that binds to the active site of the enzyme and is converted into product.

19.29 The activation energy of a reaction is the energy required for the reaction to occur.

19.31 The equilibrium constant for a chemical reaction is a reflection of the difference in energy of the reactants and products. Consider the following reaction:

$$aA + bB \rightarrow cC + dD$$

The equilibrium constant for this reaction is: $K_{eq} = [C]^c[D]^d / [A]^a[B]^b$
Because the difference in energy between reactants and products is the same regardless of what path the reaction takes, an enzyme does not alter the equilibrium constant of a reaction.

19.33 The rate of an uncatalyzed chemical reaction typically doubles every time the substrate concentration is doubled.

19.35 The rate-limiting step is that step in an enzyme-catalyzed reaction that is the slowest, and hence limits the speed with which the substrate can be converted into product.

19.37 Increasing concentration of the substrate will increase the rate of an enzyme catalyzed reaction until all of the enzyme active sites are occupied.

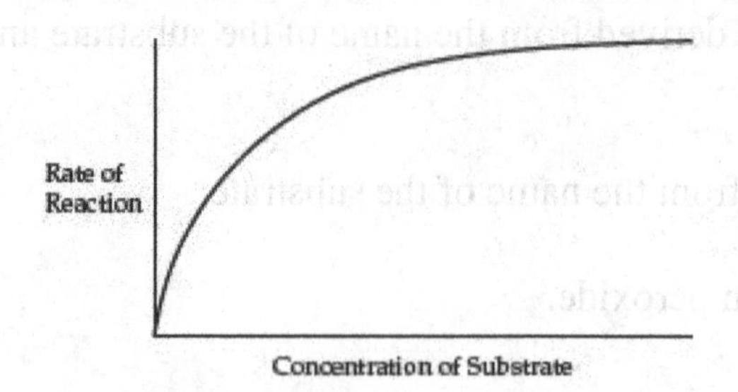

19.39 The enzyme-substrate complex is the molecular aggregate formed when the substrate binds to the active site of an enzyme.

19.41 The catalytic groups of an enzyme active site are those functional groups that are involved in carrying out catalysis.

19.43 Enzyme active sites are pockets in the surface of an enzyme that include R groups involved in binding and R groups involved in catalysis. The shape of the active site is complementary to the shape of the substrate. Thus, the conformation of the active site determines the specificity of the enzyme. Enzyme-substrate binding involves weak, noncovalent interactions.

19.45 The lock-and-key model of enzyme-substrate binding was proposed by Emil Fischer in 1894. He thought that the active site was a rigid region of the enzyme into which the substrate fit perfectly. Thus, the model purports that the substrate simply snaps into place within the active site, like two pieces of a jigsaw puzzle fitting together.

19.47 Enzyme specificity is the ability of an enzyme to bind to only one, or a very few, substrates and thus catalyze only a single reaction.

19.49 Group specificity means that an enzyme catalyzes reactions involving similar molecules having the same functional group.

19.51 Absolute specificity means that an enzyme catalyzes the reaction of only one substrate.

19.53 Hexokinase has group specificity. The advantage is that the cell does not need to encode many enzymes to carry out the phosphorylation of six-carbon sugars. Hexokinase can carry out many of these reactions.

19.55 Methionyl tRNA synthetase has absolute specificity. This is the enzyme that attaches the amino acid methionine to the transfer RNA (tRNA) that will carry the amino acid to the site of protein synthesis. If the wrong amino acid were attached to the tRNA, it could be incorporated into the protein, destroying its correct three-dimensional structure and biological function.

19.57 The first step of an enzyme-catalyzed reaction is the formation of the enzyme-substrate complex. In the second step, the transition state is formed. This is the state in which the substrate assumes a form intermediate between the original substrate and the product. In step 3, the substrate is converted to product and the enzyme-product complex is formed. Step 4 involves the release of the product and regeneration of the enzyme in its original form.

19.59 In a reaction involving bond breaking, the enzyme might distort a bond, producing a transition state in which the bond is stressed. An enzyme could bring two reactants into close proximity and in the proper orientation for the reaction to occur, producing a transition state in which the proximity of the reactants facilitates bond formation. Finally, an enzyme could alter the pH of the microenvironment of the active site, thereby serving as a transient donor or acceptor of H^+.

19.61 A cofactor helps maintain the shape of the active site of an enzyme.

19.63 It is common for coenzymes to contain modified vitamins as part of their structure. Thiamine (B_1) is found in the coenzyme thiamine pyrophosphate. Riboflavin (B_2) is found in both flavin mononucleotide and flavin adenine dinucleotide. Niacin (B_3) is found in both nicotinamide adenine dinucleotide and nicotinamide adenine dinucleotide phosphate. Pyridoxine (B_6) is found in both pyridoxyal phosphate and pyridoxamine phosphate. Cyanocobalamin (B_{12}) is found in coenzyme deoxyadenosyl cobalamin. Folic acid is found in tetrahydrofolic acid. Pantothenic acid is found in coenzyme A. Biotin is found in biocytin.

19.65 At the temperature optimum, the enzyme is functioning optimally and the rate of the reaction is maximal. Above the temperature optimum, increasing temperature begins to denature the enzyme and stop the reaction.

19.67 Each of the following answers assumes that the enzyme was purified from an organism with optimal conditions for life near 37°C, pH 7.

a. Decreasing the temperature from 37°C to 10°C will cause the rate of an enzyme-catalyzed reaction to decrease because the frequency of collisions between enzyme and substrate will decrease as the rate of molecular movement decreases.

b. Increasing the pH from 7 to 11 will generally cause a decrease in the rate of an enzyme-catalyzed reaction. In fact, most enzymes would be denatured by a pH of 11 and enzyme activity would cease.

c. Heating an enzyme from 37°C to 100°C will destroy enzyme activity because the enzyme would be denatured by the extreme heat.

19.69 High temperature denatures bacterial enzymes and structural proteins. Because the life of the cell is dependent on the function of these proteins, the cell dies.

19.71 A lysosome is a membrane-bound vesicle in the cytoplasm of cells that contains approximately fifty hydrolytic enzymes. Some of the enzymes in the lysosomes can degrade proteins to amino acids, others hydrolyze polysaccharides into monosaccharides, and some degrade lipids and nucleic acids. The lysosome contains these enzymes to prevent degradation of large biological molecules that are important to maintain cell integrity.

19.73 Enzymes used for clinical assays in hospitals are typically stored at refrigerator temperatures to ensure that they are not denatured by heat. In this way, they retain their activity for long periods.

19.75 a. Cells regulate the level of enzyme activity to conserve energy. It is a waste of cellular energy to produce an enzyme if its substrate is not present or if its product is in excess.

b. Production of proteolytic digestive enzymes must be carefully controlled because the active enzyme could destroy the cell that produces it. Thus, they are produced in an inactive form in the cell and are only activated at the site where they carry out digestion.

19.77 In positive allosterism, binding of the effector molecule turns the enzyme on. In negative allosterism, binding of the effector molecule turns the enzyme off.

19.79 A proenzyme is the inactive form of an enzyme that is converted to the active form at the site of its activity.

19.81 Blood clotting is a critical protective mechanism in the body, preventing excessive loss of blood following an injury. However, it can be a dangerous mechanism if it is triggered inappropriately. The resulting clot could cause a heart attack or stroke. By having a cascade of proteolytic reactions leading to the final formation of the clot, there are many steps at which the process can be regulated. This ensures that it will be activated only under the appropriate conditions.

19.83 Competitive enzyme inhibition occurs when a structural analog of the normal substrate occupies the enzyme active site so that the reaction cannot occur. The structural analog and the normal substrate compete for the active site. Thus, the rate of the reaction will depend on the relative concentrations of the two molecules.

19.85 A structural analog has a shape and charge distribution that is very similar to those of the normal substrate for an enzyme.

19.87 Irreversible inhibitors bind tightly to and block the active site of an enzyme and eliminate catalysis at the site.

19.89 The compound would be a competitive inhibitor of the enzyme.

19.91 A proteolytic enzyme catalyzes the cleavage of the peptide bond that maintains the primary protein structure.

19.93 The structural similarities among chymotrypsin, trypsin, and elastase suggest that these enzymes evolved from a single ancestral gene that was duplicated. Each copy then evolved independently.

19.95

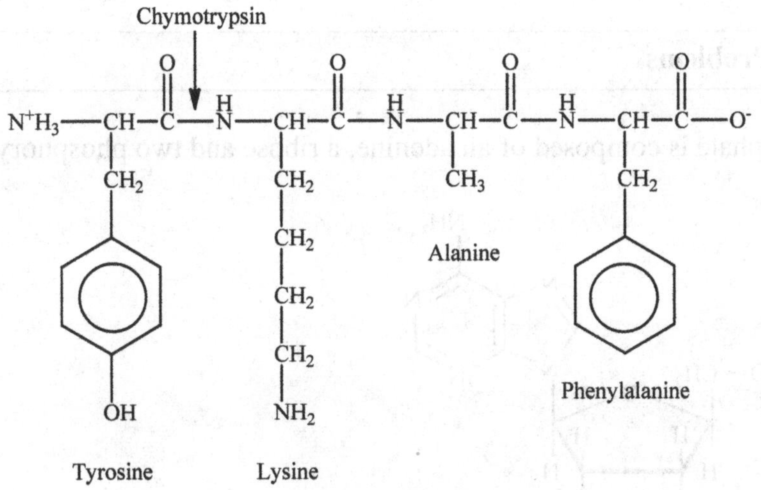

Chymotrypsin

$N^{+}H_3—CH(CH_2C_6H_4OH)—CO—NH—CH(CH_2CH_2CH_2CH_2NH_2)—CO—NH—CH(CH_3)—CO—NH—CH(CH_2C_6H_5)—COO^{-}$

Tyrosine Lysine Alanine Phenylalanine

19.97 Elastase will cleave the peptide bonds on the carbonyl side of alanine and glycine. Trypsin will cleave the peptide bonds on the carbonyl side of lysine and arginine. Chymotrypsin will cleave the peptide bonds on the carbonyl side of tryptophan and phenylalanine.

19.99 Analysis of blood serum for levels of certain enzymes can confirm a preliminary diagnosis that was made based on disease symptoms or a clinical picture. When cells die, they release their enzymes into the bloodstream. Enzyme assays can measure amounts of certain enzymes in the blood.

19.101 Creatine kinase-MB (CK-MB) and aspartate aminotransferase (AST/SGOT)

19.103 Urease is used in the clinical analysis of urea in blood. In a test called the blood urea nitrogen test (BUN), urea is converted to ammonia using the enzyme urease, the ammonia becomes an indicator of urea. This allows for the levels of urea to be measured. This measurement is useful in the diagnosis of kidney malfunction.

Chapter 20
Introduction to Molecular Genetics
Solutions to the Odd-Numbered Questions and Problems

In-Chapter Questions and Problems

20.1 a. Adenosine diphosphate is composed of an adenine, a ribose and two phosphoryl groups.

$^{-}O-P(=O)(O^{-})-O-P(=O)(O^{-})-O-CH_2$ (ribose with H, H, H, H, OH, OH; attached to adenine with NH_2)

b. Deoxyguanosine triphosphate is composed of a guanine, a deoxyribose and three phosphoryl groups.

$^{-}O-P(=O)(O^{-})-O-P(=O)(O^{-})-O-P(=O)(O^{-})-O-CH_2$ (deoxyribose with H, H, H, H, OH, H; attached to guanine with O, H, NH_2)

20.3 The RNA polymerase recognizes the promoter site for a gene, separates the strands of DNA, and catalyzes the polymerization of an RNA strand complementary to the DNA strand that carries the genetic code for a protein. It recognizes a termination site at the end of the gene and releases the RNA molecule.

20.5 The genetic code is said to be degenerate because several different triplet codons may serve as code words for a single amino acid.

20.7 The nitrogenous bases of the codons are complementary to those of the anticodons. As a result they are able to hydrogen bond to one another according to the base pairing rules.

20.9 The ribosomal P-site holds the peptidyl tRNA during protein synthesis. The peptidyl tRNA is the tRNA carrying the growing peptide chain. The only exception to this is during initiation of translation when the P-site holds the initiator tRNA.

20.11 The normal mRNA sequence, AUG-CCC-GAC-UUU, would encode the peptide sequence, methionine-proline-aspartate-phenylalanine. The mutant mRNA sequence, AUG-CGC-GAC-UUU, would encode the mutant peptide sequence, methionine-arginine-aspartate-phenylalanine. This would not be a silent mutation because a hydrophobic amino acid (proline) has been replaced by a positively charged amino acid (arginine).

End-of-Chapter Questions and Problems

20.13 A heterocyclic amine is a heterocyclic compound that contains nitrogen in at least one position of the ring skeleton.

20.15 It is the N-9 of the purine that forms the *N*-glycosidic bond with C-1 of the five-carbon sugar. The general structure of the purine ring is shown below:

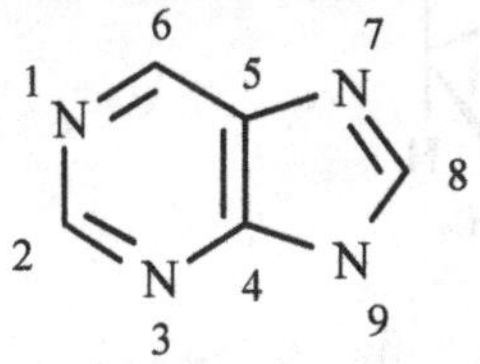

20.17 The ATP nucleotide is composed of the five-carbon sugar ribose, the purine adenine, and a triphosphate group.

20.19 The two strands of DNA in the double helix are said to be *antiparallel* because they run in opposite directions. One strand progresses in the 5'→ 3' direction, and the opposite strand progresses in the 3' → 5' direction.

20.21 The DNA double helix is 2 nm in width. The nitrogenous bases are stacked at a distance of 0.34 nm from one another. One complete turn of the helix is 3.4 nm, or ten base pairs.

20.23 Two hydrogen bonds link the adenine-thymine base pair.

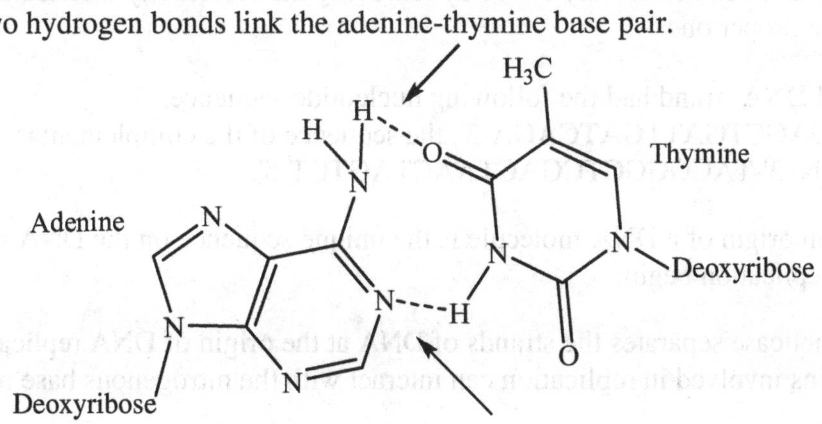

20.25 In this structure, deoxycytosine-5'-monophosphate is linked by a 3' → 5' phosphodiester bond to thymidine-5'-monophosphate.

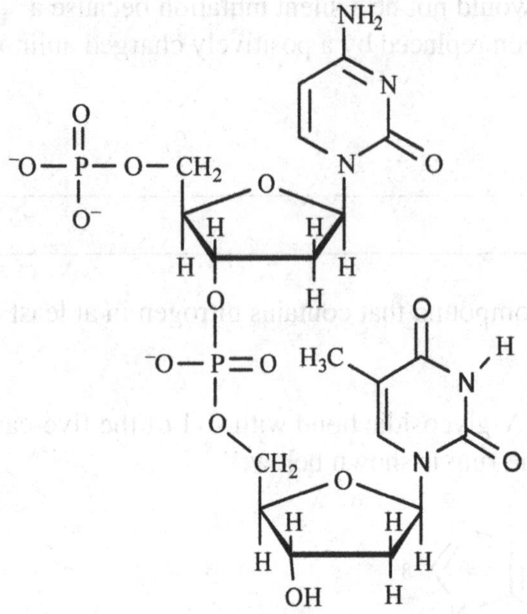

20.27 The prokaryotic chromosome is a circular DNA molecule that is supercoiled, that is, the helix is coiled on itself.

20.29 The term semiconservative DNA replication refers to the fact that each parental DNA strand serves as the template for the synthesis of a daughter strand. As a result, each of the daughter DNA molecules is made up of one strand of the original parental DNA and one strand of newly synthesized DNA.

20.31 The two primary functions of DNA polymerase III are to read a template DNA strand and catalyze the polymerization of a new daughter strand, and to proofread the newly synthesized strand and correct any errors by removing the incorrectly inserted nucleotide and adding the proper one.

20.33 If the parental DNA strand had the following nucleotide sequence: 5'-ATGCCCGAGCTGATTGATCAGA-3', the sequence of the complementary daughter strand would be 3'-TACGGGCTCGACTAACTAGTCT-5'.

20.35 The replication origin of a DNA molecule is the unique sequence on the DNA molecule where DNA replication begins.

20.37 The enzyme helicase separates the strands of DNA at the origin of DNA replication so that the proteins involved in replication can interact with the nitrogenous base pairs.

20.39 The RNA primer "primes" DNA replication by providing a 3'-OH, which can be used by DNA polymerase III for the addition of the next nucleotide in the growing DNA chain.

20.41 The central dogma of molecular biology states that information flow in cellular biological systems is unidirectional: DNA → RNA → Protein. The DNA carries the genetic information; RNA molecules carry out the expression of the genetic information to produce proteins; the final products are proteins that carry out the work of the cell and serve as cellular structural components.

20.43 Anticodons are found on transfer RNA molecules.

20.45 If a gene had the following nucleotide sequence:
5'-TACGGGCATAGGCCTTAAAGCTAGCTT-3',
the mRNA sequence would be: 3'-AUGCCCGUAUCCGGAAUUUCGAUCGAA-5'.

20.47 RNA splicing is the process by which the noncoding sequences (introns) of the primary transcript of a eukaryotic mRNA are removed and the protein coding sequences (exons) are spliced together.

20.49 The three classes of RNA molecules are messenger RNA (mRNA), transfer RNA (tRNA), and ribosomal RNA (rRNA).

20.51 Spliceosomes are small ribonucleoprotein complexes in the nucleus of the cell that are responsible for RNA splicing. By hydrogen bonding, they recognize the boundaries between exons and introns and bring together the RNA sequences involved in splicing.

20.53 The poly(A) tail is a stretch of 100 - 200 adenosine nucleotides polymerized onto the 3' end of a mRNA by the enzyme poly(A) polymerase.

20.55 The cap structure is made up of the nucleotide 7-methylguanosine attached to the 5' end of an mRNA by a 5'-5' triphosphate bridge. Generally, the first two nucleotides of the mRNA are also methylated. The cap structure is required for efficient translation of the mRNA.

20.57 There are 64 codons in the genetic code.

20.59 The reading frame of a gene is the sequential set of triplet codons that carries the genetic code for the primary structure of a protein. Each triplet specifies the addition of a particular amino acid to the growing peptide chain.

20.61 Methionine (AUG) and tryptophan (UGG) are encoded by only one codon.

20.63 The codon 5'-UUU-3' encodes the amino acid phenylalanine. The mutant codon 5'-UUA-3' encodes the amino acid leucine. Both leucine and phenylalanine are hydrophobic amino acids, however, leucine has a smaller R group. It is possible that the smaller R group would disrupt the structure of the protein.

20.65 The ribosomes serve as a platform on which protein synthesis can occur. They also carry the enzymatic activity that forms peptide bonds.

20.67 The following peptide:

Ala-gly-leu-cys-met-trp-tyr-ser-ile-gly

may have been coded by the mRNA sequence:

5'-AUG GCU GGG CUU UGU AUG UGG UAU UCU AUU GGG UAA-3'.

The codon AUG is a start codon and the codon UAA specifies a termination signal for the process of translation. The codons for the peptide amino acids are as follows:

GCU GGG CUU UGU AUG UGG UAU UCU AUU GGG.

20.69 The sequence of DNA nucleotides in a gene is transcribed to produce a complementary sequence of RNA nucleotides in a messenger RNA (mRNA). In the process of translation the sequence of the mRNA is read sequentially in words of three nucleotides (codons) to produce a protein. Each codon calls for the addition of a particular amino acid to the growing peptide chain. If one of those codons has been altered by mutation, it may now call for the addition of the wrong amino acid to the growing peptide chain. This could result in improper folding of the protein and in loss of biological function.

20.71 The bond between an amino acid and a tRNA is an ester bond formed between the carboxylate group of the amino acid and the 3'-OH of the sugar ribose in the tRNA molecule.

20.73 A point mutation is the substitution of one nucleotide pair for another in a gene.

20.75 Some mutations are silent because the change in the nucleotide sequence does not alter the amino acid sequence of the protein. This can happen because there are many amino acids encoded by multiple codons.

20.77 UV light causes the formation of pyrimidine dimers, the covalent bonding of two adjacent pyrimidine bases. Mutations occur when the UV damage repair system makes an error during the repair process. This causes a change in the nucleotide sequence of the DNA.

20.79 a. A carcinogen is a compound that causes cancer. Cancers are caused by mutations in the genes responsible for controlling cell division.

b. Carcinogens cause DNA damage that result in changes in the nucleotide sequence of the gene. Thus, carcinogens are also mutagens.

20.81 A restriction enzyme is a bacterial enzyme that "cuts" the sugar-phosphate backbone of DNA molecules at a specific nucleotide sequence.

20.83 A selectable marker is a genetic trait that can be used to detect the presence of a plasmid in a bacterium. Many plasmids have antibiotic resistance genes as selectable markers. Bacteria containing the plasmid will be able to grow in the presence of the antibiotic; those without the plasmid will be killed.

20.85 Human insulin, interferon, human growth hormone, and human blood clotting factor VIII are protein products of recombinant DNA technology that are of great value in the field of medicine.

20.87 Each round of polymerase chain reaction doubles the number of target DNA molecules. Thus, after twelve cycles of PCR, there would be 2^{12} or 4096 copies of each molecule of target DNA.

20.89 The goals of the Human Genome Project were to identify and map all of the genes of the human genome and to determine the DNA sequences of the complete three billion nucleotide pairs.

20.91 A genome library is a set of clones that represents all of the DNA sequences in the genome of an organism.

20.93 A dideoxynucleotide has hydrogen atoms rather than hydroxyl groups bonded to both the 2' and the 3' carbons of the five-carbon sugar.

20.95 Sequences that these DNA sequences have in common are highlighted in bold.

a. 5' – **AGCTCCT**GATTTCATACAGTTTCTACT**ACCTACTA** - 3'

b. 5' - AGACATTCTATCTACCTAGACTATG**TTCAGAA** - 3'

c. 5' - **TTCAGAA**CTCATTCAGACCTACTACTATACCTTGGG**AGCTCCT** - 3'

d. 5' - **ACCTACTA**GACTATACTACTACTAAGGGGACTATTCCAGACTT - 3'

The 5' end of sequence (a) is identical to the 3' end of sequence (c).

The 3' end of sequence (a) is identical to the 5' end of sequence (d).

The 3' end of sequence (b) is identical to the 5' end of sequence (c).

From 5' to 3', the sequences would form the following map:

5'_____b_____

_____c_____

_____a_____

_____d_____3'

Chapter 21
Carbohydrate Metabolism
Solutions to the Odd-Numbered Questions and Problems

In-Chapter Questions and Problems

21.1 ATP is called the universal energy currency because it is the major molecule used by all organisms to store energy. Breaking the high-energy phosphoanhydride bonds releases energy that is used for cellular work.

21.3 The first stage of catabolism is the digestion (hydrolysis) of dietary macromolecules in the stomach and intestine. Polysaccharides are hydrolyzed to monosaccharides; proteins are degraded to amino acids; and triglycerides are broken down into glycerol and fatty acids. The small molecules produced by digestion are taken into the cells lining the intestine by active or passive transport.

In the second stage of catabolism, monosaccharides, amino acids, fatty acids, and glycerol are converted by metabolic reactions into molecules that can be completely oxidized. Often they are converted into acetyl CoA.

In the third stage of catabolism, the two-carbon acetyl group of acetyl CoA is completely oxidized by the reactions of the citric acid cycle. The energy of the electrons harvested in these oxidation reactions is used to make ATP.

21.5 Substrate-level phosphorylation is one way the cell can make ATP. In this reaction, a high-energy phosphoryl group of a substrate in the reaction is transferred to ADP to produce ATP. Substrate-level phosphorylation can be summarized as follows:

$$\text{Substrate}\sim\text{P} + \text{ADP} \rightarrow \text{Product} + \text{ATP}$$

21.7 Glycolysis is a pathway involving ten reactions. In reactions 1 - 3, energy is invested in the beginning substrate, glucose. This is done by transferring high-energy phosphoryl groups from ATP to the intermediates in the pathway. The product is fructose-1,6-bisphosphate. In the energy harvesting reactions of glycolysis, fructose-1,6-bisphosphate is split into two three-carbon molecules that begin a series of rearrangement, oxidation-reduction, and substrate-level phosphorylation reactions that produce 4 ATP, 2 NADH, and 2 pyruvate molecules. Because of the investment of two ATP in the early steps of glycolysis, the net yield of ATP is two.

21.9 Both the alcohol and lactate fermentations are anaerobic reactions that use the pyruvate and reoxidize the NADH produced in glycolysis. In the alcohol fermentation, pyruvate is first decarboxylated to produce acetaldehyde. The acetaldehyde is then reduced as NADH is oxidized. The products are CO_2, ethanol, and NAD^+. In the lactate fermentation, pyruvate is reduced to lactate and NADH is oxidized to NAD^+.

21.11 Gluconeogenesis (synthesis of glucose from noncarbohydrate sources) appears to be the reverse of glycolysis (the first stage of carbohydrate degradation) because the intermediates in the two pathways are the same. However, reactions 1, 3, and 10 of glycolysis are not reversible reactions. Thus, the reverse reactions must be carried out by different enzymes.

Reaction 1 of glycolysis, the transfer of a high-energy phosphoryl group from ATP to glucose, is carried out by the enzyme hexokinase. The reverse reaction in gluconeogenesis is catalyzed by glucose-6-phosphatase.

Reaction 3 of glycolysis, the transfer of a high-energy phosphoryl group from ATP to fructose-6-phosphate, is catalyzed by phosphofructokinase. The reverse reaction of gluconeogenesis is carried out by fructose bisphosphatase.

The final reaction of glycolysis, the transfer of a high-energy phosphoryl group from phosphoenolpyruvate to ADP, is catalyzed by pyruvate kinase. This is reversed in gluconeogenesis by the action of two enzymes. Pyruvate carboxylase adds CO_2 to pyruvate to produce oxaloacetate and phosphoenolpyruvate carboxykinase removes the CO_2 and transfers a high-energy phosphoryl group from GTP to produce phosphoenolpyruvate.

21.13 The enzyme glycogen phosphorylase catalyzes the phosphorolysis of a glucose unit at one end of a glycogen molecule. The reaction involves the displacement of the glucose by a phosphate group. The products are glucose-1-phosphate and a glycogen molecule that is one glucose unit shorter.

21.15 Glucokinase traps glucose within the liver cell by phosphorylating it. Because the product, glucose-6-phosphate is charged, it cannot be exported from the cell.

21.17 Glucagon indirectly stimulates glycogen phosphorylase, the first enzyme of glycogenolysis. This speeds up glycogen degradation. Glucagon also inhibits glycogen synthase, the first enzyme in glycogenesis. This inhibits glycogen synthesis.

End-of-Chapter Questions and Problems

21.19 ATP is the molecule that is primarily responsible for conserving the energy released in catabolism.

21.21 The terminal phosphoanhydride bond of ATP is broken to produce ADP, an inorganic phosphate group, and energy (7kcal/mol).

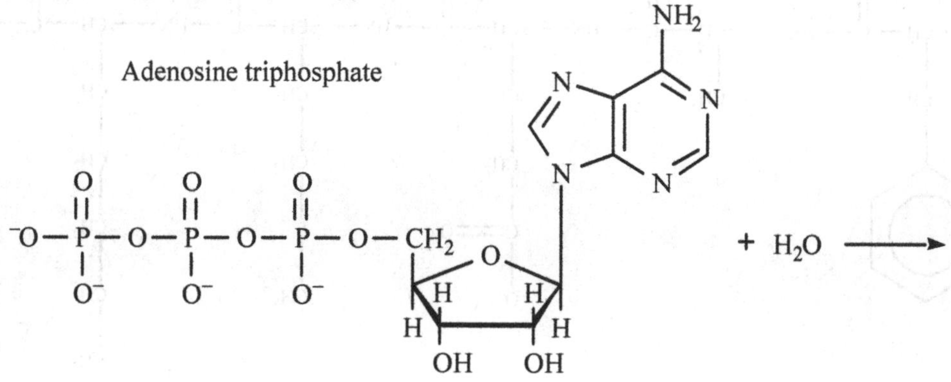

Adenosine triphosphate

$+ H_2O \longrightarrow$

Inorganic phosphate group

Adenosine diphosphate

21.23 A coupled reaction is one that can be thought of as a two-step process. In a coupled reaction, two reactions occur simultaneously. Frequently, one of the reactions releases the energy that drives the second, energy-requiring, reaction.

21.25 Carbohydrates are the most readily used energy source in the diet.

21.27 The following equation represents the hydrolysis of maltose:

β-Maltose $+ H_2O \longrightarrow 2$ β-D-Glucose

21.29 The peptide phe-ala-glu-met-lys can be hydrolyzed to produce the amino acids phe, ala, glu, met and lys.

$+ 4H_2O \longrightarrow$

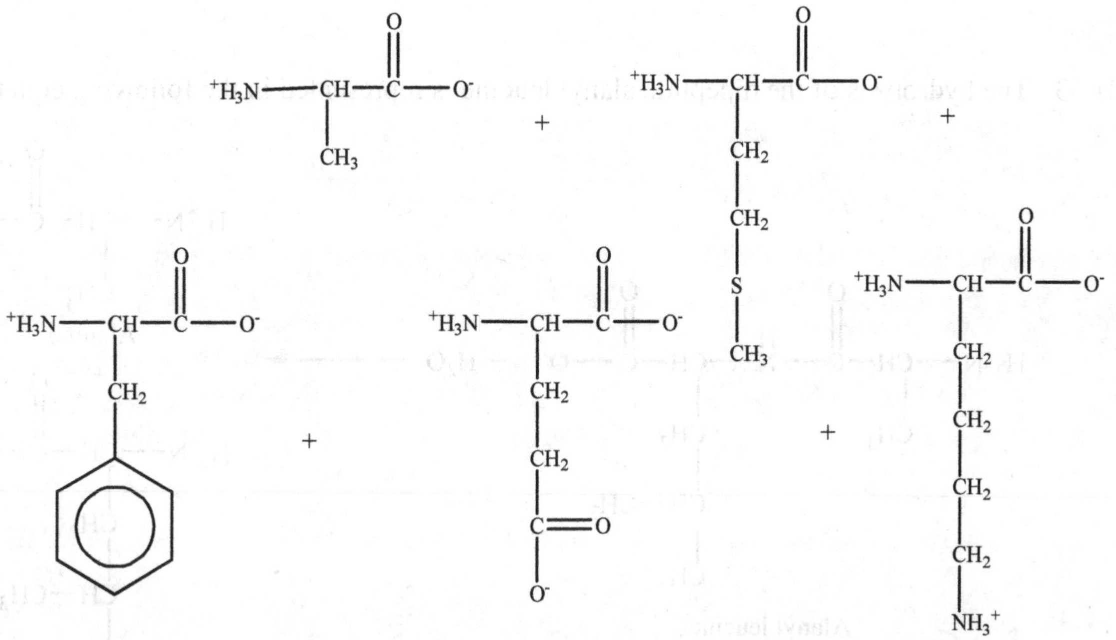

$^{+}H_3N-CH(CH_3)-C(=O)-O^-$ + $^{+}H_3N-CH(CH_2CH_2-S-CH_3)-C(=O)-O^-$ +

$^{+}H_3N-CH(CH_2-C_6H_5)-C(=O)-O^-$ + $^{+}H_3N-CH(CH_2CH_2-C(=O)-O^-)-C(=O)-O^-$ + $^{+}H_3N-CH(CH_2CH_2CH_2CH_2-NH_3^+)-C(=O)-O^-$

21.31 The hydrolysis of a triglyceride containing oleic acid, stearic acid, and linoleic acid is represented in the following equations:

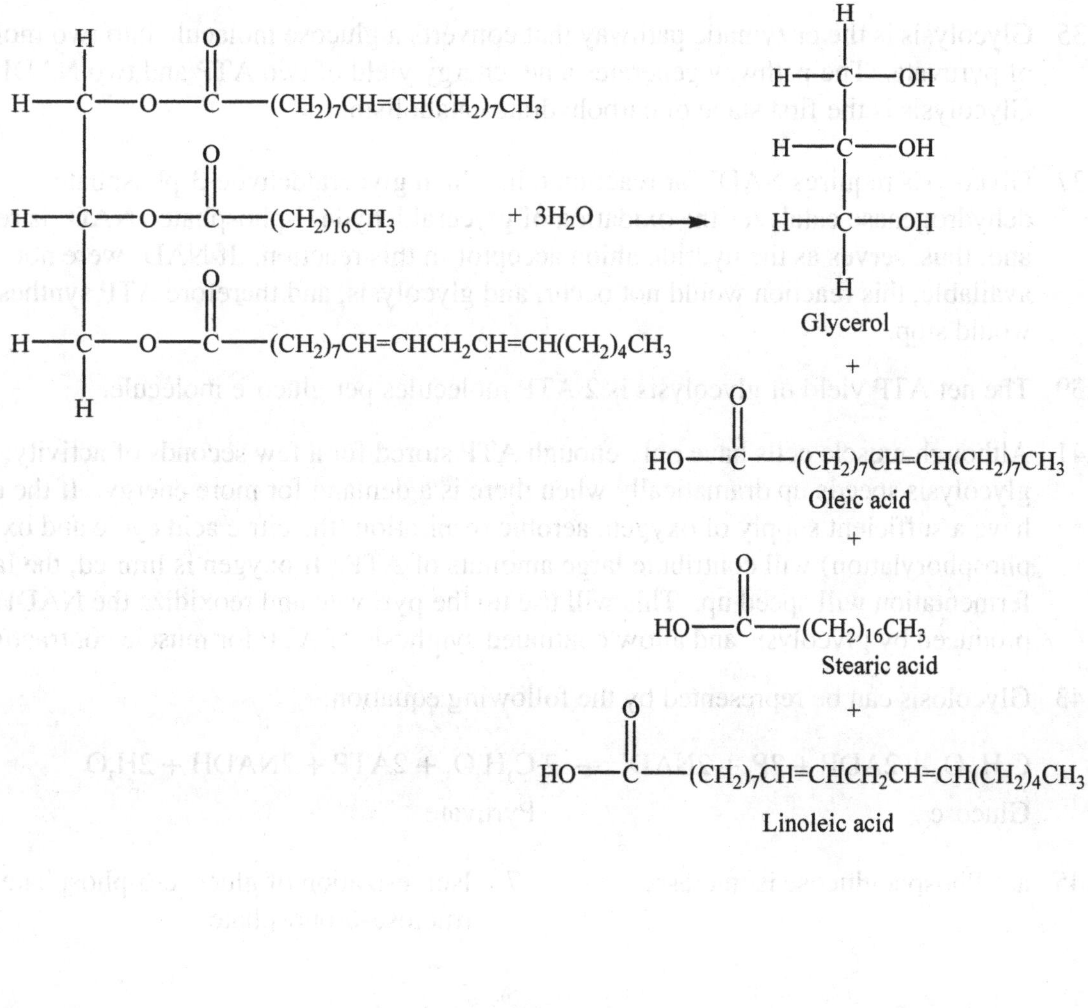

$$\begin{array}{l} H-\overset{H}{\underset{|}{C}}-O-\overset{O}{\overset{\|}{C}}-(CH_2)_7CH{=}CH(CH_2)_7CH_3 \\ H-\underset{|}{C}-O-\overset{O}{\overset{\|}{C}}-(CH_2)_{16}CH_3 \\ H-\underset{H}{\underset{|}{C}}-O-\overset{O}{\overset{\|}{C}}-(CH_2)_7CH{=}CHCH_2CH{=}CH(CH_2)_4CH_3 \end{array} + 3H_2O \longrightarrow$$

$$\begin{array}{c} H \\ | \\ H-C-OH \\ | \\ H-C-OH \\ | \\ H-C-OH \\ | \\ H \end{array}$$

Glycerol

+

$HO-\overset{O}{\overset{\|}{C}}-(CH_2)_7CH{=}CH(CH_2)_7CH_3$

Oleic acid

+

$HO-\overset{O}{\overset{\|}{C}}-(CH_2)_{16}CH_3$

Stearic acid

+

$HO-\overset{O}{\overset{\|}{C}}-(CH_2)_7CH{=}CHCH_2CH{=}CH(CH_2)_4CH_3$

Linoleic acid

21.33 The hydrolysis of the dipeptide alanyl leucine is represented in the following equation:

$$H_3^+N{-}CH(CH_3){-}C(=O){-}NH{-}CH(CH_2{-}CH(CH_3){-}CH_3){-}C(=O){-}O^- + H_2O \longrightarrow H_3^+N{-}CH(CH_3){-}C(=O){-}O^- + H_3^+N{-}CH(CH_2{-}CH(CH_3){-}CH_3){-}C(=O){-}O^-$$

Alanyl leucine → Alanine + Leucine

21.35 Glycolysis is the enzymatic pathway that converts a glucose molecule into two molecules of pyruvate. The pathway generates a net energy yield of two ATP and two NADH. Glycolysis is the first stage of carbohydrate catabolism.

21.37 Glycolysis requires NAD^+ for reaction 6 in which glyceraldehyde-3-phosphate dehydrogenase catalyzes the oxidation of glyceraldehyde-3-phosphate. NAD^+ is reduced and, thus, serves as the hydride anion acceptor in this reaction. If NAD^+ were not available, this reaction would not occur, and glycolysis, and therefore ATP synthesis, would stop.

21.39 The net ATP yield of glycolysis is 2 ATP molecules per glucose molecule.

21.41 Although muscle cells have only enough ATP stored for a few seconds of activity, glycolysis speeds up dramatically when there is a demand for more energy. If the cells have a sufficient supply of oxygen, aerobic respiration (the citric acid cycle and oxidative phosphorylation) will contribute large amounts of ATP. If oxygen is limited, the lactate fermentation will speed up. This will use up the pyruvate and reoxidize the NADH produced by glycolysis and allow continued synthesis of ATP for muscle contraction.

21.43 Glycolosis can be represented by the following equation:

$$\underset{\text{Glucose}}{C_6H_{12}O_6} + 2ADP + 2P_i + 2NAD^+ \rightarrow \underset{\text{Pyruvate}}{2\,C_3H_3O_3} + 2ATP + 2NADH + 2H_2O$$

21.45 a. Phosphoglucose isomerase. 7. Isomerization of glucose-6-phosphate into fructose-6-phosphate.

b. Phosphofructokinase.	2. Phosphorylation of fructose-6-phosphate
c. Triose phosphate isomerase.	6. Conversion of dihydroxyacetone phosphate into glyceraldehyde-3-phosphate
d. Aldolase.	4. Conversion of fructose-1,6-bisphosphate to dihydroxyacetone phosphate and glyceraldehyde-3-phosphate.
e. Hexokinase.	1. Phosphorylation of glucose.
f. Enolase.	9. Dehydration of 2-phosphoglycerate to produce phosphoenolpyruvate
g. Glyceraldehyde-3-phosphate dehydrogenase.	5. Phosphorylation and oxidation of glyceraldehyde-3-phosphate to produce 1,3-bisphosphoglycerate and NADH.
h. Phosphoglycerate kinase.	10. Substrate level phosphorylation involving transfer of a phosphoryl group from 1, 3-bisphosphoglycerate into 3-phosphoglycerate.
i. Pyruvate kinase.	3. Dephosphorylation of pyruvate.
j. Phospohglycerate mutase.	8. Isomerization of 3-phosphoglycerate into 2-phosphoglycerate.

21.47 Myopathy and hemolytic anemia are symptoms associated with a genetic defect in some of the enzymes of glycolysis. Myopathy can lead to exercise intolerance, muscle breakdown, and blood in the urine. Tarui's disease is also caused by a deficiency in one of the enzymes of glycolysis. Its symptoms include muscle pain, exercise intolerance, respiratory failure, heart muscle disease, seizures and blindness.

21.49 If a person is deficient in some of the enzymes of glycolysis, muscle cells may begin to die which can lead to the release of myoglobin into the blood and the urine. This condition is called myoglobinuria, and it results in urine that is the color of cola soft drinks.

21.51 Since the reactants and products are isomers, the enzymes that catalyze these reactions must be isomerases.

21.53 Enediol is the type of intermediate formed.

21.55 A kinase transfers a phosphoryl group from one molecule to another.

21.57 NAD^+ is reduced, accepting a hydride anion ($H:^-$).

21.59 To optimize efficiency and minimize waste, it is important that energy-harvesting pathways, such as glycolysis, respond to the energy demands of the cell. If energy in the form of ATP is abundant, there is no need for the pathway to continue at a rapid rate. When this is the case, allosteric enzymes that catalyze the reactions of the pathway are inhibited by binding to their negative effectors. Similarly, when there is a great demand for ATP, the pathway speeds up as a result of the action of allosteric enzymes binding to positive effectors.

21.61 ATP and citrate are allosteric inhibitors of phosphofructokinase, while AMP and ADP are allosteric activators.

21.63 Citrate, which is the first intermediate in the citric acid cycle, is an allosteric inhibitor of phosphofructokinase. The citric acid cycle is a pathway that results in the complete oxidation of the pyruvate produced by glycolysis. A high concentration of citrate signals that sufficient substrate is entering the citric acid cycle. The inhibition of phosphofructokinase by citrate is an example of *feedback inhibition*: the product, citrate, allosterically inhibits the activity of an enzyme early in the pathway.

21.65 The enzyme alcohol dehydrogenase catalyzes the conversion of acetaldehyde to ethanol.

$$H_3C-\overset{\overset{\displaystyle O}{||}}{C}-H \xrightarrow{NADH \quad NAD^+} CH_3CH_2OH$$

21.67 Lactate fermentation produces lactate from pyruvate when the amount of oxygen is limiting. Over time, lactate will accumulate in the muscle.

21.69 The tangy flavor of yogurt and some cheeses is the result of lactate produced by the lactate fermentation. It is the pH decrease caused by the lactate build-up that causes milk protein to coagulate. This coagulation produces the soft curd of yogurt and the hard curd of some cheeses.

21.71 Lactate dehydrogenase catalyzes the reduction of pyruvate to lactate.

21.73 This child must have the enzymes to carry out the alcohol fermentation. When the child exercised hard, there was not enough oxygen in the cells to maintain aerobic respiration. As a result, glycolysis and the alcohol fermentation were responsible for the majority of the ATP production by the child. The accumulation of alcohol (ethanol) in the child caused the symptoms of drunkenness.

21.75 The ribose-5-phosphate produced in the pentose phosphate pathway is used for the biosynthesis of nucleotides, such as ATP. The erythrose-4-phosphate is used for the biosynthesis of aromatic amino acids, such as phenylalanine, tyrosine, and tryptophan.

21.77 Gluconeogenesis is production of glucose from noncarbohydrate starting materials. This pathway can provide glucose when starvation or strenuous exercise leads to a depletion of glucose from the body.

21.79 The liver is primarily responsible for gluconeogenesis.

21.81 Lactate is first converted to pyruvate. Pyruvate is the starting substrate for gluconeogenesis.

21.83 Because steps 1, 3, and 10 of glycolysis are irreversible, gluconeogenesis is not simply the reverse of glycolysis. The reverse reactions must be carried out by different enzymes.

Reaction 1 of glycolysis, the transfer of a high-energy phosphoryl group from ATP to glucose, is carried out by the enzyme hexokinase. The reverse reaction in gluconeogenesis is catalyzed by glucose-6-phosphatase.

Reaction 3 of glycolysis, the transfer of a high-energy phosphoryl group from ATP to fructose-6-phosphate, is catalyzed by phosphofructokinase. The reverse reaction of gluconeogenesis is carried out by fructose bisphosphatase.

The final reaction of glycolysis, the transfer of a high-energy phosphoryl group from phosphoenolpyruvate to ADP, is catalyzed by pyruvate kinase. This is reversed in gluconeogenesis by the action of two enzymes. Pyruvate carboxylase adds CO_2 to pyruvate to produce oxaloacetate and phosphoenolpyruvate carboxykinase removes the CO_2 and transfers a high-energy phosphoryl group from GTP to produce phosphoenolpyruvate.

21.85 Steps 1, 3, and 10 of glycolysis are irreversible. Step 1 is the transfer of a phosphoryl group from ATP to carbon-6 of glucose and is catalyzed by hexokinase. Step 3 is the transfer of a phosphoryl group from ATP to carbon-1 of fructose-6-phosphate and is catalyzed by phosphofructokinase. Step 10 is the substrate-level phosphorylation in which a phosphoryl group is transferred from phosphoenolpyruvate to ADP and is catalyzed by pyruvate kinase.

21.87 The liver is instrumental in the maintenance of blood glucose levels by serving as a reservoir for glycogen, a branched homopolymer of D-glucose. The pancreas is also critical to the control of blood glucose. When blood glucose levels are too high, the pancreas secretes the hormone insulin, which stimulates uptake and storage of glucose from the blood. When blood glucose levels are too low, the pancreas secretes the hormone glucagon, which stimulates the breakdown of liver glycogen and the release of glucose into the blood.

21.89 Hypoglycemia is the condition in which blood glucose levels are too low.

21.91
a. Insulin stimulates glycogen synthase, the first enzyme in glycogen synthesis. It also stimulates uptake of glucose from the bloodstream into cells and phosphorylation of glucose by the enzyme glucokinase.
b. This traps glucose within liver cells and increases the storage of glucose in the form of glycogen.
c. The ultimate result is a decrease of blood glucose levels.

21.93 Any defect in the enzymes required to degrade glycogen or export glucose from liver cells will result in a reduced ability of the liver to provide glucose at times when blood glucose levels are low. This will cause hypoglycemia.

21.95 Glycogen phosphorylase catalyzes phosphorylysis of a glucose at one end of a glycogen polymer. The reaction involves the displacement of a glucose unit of glycogen by a phosphate group. As a result, glucose-1-phosphate is produced.

$$\text{Glycogen (glucose)}_x + n\,HPO_4^{2-} \rightarrow \text{Glycogen (glucose)}_{x-n} + n \text{ glucose-1-phosphate}$$

21.97 In glycogen degradation, phosphoglucomutase converts glucose-1-phosphate to glucose-6-phosphate. This allows for glucose to be released as a product of glycogen degradation. In glycogen synthesis, phosphoglucomutase converts glucose-6-phosphate to glucose-1-phosphate.

Phosphoglucomutase

21.99 Glucokinase converts glucose to glucose-6-phosphate as the first reaction of glycogen synthesis. In this reaction, ATP serves as the phosphoryl donor.

+ ATP Glucokinase + ADP + H^+

Chapter 22
Aerobic Respiration and Energy Production
Solutions to the Odd-Numbered Questions and Problems

In-Chapter Questions and Problems

22.1 Mitochondria are the organelles responsible for aerobic respiration. They have enzymes that carry out the final oxidation of carbohydrates, amino acids, and fatty acids. They produce the majority of the ATP for the cell.

22.3

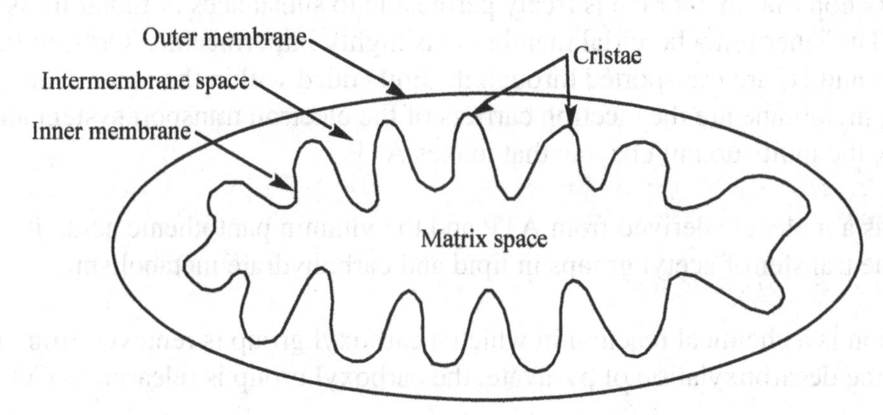

22.5 Pyruvate is converted to acetyl CoA by the pyruvate dehydrogenase complex. This huge enzyme complex requires four coenzymes, each of which is made from a different vitamin. The four coenzymes are thiamine pyrophosphate (made from thiamine), FAD (made from riboflavin), NAD^+ (made from niacin), and coenzyme A (made from the vitamin pantothenic acid). The coenzyme lipoamide is also involved in this reaction.

22.7 Oxidative phosphorylation is the process by which the energy of electrons harvested from oxidation of a fuel molecule is used to phosphorylate ADP to produce ATP.

22.9 A balanced equation for the reduction of NAD^+ is $NAD^+ + H\!:^- \rightarrow NADH$.

22.11 Pyridoxal phosphate is a coenzyme required by all transaminases. During transamination reactions, the α-amino group is transferred to pyridoxal phosphate. In the last part of the reaction, the α-amino group is transferred from pyridoxal phosphate to an α-keto acid.

22.13 The urea cycle converts toxic ammonium ions to urea, which is excreted in the urine of land animals. This keeps toxic ammonium ions out of the bloodstream.

22.15 An amphibolic pathway is a metabolic pathway that functions both in anabolism and catabolism. The citric acid cycle is amphibolic because it has a catabolic function - it completely oxidizes the acetyl group carried by acetyl CoA to provide electrons for ATP synthesis. Because citric acid cycle intermediates are precursors for the biosynthesis of many other molecules, it also serves a function in anabolism.

End-of-Chapter Questions and Problems

22.17 The mitochondrion is an organelle that serves as the cellular power plant. The reactions of the citric acid cycle, the electron transport system, and ATP synthase function together within the mitochondrion to harvest ATP energy for the cell.

22.19 The intermembrane compartment is the location of the high-energy proton (H^+) reservoir produced by the electron transport system. The energy of this H^+ reservoir is used to make ATP.

22.21 The outer mitochondrial membrane is freely permeable to substances of molar mass less than 10,000. The inner mitochondrial membrane is highly impermeable. Only certain fuel molecules and H^+ are transported through it. Embedded within the inner mitochondrial membrane are the electron carriers of the electron transport system and ATP synthase, the multisubunit enzyme that makes ATP.

22.23 Coenzyme A is a molecule derived from ATP and the vitamin pantothenic acid. It functions in the transfer of acetyl groups in lipid and carbohydrate metabolism.

22.25 Decarboxylation is a chemical reaction in which a carboxyl group is removed from a molecule. In the decarboxylation of pyruvate, the carboxyl group is released as CO_2.

22.27 Under aerobic conditions, pyruvate is converted to acetyl CoA.

22.29 The coenzymes NAD^+, FAD, thiamine pyrophosphate, and coenzyme A are required by the pyruvate dehydrogenase complex for the conversion of pyruvate to acetyl CoA. These coenzymes are synthesized from the vitamins niacin, riboflavin, thiamine, and pantothenic acid, respectively. If the vitamins are not available, the coenzymes will not be available and pyruvate cannot be converted to acetyl CoA. Since the complete oxidation of the acetyl group of acetyl CoA produces the vast majority of the ATP for the body, ATP production would be severely inhibited by a deficiency of any of these vitamins.

22.31 An aldol condensation is a reaction in which aldehydes or ketones react to form larger molecules.

22.33 The conversion of isocitrate to α-ketoglutarate is an oxidation reduction reaction. The hydroxyl group of isocitrate is oxidized to a ketone, carbon dioxide is released, and NAD^+ is reduced to NADH.

22.35 A dehydrogenation reaction is an oxidation reaction in which protons and electrons are removed from a molecule.

22.37 a. Citrate synthase — 1. Transferase
b. Aconitase — 6. Isomerase
c. Isocitrate dehydrogenase — 2. Oxidoreductase

d. α-Ketoglutarate dehydrogenase	2. Oxidoreductase
e. Succinyl CoA synthase	4. Hydrolase
f. Succinate dehydrogenase	2. Oxidoreductase
g. Fumarase	5. Lyase
h. Malate dehydrogenase	2. Oxidoreductase

22.39 The statement is true.

22.41 Three NAD^+ are reduced to three NADH in one turn of the citric acid cycle.

22.43 The net yield of ATP for anaerobic glycolysis is two ATP per glucose.

22.45 The function of acetyl CoA in the citric acid cycle is to bring the two-carbon remnant (acetyl group) of pyruvate from glycolysis and transfer it to oxaloacetate. In this way, the acetyl group enters the citric acid cycle for the final stages of oxidation.

22.47 GTP is produced in the citric acid cycle. The high-energy phosphoryl group of the GTP is transferred to ADP to produce ATP. This reaction is catalyzed by the enzyme dinucleotide diphosphokinase.

22.49 Fumarate contains an alkene carbon-carbon double bond. Addition of water to the double bond of fumarate gives malate. The enzyme fumarase catalyzes this reaction. When water is added to the alkene double bond, one of the carbons forms a new bond to –OH, and the other carbon forms a new bond to –H. As a result, the alkene becomes an alcohol.

$$\underset{\text{Fumarate}}{{}^{-}OOC\text{—}CH{=}CH\text{—}COO^{-}} \xrightarrow{\text{Fumarase}} \underset{\text{Malate}}{{}^{-}OOC\text{—}CH(OH)\text{—}CH_2\text{—}COO^{-}}$$

(*alkene* circled in fumarate; *alcohol* circled in malate)

22.51 First, write an equation representing the conversion of pyruvate to acetyl CoA to determine which carbon in acetyl CoA is labeled:

$$\underset{\text{Coenzyme A}}{\text{H-S-CoA}} + \underset{\text{Pyruvate}}{H_3\overset{*}{C}\text{—}\overset{O}{\overset{\|}{C}}\text{—}COO^-} \longrightarrow \underset{\text{Acetyl CoA}}{H_3\overset{*}{C}\text{—}\overset{O}{\overset{\|}{C}}\text{\textasciitilde}S\text{—}CoA} + CO_2$$

Next, write out the aldol condensation reaction of acetyl CoA and oxaloacetate to show the location of the radiolabeled carbon in citrate:

$$\text{Oxaloacetate} + H_3\overset{*}{C}-C(=O)\sim S-CoA \;(\text{Acetyl CoA}) \longrightarrow \text{Citrate} + \text{H-S-CoA} \;(\text{Coenzyme A})$$

Oxaloacetate: $^-OOC-C(=O)-CH_2-COO^-$

Acetyl CoA: $H_3\overset{*}{C}-C(=O)\sim S-CoA$

Citrate: $^-OOC-{*}CH_2-C(OH)(COO^-)-CH_2-COO^-$

Now draw out the intermediates of the citric acid cycle. Place an asterisk on the radio-labeled carbon and circle the $-COO^-$ groups that are released as CO_2.

Citrate → Isocitrate → α–Ketoglutarate → Succinyl CoA → Succinate → Fumarate → Malate → Oxaloacetate

- Citrate: $^-OOC-{*}CH_2-C(OH)(COO^-)-CH-COO^-$ (HO—C—COO⁻, H—CH)
- Isocitrate: $^-OOC-{*}CH_2-CH(\text{circled } COO^-)-CH(OH)-COO^-$
- α–Ketoglutarate: $^-OOC-{*}CH_2-CH_2-C(=O)-(\text{circled } COO^-)$
- Succinyl CoA: $O=C(\sim S-CoA)-{*}CH_2-CH_2-COO^-$
- Succinate: $^-OOC-{*}CH_2-CH_2-COO^-$
- Fumarate: $^-OOC-{*}CH=CH-COO^-$
- Malate: $^-OOC-\overset{*}{C}H(OH)-CH_2-COO^-$
- Oxaloacetate: $^-OOC-{*}C(=O)-CH_2-COO^-$

22.53 This reaction is an example of the oxidation of a secondary alcohol to a ketone. The two functional groups are the hydroxyl group of the alcohol and the carbonyl group of the ketone.

22.55 It is a kinase because it transfers a phosphoryl group from one molecule to another. Kinases are a specific type of transferase.

22.57 Symptoms of mutations of the citric acid cycle enzymes frequently appear first in the central nervous system because of the high energy (ATP) demands of this tissue. As a result, many of these mutations cause neurological problems including encephalopathy which has a variety of neurological symptoms. Neurological symptoms draw attention to metabolic disorders in infants because the brain has such high energy requirements.

22.59 Hypotonia is muscle weakness. Deficiencies of citric acid cycle enzymes cause hypotonia because there is insufficient ATP.

22.61 An allosteric enzyme is one that has an effector binding site and an active site. Effector binding can change the shape of the active site, causing it to be active or inactive.

22.63 Allosteric enzymes are an efficient means to regulate a biochemical pathway because they bind to effectors, such as ATP or ADP, that alter the shape of the enzyme active site, either stimulating the rate of the reaction or inhibiting the reaction.

22.65 The citric acid cycle is regulated by the following four enzymes or enzyme complexes: pyruvate dehydrogenase complex, citrate sythase, isocitrate dehydrogenase, and the α-ketoglutarate dehydrogenase complex.

22.67 Energy-harvesting pathways, such as the citric acid cycle, must be responsive to the energy needs of the cell. If the energy requirements are high, as during exercise, the reactions must speed up. If energy demands are low and ATP is in excess, the reactions of the pathway slow down.

22.69 ADP serves as a signal to increase the rate of the reactions of the citric acid cycle. If ADP is found in high concentration, then ATP levels must be low. This is a signal that the cell requires more ATP energy. The enzyme isocitrate dehydrogenase is an allosteric enzyme of the citric acid cycle that is stimulated by ADP.

22.71 The electron transport system is series of electron transport proteins embedded in the inner mitochondrial membrane that accept high-energy electrons from NADH and $FADH_2$ and transfer them in stepwise fashion to molecular oxygen (O_2).

22.73 The oxidation of NADH via oxidative phosphorylation yields 3 ATP.

22.75 The oxidation of a variety of fuel molecules, including carbohydrates, the carbon skeletons of amino acids, and fatty acids provides the electrons. The energy of these electrons is used to produce an H^+ reservoir. The energy of this proton reservoir is used for ATP synthesis.

22.77 The electron transport system passes electrons harvested during oxidation of fuel molecules to molecular oxygen. At three sites, protons are pumped from the mitochondrial matrix into the intermembrane compartment. Thus, the electron transport system builds the high-energy H^+ reservoir that provides energy for ATP synthesis.

22.79 a. Two ATP per glucose (net yield) are produced in glycolysis, while the complete oxidation of glucose in aerobic respiration (glycolysis, the citric acid cycle, and oxidative phosphorylation) results in the production of 36 ATP per glucose.
b. Aerobic respiration harvests nearly 40% of the potential energy of glucose, while anaerobic glycolysis harvests only about 2% of the potential energy of glucose.

22.81 Transaminases transfer amino groups from amino acids to ketoacids.

22.83 The glutamate family of transaminases is very important because the ketoacid corresponding to glutamate is α-ketoglutarate, one of the citric acid cycle intermediates. This provides a link between the citric acid cycle and amino acid metabolism. These transaminases provide amino groups for amino acid synthesis and collect amino groups during catabolism of amino acids.

22.85 a. Alanine is converted to pyruvate.
b. Glutamate is converted to α-ketoglutarate.
c. Aspartate is converted to oxaloacetate.
d. Phenylalanine is converted to acetyl CoA.
e. Threonine is converted to succinate.
f. Arginine is converted to α-ketoglutarate.

22.87 Transaminase binds to the amino acid in its active site. Then, the α-amino group is transferred to pyridoxal phosphate, producing pyridoxamine phosphate. The amino group is then transferred to an α-keto acid.

22.89 Hyperammonemia, an elevation of the concentration of ammonium ions in the body, results when the urea cycle is not functioning. A complete deficiency of a urea cycle enzyme results in death in early infancy. A partial deficiency causes retardation, convulsions, and vomiting and can be treated with a low protein diet.

22.91 a. The source of one amino group of urea is the ammonium ion and the source of the other is the α-amino group of the amino acid aspartate.
b. The carbonyl group of urea is derived from CO_2.

22.93 Anabolism is a term used to describe all of the cellular energy-requiring biosynthetic pathways.

22.95 The amino acid glutamate is synthesized from the citric acid cycle intermediate α- ketoglutarate.

22.97 Citric acid cycle intermediates are the starting materials for the biosynthesis of many biological molecules.

22.99 An essential amino acid is one that cannot be synthesized by the body and must be provided in the diet.

22.101 Pyruvate carboxylase catalyzes the conversion of pyruvate to oxaloacetate.

$$\underset{\text{Pyruvate}}{\begin{array}{c}COO^-\\|\\C{=}O\\|\\CH_3\end{array}} + CO_2 + ATP \longrightarrow \underset{\text{Oxaloacetate}}{\begin{array}{c}COO^-\\|\\C{=}O\\|\\CH_2\\|\\COO^-\end{array}} + ADP + P_i$$

Chapter 23
Fatty Acid Metabolism
Solutions to the Odd-Numbered Questions and Problems

In-Chapter Questions and Problems

23.1 Because dietary lipids are hydrophobic, they arrive in the small intestine as large fat globules. The bile salts emulsify these fat globules into tiny fat droplets. This greatly increases the surface area of the lipids, allowing them to be more accessible to pancreatic lipases and thus more easily digested.

23.3 Starvation, a diet low in carbohydrates, and diabetes mellitus are conditions that lead to the production of ketone bodies. Lack of carbohydrates causes a decreased amount of oxaloacetate. This slows the citric acid cycle so that less acetyl CoA is oxidized. The excess acetyl CoA is converted to ketone bodies to recover coenzyme A.

23.5 The following are four differences between fatty acid biosynthesis and β-oxidation:

a. Fatty acid biosynthesis occurs in the cytoplasm, and β-oxidation occurs in the mitochondria.

b. The acyl group carrier in fatty acid biosynthesis is Acyl Carrier Protein (ACP), and the acyl group carrier in β-oxidation is coenzyme A.

c. The seven enzymes of fatty acid biosynthesis are associated as a multienzyme complex called fatty acid synthase. The enzymes involved in β-oxidation are not physically associated with one another.

d. NADPH is the reducing agent used in fatty acid biosynthesis. NADH and $FADH_2$ are produced by β-oxidation.

23.7 When excess fuel is available, the liver synthesizes fatty acids and triglycerides.

23.9 Insulin stimulates uptake of glucose and amino acids by cells, glycogen and protein synthesis, and storage of lipids. It inhibits glycogenolysis, gluconeogenesis, breakdown of stored triglycerides, and ketogenesis.

End-of-Chapter Questions and Problems

23.11 Cholate and chenodeoxycholate are the most common bile salts in human bile.

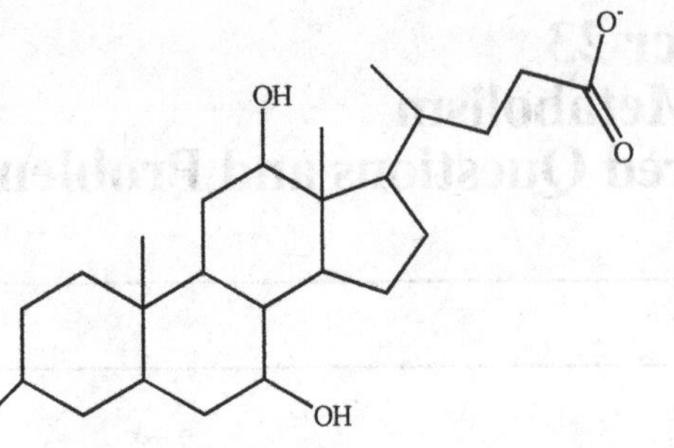

OH

HO OH

Cholate

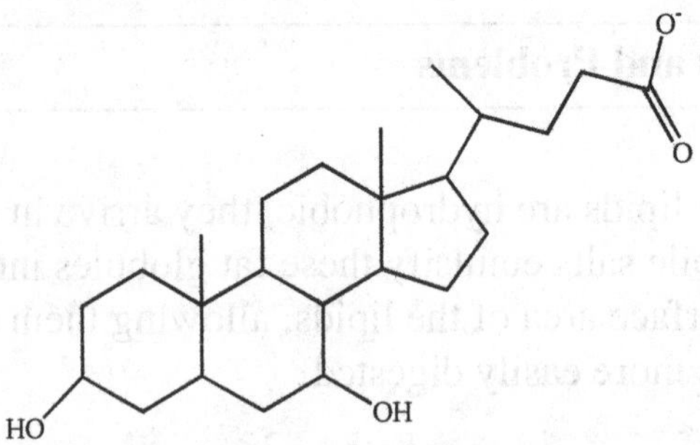

HO OH

Chenodeoxycholate

23.13 A micelle is an aggregation of molecules having nonpolar and polar regions. The nonpolar regions of the molecules aggregate, leaving the polar regions facing the surrounding water.

23.15 A triglyceride is a molecule composed of glycerol esterified to three fatty acids.

23.17 Triglycerides, cholesterol, and phospholipids are packaged into a protein-coated shell to produce the class of plasma lipoproteins called chylomicrons.

22.19 Triglycerides in adipose tissue are the major storage form of lipids.

23.21 The most outstanding feature of an adipocyte is the large fat globule that takes up nearly the entire cytoplasm.

23.23 Lipases catalyze the hydrolysis of the ester bonds of triglycerides, as seen in the following equation:

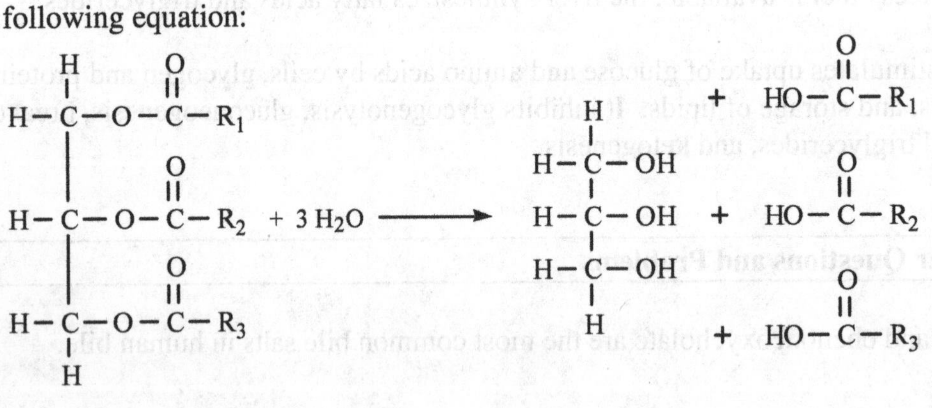

$$\begin{array}{l} H-\overset{\displaystyle H}{\underset{|}{C}}-O-\overset{\displaystyle O}{\overset{\|}{C}}-R_1 \\ H-\underset{|}{C}-O-\overset{\displaystyle O}{\overset{\|}{C}}-R_2 \\ H-\underset{\displaystyle H}{C}-O-\overset{\displaystyle O}{\overset{\|}{C}}-R_3 \end{array} + 3\,H_2O \longrightarrow \begin{array}{l} H \\ H-C-OH \\ H-C-OH \\ H-C-OH \\ H \end{array} \begin{array}{l} + \; HO-\overset{\displaystyle O}{\overset{\|}{C}}-R_1 \\ + \; HO-\overset{\displaystyle O}{\overset{\|}{C}}-R_2 \\ + \; HO-\overset{\displaystyle O}{\overset{\|}{C}}-R_3 \end{array}$$

23.25 Acetyl CoA is the precursor for fatty acids, several amino acids, cholesterol, and other steroids.

23.27 Chylomicrons are plasma lipoproteins (aggregates of protein and triglycerides) that carry dietary triglycerides from the intestine to all tissues via the bloodstream.

23.29 Bile salts serve as detergents. Fat globules stimulate their release from the gall bladder. The bile salts then emulsify the lipids, increasing the surface area and making them more accessible to digestive enzymes (pancreatic lipases).

23.31 When dietary lipids in the form of fat globules reach the duodenum, they are emulsified by bile salts. The triglycerides in the resulting tiny fat droplets are hydrolyzed into monoglycerides and fatty acids by the action of pancreatic lipases, assisted by colipase. The monoglycerides and fatty acids are absorbed by cells lining the intestine. Within intestinal cells, triglycerides are reassembled and are packaged into chylomicrons (lipoprotein particles made up of protein and dietary triglycerides). Chylomicrons are secreted into the lymphatic vessels and eventually reach the bloodstream. In the bloodstream, the triglycerides are hydrolyzed once again and the products (glycerol and free fatty acids) are absorbed by the cells of the body.

23.33 The energy source for the activation of a fatty acid entering β-oxidation is the breakdown of ATP into AMP and PP_i (pyrophosphate group), an energy expense of two high-energy phosphoester bonds.

23.35 Carnitine is a carrier molecule that brings fatty acyl groups into the mitochondrial matrix.

23.37 The following equation represents the reaction catalyzed by acyl-CoA dehydrogenase. Notice that the reaction involves the loss of two hydrogen atoms. Thus, this is an oxidation reaction.

$$H_3C-CH_2-CH_2-C(=O)\sim S\text{-}CoA$$

FAD → $FADH_2$ (Acyl-CoA dehydrogenase)

$$H_3C-CH=CH-C(=O)\sim S\text{-}CoA$$

23.39 An alcohol is the product of the hydration of an alkene.

23.41 The β-oxidation of 14-phenyltetradecanoic acid yields 6 acetyl CoA, 1 phenyl acetate, 6 NADH, and 6 $FADH_2$.

23.43 The following diagram summarizes the energy harvested by the β-oxidation of tetradecanoic acid:

Step 1 (Activation):	– 2 ATP
Steps 2 - 6 (repeated six times):	
6 $FADH_2$ x 2 ATP/$FADH_2$	12 ATP
6 NADH x 3 ATP/NADH	18 ATP
7 acetyl CoA to citric acid cycle:	
7 x 1 GTP x 1 ATP/GTP	7 ATP
7 x 3 NADH x 3 ATP/NADH	63 ATP
7 x 1 $FADH_2$ x 2 ATP/$FADH_2$	14 ATP
	112 ATP

23.45 Lauric acid is a dodecanoic acid. This twelve-carbon fatty acid would be broken down into six acetyl CoA molecules. Because five cycles through β-oxidation are required to break down a twelve-carbon fatty acid, five NADH molecules and five $FADH_2$ molecules would also be produced.

This summarizes the energy harvested by the β-oxidation of lauric acid:

Step 1 (Activation):	– 2 ATP
Steps 2 - 6 (repeated five times):	
5 $FADH_2$ x 2 ATP/$FADH_2$	10 ATP
5 NADH x 3 ATP/NADH	15 ATP
6 acetyl CoA to citric acid cycle:	
6 x 1 GTP x 1 ATP/GTP	6 ATP
6 x 3 NADH x 3 ATP/NADH	54 ATP
6 x 1 $FADH_2$ x 2 ATP/$FADH_2$	12 ATP
	95 ATP

23.47 The acetyl CoA produced by β-oxidation will enter the citric acid cycle unless there is too little oxaloacetate available. If that is the case, the acetyl CoA will be used in ketogenesis.

23.49 Ketone bodies include the compounds acetone, acetoacetone, and β-hydroxybutyrate, which are produced from fatty acids in the liver via acetyl CoA.

23.51 Ketosis is an abnormal rise in the level of ketone bodies in the blood.

23.53 Ketosis occurs in the matrix of the mitochondrion.

23.55 Acetoacetate β-Hydroxybutyrate

$$CH_3-\overset{\overset{O}{\parallel}}{C}-CH_2-\overset{\overset{O}{\parallel}}{C}-O^- \qquad CH_3-\overset{\overset{OH}{|}}{CH}-CH_2-\overset{\overset{O}{\parallel}}{C}-O^-$$

23.57 In those suffering from uncontrolled diabetes, the glucose in the blood cannot get into the cells of the body. The excess glucose is excreted in the urine. Body cells degrade fatty acids because glucose is not available. β-Oxidation of fatty acids yields enormous quantities of acetyl CoA, so much acetyl CoA, in fact, that it cannot all enter the citric acid cycle because there is not enough oxaloacetate available. Excess acetyl CoA is used for ketogenesis.

23.59 Ketone bodies are the preferred energy source of the heart.

23.61 Fatty acid biosynthesis occurs in the cytoplasm of the cell.

23.63 The structure of NADPH is:

23.65 The phosphopantetheine group allows formation of a high-energy thioester bond with a fatty acid. It is derived from the vitamin pantothenic acid and β-mercaptoethylamine.

23.67 Fatty acid synthase is a huge multienzyme complex consisting of the seven enzymes involved in fatty acid synthesis. It is found in the cell cytoplasm. The enzymes involved in β-oxidation are not physically associated with one another. They are free in the mitochondrial matrix space.

23.69 The liver produces ketone bodies under conditions of starvation or fasting.

23.71 Working muscle obtains most of its energy from the degradation of its supply of glycogen.

23.73 The energy requirements of resting muscle are generally met by β-oxidation of fatty acids.

23.75 Because ketone bodies have a free carboxylate group and are soluble in the blood, they can enter the brain and be used as an energy source.

23.77 Ketone bodies are the major fuel for the heart. Glucose is the major energy source of the brain and the liver obtains most of its energy from the oxidation of amino acid carbon skeletons.

23.79 Fatty acids are absorbed from the bloodstream by adipocytes. Using glycerol-3-phosphate, produced as a by-product of glycolysis, triglycerides are synthesized. Triglycerides are constantly being hydrolyzed and resynthesized in adipocytes. The rates of hydrolysis and synthesis are determined by lipases that are under hormonal control.

23.81 In general, insulin stimulates anabolic processes, including glycogen synthesis, uptake of amino acids and protein synthesis, and triglyceride synthesis. At the same time, catabolic processes such as glycogenolysis are inhibited.

23.83 A target cell is one that has a receptor for a particular hormone.

23.85 Decreased blood glucose levels trigger the secretion of glucagon into the bloodstream.

23.87 Insulin is produced in the β-cells of the islets of Langerhans in the pancreas.

23.89 Insulin stimulates the uptake of glucose from the blood into cells. It enhances glucose storage by stimulating glycogenesis and inhibiting glycogen degradation and gluconeogenesis.

23.91 Insulin stimulates synthesis and storage of triglycerides.

23.93 Insulin is secreted when blood glucose levels are high. It facilitates the uptake and storage of glucose by target cells to restore normal blood glucose levels. Glucagon is secreted when blood glucose levels are too low. It stimulates release of glucose into the blood to restore normal levels.

APPENDIX B

Answers to Chapter Self Tests

Chapter 1

1. Review your notes as soon as possible after class. Fill in any gaps that exist, and note any additional questions that arise.
2. In-chapter and end-of-chapter questions and problems can be used as your own personal quiz.
3. A hypothesis is an educated guess while a theory is a hypothesis that has been supported by experimentation.
4. solid, liquid, gas
5. gas
6. solid
7. b, c, d, f, g
8. extensive
9. b, c, d, e, f
10. a. heterogeneous
    b. heterogeneous
    c. homogeneous
11. Metric
12. 5.8×10^{-3} m^2
13. a. 2 b. 4 c. 3
14. a. 5.64×10^{6} b. 4.90×10^{-4}
    c. 1.0090×10^{1}
15. 0.0037
16. a. 1.48×10^{-1} b. 3.2×10^{-6}
    c. 104.1 d. 104.
17. 10,000
18. 3.95×10^{-22} g
19. $2.24 \text{x} 10^{3}$cm/s
20. -40.0 °F
21. -20.3 °C
22. 265.2 K
23. 1.02 g/mL
24. 4.21 kg
25. 0.534
26. 5.57 cm^3
27. Energy
28. 1.5×10^{6} cal.

Chapter 2

1. Electron
2. 12 protons, 12 electrons
3. $^{32}_{16}S$
4. $^{1}_{1}H$ $^{2}_{1}H$ $^{3}_{1}H$
5. It is still believed that all matter consists of tiny particles called atoms; atoms of different elements have different properties; atoms combine in simple, whole-number ratios; and chemical change involves the joining, separating or rearranging of atoms.

6. This experiment showed the presence of negatively charged electrons in the atom.
7. X-rays are higher in energy; shorter wavelengths correspond to higher energy.
8. Emits a photon.
9. Boron, silicon
10. 2s, 2p
11. s<p<d
12. 8
13. s
14. *s*, 1; *p*, 3; *d*, 5
15. Opposite directions
16. $1s^2 2s^2 2p^2$
17.

↑↓	↑↓	↑		
1*s*	2*s*	$2p_x$	$2p_y$	$2p_z$

18. $[Xe]6s^2$
19. (v.e.= valence electrons) IA(1): 1 v.e.; IIA(2): 2 v.e.; IIIA(13): 3 v.e.; IVA(14): 4 v.e.; VA(15): 5 v.e.; VIA(16): 6 v.e.; VIIA(17): 7 v.e.; VIIIA(18): 8 v.e.
20. 6
21. Helium
22. VIIA (or 17)
23. Isoelectronic ion
24. Al^{3+}
25. 17 protons, 18 electrons
26. Li, 3 protons, 3 electrons; Li^+, 3 protons, 2 electrons; $Li \rightarrow Li^+ + e^-$
27. $P + 3e^- \rightarrow P^{3-}$
28. +2
29. S< P<Mg
30. S^{2-}
31. Br
32. Energy released when a single electron is added to a neutral atom in the gaseous state.

Chapter 3

1. Ionic bond
2. Electrons are shared to hold the atoms of a covalent bond together.
3. Cations and anions
4. a. Polar covalent b. Ionic
   c. Polar covalent
   d. Nonpolar covalent
5. a. Sodium oxide b. Lithium sulfide
   c. Iron(II) sulfate
   d. Nitrogen trichloride
6. a. Cu_2O b. $Cu(NO_3)_2$
   c. $(NH_4)_2S$ d. SF_6
7. Covalent
8. NaCl
9. H–Ö: with H bonded above O

   4 bonding, 4 nonbonding
10. $[H_3O]^+$: H–O–H with H bonded above O, one lone pair on O

    6 bonding, 2 nonbonding

11. $\left[\begin{array}{c} :\ddot{O}: \\ | \\ :\ddot{O}=N-\ddot{O}: \end{array}\right]^{-}$
    1 double bond, 2 single bonds
12. Odd electron: NO
    Incomplete octet: BeH_2
    Expanded octet: SF_6
13. Two. No
14. More
15. Longer
16. Trigonal planar
17. Tetrahedron
18. BeH_2
19. Trigonal planar: 120°
    Tetrahedral: 109.4°
20. Polar
21. Nonpolar
22. Ionic
23. H_2O
24. ICl

Chapter 4

1. 19.00 amu, 19.00 g
2. 1.5×10^{24} atoms Fe
3. 4.81 g S
4. 18.02 g
5. 18.02 amu
6. 179.87 g/mol
7. Reactants
8. a. single-replacement
   b. combination
   c. double-replacement
9. $Ca + 2HCl \rightarrow CaCl_2 + H_2$
10. $H_3PO_4(aq) + 3NaOH(aq) \rightarrow Na_3PO_4(aq) + 3H_2O(l)$
11. $2CH_3OH + 3O_2 \rightarrow 2CO_2 + 4H_2O$
12. AgBr
13. Precipitate is $BaSO_4$
    Molecular equation:
    $BaCl_2(aq) + CuSO_4(aq) \rightarrow BaSO_4(s) + CuCl_2(aq)$
    Net ionic equation:
    $Ba^{2+}(aq) + SO_4^{2-}(aq) \rightarrow BaSO_4(s)$
14. Precipitate is AgBr
    Molecular equation:
    $AgNO_3(aq) + KBr(aq) \rightarrow AgBr(s) + KNO_3(aq)$
    Net ionic equation:
    $Ag^{+}(aq) + Br^{-}(aq) \rightarrow AgBr(s)$
15. 2 mol H_2
16. 1.00 mole
17. 6.0×10^{22} molecules NCl_3
18. 1.8×10^{2} mL
19. 80.08 g NaOH
20. 84.23 g H_2O
21. 7.15 g Fe_2O_3
22. 8.26 g NaCl
23. 83.8%
24. No, most reactions do not go to 100% due to experimental limitations in isolation, transfer, and recovery of the product. Many reactions do not convert completely to products.

Chapter 5

1. a. random motion.
   b. mostly empty space.
   c. or repulsive forces exist between atoms or molecules in a gas.
   d. losing energy.
   e. proportion to absolute temperature.
2. 0.770 atm
3. 761.0 torr

4. 5.0 atm
5. 2.09 L
6. 20.0 L
7. 5.97×10^{-2} mol
8. 5.5 mol
9. 0.778 L
10. N_2 and O_2
11. Resistance to flow
12. Evaporation: process of converting of a liquid to a gas.
    Vapor pressure is the pressure exerted by a gas vapor at equilibrium above its liquid.
13. The boiling point is the temperature at which the vapor pressure is equal to the atmospheric pressure (under "normal" conditions this is 1 atm)
14. Since the boiling point of a liquid depends on the atmospheric pressure, changes in the atmospheric pressure (such as during changes in altitude) change the boiling point.
15. Surface tension
16. Surfactants
17. London dispersion forces
18. δ^+ H–F δ^- ------- δ^+ H–F δ^-
19. $CH_4 < HCl < HF$
20. As intermolecular forces increases, vapor pressure decreases.
21. Melting point
22. High
23. c, a, b, d

Chapter 6

1. The solute is sucrose, the solvent is water.
2. To be an aqueous solution the solvent must be water.
3. Electrolytic
4. The particles in colloidal suspensions are not identical in size and are not homogeneously mixed throughout the solution. Additionally, colloidal suspensions scatter light while homogeneous solutions do not.
5. Decreasing the temperature will decrease the solubility of a solid dissolved in a liquid.
6. $CaCl_2$ is an ionic compound and will be more soluble in water, which is a polar solvent.
7. In a saturated solution particles are constantly dissolving and precipitating, but the total number of dissolved particles and undissolved particles remains the same.
8. High pressure and low temperature.
9. 44 g Au
10. 1.10 %
11. 800.0 mL solution
12. 1.25 L solution
13. 0.0050 g
14. 0.015 g Pb
15. 0.055 mol
16. 0.466 g
17. 0.19 *M*
18. 0.185 *M*
19. 52.1 mL
20. 0.10 *M*
21. 0.75 *m*
22. -2.1 °C
23. 0.30 mol particles/L
24. 194 atm

25. Osmosis

26. Hypertonic

27. Water is the "universal solvent"; despite being small, water has a high boiling point (due to strong intermolecular forces); water is the principle biological solvent

28. 1.0 *M*

29. During dialysis, blood is pumped through a semipermeable membrane. By the process of osmosis, waste materials pass from the blood to a dialyzing fluid.

Chapter 7

1. a. exothermic b. higher

2. a. decreased b. equal to

3. Usually nonspontaneous. An endothermic reaction may be spontaneous at high temperatures if the reaction leads to an increase in entropy (positive ΔS).

4. Large positive values

5. Gas

6. Positive ΔS. There is an increase in moles of gas.

7. Spontaneous

8. Nonspontaneous

9. Cannot be determined without the ΔS of the surroundings.

10. A calorimeter is used to measure the energy demand or energy release in a chemical reaction.

11. 1.4×10^3 cal

12. Speed

13. a. Rate decrease b. rate increase c. rate increase

14. Activated complex

15. Rate=$k[H_2]^n[I_2]^{n'}$, experimentally

16. The system is at equilibrium.

17. a. $K_{eq} = \dfrac{[N_2O_4]}{[NO_2]^2}$

b. $K_{eq} = \dfrac{[N_2]}{[NO]^2[H_2]^2}$

18. Right

19. Left

20. Right

Chapter 8

1. Acid: a substance that produces H^+ ions when added to water
Base: a substance that produces OH^- ions when added to water

2. Acid: proton donor
Base: proton acceptor

3. Acid: H_2O Base: CH_3NH_2
Conj. acid: $CH_3NH_3^+$ Conj. base: OH^-

4. a. $HNO_2 + H_2O \rightleftharpoons NO_2^- + H_3O^+$
b. $HBr + H_2O \rightarrow Br^- + H_3O^+$
c. $NH_3 + H_2O \rightleftharpoons NH_4^+ + OH^-$
d. $NaOH \rightarrow Na^+ + OH^-$

5. $2H_2O \rightleftharpoons H_3O^+ + OH^-$

6. 2.0×10^{-10} *M*
7. Acidic
8. 1.0×10^{-11} *M*
9. pH
10. 4.19
11. 5.40
12. 1.0×10^{-5} *M*
13. Basic
14. 1.0×10^{-5} *M*
15. Acidic
16. 3.00
17. Neutralization
18. 0.0592 *M*
19. 3.52
20. 3.22
21. a. Oxidation b. Oxidation
    c. Reduction d. Reduction
22. b
23. voltaic cell: a, d, e
    electrolysis: b, c, e
24. cathode

Chapter 9

1. $^{23}_{11}Na$
2. Gamma ray
3. Alpha particle
4. Alpha particle
5. Natural radioactivity
6. Gamma rays
7. Alpha particles
8. Boron-11 has a p:n ratio closer to 1:1 than does Boron-13
9. a. $^{1}_{0}n$ b. $^{210}_{83}Bi$
10. $^{210}_{82}Pb$
11. 3.125 mg
12. Radiocarbon dating can be used to determine the approximate age of fossils by comparing the relative amounts of carbon-14 and carbon-12 in the object.
13. Nuclear fission
14. Cancer or malignant cells
15. A small amount of a radioactive substance, or tracer, is given to a patient and photographs of the organ of interest are obtained to investigate the condition of the organ.
16. Background radiation
17. Film badges are sensitive to the energies of radioactive emissions and are used to measure an individual's exposure to radiation.
18. Natural radioactivity is radioactivity produced by unstable isotopes. Artificial radioactivity is radioactivity produced by bombarding a stable nucleus with protons, neutrons, or alpha particles to make it radioactive.
19. Shielding, maximum distance from a radioactive source, and minimum time of exposure all minimize the possible damaging effects of radiation on the body.
20. The unit of rem measures the biological damage caused by absorption of radiation by the body.

Chapter 10

1. a. flammable
2. Aromatics
3. e. hydrocarbons
4. Functional group
5. Carbon-to-carbon double bond
6. Hydroxyl
7. C_9H_{20}
8. C_6H_{14}, $CH_3CH_2CH_2CH(CH_3)CH_3$,

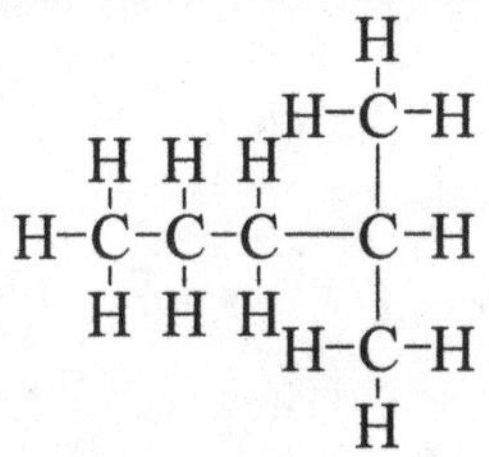

```
                H
               H-C-H
      H H H     |
    H-C-C-C—---C-H
      H H H     |
               H-C-H
                H
```

9. Tetrahedral
10. Lower
11. None or little
12. a. Methylbutane
    b. 2,2-Dimethylpentane
    c. 2,2,3-Trimethylpentane
13. 1-Bromo-4-chloro-2-methylpentane
14. 2-Bromo-2-methylpropane
15. 2-Bromo-4,4-dimethylhexane
16. Parent compound
17. Trichloromethane
18. Primary: 4, secondary: 3, tertiary: 1
19. b, c, and d
20. a. Br_2CHCH_3
21. c. Cyclobutane
22. a. CH_3CH_2OH
23. Ring
24. Stereoisomers
25. *cis*-1,3-dichlorocyclopentane; *trans*-1,3-dichlorocyclopentane
26. The bonding electrons are farther apart in the *staggered* conformation.
27. Boat
28. Combustion
29. Two
30. $C_3H_8 + 5O_2 \rightarrow 3CO_2 + 4H_2O$
31. Two products are formed: 1-chloropropane and 2-chloropropane
32. Light, or UV, or high temperature
33.

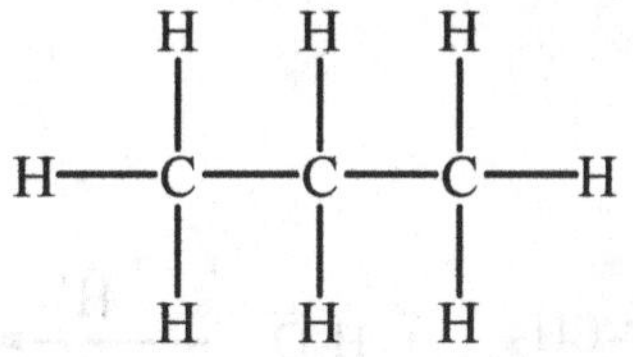

```
    H   H   H
    |   |   |
H---C---C---C---H      CH3CH2CH3
    |   |   |
    H   H   H
```

$CH_3CH_2CH_3$

Chapter 11

1. a. 1-Hexene (as with other alkenes) is nonpolar and is not soluble in water.
   b. 1-Hexene has a smaller molecular mass and thus a lower boiling point than 1-nonene.
2. a. Alkane: C_4H_{10} could be butane
   b. Alkene: C_4H_8 could be 1-butene or 2-butene
   c. Alkyne: C_4H_6 could be 1-butyne or 2-butyne
3. All C-C single bonds in the molecule
4. At least one carbon-to-carbon triple bond

5. 3-Ethyl-2-pentene
6. 4,5-Dibromo-2-hexyne
7. 2-Methyl-2-pentene: $CH_3-C(CH_3)=CH-CH_2-CH_3$
8. *cis*-2-Bromo-2-butene
9. *cis*-3-Methyl-2-pentene: $(H_3C)(H)C=C(CH_3)(CH_2CH_3)$ (H_3C and CH_3 on the same side)
10. *trans*-2-Bromo-2-butene
11. *trans*-1,2-Dibromopropene
12. Addition reactions
13. Hydrogenation
14. Palladium, platinum, and nickel
15. Butane
16. $$HC{\equiv}CCH_2CH_3 + 2H_2 \xrightarrow[\text{Heat or pressure}]{\text{Pt, Pd, or Ni}} CH_3CH_2CH_2CH_3$$
17. 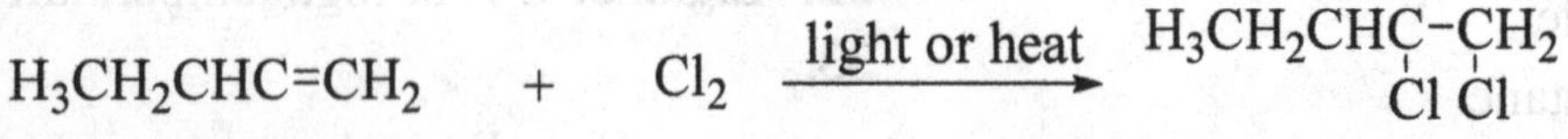 $$H_3CH_2CHC{=}CH_2 + Cl_2 \xrightarrow{\text{light or heat}} H_3CH_2CHC(Cl){-}CH_2(Cl)$$
18. 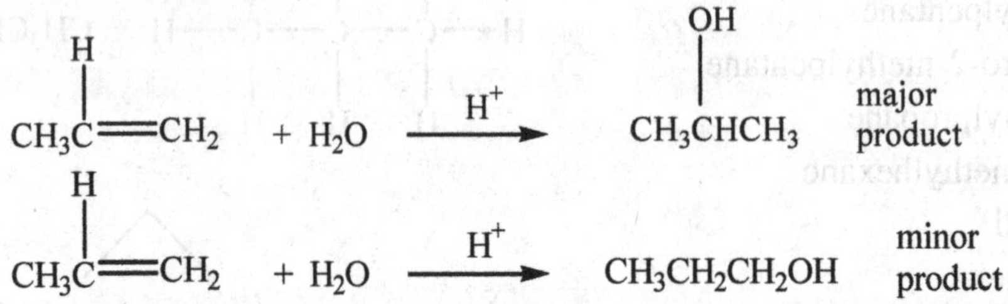 $$CH_3C(H){=}CH_2 + H_2O \xrightarrow{H^+} CH_3CH(OH)CH_3 \quad \text{major product}$$
$$CH_3C(H){=}CH_2 + H_2O \xrightarrow{H^+} CH_3CH_2CH_2OH \quad \text{minor product}$$
19. 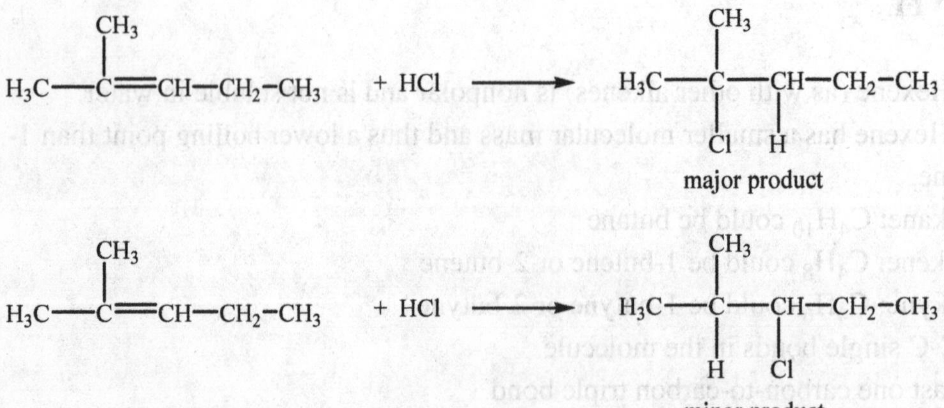 $$H_3C-C(CH_3){=}CH-CH_2-CH_3 + HCl \longrightarrow H_3C-C(CH_3)(Cl)-CH(H)-CH_2-CH_3$$
major product
$$H_3C-C(CH_3){=}CH-CH_2-CH_3 + HCl \longrightarrow H_3C-C(CH_3)(H)-CH(Cl)-CH_2-CH_3$$
minor product

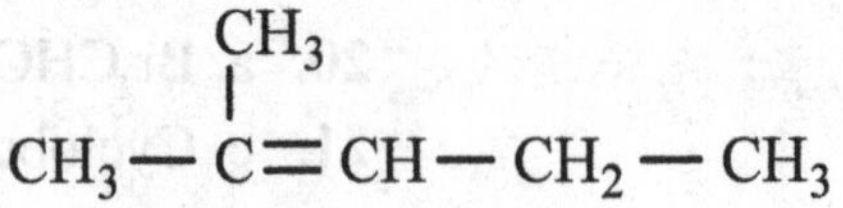

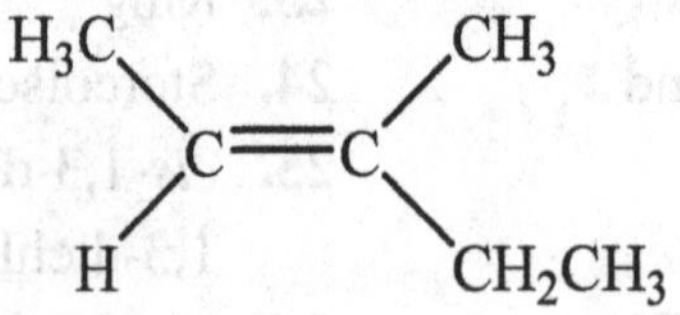

20.

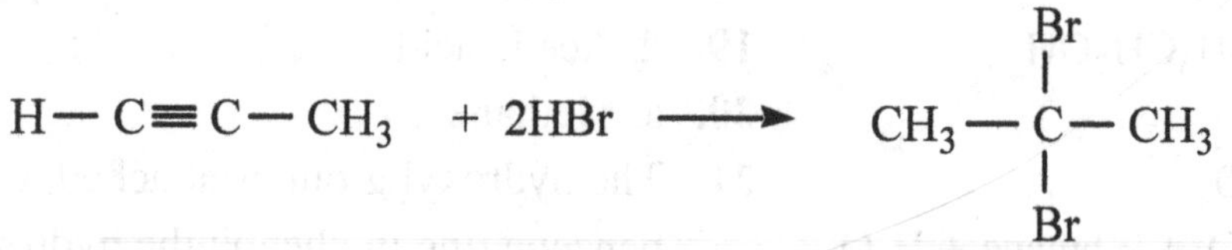

$$H-C\equiv C-CH_3 \; + 2HBr \longrightarrow CH_3-\overset{Br}{\underset{Br}{C}}-CH_3$$

21.

$$C_6H_5-CH=CH_2 \longrightarrow \left[\,-\underset{C_6H_5}{\overset{H}{C}}-\underset{H}{\overset{H}{C}}-\,\right]_n$$

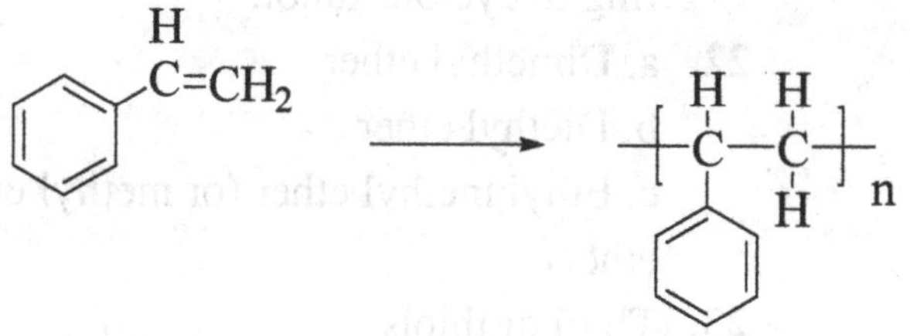

22. Benzene

23. a. Benzene
b. Phenol or hydroxybenzene
c. 1,2-Dichlorobenzene or *ortho*-dichlorobenzene or *o*-dichlorobenzene
d. 1-Bromo-2,3-dichlorobenzene
e. 1,4-Dibromobenzene or *para*-dibromobenzene or *p*-dibromobenzene

24.

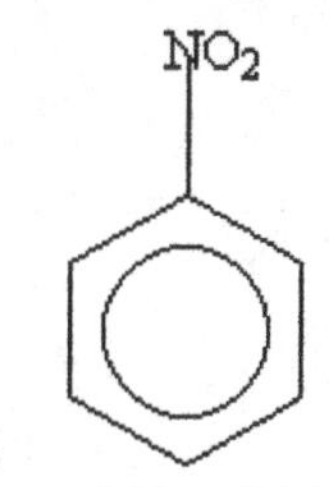

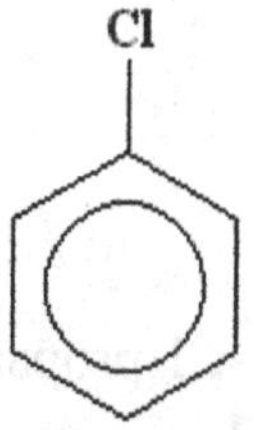

25.

$$C_6H_5-Cl$$

26. Purine, is a heterocyclic compound since it has nitrogens as part of the structure of the aromatic ring. Purines are found in RNA and DNA.

Chapter 12

1. 1

2. a. 1-Nonanol b. Methanol
c. 1-Propanol d. Dimethyl ether

3. a. 1-Propanol

4. They are polar and can form hydrogen bonds between alcohol molecules.

5. b. ethers

6. b. ethers

7. 3-Bromo-4-methyl-2-pentanol

8. 5-Chloro-2-pentanol

9. 1,2,3-propanetriol

10. a. Propyl alcohol
b. Isopropyl alcohol
c. *tert*-Butyl alcohol

11. Denatured alcohol

12. Glycerol
13. a. CH_3OH b. CH_3CH_2OH
14. a. Ethene + H_2O
    b. Propene + H_2O
    c. 1-Butene + H_2O + 2-butene + H_2O
    [major product 2-butene; minor product 1-butene]
15. a. $H-\overset{\overset{\displaystyle O}{||}}{C}-H$

    b. $CH_3\overset{\overset{\displaystyle O}{||}}{C}-H$
16. a. Formaldehyde (Methanal)
    b. Acetaldehyde (Ethanal)
    c. Acetone (Propanone)
    d. No reaction
17. Tertiary
18. Oxidation
19. d. Acetic acid
20. a. Methane
21. The hydroxyl group is attached to a benzene ring in phenol; the hydroxyl group is attached to a cyclohexane ring in cyclohexanol.
22. a. Dimethyl ether
    b. Diethyl ether
    c. Ethyl methyl ether (or methyl ethyl ether)
23. Thiol or thiols
24. $CH_3CH_2CH_2CH_2\underset{\underset{\displaystyle CH_3}{|}}{CH}-\overset{\overset{\displaystyle SH}{|}}{\underset{\underset{\displaystyle CH_3}{|}}{C}}CH_3$
25. 2-Butanethiol

Chapter 13

1. a. propane < propanone < 1-propanol
   b. pentane < methoxyethane < propanal
2. The electronegativity difference between the C and O of the C=O double bond makes the carbonyl group polar.
3. H_3C and CH_2CH_3 on $C^{\delta+}$, $C=O^{\delta-}$; dipole–dipole attraction between $O^{\delta-}$ of one molecule and $C^{\delta+}$ of another:

   $$\overset{\displaystyle H_3C\quad CH_2CH_3}{\underset{\displaystyle O^{\delta-}}{\overset{||}{C^{\delta+}}}} \cdots\cdots \overset{\displaystyle O^{\delta-}}{\underset{\displaystyle H_3C\quad CH_2CH_3}{\overset{||}{C^{\delta+}}}}$$
4. The numbers are not necessary.
5. a. $CH_3CH_2\underset{\underset{\displaystyle Br}{|}}{CH}-\overset{\overset{\displaystyle O}{||}}{C}-H$

   b. $CH_3-\overset{\overset{\displaystyle CH_3}{|}}{CH}-\overset{\overset{\displaystyle O}{||}}{C}-CH_2CH_3$
6. Acetaldehyde
7. a. Ethanal
   b. Propanone
   c. Propanal
   d. 3-Pentanone
   e. 4-Bromohexanal
8. $H-\overset{\overset{\displaystyle O}{||}}{C}-H$ Aqueous solutions of formaldehyde are used to preserve tissues.
9. Ethanol is produced from the oxidation of ethanol in the liver.
10. a. Oxidation b. Oxidation
    c. Reduction d. Oxidation

11. a. $HCOOH$ b. CH_3COOH
c. No reaction d. CH_3CH_2COOH
e. No reaction

12. Ethanal (on further oxidation, ethanoic acid)
b. Propanal (on further oxidation, propanoic acid)
c. Propanone
d. No reaction

13. tertiary

14. a. $CH_3COO^- + Ag^0$
b. No reaction: CH_3COCH_3 is a ketone.
c. No reaction: $CH_3CH_2COCH_3$ is a ketone.

15. Carboxylic acids

16.

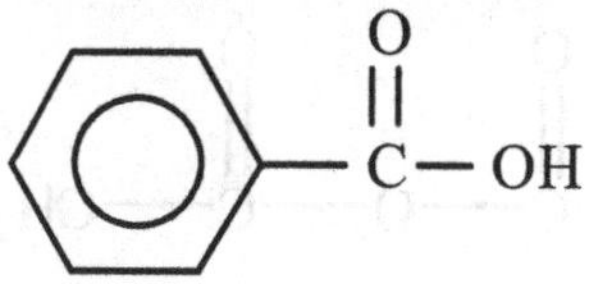

C_6H_5—C(=O)—OH

Benzoic acid

17. In the Tollens' test, aldehydes react with $Ag(NH_3)_2^+$ to produce carboxylic acids and Ag^0. Ketones are not further oxidized and thus do not react by the Tollens' test. Similarly, aldehydes (but not ketones) react with Benedict's reagent (Cu^{2+}, OH^-) to give a red solid, Cu_2O.

18.

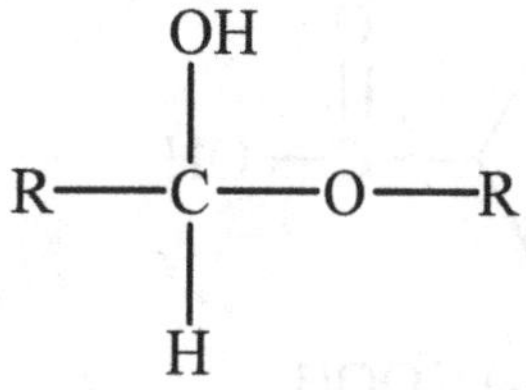

R—C(OH)(H)—O—R

19. Acetal

20. Methanol, CH_3OH

21. Cyclohexanone,

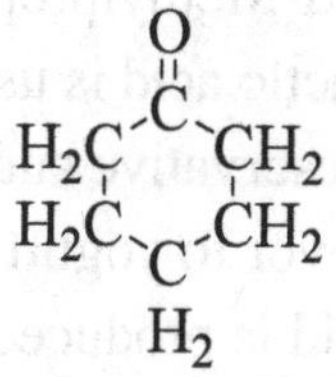

$C(=O)$ in a six-membered ring with H_2C, CH_2, H_2C, CH_2, C / H_2

22.

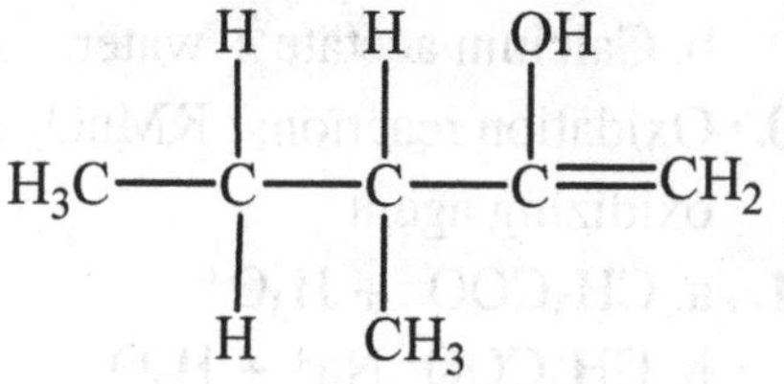

$H_3C—CH(H)—CH(CH_3)—C(OH)=CH_2$

or

$H_3C—CH_2—C(CH_3)=C(OH)—CH_3$

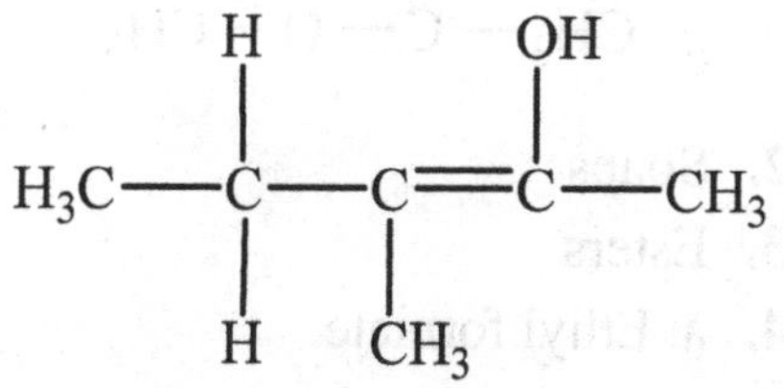

23. keto

Chapter 14

1. a. ROH b. ROR c. RCOOR
d. RCHO e. RCOOH f. RCOR

2. a. Acetic acid

3. No

4. a. Methanoic acid
b. Hexanoic acid
c. Ethanoic acid
d. 3-Methylbutanoic acid

5. a. CH_3COOH b. $HCOOH$

c. 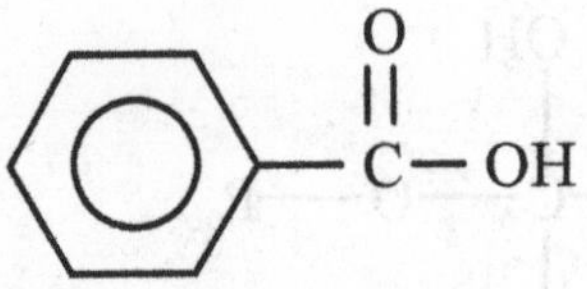 Benzoic acid structure: $C_6H_5-\overset{O}{\overset{||}{C}}-OH$

d. $CH_3(CH_2)_8COOH$

6. a. Butyric acid
   b. Propionic acid
   c. α-Methylpropionic acid
7. Lactic acid is used as a food preservative and gives the tangy flavor to yogurt and buttermilk. Lactic acid is produced in the muscle cells of the body during strenuous exercise.
8. Carboxylate ion + hydronium ion
9. a. Sodium butyrate + water
   b. Calcium acetate + water
10. Oxidation reaction; $KMnO_4$ is an oxidizing agent.
11. a. $CH_3COO^- + H_3O^+$
    b. $CH_3COO^- Na^+ + H_2O$
    c. $CH_3-\overset{O}{\overset{||}{C}}-O-CH_3$
12. Soaps
13. Esters
14. a. Ethyl formate
    b. Methyl propionate
    c. Pentyl acetate
15. a. Methyl ethanoate
    b. Ethyl propanoate
16. Methanol + propanoic acid
17. Methyl ethanoate
18. CH_3OH
19. Alcohol + carboxylic acid
20. Methanol and acetic acid
21. Polyester
22. Saponification
23. The nonpolar hydrocarbon chain dissolves oil and grease. The polar carboxylate functional group dissolves in water.
24. Propionoyl chloride, Propanoyl chloride
25. Propanoic anhydride

$$CH_3-CH_2-\overset{O}{\overset{||}{C}}-O-\overset{O}{\overset{||}{C}}-CH_2-CH_3$$

26. Two acetic acid molecules
27. Phosphate ester or phosphoester
28. Phosphoric anhydride bond or phosphoanhydride bond
29. Adenosine triphosphate (ATP)
30. Thioester
31. Coenzyme A

Chapter 15

1. Ammonia
2. a. Tertiary amine
   b. Quaternary ammonium salt
   c. Primary amine
   d. Secondary amine
3. a. RNH_2 b. R_3N
4. Diethylamine should have a higher boiling point because of hydrogen bonding.
5. a. Ethanamine
   b. 2-Propanamine
   c. *N,N*-dimethylmethanamine
6. Aniline

7. $H_3C-\overset{\overset{\displaystyle O}{\|}}{C}-\underset{\underset{\displaystyle H}{|}}{N}-CH_2-CH_3$

8. Methylammonium chloride
9. a. $(CH_3)_2NH_2^+ + OH^-$
   b. $(CH_3)_3NH^+Cl^-$
10. Because the alkylammonium salts are more soluble in water and body fluids
11. Because they have long carbon chains and a charged quaternary amine, they can act as detergents.
12. Nucleic acids or DNA + RNA
13. Solid
14. a. Ethanamide
    b. *N*-propylhexanamide
    c. *N*-methylbutanamide
15. $H_3CCOCl + 2\ NH_3 \rightarrow H_3CCONH_2 + NH_4Cl$
16. a. $CH_3COOH + NH_4^+$
    b. $CH_3CH_2COOH + CH_3N^+H_3$
17. Proteins
18. Carboxyl groups and amino groups
19. $H_3\overset{+}{N}-\underset{\underset{\displaystyle H}{|}}{\overset{\displaystyle H}{\overset{|}{C}}}-\overset{\overset{\displaystyle O}{\|}}{C}-\underset{\underset{\displaystyle H}{|}}{N}-\underset{\underset{\displaystyle CH_3}{|}}{\overset{\displaystyle H}{\overset{|}{C}}}-\overset{\overset{\displaystyle O}{\|}}{C}-O$
20. Analgesic
21. To carry a message, or signal, from a nerve cell to a target cell
22. Parkinson's disease
23. Acetylcholine
24. Histamine

Chapter 16

1. Simple carbohydrates, such as glucose, are the smallest type of sugars which are polymerized together to form complex carbohydates, such as starch.
2. It is currently recommended that 45-65% of the calories in the diet should be from carbohydrates. It is recommended that the intake of sucrose should be minimized and not account for more than 5% of daily calories.
3. The monosaccharides are a, b, c and f. Sucrose is a disaccharide, and starch is a polysaccharide.
4. A triose contains three carbon atoms and a pentose contains five carbon
5. a. Aldohexose b. Ketohexose
   c. Aldohexose d. Aldopentose
   e. Aldopentose f. Aldotriose
6. Any carbon of a sugar which has four different groups attached to it is chiral.
7. A pair of stereoisomers have the same molecular formula and the same connectivity of the atoms. They differ in the spatial arrangement of the atoms.
8. Enantiomers
9. a. D-Glyceraldehyde b. D-Ribose
   c. L-Ribose d. D-2-Deoxyribose
10. D-Glucose
11. The sugar ribose is found in RNA.
12. Fructose, a ketohexose, is the sweetest of all the sugars.
13. A hemiacetal
14. α-D-glucose
15. Reducing sugars
16. Glucose is a reducing sugar and thus if glucose is present in the urine, the

urine should show a positive Benedict's test. Benedict's reagent will turn a variety of colors depending on the amount of glucose present. The presence and amount of glucose in the urine is monitored for individuals with Type I diabetes to determine if they are taking the correct amount of insulin.

17. Glycosidic bond

18. In an $\alpha(1 \rightarrow 6)$ glycosidic linkage the oxygen joining the two monosaccharides links C-1 of the first sugar to C-6 of the second sugar. Also, the oxygen is located below the first ring. In a $\beta(1 \rightarrow 4)$ glycosidic linkage the oxygen joining the two monosaccharides is above the ring and is connecting C-1 on the first sugar with C-4 on the second sugar.

19. Maltose, lactose, and sucrose are all disaccharides and all contain at least one glucose. Maltose is composed of two glucose monosaccharides joined by an $\alpha(1 \rightarrow 4)$ glycosidic bond. Lactose is composed of glucose and galactose joined by a $\beta(1 \rightarrow 4)$ glycosidic bond. Sucrose is composed of glucose and fructose joined by a $(\alpha 1 \rightarrow \beta 2)$ glycosidic bond. Maltose and lactose are reducing sugars, sucrose is a nonreducing sugar.

20. Lactose intolerance

21. Galactosemia

22. Sucrose will not react with Benedict's solution and thus is a non-reducing sugar. A hemiacetal functional group must be present in the molecule to give a reducing sugar. These functional groups are lacking in sucrose and therefore sucrose cannot be further oxidized using Benedict's reagent.

23. α-Amylase cleaves the glycosidic bonds of amylose chains. β-Amylase removes maltose molecules from the reducing end of the amylose chain.

24. Amylose and amylopectin are both components of starch and are both polymers of glucose. Amylose is a linear chain of glucose units joined by $\alpha(1 \rightarrow 4)$ glycosidic bonds. Amylopectin contains the amylose chain but has branches formed by $\alpha(1 \rightarrow 6)$ bonds off the chain.

25. Amylopectin is a component of starch which is present in plants. Glycogen is the main glucose storage molecule in animals. Glycogen is very similar to amylopectin but contains more, and shorter, branches.

26. The amount of glucose in the blood must be maintained at a fairly constant level. The body uses glycogen to help maintain this level. When the glucose level in the blood gets too low the body signals the breakdown of glycogen and the glucose units are released into the blood. When the blood glucose level gets too high, the body signals the synthesis of glycogen which takes glucose from the blood and polymerizes it together to form glycogen.

27. Cellulose

28. Humans cannot make the enzyme cellulase, an enzyme needed to hydrolyze cellulose.

29. a. No reaction
b. glucose + galactose
c. glucose + fructose
d. many glucose molecules

Chapter 17

1. They are all insoluble in water.
2. They are nonpolar molecules.
3. Fatty acids
4. They are long-chain monocarboxylic acids with an even number of carbon atoms.
5. Solid
6. The saturated fatty acid contains all C-C single bonds, while the unsaturated fatty acid contains at least one C=C double bond. The unsaturated fatty acid has a lower melting point.
7. Prostaglandins, leukotrienes, and thromboxanes
8. Prostaglandins are formed from the essential fatty acid, linoleic acid.
9. Dilation
10. Aspirin inhibits the enzyme cyclooxygenase, which in turn, inhibits the synthesis of prostaglandins.
11. Triglycerides (triacylglycerols)
12. Nonionic and nonpolar
13. Ester + water
14. Glycerol + two fatty acid molecules
15. Glycerol + 3 fatty acids
16. Saturated solid fats
17. Ca^{2+} or Mg^{2+}
18. When triglycerides react with water in the presence of base soap molecules are formed. This is called a saponification reaction.
19. polar and nonpolar
20. A sphingolipid is based on the alcohol, sphingosine, while a glyceride is based on the alcohol glycerol.
21. Sphingomyelin surrounds and insulates the cells of the central nervous system.
22. The four fused rings known as the steroid nucleus
23. Egg yolks, dairy products, liver, and other animal meats
24. lecithin
25. Progesterone
26. Cortisone
27. Birth control
28. Long-chain alcohol + fatty acid
29. Chylomicrons, very low-density lipoproteins (VLDL), low-density lipoproteins (LDL), high-density lipoproteins (HDL)
30. HDL
31. High-density lipoproteins
32. Cholesterol
33. Hydrophobic (nonpolar) fatty acid tail
34. Cholesterol

Chapter 18

1. nitrogen and sulfur
2. Transport proteins carry materials around the body
3. Structural proteins provide mechanical support and outer coverings to animals. One example of a structural protein is keratin, found in hair and fingernails.
4. A zwitterion is a molecule that exists as a dipolar ion. Amino acids form a negatively charged carboxyl group and a positively charged amino group.
5. A hydrophobic molecule is nonpolar and thus is not water soluble.
6. Amino acids which contain COO^- in their side chain: aspartate and glutamate
7. Amino acids which contain NH_3^+ in their side chain: lysine, arginine, and histidine
8. This is a hydrophobic R group
9.

```
          H   O
          |   ||
  H3+N— C — C — O⁻
          |
          CH2
          |
          OH
```

10. Peptide bond or amide bond
11.

```
       H  O       H   O        H  O       H  O
       |  ||      |   ||       |  ||      |  ||
H3+N— C — C — N — C — C — N — C — C — N — C — C — O⁻
       |      |   |       |   |       |   |
       CH3    H   CH      H   H       H   CH2
                 /  \                     |
               H3C   CH3                  SH
```

12. Primary structure
13. Hydrogen bonding between the carbonyl oxygens and the amide hydrogens of the peptide maintain secondary structure.
14. α-Helix or helices
15. Hydrogen bonding, ionic bonding, and van der Waals forces between the side chains, as well as covalent disulfide bridges maintain tertiary structure.
16. Tertiary and quaternary structure involve the same interactions between side chains, however, in tertiary structure the amino acid residues are within the same peptide chain, while in quaternary structure, separate peptide chains aggregate together to form the final active protein.
17. Myoglobin stores oxygen in the muscle cells while hemoglobin transports oxygen through the body.
18. The iron containing heme group is found in both myoglobin and hemoglobin and allows oxygen to be stored or transported in the body.
19. The final three-dimensional structure of a protein defines its biological function.
20. If the temperature of a protein rises too high the interactions that maintain the three dimensional structure of a protein are interrupted and the protein is denatured.

21. When the pH drops too low, a protein becomes a polycation. This interrupts interactions in the protein and leads to denaturation.
22. A mixture of the individual amino acids
23. Pepsin
24. Essential amino acids

Chapter 19

1. They catalyze the transfer of phosphate groups.
2. Lipase
3. Lactate dehydrogenase removes hydrogens from the substrate lactate.
4. a. Succinate b. Sucrose
   c. Glycogen
5. The enzyme lowers the activation energy of the reaction.
6. No effect
7. Substrate
8. The rate of reaction increases until, at a certain substrate concentration, the rate reaches a maximum.
9. The active sites of all the enzyme molecules in solution are occupied by substrate molecules.
10. Formation of enzyme-substrate complex
11. In the lock and key model, the substrate fits exactly into the active site. In the induced-fit model, the substrate approximately fits the active site and the active site changes shape to conform to the substrate.
12. Step 2: The shape of the substrate is altered to form a transition state. Step 3: The substrate is converted into product. Step 4: The product is released from the active site.
13. a. Absolute specificity
    b. Group specificity
    c. Linkage specificity
    d. Stereochemical specificity
14. Absolute specificity
15. For an enzyme that requires a nonprotein component in order to function in catalysis, the protein portion is called its apoenzyme and the nonprotein part is the cofactor.
16. Usually it is nicotinamide adenine dinucleotide, as either the oxidized form (NAD^+) or its reduced form ($NADH + H^+ = NADH$).
17. Most are used in some form as coenzymes.
18. pH optimum
19. Denaturation
20. The optimum temperature (37°C for most enzymes found in the human body)
21. Irreversible inhibitors and competitive inhibitors
22. The sulfa drugs are competitive inhibitors of a bacterial enzyme required for the synthesis of the required vitamin folic acid.
23. Effector binding
24. Generally, the end-product of the pathway
25. Proenzyme
26. They are proteases, digestive enzymes that hydrolyze peptide bonds in proteins.

27. Acute myocardial infarction can cause some of the cells of the heart muscle to die, which releases these enzymes into the bloodstream.

Chapter 20

1. DNA
2. DNA
3. a. dATP b. CTP c. TMP
4. Hydrogen bonding between A and T, as well as G and C.
5. Thymine
6. 3 ' → 5' direction
7. The two strands are complementary. One strand specifies the sequence of bases on the other.
8. RNA molecules are single stranded. The sugar in RNA is ribose rather than 2'-deoxyribose, and uracil (U) replaces thymine (T).
9. The two strands of the double helix are separated by the enzyme helicase.
10. Replication origin
11. Replication fork
12. a. It reads the parental strand and produces the daughter strand.
    b. It proofreads the newly synthesized daughter strand to ensure that no errors have been made. If an error is detected, it is corrected.
13. DNA → RNA → protein
14. mRNA carries the genetic information for a protein from DNA to the ribosomes.
15. transfer RNA
16. Promoter sequence
17. The poly(A) tail protects the mRNA from enzymatic degradation.
18. Intervening sequences, or introns
19. Exons
20. Codons
21. The genetic code is said to be degenerate because often several triplet codons code for the same amino acid.
22. Ribosomes are "platforms" on which protein synthesis occurs.
23. A point mutation results when a single nucleotide is substituted for another one.
24. Silent mutation
25. Pyrimidine dimers
26. Mutations may lead to different amino acid sequences resulting in nonfunctional or improperly functioning proteins, which can kill a cell.
27. Cancer
28. Restriction enzyme
29. Phage vectors and plasmid vectors
30. The polymerase chain reaction is used to produce many copies of a particular DNA sequence.
31. In PCR, the two strands of a small piece of DNA are separated with heat. With a reduction in temperature, a primer binds to the target DNA, and finally the temperature is again increased to allow Taq polymerase to polymerize a daughter strand beginning from the primer. This is repeated many times, each time doubling the amount of gene produced.

32. Dideoxynucleotides cause chain termination, this results in a family of different size DNA fragments, which can be separated by gels to determine the DNA sequence.

Chapter 21

1. Adenosine triphosphate, or ATP
2. Energy is stored by the formation of high energy phosphoanhydride bonds in ATP. Energy is released from ATP when these high energy phosphoanhydride bonds are broken.
3. Anabolic
4. Catabolism
5. Carbohydrates
6. The first stage of catabolism converts carbohydrates in monosaccharides; proteins into oligopeptides and amino acids; and lipids into fatty acids, glycerol, and monoglycerides.
7. Hydrolysis
8. Amylase
9. Pepsin
10. Small intestine
11. In the second stage of catabolism monomers are converted into forms that can enter one of the later energy yielding catabolic pathways. In the third stage of catabolism nutrients are completely oxidized to produce ATP.
12. Glycolysis is thought to be the first metabolic pathway because it does not require oxygen and is carried out by enzymes in the cytoplasm of the cell.
13. The first segment of glycolysis involves in investment of energy. The second segment of glycolysis involves the net yield of energy.
14. Anaerobic, glycolysis does not require oxygen.
15. Cytoplasm of the cell.
16. Two
17. Glucose-6-phosphate
18. ATP is also needed. A kinase enzyme catalyzes the transfer of the phosphate group (specifically phosphofructokinase).
19. Substrate-level phosphorylation
20. This step yields NADH which is used in later pathways to generate energy for the cell.
21. ATP inhibits glycolysis. The abundance of ATP in the cell indicates that the cell has sufficient energy and glycolysis, and the production of more ATP, is inhibited.
22. Fermentation reactions occur to reoxidize NADH to NAD^+ to allow glycolysis to continue and to consume the pyruvate generated by glycolysis.
23. Ethanol
24. Lactate
25. The pentose pathway is another degradation pathway for glucose and serves to produce molecules needed for the biosynthesis of more complex molecules. For example, the pentose pathway produces NADPH, ribose-5-phosphate, and erythrose-4-phosphate.

26. Gluconeogenesis is a pathway that synthesizes glucose while glycolysis is a pathway that breaks down glucose. Gluconeogenesis is almost the reverse of the glycolysis pathway with the exception of three irreversible steps which are bypassed by using other enzymes.

27. The Cori Cycle occurs in the liver and converts lactate into glucose.
28. The body responds to low blood glucose levels by producing the hormone glucagon. Glucagon initiates glycogenolysis and inhibits glycogenesis.

Chapter 22

1. Mitochondria
2. Energy need
3. Oxaloacetate
4. The formation of acetyl CoA is catalyzed by the pyruvate dehydrogenase complex. This complex requires coenzymes derived from thiamine (vitamin B1), riboflavin (vitamin B2), niacin and pantothenic acid. A deficiency in the B vitamins would result in a deficiency in the coenzymes for the pyruvate dehydrogenase complex and as a result, a decrease in the activity of this enzyme.
5. Matrix
6. This final step forms oxaloacetate, which can react with more acetyl CoA, continuing the citric acid cycle.
7. Two
8. Three
9. An abundance of ATP signals that the cell has sufficient energy, thus ATP serves as a negative effector and inhibits the activity of several enzymes of the citric acid cycle.
10. Electron transport system
11. The protons are pumped into the intermembrane space.
12. NAD+
13. It returns to be used again in the citric acid cycle.
14. Protons
15. F_1 catalyzes the phosphorylation of ADP to produce ATP.
16. Oxygen, or O_2
17. $FADH_2$ donates is electrons to a carrier later in the electron transport system, and thus fewer protons are pumped into the intermembrane space.
18. Three
19. Two
20. The diet
21. Liver
22. Removal of the amino group
23. A new α-amino acid and a new α-ketoacid.
24. Oxalocetate and glutamate
25. Citric acid cycle, or Krebs cycle
26. Liver
27. 2 ATP
28. Hyperammonemia is caused by a deficiency in one of the enzymes of the urea cycle, resulting in accumulation of toxic ammonium ions in the body.
29. Amphibolic

30. The citric acid cycle allows for the oxidation of carbohydrates, lipids, and amino acids into species that can be sent to the electron transport system to result in ATP for the cell. The citric acid cycle also provides the intermediates necessary for the synthesis of many biomolecules.

Chapter 23

1. Fat globules
2. Cholate and chenodeoxycholate
3. Bile contains bile salts which emulsify the fat globules into tiny droplets that can be hydrolyzed to monoglycerides and free fatty acids.
4. A spherical micelle is an aggregate of molecules having polar and nonpolar regions.
5. Colipase
6. Chylomicrons
7. Adipose tissue
8. Acetyl coenzyme A
9. A 12-carbon fatty acid would go through five cycles of β-oxidation.
10. 1 $FADH_2$ and 1 NADH
11. Acyl CoA
12. Acetyl CoA
13. 129 ATP molecules
14. The abnormal rise in concentration of blood ketone bodies.
15. Starvation, a diet that is extremely low in carbohydrates, and the disease diabetes mellitus.
16. When there is not enough oxaloacetate to bring acetyl CoA molecules into the citric acid cycle, the excess acetyl CoA molecules are converted into ketone bodies.
17. Ketoacidosis
18. The ketone bodies acetoacetate and β-hydroxybutyrate can be circulated to other tissues through the blood and reconverted to acetyl CoA and used to produce ATP.
19. Heart muscle
20. Acetyl CoA
21. Acetyl CoA is produced by the degradation of fatty acids through β-oxidation and allows the carbons of the fatty acid to be oxidized via the citric acid cycle. Acetyl CoA is also used as the starting material for the biosynthesis of fatty acids.
22. Fatty acid synthase
23. NADPH
24. Ketone bodies
25. VLDL
26. Glycerol-3-phosphate is required for the synthesis of triglycerides in adipose tissue. This is synthesized from glyceraldehyde-3-phosphate produced during glycolysis.
27. Fatty acids
28. Liver, adipose, and muscle cells
29. Insulin increases glycogen synthesis.
30. Glucagon increases lipolysis.

Notes

Notes

Notes

Notes

Notes